T0212644

Lecture Notes in Computer Science   9241

Commenced Publication in 1973
Founding and Former Series Editors:
Gerhard Goos, Juris Hartmanis, and Jan van Leeuwen

More information about this series at http://www.springer.com/series/7410

Keisuke Tanaka · Yuji Suga (Eds.)

Advances in Information and Computer Security

10th International Workshop on Security, IWSEC 2015
Nara, Japan, August 26–28, 2015
Proceedings

 Springer

Editors
Keisuke Tanaka
Tokyo Institute of Technology
Tokyo
Japan

Yuji Suga
Internet Initiative Japan Inc.
Tokyo
Japan

ISSN 0302-9743 ISSN 1611-3349 (electronic)
Lecture Notes in Computer Science
ISBN 978-3-319-22424-4 ISBN 978-3-319-22425-1 (eBook)
DOI 10.1007/978-3-319-22425-1

Library of Congress Control Number: 2015945151

LNCS Sublibrary: SL4 – Security and Cryptology

Springer Cham Heidelberg New York Dordrecht London

Printed on acid-free paper

Springer International Publishing AG Switzerland is part of Springer Science+Business Media
(www.springer.com)

Preface

The 10th International Workshop on Security (IWSEC 2015) was held at Todaiji Cultural Center, in Nara, Japan, during August 26-28, 2015. The workshop was organized by CSEC of IPSJ (Special Interest Group on Computer Security of Information Processing Society of Japan) and supported by ISEC in ESS of IEICE (Technical Committee on Information and Communication Engineers).

This year, the workshop received 58 submissions. Finally, 18 papers were accepted as regular papers, and 3 papers were accepted as short papers. Each submission was anonymously reviewed by at least four reviewers, and these proceedings contain the revised versions of the accepted papers. In addition to the presentations of the papers, the workshop also featured a poster session.

The best paper award was given to "Identity-based Lossy Encryption from Learning with Errors" by Jingnan He, Bao Li, Xianhui Lu, Dingding Jia, Haiyang Xue, and Xiaochao Sun, and the best student paper award was given to "Improved (Pseudo) Preimage Attacks on Reduced-Round GOST and Grøstl-256 and Studies on Several Truncation Patterns for AES-like Compression Functions" by Bingke Ma, Bao Li, Ronglin Hao, and Xiaoqian Li.

A number of people contributed to the success of IWSEC 2015. We would like to thank the authors for submitting their papers to the workshop. The selection of the papers was a challenging and dedicated task, and we are deeply grateful to the members of Program Committee and the external reviewers for their in-depth reviews and detailed discussions. We are also grateful to Andrei Voronkov for developing Easy-Chair, which was used for the paper submission, reviews, discussions, and preparation of these proceedings.

Last but not least, we would like to thank the general co-chairs, Yukiyasu Tsunoo and Satoru Torii, for leading the Local Organizing Committee, and we also would like to thank the members of the Local Organizing Committee for their efforts to ensure the smooth running of the workshop.

June 2015

Keisuke Tanaka
Yuji Suga

IWSEC 2015
10th International Workshop on Security

Organization

Nara, Japan, August 26–28, 2015

Organized by
CSEC of IPSJ
(Special Interest Group on Computer Security of Information Processing
Society of Japan)

and

Supported by
ISEC in ESS of IEICE
(Technical Committee on Information Security in Engineering Sciences Society
of the Institute of Electronics, Information and Communication Engineers)

General Co-chairs

Yukiyasu Tsunoo	NEC Corporation, Japan
Satoru Torii	Fujitsu Laboratories Ltd., Japan

Advisory Committee

Hideki Imai	University of Tokyo, Japan
Kwangjo Kim	Korea Advanced Institute of Science and Technology, Korea
Günter Müeller	University of Freiburg, Germany
Yuko Murayama	Iwate Prefectural University, Japan
Koji Nakao	National Institute of Information and Communications Technology, Japan
Eiji Okamoto	University of Tsukuba, Japan
C. Pandu Rangan	Indian National Academy of Engineering, India
Kai Rannenberg	Goethe University Frankfurt, Germany
Ryoichi Sasaki	Tokyo Denki University, Japan

Program Co-chairs

Keisuke Tanaka	Tokyo Institute of Technology, Japan
Yuji Suga	Internet Initiative Japan Inc., Japan

Local Organizing Committee

Yuki Ashino	NEC Corporation, Japan
Keita Emura	National Institute of Information and Communications Technology, Japan
Atsuo Inomata	Nara Institute of Science and Technology, Japan
Yu-ichi Hayashi	Tohoku Gakuen University, Japan
Yoshiaki Hori	Saga University, Japan
Masaki Inamura	Tokyo Denki University, Japan
Masaki Kamizono	PricewaterhouseCoopers Co., Ltd., Japan
Akira Kanaoka	Toho University, Japan
Masahiro Mambo	Kanazawa University, Japan
Koichi Mouri	Ritsumeikan University, Japan
Masakazu Nishigaki	Shizuoka University, Japan
Toshihiro Ohigashi	Tokai University, Japan
Kazuto Ogawa	Japan Broadcasting Corporation, Japan
Kazumasa Omote	Japan Advanced Institute of Science and Technology, Japan
Kazuo Sakiyama	The University of Electro-Communications, Japan
Kouichi Sakurai	Kyushu University, Japan
Toshiaki Tanaka	KDDI R&D Laboratories Inc., Japan
Masayuki Terada	NTT Docomo, Japan
Takeaki Terada	Fujitsu Laboratories Ltd., Japan
Yuu Tsuda	National Institute of Information and Communications Technology, Japan
Noritaka Yamashita	NEC Corporation, Japan
Naoto Yanai	Osaka University, Japan
Maki Yoshida	National Institute of Information and Communications Technology, Japan

Program Committee

Nuttapong Attrapadung	National Institute of Advanced Industrial Science and Technology, Japan
Chien-Ning Chen	Nanyang Technological University, Singapore
Koji Chida	NTT, Japan
Sabrina De Capitani di Vimercati	DI - Università degli Studi di Milano, Italy
David Galindo	SCYTL Secure Electronic Voting, Spain
Benedikt Gierlichs	KU Leuven, Belgium
Goichiro Hanaoka	National Institute of Advanced Industrial Science and Technology, Japan
Yoshiaki Hori	Saga University, Japan
Toshiyuki Isshiki	NEC Corporation, Japan
Mitsugu Iwamoto	University of Electro-Communications, Japan
Tetsu Iwata	Nagoya University, Japan
Akinori Kawachi	The University of Tokushima, Japan

Hyung Chan Kim	The Attached Institute of ETRI, South Korea
Takeshi Koshiba	Saitama University, Japan
Noboru Kunihiro	The University of Tokyo, Japan
Hyung Tae Lee	Nanyang Technological University, Singapore
Kazuhiko Minematsu	NEC Corporation, Japan
Kirill Morozov	Kyushu University, Japan
Koichi Mouri	Ritsumeikan University, Japan
Ken Naganuma	Hitachi Ltd., Japan
Takashi Nishide	University of Tsukuba, Japan
Wakaha Ogata	Tokyo Institute of Technology, Japan
Kazuto Ogawa	Japan Broadcasting Corporation, Japan
Takao Okubo	Institute of Information Security, Japan
Thomas Peyrin	Nanyang Technological University, Singapore
Kai Rannenberg	Goethe University Frankfurt, Germany
Kazuo Sakiyama	The University of Electro-Communications, Japan
Yu Sasaki	NTT, Japan
Yuji Suga	Internet Initiative Japan Inc., Japan
Willy Susilo	University of Wollongong, Australia
Mio Suzuki	National Institute of Information and Communications Technology, Japan
Katsuyuki Takashima	Mitsubishi Electric Corporation, Japan
Keisuke Tanaka	Tokyo Institute of Technology, Japan
Masayuki Terada	NTT Docomo, Japan
Takeaki Terada	Fujitsu Laboratories Ltd., Japan
Damien Vergnaud	École normale supérieure, France
Sven Wohlgemuth	
Chung-Huang Yang	National Kaohsiung Normal University, Taiwan
Kenji Yasunaga	Kanazawa University, Japan
Kazuki Yoneyama	Ibaraki University, Japan
Maki Yoshida	National Institute of Information and Communications Technology, Japan
Katsunari Yoshioka	Yokohama National University, Japan
Rui Zhang	Chinese Academy of Sciences, China

External Reviewers

Kazumaro Aoki	Yoshikazu Hanatani
Yoshinori Aono	Marvin Hegen
Shivam Bhasin	Alexander Herrmann
Jeroen Delvaux	Haruna Higo
Keita Emura	Satomi Honda
Daisuke Fujimoto	Masaki Inamura
Koki Hamada	Takanori Isobe

Contents

Security in Hardware

Identity-Based Encryption

Identity-Based Lossy Encryption from Learning with Errors

Jingnan He[1,2,3(✉)], Bao Li[1,2], Xianhui Lu[1,2], Dingding Jia[1,2], Haiyang Xue[1,2], and Xiaochao Sun[1,2]

[1] State Key Laboratory of Information Security,
Institute of Information Engineering of Chinese Academy of Sciences,
Beijing, China
{jnhe13,lb,xhlu,ddjia,hyxue12,xchsun}@is.ac.cn
[2] Data Assurance and Communication Security Research Center
of Chinese Academy of Sciences, Beijing, China
[3] University of Chinese Academy of Sciences, Beijing, China

Abstract. We extend the notion of lossy encryption to the scenario of identity-based encryption (IBE), and propose a new primitive called identity-based lossy encryption (IBLE). Similar as the case of lossy encryption, we show that IBLE can also achieve selective opening security. Finally, we present a construction of IBLE from the assumption of learning with errors.

Keywords: Lossy encryption · Learning with errors · Identity-based lossy encryption

1 Introduction

1.1 Background

Lossy encryption was proposed by Bellare, Hofheinz and Yilek [3] to achieve selective opening security. Briefly, the key generation algorithm of lossy encryption runs in two indistinguishable modes, the real mode and the lossy mode. In the real mode, a real pubic key PK_{real} is generated and scheme works just as standard public key encryption scheme. In the lossy mode, a lossy public key PK_{loss} is generated, and the plaintext is information-theoretically hidden.

Lossy encryption can be constructed from several primitives, such as lossy trapdoor functions (LTDF) [3], re-randomizable encryption [18] and oblivious transfer [18]. It also can be constructed from concrete assumptions, such as decision Diffie-Hellman (DDH) [3,19], quadratic residuosity (QR) [3], learning with errors (LWE) [22] and so on.

This research is supported by the National Nature Science Foundation of China (No. 61379137 and No. 61272040), the National Basic Research Program of China (973 project)(No. 2013CB338002), and IIE's Cryptography Research Project (No. Y4Z0061403 and No. Y4Z0061D03).

K. Tanaka and Y. Suga (Eds.): IWSEC 2015, LNCS 9241, pp. 3–20, 2015.
DOI: 10.1007/978-3-319-22425-1_1

As a strengthen version of public key encryption, identity-based encryption (IBE), proposed by Shamir [25], is a powerful primitive in which the public key can be an arbitrary string. Currently, IBE schemes can be constructed from pairings [7–9,24,26,27], lattices (LWE) [1,2,11,16] and QR [10,12,20].

Motivation. Currently the most important application of lossy encryption is to achieve selective opening security. However, in the scenario of IBE, the selective opening security is achieved by using one-sided public openability [5,21]. Whether the selective opening secure IBE scheme can be constructed via the idea of lossy encryption is an interesting problem.

1.2   Our Contributions

New Definition and Its Application. We give the definition of identity-based lossy encryption (IBLE). Similar to lossy encryption, there are also two indistinguishable modes in identity-based lossy encryption, the real mode and the lossy mode. The real mode is akin to a normal IBE, but the case of lossy mode is more delicate. Specifically, in the lossy mode the lossiness of the master public key $\mathsf{MPK_{lossy}}$ can be triggered by a particular identity $\mathsf{id_{lossy}}$ only. The reason is that in IBE the adversary can obtain the identity private keys $\mathsf{SK_{id}}$ for arbitrary identities except the challenge identity by a series of extraction queries, it can distinguish $\mathsf{MPK_{real}}$ from $\mathsf{MPK_{lossy}}$ with the help of $\mathsf{SK_{id}}$. With IBLE, we obtain indistinguishability-based selective opening security in the selective identity setting (IND-sID-SO).

Construction from LWE. Inspired by [1,2,11,16], we start the construction of our IBLE scheme by designing a dual Regev type lossy encryption. Specifically let $(\mathbf{A_1 s} + \mathbf{e_1}, \mathbf{A_2 s} + \mathbf{e_2} + \mathbf{m}\lfloor \frac{q}{2} \rfloor)$ be the ciphertext of a dual Regev type encryption scheme, where $(\mathbf{A_1}, \mathbf{A_2})$ is the public key, $\mathbf{s}, \mathbf{e_1}, \mathbf{e_2}$ are random numbers, and $\mathbf{m}$ is the message. The main technical difficulty of constructing lossy encryption is to information-theoretically hide the plaintext message $\mathbf{m}$. However, the random number $\mathbf{s}$ is completely determined by the first item of the ciphertext, consequently, $\mathbf{m}$ is fixed by the second item of the ciphertext. Our solution is to lose the information of $\mathbf{s}$ with the technique proposed in [6,17]. Concretely, the randomly selected $\mathbf{A_1} \in \mathbb{Z}_q^{\mathbf{m} \times \mathbf{n}}$ is replaced by an LWE sample $(\mathbf{BC} + \mathbf{Z})$ where $\mathbf{B} \in \mathbb{Z}_q^{m \times n_1}, \mathbf{C} \in \mathbb{Z}_q^{n_1 \times n}, \mathbf{Z} \in \mathbb{Z}^{m \times n}$ sampled from the discrete Gaussians distribution. If $n_1 < n$ and the element of $\mathbf{Z}$ is small enough, then $\mathbf{s}$ is information-theoretically undetermined given $(\mathbf{BC} + \mathbf{Z})\mathbf{s} + \mathbf{e}$ [6,17].

Combining our dual Regev type lossy encryption and the technique for constructing IBE scheme in [1], we obtain an IBLE scheme. To hide the plaintext message information-theoretically for $\mathsf{id_{lossy}}$ and simultaneously extract the identity private key for other identities, the main technical challenge is to guarantee that $\mathbf{s}$ is still information-theoretically undetermined given $(\begin{bmatrix} \mathbf{B} \\ \mathbf{R}^t \mathbf{B} \end{bmatrix} \mathbf{C} + \begin{bmatrix} \mathbf{Z} \\ \mathbf{R}^t \mathbf{Z} \end{bmatrix})\mathbf{s} + \mathbf{e}$, where $\mathbf{R} \in \{-1, 1\}^{m \times m}$. Luckily, we prove that it still holds when

$n_1 < n$ and the element of $\mathbf{Z}$ is small enough. From another point of view, it is not only an extension of the result proved in [6], but also provides another choice for constructing lossy branch.

1.3 Related Work

LTDF, which is closely related with lossy encryption, has been extended to the identity-based scenario by Bellare, Kiltz, Peikert and Waters [4]. Escala, Herranz, Libert, and Ráfols [15] further studied hierarchical identity-based LTDF. Similar to the construction of lossy encryption from LTDF, the primitive IBLE can also be obtained from identity-based LTDF.

1.4 Organization

The rest of this paper is organized as follows. In Sect. 2 we introduce some notations, definitions and previous results. In Sect. 3, we give the definition of IBLE, prove that IBLE scheme implies selective opening security, and propose a construction of IBLE from LWE.

2 Preliminaries

2.1 Notations

Unless otherwise noted, all operations in this paper are under the modulo operation of q, and log means $\log_2$. Throughout, we use λ to denote our security parameter. We use bold lower-case letters (e.g. $\mathbf{s}$) to denote vectors, and bold upper-case letters (e.g. $\mathbf{A}$) to denote matrices. We use $x \xleftarrow{\$} X$ to denote that x is drawn uniformly at random over a set X. We use $x \leftarrow \mathcal{X}$ to denote that x is drawn from a distribution $\mathcal{X}$. To denote the statistical distance between two distributions, we write $\Delta(\mathcal{X}, \mathcal{Y})$. For two distribution ensembles $\mathcal{X} = \mathcal{X}_\lambda, \mathcal{Y} = \mathcal{Y}_\lambda$, we write $\mathcal{X} \approx_s \mathcal{Y}$ if $\Delta(\mathcal{X}, \mathcal{Y})$ is a negligible function of λ, and we write $\mathcal{X} \approx_c \mathcal{Y}$ if for all probabilistic polynomial time (PPT) distinguishers D there is a negligible function $negl(\cdot)$ such that: $|\Pr[D(1^\lambda, \mathcal{X}_\lambda) = 1] - \Pr[D(1^\lambda, \mathcal{Y}_\lambda) = 1]| \leq negl(\lambda)$. We let $\lfloor x \rceil$ be the closest integer to x. We use $\|\mathbf{S}\|$ to denote the L_2 length of the longest vector in $\mathbf{S}$, and $\|\tilde{\mathbf{S}}\|$ to denote the Gram-Schmidt norm of $\mathbf{S}$. Let $\Lambda_q^\perp(\mathbf{A}) = \{\mathbf{e} \in \mathbb{Z}^m \; s.t. \; \mathbf{Ae} = \mathbf{0} \mod q\}$, and $\Lambda_q^{\mathbf{u}} = \{\mathbf{e} \in \mathbb{Z}^m \; s.t. \; \mathbf{Ae} = \mathbf{u} \mod q\}$, given a matrix $\mathbf{A} \in \mathbb{Z}_q^{n \times m}$ and a vector $\mathbf{u} \in \mathbb{Z}_q^n$.

2.2 Min-Entropy

The **min-entropy** of a random variable X is $\tilde{H}_\infty(X) = -\log(\max_x \Pr[X = x])$, and the **average min-entropy** of X conditioned on Y, defined by [13], is $H_\infty(X|Y) = -\log(\mathbf{E}_{y \leftarrow Y}[\max_x \Pr[X = x|Y = y]]) = -\log(\mathbf{E}_{y \leftarrow Y}[2^{-H_\infty(X|Y=y)}])$.

Definition 1 ([13]). *For two random variables X and Y, the ϵ-smooth average min-entropy of X conditioned on Y, denoted $\tilde{H}_\infty^\epsilon(X|Y)$ is $\tilde{H}_\infty^\epsilon(X|Y) =$*
$$\max_{(X',Y'):\Delta((X,Y),(X',Y'))<\epsilon} \tilde{H}_\infty(X'|Y').$$

2.3 Learning with Errors

Learning with errors (LWE) problem initially stated in [23]. Here we recall the concepts and the hardness of LWE.

Learning with Errors (LWE). Let $m = m(n)$, $q = q(n)$ be integers, and χ be a distribution on $\mathbb{Z}_q$. Let $\mathbf{A} \xleftarrow{\$} \mathbb{Z}_q^{m \times n}$, $\mathbf{s} \xleftarrow{\$} \mathbb{Z}_q^n$, $\mathbf{e} \leftarrow \chi^m$, then the LWE$(m, n, q, \chi)$ problem is to find $\mathbf{s}$, given $(\mathbf{A}, \mathbf{As} + \mathbf{e})$.

This is the search version of the LWE problem, and there is a decisional version of the LWE problem.

(Decisional) Learning with Errors (DLWE). Let $m = m(n)$, $q = q(n)$ be integers, and χ be a distribution on $\mathbb{Z}_q$. Let $\mathbf{A} \xleftarrow{\$} \mathbb{Z}_q^{m \times n}$, $\mathbf{s} \xleftarrow{\$} \mathbb{Z}_q^n$, $\mathbf{e} \leftarrow \chi^m$, then the DLWE$(m, n, q, \chi)$ problem is that given $(\mathbf{A}, \mathbf{b})$, decide whether $\mathbf{b}$ is distributed by $\mathbf{As} + \mathbf{e}$ or chosen uniformly at random from $\mathbb{Z}_q^m$.

The hardness of the matrix-version of the DLWE problem is as below.

Lemma 1 ([14]). *Let $m(n), k(n) = \mathsf{poly}(n)$. Assume that DLWE$(m, n, q, \chi)$ is pseudorandom [23]. Then the distribution $(\mathbf{A}, \mathbf{AX} + \mathbf{E})$ is also pseudorandom, where $\mathbf{A} \in \mathbb{Z}_q^{m \times n}$ and $\mathbf{X} \in \mathbb{Z}_q^{n \times k}$ are chosen uniformly at random and $\mathbf{E}$ is chosen according to $\mathcal{D}_{\mathbb{Z}, \alpha q}^{m \times k}$.*

2.4 Discrete Gaussians

For any $s > 0$ and $\mathbf{c} \in \mathbb{R}^n$, define the Gaussian function: $\forall \mathbf{x} \in \mathbb{R}^n, \rho_{s,\mathbf{c}}(\mathbf{x}) = \exp(-\pi \|\mathbf{x} - \mathbf{c}\|^2 / s^2)$.

For any $\mathbf{c} \in \mathbb{R}^n$, real $s > 0$, and n-dimensional lattice Λ, define the discrete Gaussian distribution over Λ as: $\forall \mathbf{x} \in \Lambda, \mathcal{D}_{\Lambda, s, \mathbf{c}}(\mathbf{x}) = \frac{\rho_{s,\mathbf{c}}(\mathbf{x})}{\rho_{s,\mathbf{c}}(\Lambda)}$ where $\rho_{s,\mathbf{c}}(\Lambda) = \sum_{\mathbf{y} \in \Lambda} \rho_{s,\mathbf{c}}(\mathbf{y})$. We omit the parameter $\mathbf{c}$ when $\mathbf{c} = \mathbf{0}$.

2.5 Lossy Encryption

Lossy Encryption Scheme is defined in [3]. It is given by a tuple of PPT algorithms {KeyGen$_{real}$, KeyGen$_{loss}$, Enc, Dec}. The details are as below.

- **KeyGen$_{real}$**(1^λ): a key generation algorithm takes a security parameter λ as input, and outputs a pair of real public key and corresponding secret key (PK$_{real}$, SK).
- **KeyGen$_{loss}$**(1^λ): a key generation algorithm takes a security parameter λ as input, and outputs a pair of lossy public key and $\perp$ instead of SK (PK$_{lossy}$, $\perp$).
- **Enc**(PK, m): an encryption algorithm takes either PK$_{real}$ or PK$_{loss}$ and message m as input, and outputs a ciphertext C.
- **Dec**(SK, C): a decryption algorithm takes a secret key SK and a ciphertext C as input, and outputs either a message m or $\perp$ in the case of failure.

A *Lossy Encryption Scheme* should have the properties below.

1. *Correctness on Real Keys.* For all $(\mathsf{PK_{real}}, \mathsf{SK})$ generated by $\mathbf{KeyGen_{real}}(1^k)$ and all message m, $\mathbf{Dec}(\mathsf{SK}, \mathbf{Enc}(\mathsf{PK_{real}}, m)) = m$.
2. *Lossiness of Encryption with Lossy Keys.* For any lossy keys $\mathsf{PK_{loss}}$ generated by $\mathbf{KeyGen_{loss}}(1^k)$ and any two messages $m_0 \neq m_1$, there is $\mathbf{Enc}(\mathsf{PK_{loss}}, m_0) \approx_s \mathbf{Enc}(\mathsf{PK_{loss}}, m_1)$.
3. *Indistinguishability Between Real Public Key and Lossy Public Key.* For any $\mathsf{PK_{real}}$ generated by $\mathbf{KeyGen_{real}}$ and any $\mathsf{PK_{loss}}$ generated by $\mathbf{KeyGen_{loss}}$, there is $\mathsf{PK_{real}} \approx_c \mathsf{PK_{loss}}$.

2.6 Some Results About Randomness

Randomness plays an important role in constructing lossy encryption schemes, so we introduce some results about randomness which will be used as tools in the later section.

Lemma 2 ([17]). *Let $\mathcal{D}$ be a distribution over $\mathbb{Z}_q^n$ with min-entropy k. For any $\varepsilon > 0$ and $l \leq (k - 2\log(1/\varepsilon) - O(1))/\log q$, the joint distribution of $(\mathbf{C}, \mathbf{C} \cdot \mathbf{s})$ where $\mathbf{C} \leftarrow \mathbb{Z}_q^{l \times n}$ is uniformly random and $\mathbf{s} \in \mathbb{Z}_q^{l \times n}$ is drawn from the distribution $\mathcal{D}$ is ε-close to the uniform distribution over $\mathbb{Z}_q^{l \times n} \times \mathbb{Z}_q^l$.*

Lemma 3 ([16]). *Let n and q be positive integers with q prime, and let $m \geq 2n\log q$. Then for all but a $2q^{-n}$ fraction of all $\mathbf{A} \in \mathbb{Z}_q^{m \times n}$ and for any $s \geq \omega(\sqrt{\log m})$, the distribution of the syndrome $\mathbf{u}^t = \mathbf{e}^t \mathbf{A} \mod q$ is statistically close to uniform over $\mathbb{Z}_q^n$, where $\mathbf{e} \sim \mathcal{D}_{\mathbb{Z},s}^m$.*

Lemma 4 ([6]). *There exists a distribution Lossy such that $\bar{\mathbf{A}} \leftarrow \mathsf{Lossy} \approx_c \mathbf{U} \xleftarrow{\$} \mathbb{Z}_q^{m \times n}$ and given $\mathbf{s} \xleftarrow{\$} \mathbb{Z}_q^n$, and $\mathbf{e} \leftarrow \mathcal{D}_{\mathbb{Z}, \beta q}^{m \times n}$, $\tilde{H}_\infty^\epsilon(\mathbf{s} | \bar{\mathbf{A}}, \bar{\mathbf{A}}\mathbf{s} + \mathbf{x}) \geq n$, where $\epsilon = negl(\lambda)$. Lossy is as follows,*

- *Choose $\mathbf{B} \xleftarrow{\$} \mathbb{Z}_q^{m \times k}$, $\mathbf{C} \xleftarrow{\$} \mathbb{Z}_q^{k \times n}$, and $\mathbf{Z} \leftarrow \mathcal{D}_{\mathbb{Z}, \alpha q}^{m \times n}$, where $\frac{\alpha}{\beta} = negl(\lambda)$ and $k\log q \leq n - 2\lambda + 2$.*
- *Let $\bar{\mathbf{A}} = \mathbf{B}\mathbf{C} + \mathbf{Z}$.*
- *Output $\bar{\mathbf{A}}$.*

3 Identity-Based Lossy Encryption

In this section, we give the definition of IBLE. An IBLE scheme works in two modes. One is the real mode which is the same as an IBE scheme with standard master key generation algorithm and extraction algorithm. The other is the lossy mode with a lossy master key generation algorithm, and the corresponding extraction algorithm. The two modes share the same encryption and decryption algorithms. For identities $\mathsf{id} \neq \mathsf{id_{lossy}}$, encryptions with the lossy master public key $\mathsf{MPK_{lossy}}$ are committing as the same in the real mode. For $\mathsf{id_{lossy}}$, encryptions are not committing.

Formally, the real mode is a tuple of PPT algorithms $\{\mathsf{Setup_{real}}, \mathsf{Extract_{real}}, \mathsf{Enc}, \mathsf{Dec}\}$:

- **Setup$_{real}$(1^λ):** a master key generation algorithm takes a security parameter λ as input, and outputs a pair of real master public key and corresponding master secret key (MPK$_{real}$, MSK).
- **Extract$_{real}$(id, MPK$_{real}$, MSK):** a user secret key generation algorithm takes an identity **id**, the master public key MPK$_{real}$ and the master secret key MSK as inputs, and outputs a user secret key SK$_{id}$ for the identity.
- **Enc(id, MPK, m):** a user encryption algorithm takes an identity **id**, the master public key MPK and a message **m** as inputs, and outputs a ciphertext C.
- **Dec(id, SK$_{id}$, C):** a user decryption algorithm takes an identity **id**, the user secret key SK$_{id}$ and a ciphertext C as inputs, and outputs either a message **m** or $\perp$ in the case of failure.

The lossy mode is a tuple of PPT algorithms {Setup$_{lossy}$, Extract$_{lossy}$, Enc, Dec}:

- **Setup$_{lossy}$(id$_{lossy}$):** a master key generation algorithm takes an identity **id$_{lossy}$** as input, and outputs a pair of lossy master public key and corresponding master secret key (MPK$_{lossy}$, MSK).
- **Extract$_{lossy}$(id, MPK$_{lossy}$, MSK):** a user secret key generation algorithm takes an identity **id**, the master public key MPK$_{lossy}$ and the master secret key MSK as inputs, and outputs either a user secret key SK$_{id}$ when $\mathbf{id} \neq \mathbf{id}_{lossy}$ or $\perp$ when $\mathbf{id} = \mathbf{id}_{lossy}$.
- **Enc** and **Dec** algorithms are the same as those in the real mode.

An *Identity-based Lossy Encryption Scheme* should have the properties as below.

1. *Correctness on Keys for All* $\mathbf{id} \neq \mathbf{id}_{lossy}$. For any (MPK, MSK) generated by **Setup$_{real}$(1^k)** or **Setup$_{lossy}$(id$_{lossy}$)**, any SK$_{id}$ generated by **Extract$_{real/lossy}$(id, MPK, MSK)**, and any message **m**, **Dec(id, SK$_{id}$, Enc(id, MPK, m)) = m**.
2. *Lossiness of Encryption with Lossy Keys for* $\mathbf{id} = \mathbf{id}_{lossy}$. For any lossy keys MPK$_{lossy}$ generated by **Setup$_{lossy}$(id$_{lossy}$)** and any two messages $\mathbf{m}_0 \neq \mathbf{m}_1$, there is **Enc(id$_{lossy}$, MPK$_{lossy}$, m$_0$)** $\approx_s$ **Enc(id$_{lossy}$, MPK$_{lossy}$, m$_1$)**. The advantage of $\mathcal{A}$ whose target is to distinguish those two ciphertexts (i.e. $Adv_{\mathcal{A}, \mathcal{IBLE}}^{lossy}$-ind) means the advantage of $\mathcal{A}$ in the standard IND-CPA game when the public key in the IND-CPA game is lossy.
3. *Indistinguishability Between Real Keys and Lossy Keys.* We use a game to describe this property.
   The advantage of the adversary is $Adv_{\mathcal{A}, \mathcal{IBLE}}^{lossy\text{-}keys} = |2Pr[b' = b] - 1|$. If for all PPT adversaries $\mathcal{A}$ we have that $Adv_{\mathcal{A}, \mathcal{IBLE}}^{lossy-keys}$ is a negligible function, then we say that the real keys generated in the real mode is indistinguishable with the lossy keys generated in the lossy mode.

Obviously the definition of IBLE implies IND-CPA security of IBE.

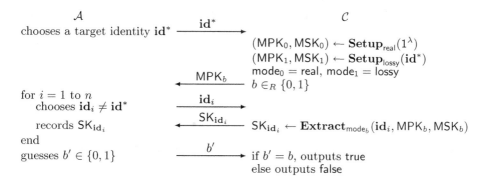

Fig. 1. Game of indistinguishability between real keys and lossy keys

3.1 Selective Opening Security

Here we prove that the notion of IBLE implies indistinguishability-based selective opening secure (IND-sID-SO) under chosen-plaintext attack. Firstly, we use a game to define IND-sID-SO. Let $\mathcal{D}_{\mathcal{M}}$ be any message sampler.

Init: The adversary outputs a target identity $\mathbf{id}^*$.

Setup: The challenger runs $\mathbf{Setup}(1^\lambda)$ and keeps the master secret key MSK. The challenger samples n messages $\{\mathbf{m}_0^i\}_{i=1..n}$ from $\mathcal{D}_{\mathcal{M}}$ and gets n ciphertexts by using algorithm $\mathbf{Enc}(\mathbf{id}^*, \mathsf{MPK}, \mathbf{m}_0^i)$, $i = 1..n$. The master public key MPK and the n ciphertexts $\mathbf{c}_1, \mathbf{c}_2, ..., \mathbf{c}_n$ are sent to the adversary.

Phase 1: The adversary issues queries $q_1, ..., q_k$ where the i-th query q_i is a query on $\mathbf{id}_i$. We require that $\mathbf{id}_i \neq \mathbf{id}^*$. The challenger responds by using algorithm Extract to obtain a private key $\mathsf{SK}_{\mathbf{id}_i}$ for $\mathbf{id}_i$, and sends $\mathsf{SK}_{\mathbf{id}_i}$ to the adversary. All queries may be made adaptively, that is, the adversary may ask q_i with knowledge of the challenger's responses to $q_1, ..., q_{i-1}$.

Open and Challenge: Once the adversary decides that Phase 1 is over it specifies a set J and sends it to the challenger. Then the challenger resamples n messages $\{\mathbf{m}_1^i\}_{i=1..n}$ from $\mathcal{D}_{\mathcal{M}}$ such that $\mathbf{m}_1^{[J]} = \mathbf{m}_0^{[J]}$. The challenger picks a random bit $b \in \{0, 1\}$ and sends the adversary the messages $\mathbf{m}_b$ and the randomnesses $\mathbf{r}[J]$ used in ciphertexts $\mathbf{c}[J]$.

Phase 2: The adversary issues additional adaptive queries $q_{k+1}, ..., q_m$ where q_i is a private-key extraction query on $\mathbf{id}_i$, where $\mathbf{id}_i \neq \mathbf{id}^*$. The challenger responds the same as in **Phase 1**.

Guess: Finally, the adversary outputs a guess $b' \in \{0, 1\}$ and wins if $b' = b$. The advantage of $\mathcal{A}$ in attacking an IBE scheme $\mathcal{E}$ is $Adv_{\mathcal{A},\mathcal{E},\mathcal{D}_{\mathcal{M}},n}^{\mathsf{IND}}\text{-s}ID\text{-SO}(\lambda) = |2 \cdot \Pr[b = b'] - 1|$. The probability is over the random bits used by the challenger and the adversary.

Definition 2. *We say that an IBE system $\mathcal{E}$ is* IND-sID-SO *secure if for all* IND-sID-SO *PPT adversaries $\mathcal{A}$ we have that $Adv_{\mathcal{A},\mathcal{E},\mathcal{M},n}^{\mathsf{IND\text{-}sID\text{-}SO}}(\lambda)$ is a negligible function.*

Theorem 1 (Identity-Based Lossy Encryption Implies IND-sID-SO Security). *Let λ be a security parameter. If $\mathcal{IBLE}$ is any identity-based lossy encryption scheme, then for all* IND-sID-SO *PPT adversaries $\mathcal{A}$, $Adv_{\mathcal{A},\mathcal{IBLE}}^{\mathsf{IND\text{-}sID\text{-}SO}}(\lambda)$ is a negligible function.*

Proof. At the beginning, we describe an algorithm called Opener. By the properties of IBLE, a ciphertext can be explained to any message with high probability. It means that, given a ciphertext C and a message $\mathbf{m}$, the algorithm Opener can find a set of random numbers r such that $\mathbf{Enc}(\mathsf{id}_{\mathsf{lossy}}, \mathsf{MPK}_{\mathsf{lossy}}, \mathbf{m}; r) = C$ and outputs a random element of that set by traversing all values of the random number. The distribution of randomness is correct and the algorithm Opener is unbounded. Let $\mathcal{A}$ be any IND-sID-SO PPT adversary of $\mathcal{IBLE}$. The game sequence is as below.

$\mathbf{G_0}$: The IND-sID-SO original game as the definition.

$\mathbf{G_1}$: $\mathbf{Setup_{real}}$ in $\mathbf{G_0}$ is replaced by $\mathbf{Setup_{lossy}}$, and correspondingly, $\mathbf{Extra\text{-}ct_{real}}$ in $\mathbf{G_0}$ is replaced by $\mathbf{Extract_{lossy}}$.

$\mathbf{H_0}$: Based on $\mathbf{G_1}$, in the process of Open & Challenge, the challenger uses the algorithm $\mathsf{Opener}(\mathbf{id}^*, \mathsf{MPK}_{\mathsf{lossy}}, \mathbf{c}_i, \mathbf{m}_0^i)$ to generate the random numbers corresponding to the $|J|$ ciphertexts.

$\mathbf{H_k}$: We generalize $\mathbf{H_0}$ with a sequence of hybrid games. In this game, besides the true messages $\mathbf{m}_0$ sampled from $\mathcal{D}_{\mathcal{M}}$, the challenger randomly chooses another k messages $\mathbf{m}_0'^1, \mathbf{m}_0'^2, ..., \mathbf{m}_0'^k$ from the plaintext space $\mathcal{M}$ as fake messages and encrypts them in the lossy mode $(\mathbf{c}_1', \mathbf{c}_2', ..., \mathbf{c}_k')$ to replace the first k ciphertexts. The challenger sends $\mathbf{c}_1', ..., \mathbf{c}_k', \mathbf{c}_{k+1}, ..., \mathbf{c}_n$ to the adversary in the Setup process. In the Open & Challenge process, the challenger still uses Opener to reveal the random numbers by the true messages $\mathbf{m}_0$, i.e. $\mathbf{r}_i = \mathsf{Opener}(\mathbf{id}^*, \mathsf{MPK}_{\mathsf{lossy}}, \mathbf{c}_i', \mathbf{m}_0^i)$ when $i \leq k$ and $\mathbf{r}_i = \mathsf{Opener}(\mathbf{id}^*, \mathsf{MPK}_{\mathsf{lossy}}, \mathbf{c}_i, \mathbf{m}_0^i)$ when $i > k$.

$\mathbf{H_n}$: In this game, the n ciphertexts sent to $\mathcal{A}$ are all replaced by encryptions of other n fake messages $\{\mathbf{m}_0'^i\}_{i=1..n}$. The revealed random numbers are opened by $\mathsf{Opener}(\mathbf{id}^*, \mathsf{MPK}_{\mathsf{lossy}}, \mathbf{c}_i', \mathbf{m}_0^i)$.

Then we will analyze this game sequence. First, the change of $\mathbf{G_1}$ is that the real keys $(\mathsf{MPK}_{\mathsf{real}}, \mathsf{SK}_{\mathbf{id}\_\mathsf{real}})$ are replaced by lossy keys $(\mathsf{MPK}_{\mathsf{lossy}}, \mathsf{SK}_{\mathbf{id}\_\mathsf{lossy}})$. Then if an adversary can distinguish $\mathbf{G_0}$ and $\mathbf{G_1}$, there is an adversary can distinguish the real keys and the lossy keys in $\mathcal{IBLE}$. It means that there is an PPT adversary $\mathcal{B}_1$ such that

$$\Pr[\mathbf{G_0}] - \Pr[\mathbf{G_1}] = Adv_{\mathcal{B}_1, \mathcal{IBLE}}^{\mathsf{lossy\text{-}key}}(\lambda).$$

By the third property of IBLE, $Adv^{\text{lossy-key}}_{\mathcal{B}_1,\mathcal{IBLE}}(\lambda)$ is $negl(\lambda)$.

Second, the algorithm Opener in $\mathbf{H}_0$ uses the true message m_0^i and its corresponding ciphertext c_i, so the distribution of random number revealed by Opener is the same as in $\mathbf{G}_1$. Then there is

$$\Pr[\mathbf{G}_1] = \Pr[\mathbf{H}_0].$$

Third, compared with $\mathbf{H}_0$, the change of $\mathbf{H}_1$ is that the first ciphertext $\mathbf{c}_1'$ sent to $\mathcal{A}$ is encrypted by the fake message $\mathbf{m'}_0^1$ instead of the true message $\mathbf{m}_0^1$. However, $\mathbf{H}_1$ still uses the true message $\mathbf{m}_0^1$ to open the random number of the ciphertext $\mathbf{c}_1'$. In other words, $\mathbf{r}_1' = \mathsf{Opener}(\mathbf{id}^*, \mathsf{MPK}_{\text{lossy}}, \mathbf{c}_1', \mathbf{m}_0^1)$, satisfies $\mathbf{Enc}(\mathbf{id}^*, \mathsf{MPK}_{\text{lossy}}, \mathbf{m}_0^1; \mathbf{r}_1') = \mathbf{c}_1' = \mathbf{Enc}(\mathbf{id}^*, \mathsf{MPK}_{\text{lossy}}, \mathbf{m'}_0^1; \mathbf{r}_1)$. Therefore, if there is an adversary can distinguish $\mathbf{H}_0$ from $\mathbf{H}_1$, there is an unbounded (because the algorithm Opener is unbounded) adversary $\mathcal{B}_2$ can distinguish the ciphertexts in $\mathcal{IBLE}$. That is

$$\Pr[\mathbf{H}_0] - \Pr[\mathbf{H}_1] = Adv^{\text{lossy-ind}}_{\mathcal{B}_2,\mathcal{IBLE}}(\lambda).$$

$\mathbf{H}$ is a hybrid sequence, and the only difference between $\mathbf{H}_i$ and $\mathbf{H}_{i+1}$ is the ciphertext $\mathbf{c}_{i+1}'$. Therefore,

$$\Pr[\mathbf{H}_0] - \Pr[\mathbf{H}_n] = n \cdot Adv^{\text{lossy-ind}}_{\mathcal{B}_2,\mathcal{IBLE}}(\lambda).$$

Because the distribution of ciphertexts encrypted by any messages in identity-based lossy encryption are statistically close, $Adv^{\text{lossy-ind}}_{\mathcal{B}_2,\mathcal{IBLE}}(\lambda)$ is $negl(\lambda)$.

Last, in $\mathbf{H}_n$, all n ciphertexts $\mathbf{c}_1', \mathbf{c}_2', ..., \mathbf{c}_n'$ are encrypted by fake messages $\mathbf{m'}_0^1, \mathbf{m'}_0^2, ..., \mathbf{m'}_0^n$ which has no information of the true messages $\mathbf{m}_0^1, \mathbf{m}_0^2, ..., \mathbf{m}_0^n$. So the adversary can just randomly guess b. That is,

$$\Pr[\mathbf{H}_n] = \frac{1}{2}.$$

Above all, $Adv^{\text{ind-sid-so}}_{\mathcal{A},\mathcal{IBLE},\mathcal{M},n}(\lambda) = |2\Pr[\mathbf{G}_0] - 1| \leq 2 \cdot Adv^{\text{lossy-key}}_{\mathcal{B}_1,\mathcal{IBLE}}(\lambda) + 2n \cdot Adv^{\text{lossy-ind}}_{\mathcal{B}_2,\mathcal{IBLE}}(\lambda) = negl(\lambda)$ states that identity-based lossy encryption implies IND-sID-SO secure.   □

3.2   Construction from LWE

The dual Regev's cryptosystem was proposed to construct IBE with random oracle in [16]. Then, Agrawal, Boneh and Boyen [1] used it to construct an IBE scheme in the standard model. Before constructing IBLE, we construct a dual Regev type lossy encryption to get some inspiration.

3.2.1   Construction of Dual Regev Type Lossy Encryption

We construct a lossy encryption based on dual Regev's cryptosystem. However, the process of encryption is different from dual Regev's cryptosystem. We only

$\mathbf{KeyGen_{real}}(1^\lambda)$	$\mathbf{KeyGen_{loss}}(1^\lambda)$	$\mathbf{Enc}(\mathsf{PK}, \mathbf{m})$	$\mathbf{Dec}(\mathsf{SK}, \mathsf{C})$
$\mathbf{A} \xleftarrow{\$} \mathbb{Z}_q^{n \times m}$	$\mathbf{B} \xleftarrow{\$} \mathbb{Z}_q^{m \times k}$	$\mathbf{s} \xleftarrow{\$} \mathbb{Z}_q^n$	$(\mathbf{c_1}, \mathbf{c_2}) := \mathsf{C}$
$\mathbf{T} \xleftarrow{\$} \mathcal{D}_{\mathbb{Z}, r}^{m \times l}$	$\mathbf{C} \xleftarrow{\$} \mathbb{Z}_q^{k \times n}$	$\mathbf{e} \longleftarrow \mathcal{D}_{\mathbb{Z}, \beta q}^m$	$\mathbf{m'} := \mathbf{c_2} - \mathsf{SK}^t \mathbf{c_1}$
$\mathsf{PK_{real}} := (\mathbf{A}, \mathbf{AT})$	$\mathbf{Z} \longleftarrow \mathcal{D}_{\mathbb{Z}, \alpha q}^{m \times n}$	$(\mathbf{PK_1}, \mathbf{PK_2}) := \mathsf{PK}$	$\mathbf{m} := decode(\mathbf{m'})$
$\mathsf{SK} := \mathbf{T}$	$\mathbf{U} \xleftarrow{\$} \mathbb{Z}_q^{l \times n}$	$\mathbf{c_1} := \mathbf{PK_1}^t \mathbf{s} + \mathbf{e}$	return $\mathbf{m}$
return $(\mathsf{PK_{real}}, \mathsf{SK})$	$\mathsf{PK_{loss}} := ((\mathbf{BC} + \mathbf{Z})^t, \mathbf{U}^t)$	$\mathbf{c_2} := \mathbf{PK_2}^t \mathbf{s} + \mathbf{m}\lfloor \frac{q}{2} \rceil$	
	return $(\mathsf{PK_{loss}}, \perp)$	return $(\mathbf{c_1}, \mathbf{c_2})$	

Fig. 2. Construction of dual Regev type lossy encryption

choose the noisy vector $\mathbf{e}$ once in our construction instead of twice in the original cryptosystem. The message space $\mathcal{M}$ is $\{0, 1\}^l$, and the concrete construction is as follows.

The $decode(\mathbf{m'})$ means that for every element m'_i of the vector $\mathbf{m'}$, outputs 0 if m'_i is closer to 0 than to $\lfloor \frac{q}{2} \rceil$ modulo q, otherwise outputs 1.

Parameters. Consider requirements of correct decryption, and the lossiness and so on, parameters are as below. $m \geq 2n \log q, k \log q \leq n - 2\lambda + 2, l \leq (k - 2\log(1/\varepsilon) - O(1))/\log q, q \geq 5rm, r \geq \omega(\sqrt{\log m}), \beta \leq 1/(r\sqrt{m}\omega(\sqrt{\log m})), \beta q > O(2\sqrt{n}), \frac{\alpha}{\beta} = negl(\lambda)$. To satisfy these requirements, q should be super-polynomial of the secure parameter λ.

Then, we show this scheme fulfills the properties of lossy encryption.

1. *Correctness on Real Keys.* For all $(\mathsf{PK_{real}}, \mathsf{SK})$ generated by $\mathbf{KeyGen_{real}}(1^\lambda)$ and all message $\mathbf{m}$,

$$\mathbf{Dec}(\mathsf{SK}, \mathbf{Enc}(\mathsf{PK_{real}}, \mathbf{m})) = \mathbf{Dec}(\mathbf{T}, \mathbf{Enc}((\mathbf{A}, \mathbf{AT}), \mathbf{m}))$$
$$= \mathbf{Dec}(\mathbf{T}, (\mathbf{A}^t \mathbf{s} + \mathbf{e}, \mathbf{T}^t \mathbf{A}^t \mathbf{s} + \mathbf{m}\lfloor \frac{q}{2} \rceil))$$
$$= decode(\mathbf{T}^t \mathbf{A}^t \mathbf{s} + \mathbf{m}\lfloor \frac{q}{2} \rceil - \mathbf{T}^t \mathbf{A}^t \mathbf{s} - \mathbf{T}^t \mathbf{e})$$
$$= decode(\mathbf{m}\lfloor \frac{q}{2} \rceil - \mathbf{T}^t \mathbf{e})$$
$$= \mathbf{m}$$

By Lemma 5, the algorithm $decode()$ will get the correct message with overwhelming probability.

2. *Lossiness of Encryption with Lossy Keys.*

$$\mathbf{Enc}(\mathsf{PK_{loss}}, \mathbf{m}) = \mathbf{Enc}(((\mathbf{BC} + \mathbf{Z})^t, \mathbf{U}^t), \mathbf{m})$$
$$= ((\mathbf{BC} + \mathbf{Z})\mathbf{s} + \mathbf{e}, \mathbf{Us} + \mathbf{m}\lfloor \frac{q}{2} \rceil)$$

By Lemma 4, $\tilde{H}_\infty(\mathbf{s}|(\mathbf{BC} + \mathbf{Z})\mathbf{s} + \mathbf{e}) \geq n$. Because $l \leq (k - 2\log(1/\varepsilon) - O(1))/\log q$, and by Lemma 2, given $(\mathbf{BC} + \mathbf{Z})\mathbf{s} + \mathbf{e}$, $\mathbf{Us}$ is ε-close to $\mathcal{U}(\mathbb{Z}_q^l)$.

When $\varepsilon = negl(\lambda)$, $\mathbf{Us} \approx_s \mathcal{U}(\mathbb{Z}_q^l)$ given $(\mathbf{BC} + \mathbf{Z})\mathbf{s} + \mathbf{e}$. Therefore, for any $\mathbf{m} \in \mathcal{M}$, $(\mathbf{Us} + \mathbf{m}\lfloor\frac{q}{2}\rfloor)$ is statistically close to $\mathcal{U}(\mathbb{Z}_q^l)$, given $(\mathbf{BC} + \mathbf{Z})\mathbf{s} + \mathbf{e}$, i.e. for any lossy keys $\mathsf{PK}_{\mathsf{loss}}$ generated by $\mathbf{KeyGen_{loss}}(1^\lambda)$ and any two messages $\mathbf{m}_0 \neq \mathbf{m}_1$, $\mathbf{Enc}(\mathsf{PK}_{\mathsf{loss}}, \mathbf{m}_0) \approx_s \mathbf{Enc}(\mathsf{PK}_{\mathsf{loss}}, \mathbf{m}_1)$ holds.

3. *Indistinguishability Between Real Public Key and Lossy Public Key.* $\mathsf{PK}_{\mathsf{real}}$ is $(\mathbf{A}, \mathbf{AT})$, and $\mathsf{PK}_{\mathsf{loss}}$ is $((\mathbf{BC} + \mathbf{Z})^t, \mathbf{U}^t)$. Because $m \geq 2n \log q$, by Lemma 3, $(\mathbf{A}^t, (\mathbf{TA})^t) \approx_s (\mathbf{U}_1, \mathbf{U}_2)$, $\mathbf{U}_1 \xleftarrow{\$} \mathbb{Z}_q^{m \times n}$ and $\mathbf{U}_2 \xleftarrow{\$} \mathbb{Z}_q^{l \times n}$. Under the hardness of LWE, $(\mathbf{BC} + \mathbf{Z}, \mathbf{U}) \approx_c (\mathbf{U}_1, \mathbf{U}_2)$. Therefore, $(\mathbf{BC} + \mathbf{Z}, \mathbf{U}) \approx_c (\mathbf{A}^t, (\mathbf{TA})^t)$, i.e. $\mathsf{PK}_{\mathsf{real}}$ and $\mathsf{PK}_{\mathsf{loss}}$ are computationally indistingushable.

Lemma 5 ([16]). *Let $q \geq 5rm$, let $\beta \leq 1/(r\sqrt{m} \cdot \omega(\sqrt{\log n}))$. Then $\mathbf{Dec}(\mathsf{SK}, \mathbf{C})$ in Fig. 2 decrypts correctly with overwhelming probability (over the random choices of $\mathbf{KeyGen_{real}}(1^\lambda)$ and $\mathbf{Enc}(\mathsf{PK}, \mathbf{m})$).*

3.2.2 Construction of IBLE

Before describing the construction, we will introduce some algorithms which will be used.

Lemma 6 ([1]). *Let $q \geq 3$ be odd and $m := \lceil 6n \log q \rceil$. There is a probabilistic polynomial-time algorithm $\mathsf{TrapGen}(q, n)$ that outputs a pair $(\mathbf{A} \in \mathbb{Z}_q^{n \times m}, \mathbf{S} \in \mathbb{Z}^{m \times m})$ such that $\mathbf{A}$ is statistically close to a uniform matrix in $\mathbb{Z}_q^{n \times m}$ and $\mathbf{S}$ is a basis for $\Lambda_q^\perp(\mathbf{A})$ satisfying $\|\tilde{\mathbf{S}}\| \leq O(\sqrt{n \log q})$ and $\|\mathbf{S}\| \leq O(n \log q)$ with all but negligible probability in n.*

Lemma 7 ([1]). *Let $q > 2$, $m > 2n \log q$ and $\sigma > \|\tilde{\mathbf{T}}_A\| \cdot \omega(\sqrt{\log(m + m_1)})$. There is a probabilistic polynomial-time algorithm $\mathsf{SampleLeft}(\mathbf{A}, \mathbf{M}_1, \mathbf{T}_A, \mathbf{u}, \sigma)$ that, given a rank n matrix $\mathbf{A}$ in $\mathbb{Z}_q^{n \times m}$, a matrix $\mathbf{M}_1$ in $\mathbb{Z}_q^{n \times m_1}$, a "short" basis $\mathbf{T}_A$ of $\Lambda_q^\perp(\mathbf{A})$ and a vector $\mathbf{u} \in \mathbb{Z}_q^n$, outputs a vector $\mathbf{e} \in \mathbb{Z}^{m+m-1}$ distributed statistically close to $\mathcal{D}_{\Lambda_q^\mathbf{u}(\mathbf{F}_1),\sigma}$ where $\mathbf{F}_1 := (\mathbf{A}\|\mathbf{M}_1)$.*

Lemma 8 ([1]). *Let $q > 2$, $m > n$ and $\sigma > \|\tilde{\mathbf{T}}_B\| \cdot \sqrt{m} \cdot \omega(\log m)$. There is a probabilistic polynomial-time algorithm $\mathsf{SampleRight}(\mathbf{A}, \mathbf{B}, \mathbf{R}, \mathbf{T}_B, \mathbf{u}, \sigma)$ that, given a matrix $\mathbf{A}$ in $\mathbb{Z}_q^{n \times m}$, a rank n matrix $\mathbf{B}$ in $\mathbb{Z}_q^{n \times m}$, a uniform random matrix $\mathbf{R} \in \{-1, 1\}^{m \times m}$, a basis $\mathbf{T}_B$ of $\Lambda_q^\perp(\mathbf{B})$ and a vector $\mathbf{u} \in \mathbb{Z}_q^n$, outputs a vector $\mathbf{e} \in \mathbb{Z}^{2m}$ distributed statistically close to $\mathcal{D}_{\Lambda_q^\mathbf{u}(\mathbf{F}_2),\sigma}$ where $\mathbf{F}_2 := (\mathbf{A}\|\mathbf{AR} + \mathbf{B})$.*

If the input vector $\mathbf{u}$ is replaced by a matrix $\mathbf{U}$, the algorithms of $\mathsf{SampleLeft}$ and $\mathsf{SampleRight}$ still work normally and the outputs of them are matrices. We will use the matrix version of them in the construction.

Next we prove an extension of Lemma 4 using the similar method of [6], which is important to our construction of IBLE.

Lemma 9. *There is a distribution Lossy such that $\bar{\mathbf{A}} \leftarrow \mathsf{Lossy} \approx_c \mathbf{U} \xleftarrow{\$} \mathbb{Z}_q^{2m \times n}$ and given $\mathbf{s} \xleftarrow{\$} \mathbb{Z}_q^n$, and $\mathbf{e} \leftarrow \mathcal{D}_{\mathbb{Z},\beta q}^{2m \times n}$, $\tilde{H}_\infty^\epsilon(\mathbf{s}|\bar{\mathbf{A}}, \bar{\mathbf{A}}\mathbf{s} + \mathbf{x}) \geq n$, where $\epsilon = negl(\lambda)$. Lossy is as follows.*

- *Choose* $\mathbf{B} \overset{\$}{\leftarrow} \mathbb{Z}_q^{m \times k}$, $\mathbf{C} \overset{\$}{\leftarrow} \mathbb{Z}_q^{k \times n}$, $\mathbf{Z} \leftarrow \mathcal{D}_{\mathbb{Z}, \alpha q}^{m \times n}$, *and* $\mathbf{R} \overset{\$}{\leftarrow} \{1, -1\}^{m \times m}$, *where* $\frac{\alpha}{\beta} = negl(\lambda)$, $k \log q \leq n - 2\lambda + 2$, *and* $n \log q \leq m - 2\lambda + 2$.
- *Let* $\bar{\mathbf{A}} = \begin{bmatrix} \mathbf{B} \\ \mathbf{R}^t \mathbf{B} \end{bmatrix} \mathbf{C} + \begin{bmatrix} \mathbf{Z} \\ \mathbf{R}^t \mathbf{Z} \end{bmatrix}$.
- *Output* $\bar{\mathbf{A}}$.

Proof. 1. $\bar{\mathbf{A}} \approx_c \mathbf{U} \overset{\$}{\leftarrow} \mathbb{Z}_q^{2m \times n}$:

$$(\mathbf{BC} + \mathbf{Z}, \mathbf{R}^t(\mathbf{BC} + \mathbf{Z})) \overset{(1)}{\approx_c} (\mathbf{U}_1, \mathbf{R}^t \mathbf{U}_1) \overset{(2)}{\approx_c} (\mathbf{U}_1, \mathbf{U}_2), \qquad \mathbf{U}_1, \mathbf{U}_2 \overset{\$}{\leftarrow} \mathbb{Z}_q^{m \times n}$$

- Under the hardness of LWE assumption, approximate formula (1) holds.
- Let $\mathbf{r}_i$ be the i-th column of $\mathbf{R}$ where $\mathbf{r}_i \leftarrow \{-1, 1\}^m$ is uniformly random. Because $n \log q \leq m - 2\lambda + 2$, by Lemma 2, $(\mathbf{U}_1^t, \mathbf{U}_1^t \mathbf{r}_i)$ is statistically close to the uniform distribution over $\mathbb{Z}_q^{n \times m} \times \mathbb{Z}_q^n$. Because the columns of $\mathbf{R} = [\mathbf{r}_1 \ \mathbf{r}_2 \ ... \ \mathbf{r}_m]$ are sampled independently, $(\mathbf{U}_1^t, \mathbf{U}_1^t \mathbf{R})$ is statistically close to the uniform distribution over $\mathbb{Z}_q^{n \times m} \times \mathbb{Z}_q^{n \times m}$. Taking the transpose, (2) holds.

2. $\tilde{H}_\infty^\epsilon(\mathbf{s} | \bar{\mathbf{A}}, \bar{\mathbf{A}}\mathbf{s} + \mathbf{x}) \geq n$, where $\epsilon = negl(\lambda)$: Let $\mathbf{s}_0 \overset{\$}{\leftarrow} \{0, 1\}^n$, $\mathbf{s}_1 \overset{\$}{\leftarrow} \mathbb{Z}_q^n$, then, think of $\mathbf{s} = \mathbf{s}_0 + \mathbf{s}_1$. Because $\tilde{H}_\infty^\epsilon(\mathbf{s} | \bar{\mathbf{A}}\mathbf{s} + \mathbf{e}) \geq \tilde{H}_\infty^\epsilon(\mathbf{s}_0 | \bar{\mathbf{A}}\mathbf{s} + \mathbf{e})$, we will consider $\tilde{H}_\infty^\epsilon(\mathbf{s}_0 | \bar{\mathbf{A}}\mathbf{s} + \mathbf{e})$.

$$\bar{\mathbf{A}}\mathbf{s} + \mathbf{e} = \begin{bmatrix} \mathbf{B} \\ \mathbf{R}^t \mathbf{B} \end{bmatrix} \mathbf{C}\mathbf{s}_0 + \begin{bmatrix} \mathbf{Z} \\ \mathbf{R}^t \mathbf{Z} \end{bmatrix} \mathbf{s}_0 + \begin{bmatrix} \mathbf{B} \\ \mathbf{R}^t \mathbf{B} \end{bmatrix} \mathbf{C}\mathbf{s}_1 + \begin{bmatrix} \mathbf{Z} \\ \mathbf{R}^t \mathbf{Z} \end{bmatrix} \mathbf{s}_1 + \mathbf{e}$$

$$\overset{(1)}{\approx_s} \begin{bmatrix} \mathbf{B} \\ \mathbf{R}^t \mathbf{B} \end{bmatrix} \mathbf{C}\mathbf{s}_0 + \begin{bmatrix} \mathbf{B} \\ \mathbf{R}^t \mathbf{B} \end{bmatrix} \mathbf{C}\mathbf{s}_1 + \begin{bmatrix} \mathbf{Z} \\ \mathbf{R}^t \mathbf{Z} \end{bmatrix} \mathbf{s}_1 + \mathbf{e}$$

$$\overset{(2)}{\approx_s} \begin{bmatrix} \mathbf{B} \\ \mathbf{R}^t \mathbf{B} \end{bmatrix} \mathbf{u}_0 + \begin{bmatrix} \mathbf{B} \\ \mathbf{R}^t \mathbf{B} \end{bmatrix} \mathbf{C}\mathbf{s}_1 + \begin{bmatrix} \mathbf{Z} \\ \mathbf{R}^t \mathbf{Z} \end{bmatrix} \mathbf{s}_1 + \mathbf{e}$$

- Since $\frac{\alpha}{\beta} = negl(\lambda)$, each element of $\mathbf{Z}\mathbf{s}_0$ is negligibly small compared to the corresponding element of $\mathbf{e}$. And $\mathbf{R}^t \mathbf{Z}\mathbf{s}_0$ is polynomial number of adds operating on elements of $\mathbf{Z}\mathbf{s}_0$ where the elements of $\mathbf{R}$ are uniformly random chosen from $\{-1, 1\}$, so each element of $\mathbf{R}^t \mathbf{Z}\mathbf{s}_0$ is negligibly small compared to the corresponding element of $\mathbf{e}$. Therefore, $\mathbf{e} + \begin{bmatrix} \mathbf{Z} \\ \mathbf{R}^t \mathbf{Z} \end{bmatrix} \mathbf{s}_0 \approx_s \mathbf{e}$, and the approximate formula (1) holds. It means that their statistical distance is some $\epsilon_1 = negl(\lambda)$.
- Since $\mathbf{s}_0 \overset{\$}{\leftarrow} \{0, 1\}^n$ and $k \log q \leq n - 2\lambda + 2$, by Lemma 2, the approximate formula (2) holds where $\mathbf{u}_0 \overset{\$}{\leftarrow} \mathbb{Z}_q^k$. It means that their statistical distance is some $\epsilon_2 = negl(\lambda)$.

Then, for $\epsilon = \epsilon_1 + \epsilon_2 = negl(\lambda)$,

$$\tilde{H}_\infty^\epsilon(\mathbf{s}_0 | \begin{bmatrix} \mathbf{B} \\ \mathbf{R}^t \mathbf{B} \end{bmatrix} \mathbf{C}\mathbf{s}_0 + \begin{bmatrix} \mathbf{Z} \\ \mathbf{R}^t \mathbf{Z} \end{bmatrix} \mathbf{s}_0 + \begin{bmatrix} \mathbf{B} \\ \mathbf{R}^t \mathbf{B} \end{bmatrix} \mathbf{C}\mathbf{s}_1 + \begin{bmatrix} \mathbf{Z} \\ \mathbf{R}^t \mathbf{Z} \end{bmatrix} \mathbf{s}_1 + \mathbf{e})$$

$$\geq \tilde{H}_\infty^\epsilon(\mathbf{s}_0 | \begin{bmatrix} \mathbf{B} \\ \mathbf{R}^t \mathbf{B} \end{bmatrix} \mathbf{C}\mathbf{s}_0 + \begin{bmatrix} \mathbf{B} \\ \mathbf{R}^t \mathbf{B} \end{bmatrix} \mathbf{C}\mathbf{s}_1 + \begin{bmatrix} \mathbf{Z} \\ \mathbf{R}^t \mathbf{Z} \end{bmatrix} \mathbf{s}_1 + \mathbf{e})$$

$$\geq \tilde{H}_\infty^\epsilon\left(\mathbf{s}_0 \Big| \begin{bmatrix} \mathbf{B} \\ \mathbf{R}^t\mathbf{B} \end{bmatrix} \mathbf{u}_0 + \begin{bmatrix} \mathbf{B} \\ \mathbf{R}^t\mathbf{B} \end{bmatrix} \mathbf{C}\mathbf{s}_1 + \begin{bmatrix} \mathbf{Z} \\ \mathbf{R}^t\mathbf{Z} \end{bmatrix} \mathbf{s}_1 + \mathbf{e}\right)$$

$$\overset{(3)}{=} \tilde{H}_\infty(\mathbf{s}_0)$$

$$= n$$

Because each of $\mathbf{B}, \mathbf{R}, \mathbf{C}, \mathbf{Z}, \mathbf{u}_0, \mathbf{s}_1, \mathbf{e}$ is independent of $\mathbf{s}_0$, (3) holds. □

Next we describe our construction of IBLE from LWE inspired by the construction of IBE in [1]. There are some changes compared with [1]. In [1], there are errors chosen from gaussian distribution in both two ciphertexts. And the error used in the second ciphertext consists of two parts, one part $\mathbf{e}$ is from gaussian distribution, and the other is $\mathbf{R}^t\mathbf{e}$. However, in our construction, only the first ciphertext needs an error from gaussian distribution. The concrete construction is as Fig. 3. $\mathbf{H} : \mathbb{Z}_q^n \rightarrow \mathbb{Z}_q^{n\times n}$ is an encoding function constructed in [1]. This encoding function has the property that, for any two distinct inputs $\mathbf{id}_1$ and $\mathbf{id}_2$, the difference between the outputs $\mathbf{H}(\mathbf{id}_1)$ and $\mathbf{H}(\mathbf{id}_2)$ is never singular.

$\mathbf{Setup_{real}}(1^\lambda)$	$\mathbf{Setup_{lossy}}(1^\lambda, \mathbf{id}^*)$
1 $(\mathbf{A}, \mathbf{S}_A) \in \mathbb{Z}_q^{n\times m} \times \mathbb{Z}_q^{m\times m} \leftarrow \mathsf{TrapGen}(q, n)$ 2 $\mathbf{A}_1 \xleftarrow{\$} \mathbb{Z}_q^{n\times m}$, $\mathbf{A}_2 \xleftarrow{\$} \mathbb{Z}_q^{n\times m}$ 3 $\mathbf{Y} \xleftarrow{\$} \mathbb{Z}_q^{n\times l}$ 4 MPK $:= (\mathbf{A}, \mathbf{A}_1, \mathbf{A}_2, \mathbf{Y})$ 5 MSK $:= \mathbf{S}_A$ 6 return (MPK, MSK)	1 $\mathbf{B} \xleftarrow{\$} \mathbb{Z}_q^{m\times k}$, $\mathbf{C} \xleftarrow{\$} \mathbb{Z}_q^{k\times n}$, $\mathbf{Z} \leftarrow \mathcal{D}_{\mathbb{Z},\alpha q}^{m\times n}$ 2 $\mathbf{A} := (\mathbf{BC} + \mathbf{Z})^t$ 3 $\mathbf{R} \xleftarrow{\$} \{-1, 1\}^{m\times m}$, $\mathbf{Y} \xleftarrow{\$} \mathbb{Z}_q^{n\times l}$ 4 $(\mathbf{A}_2, \mathbf{S}_{A_2}) \in \mathbb{Z}_q^{n\times m} \times \mathbb{Z}_q^{m\times m} \leftarrow \mathsf{TrapGen}(q, n)$ 5 $\mathbf{A}_1 := \mathbf{AR} - \mathbf{H}(\mathbf{id}^*)\mathbf{A}_2$ 6 MPK $:= (\mathbf{A}, \mathbf{A}_1, \mathbf{A}_2, \mathbf{Y})$ 7 MSK $:= (\mathbf{S}_{A_2}, \mathbf{R})$ 8 return (MPK, MSK)
$\mathbf{Extract_{real}}(\mathbf{id}, \mathsf{MPK}, \mathsf{MSK})$	$\mathbf{Extract_{lossy}}(\mathbf{id}, \mathsf{MPK}, \mathsf{MSK})$
1 $(\mathbf{A}, \mathbf{A}_1, \mathbf{A}_2, \mathbf{Y}) := \mathsf{MPK}$ 2 $\mathbf{S}_A := \mathsf{MSK}$ 3 $\mathbf{M} := \mathbf{A}_1 + \mathbf{H}(\mathbf{id})\mathbf{A}_2$ 4 $\mathbf{X_{id}} \leftarrow \mathsf{SampleLeft}(\mathbf{A}, \mathbf{M}, \mathbf{S}_A, \mathbf{Y}, \sigma)$ 5 $\mathsf{SK_{id}} := \mathbf{X_{id}}$ 6 return $\mathsf{SK_{id}}$	1 $(\mathbf{A}, \mathbf{A}_1, \mathbf{A}_2, \mathbf{Y}) := \mathsf{MPK}$ 2 $(\mathbf{S}_{A_2}, \mathbf{R}) := \mathsf{MSK}$ 3 $\mathbf{M} := \mathbf{A}_1 + \mathbf{H}(\mathbf{id})\mathbf{A}_2 - \mathbf{AR}$ 4 $\mathbf{X_{id}} \leftarrow \mathsf{SampleRight}(\mathbf{A}, \mathbf{M}, \mathbf{R}, \mathbf{S}_{A_2}, \mathbf{Y}, \sigma)$ 5 $\mathsf{SK_{id}} := \mathbf{X_{id}}$ 6 return $\mathsf{SK_{id}}$
$\mathbf{Enc}(\mathbf{id}, \mathsf{MPK}, \mathbf{m})$	$\mathbf{Dec}(\mathsf{MPK}, \mathsf{SK_{id}}, \mathsf{C})$
1 $\mathbf{s} \xleftarrow{\$} \mathbb{Z}_q^n$, $\mathbf{e} \leftarrow \mathcal{D}_{\mathbb{Z},\beta q}^{2m}$ 2 $(\mathbf{A}, \mathbf{A}_1, \mathbf{A}_2, \mathbf{Y}) := \mathsf{MPK}$ 3 $\mathbf{A}(\mathbf{id}) := (\mathbf{A}\|\mathbf{A}_1 + \mathbf{H}(\mathbf{id})\mathbf{A}_2)$ 4 $\mathbf{c}_1 := \mathbf{A}(\mathbf{id})^t\mathbf{s} + \mathbf{e}$ 5 $\mathbf{c}_2 := \mathbf{Y}^t\mathbf{s} + \mathbf{m}\lfloor\frac{q}{2}\rceil$ 6 return $(\mathbf{c}_1, \mathbf{c}_2)$	1 $(\mathbf{c}_1, \mathbf{c}_2) := \mathsf{C}$ 2 $(\mathbf{A}, \mathbf{A}_1, \mathbf{A}_2, \mathbf{Y}) := \mathsf{MPK}$ 3 $\mathbf{m}' := \mathbf{c}_2 - \mathsf{SK}_{\mathbf{id}}^t\mathbf{c}_1$ 4 $\mathbf{m} := decode(\mathbf{m}')$ 5 return $\mathbf{m}$

Fig. 3. Construction of identity-based lossy encryption

The algorithm $decode()$ is the same in Sect. 3.2.1. In the algorithm $\textbf{Extra-ct}_{\textsf{lossy}}(\textsf{id}, \textsf{MPK}, \textsf{MSK})$, the matrix $\mathbf{M}$ is $(\mathbf{H}(\textsf{id}) - \mathbf{H}(\textsf{id}^*))\mathbf{A}_2$. When $\textsf{id} \neq \textsf{id}^*$, the trapdoor $\mathbf{S}_{A_2}$ of $\Lambda_q^{\perp}(\mathbf{A}_2)$ is also a trapdoor for $\Lambda_q^{\perp}(\mathbf{M})$ since $(\mathbf{H}(\textsf{id}) - \mathbf{H}(\textsf{id}^*))$ is non-singular.

Parameters. Consider requirements of correct decryption, and the lossiness and so on, parameters are as below, $m = \lceil 6n \log q \rceil, l \log q \leq k - 2\lambda + 2, k \log q \leq n - 2\lambda + 2, n \log q \leq m - 2\lambda + 2, q \geq 10\sigma m, \sigma \geq O(\sqrt{n \log qm})\omega(\log m), \beta \leq 1/(\sigma\sqrt{2m}\omega(\sqrt{\log 2m})), \frac{\alpha}{\beta} = negl(\lambda), \beta q > O(2\sqrt{n}), \alpha q > O(2\sqrt{n})$. To satisfy these requirements, q should be super-polynomial of the secure parameter λ.

Then, we show this scheme fulfills the properties of IBLE.

1. *Correctness on Keys for All* $\textsf{id} \neq \textsf{id}_{\textsf{lossy}}$. For all $(\textsf{MPK}, \textsf{MSK})$ generated by $\textbf{Setup}_{\textsf{real}}(1^k)$ and $\textbf{Setup}_{\textsf{lossy}}(\textsf{id}_{\textsf{lossy}})$, all $\textsf{SK}_{\textsf{id}}$ generated by $\textbf{Extract}_{\textsf{real},\textsf{lossy}}(\textsf{id}, \textsf{MPK}, \textsf{MSK})$, and all messages $\mathbf{m}$,

$$\textbf{Dec}(\textsf{id}, \textsf{SK}_{\textsf{id}}, \textbf{Enc}(\textsf{id}, \textsf{MPK}, \mathbf{m}))$$
$$= \textbf{Dec}(\textsf{id}, \textsf{SK}_{\textsf{id}}, \textbf{Enc}((\mathbf{A}, \mathbf{A}_1, \mathbf{A}_2, \mathbf{Y}), \mathbf{m}))$$
$$= \textbf{Dec}(\textsf{id}, \textsf{SK}_{\textsf{id}}, ((\mathbf{A}\|\mathbf{A}_1 + \mathbf{H}(\textsf{id})\mathbf{A}_2)^t\mathbf{s} + \mathbf{e}, \mathbf{Y}^t\mathbf{s} + \mathbf{m}\lfloor\frac{q}{2}\rceil))$$
$$= decode(\mathbf{Y}^t\mathbf{s} + \mathbf{m}\lfloor\frac{q}{2}\rceil - \textsf{SK}_{\textsf{id}}^t\mathbf{A}(\textsf{id})^t\mathbf{s} - \textsf{SK}_{\textsf{id}}^t\mathbf{e})$$
$$\stackrel{(1)}{=} decode(\mathbf{Y}^t\mathbf{s} + \mathbf{m}\lfloor\frac{q}{2}\rceil - \mathbf{Y}^t\mathbf{s} - \textsf{SK}_{\textsf{id}}^t\mathbf{e})$$
$$= decode(\mathbf{m}\lfloor\frac{q}{2}\rceil - \textsf{SK}_{\textsf{id}}^t\mathbf{e})$$
$$= \mathbf{m}$$

Because $\textsf{SK}_{\textsf{id}}$ is generated by SampleLeft, $\textsf{SK}_{\textsf{id}}^t = \mathbf{Y}^t$, and (1) holds. By Lemma 5, the algorithm $decode()$ will get the correct message with overwhelming probability.

2. *Lossiness of Encryption with Lossy Keys for* $\textsf{id} = \textsf{id}_{\textsf{lossy}}$.

$$\textbf{Enc}(\textsf{id}, \textsf{MPK}_{\textsf{lossy}}, \mathbf{m})$$
$$= \textbf{Enc}(\textsf{id}, (\mathbf{BC} + \mathbf{Z})^t, (\mathbf{BC} + \mathbf{Z})^t\mathbf{R} - \mathbf{H}(\textsf{id})\mathbf{A}_2, \mathbf{A}_2, \mathbf{Y}), \mathbf{m})$$
$$= (\begin{bmatrix} \mathbf{BC} + \mathbf{Z} \\ ((\mathbf{BC} + \mathbf{Z})^t\mathbf{R} - \mathbf{H}(\textsf{id})\mathbf{A}_2 + \mathbf{H}(\textsf{id})\mathbf{A}_2)^t \end{bmatrix}\mathbf{s} + \mathbf{e}, \mathbf{Y}^t\mathbf{s} + \mathbf{m}\lfloor\frac{q}{2}\rceil)$$
$$= (\begin{bmatrix} \mathbf{BC} + \mathbf{Z} \\ \mathbf{R}^t(\mathbf{BC} + \mathbf{Z}) \end{bmatrix}\mathbf{s} + \mathbf{e}, \mathbf{Y}^t\mathbf{s} + \mathbf{m}\lfloor\frac{q}{2}\rceil)$$
$$= ((\begin{bmatrix} \mathbf{B} \\ \mathbf{R}^t\mathbf{B} \end{bmatrix}\mathbf{C} + \begin{bmatrix} \mathbf{Z} \\ \mathbf{R}^t\mathbf{Z} \end{bmatrix})\mathbf{s} + \mathbf{e}, \mathbf{Y}^t\mathbf{s} + \mathbf{m}\lfloor\frac{q}{2}\rceil)$$

Let $\mathbf{A}' = \begin{bmatrix} \mathbf{B} \\ \mathbf{R}^t\mathbf{B} \end{bmatrix}\mathbf{C} + \begin{bmatrix} \mathbf{Z} \\ \mathbf{R}^t\mathbf{Z} \end{bmatrix}$. Because the parameters satisfy the requirements of Lemma 9, we know that $\tilde{H}_{\infty}(\mathbf{s}|\mathbf{A}'\mathbf{s} + \mathbf{e}) \geq n$. Then because

$l \leq (k - 2\log(1/\varepsilon) - O(1))/\log q$, and by Lemma 2, given $\mathbf{A}'\mathbf{s} + \mathbf{e}$, $\mathbf{Y}^t\mathbf{s}$ is ε-close to $\mathcal{U}(\mathbb{Z}_q^l)$. When $\varepsilon = negl(\lambda)$, $\mathbf{Y}^t\mathbf{s} \approx_s \mathcal{U}(\mathbb{Z}_q^l)$ given $\mathbf{A}'\mathbf{s} + \mathbf{e}$. Therefore, for any $\mathbf{m} \in \mathcal{M}$, $(\mathbf{Y}^t\mathbf{s} + \mathbf{m}\lfloor\frac{q}{2}\rfloor)$ is statistically close to $\mathcal{U}(\mathbb{Z}_q^l)$, given $\mathbf{A}'\mathbf{s} + \mathbf{e}$, i.e. for the lossy identity $\mathbf{id}$, any lossy keys $\mathsf{MPK}_{\mathsf{lossy}}$ generated by $\mathbf{Setup}_{\mathbf{lossy}}(\mathbf{id})$ and any two messages $\mathbf{m}_0 \neq \mathbf{m}_1$, there is $\mathbf{Enc}(\mathbf{id}, \mathsf{MPK}_{\mathsf{lossy}}, \mathbf{m}_0) \approx_s \mathbf{Enc}(\mathbf{id}, \mathsf{MPK}_{\mathsf{lossy}}, \mathbf{m}_1)$.

3. *Indistinguishability Between Real Keys and Lossy Keys.* We use a game sequence to prove this property.

$\mathbf{G}_0$: This is the original game from the definition of the third property of identity-based lossy encryption described as Fig. 1.

$\mathbf{G}_1$: In $\mathbf{G}_0$, the master public key MPK_1 generated by the challenger is $((\mathbf{BC} + \mathbf{Z})^t,(\mathbf{BC} + \mathbf{Z})^t\mathbf{R} - \mathbf{H}(\mathbf{id}^*)\mathbf{A}_2, \mathbf{A}_2, \mathbf{Y})$. In $\mathbf{G}_1$, we use a random matrix $\mathbf{U}$ in $\mathbb{Z}_q^{n \times m}$ to replace $(\mathbf{BC} + \mathbf{Z})^t$. It means that MPK_1 in this game is $(\mathbf{U}, \mathbf{UR} - \mathbf{H}(\mathbf{id}^*)\mathbf{A}_2, \mathbf{A}_2, \mathbf{Y})$. The remainder of the game is unchanged.

Suppose there is an adversary $\mathcal{A}$ has non-negligible advantage in distinguishing $\mathbf{G}_0$ and $\mathbf{G}_1$. Then we use $\mathcal{A}$ to construct an algorithm $\mathcal{S}$ as Fig. 4 to distinguish a random matrix $\mathbf{U}$ and an LWE instance $((\mathbf{BC} + \mathbf{Z})^t)$. In words, the algorithm $\mathcal{S}$ proceeds as follows. $\mathcal{S}$ requests the oracle $\mathcal{O}$ which outputs a random matrix $\mathbf{U}$ or an LWE instance $(\mathbf{BC} + \mathbf{Z})^t$, and receives a matrix $\mathbf{A}'$. After receiving the target identity $\mathbf{id}^*$ sent by $\mathcal{A}$, $\mathcal{S}$ works as the $\mathbf{Setup}_{\mathbf{real}}$ algorithm in Fig. 3 to generate the real keys $\mathsf{MPK}_0, \mathsf{MSK}_0$, and uses $\mathbf{A}'$ to generate the lossy keys $\mathsf{MPK}_1, \mathsf{MSK}_1$. Then $\mathcal{S}$ randomly chooses b from $\{0, 1\}$ and sends MPK_b to $\mathcal{A}$. Then $\mathcal{A}$ issues private key extraction queries on $\mathbf{id}_i$ where $\mathbf{id}_i \neq \mathbf{id}^*$.

We argue that when the oracle $\mathcal{O}$ outputs an LWE instance $(\mathbf{BC} + \mathbf{Z})^t$, MPK_b is distributed exactly as in $\mathbf{G}_0$. If $b = 0$, MPK_0 is the real public key, and else $b = 1$, MPK_1 is $((\mathbf{BC} + \mathbf{Z})^t, (\mathbf{BC} + \mathbf{Z})^t\mathbf{R} - \mathbf{H}(\mathbf{id}^*)\mathbf{A}'_2, \mathbf{A}'_2, \mathbf{Y}')$. This is the same as in $\mathbf{G}_0$. When the oracle $\mathcal{O}$ outputs a random matrix $\mathbf{U}$, MPK_0 is unchanged and MPK_1 is $(\mathbf{U}, \mathbf{UR} - \mathbf{H}(\mathbf{id}^*)\mathbf{A}'_2, \mathbf{A}'_2, \mathbf{Y}')$. In this case, MPK is the same as in $\mathbf{G}_1$.

At last, $\mathcal{A}$ guesses if it is interacting with $\mathbf{G}_0$ or $\mathbf{G}_1$. $\mathcal{S}$ uses $\mathcal{A}$'s guess to answer whether $\mathbf{A}'$ is a random matrix or an LWE instance. Hence, $\mathcal{S}$'s advantage in distinguishing $\mathbf{U}$ and $(\mathbf{BC} + \mathbf{Z})^t$ is the same as $\mathcal{A}$'s advantage in distinguishing $\mathbf{G}_0$ and $\mathbf{G}_1$. Because $(\mathbf{BC} + \mathbf{Z})^t \approx_c \mathbf{U}$ by Lemma 4, $\mathbf{G}_0$ and $\mathbf{G}_1$ are computationally indistinguishable.

In $\mathbf{G}_1$, $\mathsf{MPK}_1 = (\mathbf{U}, \mathbf{UR} - \mathbf{H}(\mathbf{id}^*)\mathbf{A}'_2, \mathbf{A}'_2, \mathbf{Y}') \approx_c (\mathbf{U}_1, \mathbf{U}_2, \mathbf{A}'_2, \mathbf{Y}')$ by Lemma 2 where $\mathbf{U}_1, \mathbf{U}_2$ are random matrices. Hence, MPK_1 is statistically indistinguishable with MPK_0 which is $(\mathbf{A}, \mathbf{A}_1, \mathbf{A}_2, \mathbf{Y})$. Let $\mathbf{F}_1 = (\mathbf{A}\|\mathbf{A}_1 + \mathbf{H}(\mathbf{id})\mathbf{A}_2)$, $\mathbf{F}_2 = (\mathbf{U}\|\mathbf{UR} + (\mathbf{H}(\mathbf{id}) - \mathbf{H}(\mathbf{id}^*))\mathbf{A}'_2)$. For all $\mathbf{id} \neq \mathbf{id}^*$, $\mathbf{Extract}_{\mathbf{real}}(\mathbf{id}, \mathsf{MPK}_0, \mathsf{MSK}_0)$ uses algorithm SampleLeft to extract $\mathsf{SK}_{\mathbf{id}}$, so by Lemma 7, the distribution of $\mathsf{SK}_{\mathbf{id}}$ is statistically close to $\mathcal{D}_{\Lambda_q^{\mathbf{Y}}(\mathbf{F}_1), \sigma}$. $\mathbf{Extract}_{\mathbf{lossy}}(\mathbf{id}, \mathsf{MPK}_1, \mathsf{MPK}_1)$ uses algorithm SampleRight to extract $\mathsf{SK}_{\mathbf{id}}$, so by Lemma 8, the distribution of $\mathsf{SK}_{\mathbf{id}}$ is statistically close to $\mathcal{D}_{\Lambda_q^{\mathbf{Y}}(\mathbf{F}_2), \sigma}$. And because $\mathsf{MPK}_0 \approx_s \mathsf{MPK}_1$, $\mathbf{F}_1$ and $\mathbf{F}_2$ are statistically indistinguishable.

The process of $\mathcal{S}$

$\mathbf{A}' \leftarrow \mathcal{O}$
$\mathsf{id}^* \leftarrow \mathcal{A}$
$(\mathsf{MPK}_0, \mathsf{MSK}_0) \leftarrow \mathbf{Setup}_{\mathsf{real}}(1^\lambda)$
$(\mathbf{A}'_2, \mathbf{S}_{A'_2}) \in \mathbb{Z}_q^{n \times m} \times \mathbb{Z}_q^{m \times m} \leftarrow \mathbf{TrapGen}$
$\mathbf{R} \xleftarrow{\$} \{1, -1\}^{m \times m}, \mathbf{Y}' \xleftarrow{\$} \mathbb{Z}_q^{n \times l}$
$\mathbf{A}'_1 := \mathbf{A}'\mathbf{R} - \mathbf{H}(\mathsf{id}^*)\mathbf{A}'_2$
$\mathsf{MPK}_1 := (\mathbf{A}', \mathbf{A}'_1, \mathbf{A}'_2, \mathbf{Y}')$
$\mathsf{MSK}_1 := \mathbf{S}_{A'_2}$
$\mathsf{mode}_0 := \mathsf{real}$
$\mathsf{mode}_1 := \mathsf{lossy}$
$b \in_R \{0, 1\}$
sends MPK_b
for $i = 1$ to t
    $\mathsf{id}_i \leftarrow \mathcal{A}$
    $\mathsf{SK}_{\mathsf{id}_i} \leftarrow \mathbf{Extract}_{\mathsf{mode}_b}(\mathsf{id}_i, \mathsf{MPK}_b, \mathsf{MSK}_b)$
    sends $\mathsf{SK}_{\mathsf{id}_i}$
end

Fig. 4. The process of $\mathcal{S}$

Therefore, $\mathcal{D}_{\Lambda_q^\mathbf{Y}(\mathbf{F}_1)}$ is statistically indistinguishable with $\mathcal{D}_{\Lambda_q^\mathbf{Y}(\mathbf{F}_2)}$, i.e. any $\mathsf{SK}_{\mathsf{id}}$ generated by $\mathbf{Extract}_{\mathsf{real}}(\mathsf{id}, \mathsf{MPK}_0, \mathsf{MPK}_0)$ is statistically indistinguishable with any $\mathsf{SK}_{\mathsf{id}}$ generated by $\mathbf{Extract}_{\mathsf{lossy}}(\mathsf{id}, \mathsf{MPK}_1, \mathsf{MPK}_1)$ for all $\mathsf{id} \neq \mathsf{id}^*$. Therefore, the advantage of the adversary of $\mathbf{G}_1$ is $negl(\lambda)$.

Above all, the advantage of $\mathbf{G}_0$'s adversary is $negl(\lambda)$. This completes the proof. $\square$

4   Conclusion

In this paper, we extend the notion of lossy encryption proposed by [3] to the scenario of identity-based encryption. This new notion, identity-based lossy encryption, implies IND-sID-SO security under selective identity. And we provide a construction of identity-based lossy encryption based on LWE.

Acknowledgments. We would like to thank the anonymous reviewers for their helpful comments. We would further like to thank Yamin Liu for the helpful revision suggestion.

References

1. Agrawal, S., Boneh, D., Boyen, X.: Efficient lattice (H)IBE in the standard model. In: Gilbert, H. (ed.) EUROCRYPT 2010. LNCS, vol. 6110, pp. 553–572. Springer, Heidelberg (2010)

2. Agrawal, S., Boneh, D., Boyen, X.: Lattice basis delegation in fixed dimension and shorter-ciphertext hierarchical IBE. In: Rabin, T. (ed.) CRYPTO 2010. LNCS, vol. 6223, pp. 98–115. Springer, Heidelberg (2010)

3. Bellare, M., Hofheinz, D., Yilek, S.: Possibility and impossibility results for encryption and commitment secure under selective opening. In: Joux, A. (ed.) EUROCRYPT 2009. LNCS, vol. 5479, pp. 1–35. Springer, Heidelberg (2009)

4. Bellare, M., Kiltz, E., Peikert, C., Waters, B.: Identity-based (lossy) trapdoor functions and applications. In: Pointcheval, D., Johansson, T. (eds.) EUROCRYPT 2012. LNCS, vol. 7237, pp. 228–245. Springer, Heidelberg (2012)

5. Bellare, M., Waters, B., Yilek, S.: Identity-based encryption secure against selective opening attack. In: Ishai, Y. (ed.) TCC 2011. LNCS, vol. 6597, pp. 235–252. Springer, Heidelberg (2011)

6. Berkoff, A., Liu, F.-H.: Leakage resilient fully homomorphic encryption. In: Lindell, Y. (ed.) TCC 2014. LNCS, vol. 8349, pp. 515–539. Springer, Heidelberg (2014)

7. Boneh, D., Boyen, X.: Efficient selective-id secure identity-based encryption without random oracles. In: Cachin, C., Camenisch, J.L. (eds.) EUROCRYPT 2004. LNCS, vol. 3027, pp. 223–238. Springer, Heidelberg (2004)

8. Boneh, D., Boyen, X.: Secure identity based encryption without random oracles. In: Franklin, M. (ed.) CRYPTO 2004. LNCS, vol. 3152, pp. 443–459. Springer, Heidelberg (2004)

9. Boneh, D., Franklin, M.: Identity-based encryption from the weil pairing. In: Kilian, J. (ed.) CRYPTO 2001. LNCS, vol. 2139, pp. 213–229. Springer, Heidelberg (2001)

10. Boneh, D., Gentry, C., Hamburg, M.: Space-efficient identity based encryption-without pairings. In: 48th Annual IEEE Symposium on Foundations of Computer Science, 2007, FOCS 2007, pp. 647–657. IEEE (2007)

11. Cash, D., Hofheinz, D., Kiltz, E., Peikert, C.: Bonsai trees, or how to delegate a lattice basis. J. Cryptol. **25**(4), 601–639 (2012)

12. Cocks, C.: An identity based encryption scheme based on quadratic residues. In: Honary, B. (ed.) Cryptography and Coding 2001. LNCS, vol. 2260, pp. 360–363. Springer, Heidelberg (2001)

13. Dodis, Y., Reyzin, L., Smith, A.: Fuzzy extractors: how to generate strong keys from biometrics and other noisy data. In: Cachin, C., Camenisch, J.L. (eds.) EUROCRYPT 2004. LNCS, vol. 3027, pp. 523–540. Springer, Heidelberg (2004)

14. Döttling, N., Müller-Quade, J.: Lossy codes and a new variant of the learning-with-errors problem. In: Johansson, T., Nguyen, P.Q. (eds.) EUROCRYPT 2013. LNCS, vol. 7881, pp. 18–34. Springer, Heidelberg (2013)

15. Escala, A., Herranz, J., Libert, B., Ràfols, C.: Identity-based lossy trapdoor functions: new definitions, hierarchical extensions, and implications. In: Krawczyk, H. (ed.) PKC 2014. LNCS, vol. 8383, pp. 239–256. Springer, Heidelberg (2014)

16. Gentry, C., Peikert, C., Vaikuntanathan, V.: Trapdoors for hard lattices and new cryptographic constructions. In: Proceedings of the Fortieth Annual ACM Symposium on Theory of computing, pp. 197–206. ACM (2008)

17. Goldwasser, S., Kalai, Y., Peikert, C., Vaikuntanathan, V.: Robustness of the learning with errors assumption. In: Yao, A.C.-C. (ed.) ICS, pp. 230–240. Tsinghua University Press, Beijing (2010)

18. Hemenway, B., Libert, B., Ostrovsky, R., Vergnaud, D.: Lossy encryption: constructions from general assumptions and efficient selective opening chosen ciphertext security. In: Lee, D.H., Wang, X. (eds.) ASIACRYPT 2011. LNCS, vol. 7073, pp. 70–88. Springer, Heidelberg (2011)

19. Hemenway, B., Ostrovsky, R.: Building lossy trapdoor functions from lossy encryption. In: Sako, K., Sarkar, P. (eds.) ASIACRYPT 2013, Part II. LNCS, vol. 8270, pp. 241–260. Springer, Heidelberg (2013)
20. Jhanwar, M.P., Barua, R.: A variant of Boneh-Gentry-Hamburg's pairing-free identity based encryption scheme. In: Yung, M., Liu, P., Lin, D. (eds.) Inscrypt 2008. LNCS, vol. 5487, pp. 314–331. Springer, Heidelberg (2009)
21. Lai, J., Deng, R.H., Liu, S., Weng, J., Zhao, Y.: Identity-based encryption secure against selective opening chosen-ciphertext attack. In: Nguyen, P.Q., Oswald, E. (eds.) EUROCRYPT 2014. LNCS, vol. 8441, pp. 77–92. Springer, Heidelberg (2014)
22. Peikert, C., Vaikuntanathan, V., Waters, B.: A framework for efficient and composable oblivious transfer. In: Wagner, D. (ed.) CRYPTO 2008. LNCS, vol. 5157, pp. 554–571. Springer, Heidelberg (2008)
23. Regev, O.: On lattices, learning with errors, random linear codes, and cryptography. In: Proceedings of the Thirty-seventh Annual ACM Symposium on Theory of Computing, STOC 2005, pp. 84–93. ACM, New York, NY, USA (2005)
24. Sakai, R., Ohgishi, K., Kasahara, M.: Cryptosystems based on pairing. In: The 2000 Symposium on Cryptography and Information Security, Okinawa, Japan, pp. 135–148 (2000)
25. Shamir, A.: Identity-based cryptosystems and signature schemes. In: Blakely, G.R., Chaum, D. (eds.) CRYPTO 1984. LNCS, vol. 196, pp. 47–53. Springer, Heidelberg (1985)
26. Waters, B.: Efficient identity-based encryption without random oracles. In: Cramer, R. (ed.) EUROCRYPT 2005. LNCS, vol. 3494, pp. 114–127. Springer, Heidelberg (2005)
27. Waters, B.: Dual system encryption: realizing fully secure IBE and HIBE under simple assumptions. In: Halevi, S. (ed.) CRYPTO 2009. LNCS, vol. 5677, pp. 619–636. Springer, Heidelberg (2009)

Adaptive-ID Secure Revocable Hierarchical Identity-Based Encryption

Jae Hong Seo[1] and Keita Emura[2]($^{(\boxtimes)}$)

[1] Department of Mathematics, Myongji University, Seoul, Korea
`jaehongseo@mju.ac.kr`
[2] Security Fundamentals Lab, National Institute of Information
and Communications Technology (NICT), Tokyo, Japan
`k-emura@nict.go.jp`

Abstract. Revocable Hierarchical Identity-Based Encryption (RHIBE) is a variant of Identity-Based Encryption (IBE), which enables the dynamic user management; a Key Generation Center (KGC) of a usual IBE has a key issuing ability. In contrast, in a RHIBE, a KGC can revoke compromised secret keys and even delegate both key issuing ability and revocation ability.

Recently, Seo and Emura proposed the first construction for RHIBE (CT-RSA 2013) and then refined the security model and the construction for RHIBE (CT-RSA 2015). Nevertheless, their constructions achieve only a slightly weaker security notion, called *selective-ID security*, in the sense that the adversary has to choose and declare the target identity before she receives the system parameter of target RHIBE scheme.

In this paper, we propose the first RHIBE construction that achieves a *right* security notion, called *adaptive-ID security*. In particular, our construction still has the advantages of the Seo-Emura RHIBE schemes; that is, it is scalable and achieves history-free update, security against insiders, and short ciphertexts. We employ the dual system encryption methodology.

1 Introduction

Revocable Identity-Based Encryption: Revocation is one of the important issues in the context of Identity-Based Encryption (IBE). Boneh and Franklin [3,4] considered the first (non-scalable) revocation method in the IBE context, where for all non-revoked identities ID at a time period T, secret keys are computed for $\mathsf{ID}\|T$. The first scalable IBE with revocation, which we call RIBE, is proposed by Boldyreva, Goyal, and Kumar (BGK) [1] where a Key Generation Center (KGC) broadcasts key update information ku_T at time T, and only non-revoked users at T can compute their decryption keys $\mathsf{dk}_{\mathsf{ID},T}$ by using their (long-term) secret key $\mathsf{sk}_{\mathsf{ID}}$ and ku_T. It is worth noting that the size of ku_T is $O(R\log(N/R))$ by using the Naor-Naor-Lotspiech (NNL) framework [8] called Complete Subtree (CS) method, where R is the number of revoked users and N is the total number of users, whereas that of the Boneh-Franklin

© Springer International Publishing Switzerland 2015
K. Tanaka and Y. Suga (Eds.): IWSEC 2015, LNCS 9241, pp. 21–38, 2015.
DOI: 10.1007/978-3-319-22425-1_2

scheme is $O(N - R)$. Though the BGK scheme is selective-ID secure, Libert and Vergnaud (LV) [7] proposed the first adaptive-ID secure RIBE scheme. Next, Seo and Emura [10,12] considered a new security threat called decryption key exposure resistance. They show that the Boneh-Franklin RIBE scheme is secure against decryption key exposure attack but the BGK and LV schemes are vulnerable against this attack, and proposed the first scalable RIBE scheme with decryption key exposure resistance.

RIBE with Hierarchical Structures: Seo and Emura [9,11] proposed the first Revocable hierarchical IBE (RHIBE) scheme with $O(\ell^2 \log N)$-size secret key, where ℓ is the maximum hierarchical level. However, in their construction a low-level user must know the history of key updates performed by ancestors in the current time period. This history-preserving update methodology makes the scheme very complex. Moreover, the ciphertext size depends on the depth of the hierarchy (i.e., $O(\ell)$). In addition to this, the security model only considered outsiders, where an adversary is not allowed to access to state information used by internal users of the RHIBE system.

Recently, Seo and Emura [13] proposed a RHIBE scheme which implements (1) history-free updates where a low-level user does not need to know the history of the key updates of ancestors, (2) security against insiders where an adversary is allowed to obtain internal state information, and (3) short ciphertexts where the size of ciphertext is constant in terms of the hierarchical level. Moreover, they considered decryption key exposure attack [10,12]. Their scheme is based on the Boneh-Boyen-Goh (BBG) HIBE scheme [2], i.e., they proved that the Seo-Emura RHIBE scheme is secure if the BBG HIBE is secure.

Motivation: There is a room for improvement of the Seo-Emura RHIBE scheme [13] because it achieves selective-ID security, where an adversary is required to declare the challenge identity before the system setup phase. We remark that the RHIBE scheme in [9,11] also selective-ID secure. Though Tsai et al. [14] proposed an adaptive-ID secure RHIBE scheme, this scheme is not scalable, i.e., the size of key update linearly depends on N.

The Seo-Emura contrcution [13] is a kind of conversion from HIBE to RHIBE since their construction and security proof are based on the (non-revocable) BBG HIBE scheme. Basically, the selective-ID security of the Seo-Emura construction comes from the underlying BBG HIBE scheme, which is also proven in the selective-ID security model only. Therefore, one may think that if one changes the underlying HIBE with the adaptive-ID secure HIBE such as [6], one can directly obtain the first adaptively secure RHIBE construction. Unfortunately, In the security proof of Seo and Emura [13], the reduction algorithm itself requires the challenge identity before the setup phase so that we can easily see this trivial approach ends in failure. Therefore, we need a different technique for the adaptive-ID secure RHIBE scheme.

Contribution and Methodology: In this paper, we propose the first scalable and adaptive-ID secure RHIBE scheme which has the advantages of the Seo-Emura RHIBE scheme [13]. In the contrast to the previous approaches [9–13], we propose a scalable scheme and then directly prove its security. To this end, we modify the Lewko-Waters HIBE scheme [6] appropriately and then directly prove its adaptive security by employing dual system encryption methodology [15].

In the (usual) dual system encryption, a normal ciphertext can be decrypted by using either a normal decryption key or a semi-functional decryption key, but a semi-functional ciphertext cannot be decrypted by using a semi-functional decryption key. Since a decryption key is computed by key update information and a secret key in the RHIBE context, we need to check whether a decryption key is semi-functional or not when it is generated by normal/semi-functional key update information and a normal/semi-functional secret key. Especially, in the security proof of the Lewko-Waters HIBE scheme, in order to prevent that the simulator $\mathcal{B}$ attempts to test itself whether key is semi-functional by creating a semi-functional ciphertext and trying to decrypt, $\mathcal{B}$ is required to compute a "nominally" semi-functional key only which still works for a semi-functional ciphertext. Since there are many routes for generating decryption keys in the RHIBE context, in the security proof we need to guarantee not only a decryption key is nominally semi-functional but also a decryption key generated by key update information and a secret key is nominally semi-functional.

2 Preliminaries

In this section, we give the definitions of complexity assumptions. Let $\mathcal{G}$ be a group generator which takes as input the security parameter λ, and outputs $(n = p_1p_2p_3, \mathbb{G}, \mathbb{G}_T, e)$, where $\mathbb{G}$ and $\mathbb{G}_T$ be cyclic groups with order n, p_1, p_2, and p_3 are distinct prime numbers, and $e : \mathbb{G} \times \mathbb{G} \to \mathbb{G}_T$ is an efficient, non-degenerate bilinear map. We denote the subgroup of $\mathbb{G}$ with order m as $\mathbb{G}_m$. Here, for all $X_i \in \mathbb{G}_{p_i}$ and $X_j \in \mathbb{G}_{p_j}$ ($i, j \in \{1, 2, 3\}$), $e(X_i, X_j) = 1_T$ holds if $i \neq j$.

Definition 1 (Assumption 1 [6]). *Let* $(n, \mathbb{G}, \mathbb{G}_T, e) \overset{\$}{\leftarrow} \mathcal{G}(1^\lambda)$, $n = p_1p_2p_3$, $g \overset{\$}{\leftarrow} \mathbb{G}_{p_1}$, *and* $X_3 \overset{\$}{\leftarrow} \mathbb{G}_{p_3}$. *Set* $D := ((n, \mathbb{G}, \mathbb{G}_T, e), g, X_3)$. *We say that Assumption 1 holds if for all probabilistic polynomial-time (PPT) adversaries* $\mathcal{A}$, $|\Pr[\mathcal{A}(D, T_1) = 1] - \Pr[\mathcal{A}(D, T_2) = 1]|$ *is negligible, where* $T_1 \overset{\$}{\leftarrow} \mathbb{G}_{p_1p_2}$ *and* $T_2 \overset{\$}{\leftarrow} \mathbb{G}_{p_1}$.

Definition 2 (Assumption 2 [6]). *Let* $(n, \mathbb{G}, \mathbb{G}_T, e) \overset{\$}{\leftarrow} \mathcal{G}(1^\lambda)$, $n = p_1p_2p_3$, $g, X_1 \overset{\$}{\leftarrow} \mathbb{G}_{p_1}$, $X_2, Y_2 \overset{\$}{\leftarrow} \mathbb{G}_{p_2}$, *and* $X_3, Y_3 \overset{\$}{\leftarrow} \mathbb{G}_{p_3}$. *Set* $D := ((n, \mathbb{G}, \mathbb{G}_T, e), g, X_1X_2, X_3, Y_2Y_3)$. *We say that Assumption 2 holds if for all PPT adversaries* $\mathcal{A}$, $|\Pr[\mathcal{A}(D, T_1) = 1] - \Pr[\mathcal{A}(D, T_2) = 1]|$ *is negligible, where* $T_1 \overset{\$}{\leftarrow} \mathbb{G}$ *and* $T_2 \overset{\$}{\leftarrow} \mathbb{G}_{p_1p_3}$.

Definition 3 (Assumption 3 [6]). *Let* $(n, \mathbb{G}, \mathbb{G}_T, e) \overset{\$}{\leftarrow} \mathcal{G}(1^\lambda)$, $n = p_1p_2p_3$, $\alpha, s \overset{\$}{\leftarrow} \mathbb{Z}_n$, $g \overset{\$}{\leftarrow} \mathbb{G}_{p_1}$, $X_2, Y_2, Z_2 \overset{\$}{\leftarrow} \mathbb{G}_{p_2}$, *and* $X_3 \overset{\$}{\leftarrow} \mathbb{G}_{p_3}$. *Set* $D := ((n, \mathbb{G}, \mathbb{G}_T, e), g, g^\alpha X_2, X_3, g^s Y_2, Z_2)$. *We say that Assumption 3 holds if for all*

PPT adversaries $\mathcal{A}$, $|\Pr[\mathcal{A}(D, T_1) = 1] - \Pr[\mathcal{A}(D, T_2) = 1]|$ *is negligible, where* $T_1 = e(g, g)^{\alpha s}$ *and* $T_2 \xleftarrow{\$} \mathbb{G}_T$.

Next, we introduce the KUNode algorithm [1] which implements the CS method. If v is a non-leaf node, then v_l denotes the left child of v. Similarly, v_r is the right child of v. We assume that each user is assigned to a unique leaf node. If a user, which is assigned to v, is revoked on time T, then (v, T) is added into the revocation list RL.

KUNode(BT, RL, T) :

   X, Y $\leftarrow \emptyset$;

   For $\forall (v_i, T_i) \in RL$, if $T_i \leq T$, then add Path(v_i) to X;

   For $\forall v \in$ X, if $v_l \notin$ X, then add v_l to Y;

        if $v_r \notin$ X, then add v_r to Y;

   If $|RL| = 0$, then add root to Y;

   Return Y;

3 Revocable Hierarchical Identity-Based Encryption

In this section, we introduce the definition of RHIBE given by Seo and Emura [13] as follows. In order to achieve history-free updates, Seo and Emura pointed out that the following two situations are equivalent; (1) a user ID is not revoked at time T, and (2) the user can generate the decryption key $dk_{ID,T}$, and re-defined the syntax of RHIBE from that of [9,11].

Definition 1 [13]. *RHIBE consists of seven algorithms* Setup, SKGen, KeyUp, DKGen, Enc, Dec, *and* Revoke *defined as follows.*

Setup($1^\lambda, N, \ell$): *This algorithm takes as input the security parameter* 1^λ, *maximum number of users in each level* N, *and maximum hierarchical length* ℓ, *and outputs the public system parameter* mpk, *the master secret key* msk, *initial state information* st_0, *and empty revocation list* RL. *We assume that* mpk *contains description of message space* $\mathcal{M}$, *identity space* $\mathcal{I}$, *and time space* $\mathcal{T}$. *For simplicity, we often omit* mpk *in the input of other algorithms.*

SKGen($st_{ID|_{k-1}}, ID|_k$): *This algorithm takes as input* $st_{ID|_{k-1}}$ *and an identity* $ID|_k$, *outputs the secret key* $sk_{ID|_k}$, *and updates* $st_{ID|_{k-1}}$.

KeyUp($dk_{ID|_{k-1},T}, st_{ID|_{k-1}}, RL_{ID|_{k-1}}, T$): *This algorithm takes as input the revocation list* $RL_{ID|_{k-1}}$, *state information* $st_{ID|_{k-1}}$, *the decryption key* $dk_{ID|_{k-1},T}$, *and a time period* T. *For* $k = 1$, *we set* $dk_{ID|_{k-1},T}$ *to be* msk *disregarding* T. *Then, it outputs the key update* $ku_{ID|_{k-1},T}$.

DKGen($sk_{ID|_k}, ku_{ID|_{k-1},T}$): *This algorithm takes as input* $sk_{ID|_k}$ *of* $ID|_k$ *and* $ku_{ID|_{k-1},T}$, *and outputs the decryption key* $dk_{ID|_k,T}$ *of* $ID|_k$ *at time* T *if* $ID|_k$ *is not revoked at* T *by the parent.*

$\mathsf{Enc}(M, \mathsf{ID}, T)$: *This algorithm takes as input a message M, ID and the current time T and outputs the ciphertext CT.*

$\mathsf{Dec}(\mathsf{CT}, \mathsf{dk}_{\mathsf{ID},T})$: *This algorithm takes as input CT and $\mathsf{dk}_{\mathsf{ID},T}$, and outputs the message.*

$\mathsf{Revoke}(\mathsf{ID}|_k, T, RL_{\mathsf{ID}|_{k-1}})$: *This algorithm takes as input $\mathsf{ID}|_k$ and T, updates $RL_{\mathsf{ID}|_{k-1}}$ managed by $\mathsf{ID}|_{k-1}$, who is the parent user of $\mathsf{ID}|_k$, by adding $(\mathsf{ID}|_k, T)$.*

For any output $\mathsf{Setup}(1^\lambda, N, \ell) \to (\mathsf{mpk}, \mathsf{msk})$, any message $M \in \mathcal{M}$, any identity $\mathsf{ID}|_k \in \mathcal{I}$ where $k \in [1, \ell]$, any time $T \in \mathcal{T}$, all possible states $\{\mathsf{st}_{\mathsf{ID}|_i}\}_{i \in [1, k-1]}$, and all possible revocation lists $\{RL_{\mathsf{ID}|_i}\}_{i \in [1, k-1]}$, if $\mathsf{ID}|_k$ is not revoked on time T, the following probability should be 1; i is initialized by 1. While $i \in [1, k]$, repeatedly run

$$\begin{pmatrix} \mathsf{SKGen}(\mathsf{st}_{\mathsf{ID}|_{i-1}}, \mathsf{ID}|_i) \to \mathsf{sk}_{\mathsf{ID}|_i}; \\ \mathsf{KeyUp}(\mathsf{dk}_{\mathsf{ID}|_{i-1}, T}, \mathsf{st}_{\mathsf{ID}|_{i-1}}, RL_{\mathsf{ID}|_{i-1}}) \to \mathsf{ku}_{\mathsf{ID}|_{i-1}, T}; \\ \mathsf{DKGen}(\mathsf{sk}_{\mathsf{ID}|_i}, \mathsf{ku}_{\mathsf{ID}|_{i-1}, T}) \to \mathsf{dk}_{\mathsf{ID}|_i, T}; \\ i \leftarrow i + 1; \end{pmatrix}$$

Finally, compute $\mathsf{Enc}(M, \mathsf{ID}|_k, T) \to \mathsf{CT}$ and $\mathsf{Dec}(\mathsf{CT}, \mathsf{dk}_{\mathsf{ID}|_k, T}) \to M'$. The perfect correctness requires that the probability that $M = M'$ to be 1, where the probability is taken over the randomness used in all algorithms.

Next, we give the security model for RHIBE defined by Seo and Emura [13]. It is worth noting that the SKGen oracle returns not only $\mathsf{sk}_{\mathsf{ID}}$ but also $\mathsf{st}_{\mathsf{ID}}$. So, security against insider is considered. They also considered decryption key exposure resistance [10,12]. In this paper, we additionally consider adaptive-ID security.

$\mathsf{SKGen}(\cdot)$: The oracle takes as input ID, and outputs $\mathsf{sk}_{\mathsf{ID}}$ and $\mathsf{st}_{\mathsf{ID}}$.

$\mathsf{DKGen}(\cdot, \cdot)$: The oracle takes as input ID and T, and outputs $\mathsf{dk}_{\mathsf{ID}, T}$.

$\mathsf{KeyUp}(\cdot, \cdot)$: The oracle takes as input $\mathsf{ID}|_{k-1}$ and T, and outputs $\mathsf{ku}_{\mathsf{ID}|_{k-1}, T}$. If $k = 1$, it means that $\mathcal{A}$ asks the key updates for the first level users generated by the KGC.

$\mathsf{Revoke}(\cdot, \cdot)$: The oracle takes as input $\mathsf{ID}|_k$ and T, and adds a pair $(\mathsf{ID}|_k, T)$ into $RL_{\mathsf{ID}|_{k-1}}$, where $\mathsf{ID}|_{k-1}$ is the parent of $\mathsf{ID}|_k$.

Given a RHIBE scheme $\mathcal{RHIBE} = (\mathsf{Setup}, \mathsf{SKGen}, \mathsf{DKGen}, \mathsf{KeyUp}, \mathsf{Enc}, \mathsf{Dec}, \mathsf{Revoke})$ and an adversary $\mathcal{A} = \{\mathcal{A}_1, \mathcal{A}_2\}$, we define an experiment $\mathbf{Exp}_{\mathcal{RHIBE}, \mathcal{A}}^{\text{IND-RID-CPA}}(1^\lambda, N, \ell; \rho)$, where ρ is a random tape for all randomness used in the experiment. Let $\mathcal{O}$ be a set of oracles ($\mathsf{SKGen}(\cdot)$, $\mathsf{DKGen}(\cdot, \cdot)$, $\mathsf{KeyUp}(\cdot, \cdot)$, $\mathsf{Revoke}(\cdot, \cdot)$).

$\mathbf{Exp}_{\mathcal{RHIBE}, \mathcal{A}}^{\text{IND-RID-CPA}}(\lambda, N, \ell; \rho):$
$(\mathsf{mpk}, \mathsf{msk}) \leftarrow \mathsf{Setup}(1^\lambda, N, \ell); (\mathsf{ID}|_{k^*}^*, M_0^*, M_1^*, state) \leftarrow \mathcal{A}_1^{\mathcal{O}}(\mathsf{mpk});$
$b \xleftarrow{\$} \{0, 1\}; \mathsf{CT}^* \leftarrow \mathsf{Enc}(M_b^*, \mathsf{ID}|_{k^*}^*, T^*); b' \leftarrow \mathcal{A}_2^{\mathcal{O}}(\mathsf{mpk}, \mathsf{CT}^*, state);$
Return $\begin{cases} 1 \text{ if } b = b' \text{ and the Conditions are satisfied} \\ 0 \text{ otherwise} \end{cases}$

1. M_0^* and M_1^* have the same length.
2. $\mathcal{A}$ has to query to $\mathsf{KeyUp}(\cdot, \cdot)$ and $\mathsf{Revoke}(\cdot, \cdot)$ in increasing order of time.
3. $\mathcal{A}$ cannot query to $\mathsf{Revoke}(\cdot, \cdot)$ on time T if it already queried to $\mathsf{KeyUp}(\cdot, \cdot)$ on time T.
4. If $\mathsf{ID}|_{k'}^*$ is queried to $\mathsf{SKGen}(\cdot)$ on time T', where $k' \leq k^*$ and $T' \leq T$, then $\mathcal{A}$ must query to revoke the challenge identity $\mathsf{ID}|_{k^*}^*$ or one of its ancestors on time $T' \leq T'' \leq T$.
5. $\mathcal{A}$ cannot query decryption keys $\mathsf{dk}_{\mathsf{ID}|_{k'}^*, T^*}$ of the challenge identity or its ancestors on the challenge time T^*, where $k' \leq k^*$.

The advantage is defined as

$$\mathbf{Adv}_{\mathcal{RHIBE},\mathcal{A}}^{\text{IND-RID-CPA}}(\lambda, N, \ell) := \left| \Pr_{\rho}[\mathbf{Exp}_{\mathcal{RHIBE},\mathcal{A}}^{\text{IND-RID-CPA}}(\lambda, N, \ell; \rho) \rightarrow 1] - \frac{1}{2} \right|$$

Definition 2 (IND-RID-CPA). *We say that $\mathcal{RHIBE}$ is IND-RID-CPA secure if for any polynomials N and ℓ and probabilistic polynomial time algorithm $\mathcal{A}$, the function $\mathbf{Adv}_{\mathcal{RHIBE},\mathcal{A}}^{\text{IND-RID-CPA}}(\lambda, N, \ell)$ is a negligible function in the security parameter λ.*

4   Proposed Adaptive-ID Secure RHIBE

In this section, we give the proposed RHIBE scheme based on the Lewko-Waters HIBE scheme. A decryption key $\mathsf{dk}_{\mathsf{ID}|_k, T}$ can be computed from a (long-term) secret key $\mathsf{sk}_{\mathsf{ID}|_k}$ and $\mathsf{ku}_{\mathsf{ID}|_{k-1}, T}$ if $\mathsf{ID}|_k$ is not revoked at T. $\mathsf{sk}_{\mathsf{ID}|_k}$ is the same as that of the Lewko-Waters HIBE scheme, except that g^α is replaced to P_θ for a history-free construction as in the Seo-Emura RHIBE scheme [13]. That is, the secret key generation algorithm does not require any secret information from the ancestors but $\mathsf{sk}_{\mathsf{ID}|_k}$ is necessary to compute a decryption key, namely P_θ, called msk-shade, plays the role of a delegation key.

In the description, if $k = 1$, then $\mathsf{sk}_{\mathsf{ID}|_{k-1}}$ means msk, $\mathsf{BT}_{\mathsf{ID}|_{k-1}}$ means BT_0, and $\mathsf{ku}_{\mathsf{ID}|_{k-1}, T}$ means $\mathsf{ku}_{0, T}$. We remark that $\mathsf{ku}_{0, T}$ in [13] has the special form such that $\mathsf{ku}_{0, T} = \{(P_\theta^{-1} g_2^\alpha (u'^T h')^{t_\theta}, g^{t_\theta})\}_\theta$, i.e., no g^{r_θ} part is contained. We explicitly define $\mathsf{ID}_0 = \mathsf{I}_0$ as the identity of KGC and add the g^{r_θ} part to $\mathsf{ku}_{0, T}$ in order to fix the form of key update information. This is necessary for employing dual system encryption since it is not guaranteed that a decryption key generated by semi-functional $\mathsf{ku}_{0, T}$ is also semi-functional if $\mathsf{ku}_{0, T}$ has a special form.

$\mathsf{Setup}(1^\lambda, N, \ell)$: Run $\mathcal{G}(1^\lambda) \rightarrow (n = p_1 p_2 p_3, \mathbb{G}, \mathbb{G}_T, e)$. Choose random elements $g, h, v, u_0, u_1 \ldots, u_\ell, u', h' \xleftarrow{\$} \mathbb{G}_{p_1}$, $X_3 \xleftarrow{\$} \mathbb{G}_{p_3}$, and a random integer $\alpha \xleftarrow{\$} \mathbb{Z}_n$. Publish

$$\mathsf{mpk} = \{g, h, u_0, u_1, \ldots, u_\ell, e(g, g)^\alpha, u', h', X_3\}$$

and keep $\mathsf{msk} = \alpha$ in a secure storage.

$\mathsf{SKGen}(\mathsf{st}_{\mathsf{ID}|_{k-1}}, \mathsf{ID}|_k)$: The state information $\mathsf{st}_{\mathsf{ID}|_{k-1}}$, which is kept by $\mathsf{ID}|_{k-1}$, contains the binary tree $\mathsf{BT}_{\mathsf{ID}|_{k-1}}$. For $\mathsf{ID}|_k = (\mathsf{I}_0, \mathsf{I}_1, \ldots, \mathsf{I}_k)$, assign a random leaf node of $\mathsf{BT}_{\mathsf{ID}|_{k-1}}$ to $\mathsf{ID}|_k$. For each θ in $\mathsf{path}(\mathsf{ID}|_k) \subset \mathsf{BT}_{\mathsf{ID}|_{k-1}}$, recall P_θ if it is stored. Otherwise, choose $P_\theta \xleftarrow{\$} \mathbb{G}_{p_1}$ using g, assign and store it in the corresponding node in the tree. For each $\theta \in \mathsf{Path}(\mathsf{ID}|_k)$ choose $R'_{3,\theta}, R_{3,\theta}, R_{k+1,\theta}, \ldots, R_{\ell,\theta} \xleftarrow{\$} \mathbb{G}_{p_3}$ using X_3 and $r_\theta \xleftarrow{\$} \mathbb{Z}_n$. Compute and output $\mathsf{sk}_{\mathsf{ID}|_k}$ defined as

$$\left\{ (P_\theta(u_0^{\mathsf{I}_0} \cdots u_k^{\mathsf{I}_k} h)^{r_\theta} R'_{3,\theta}, g^{r_\theta} R_{3,\theta}, u_{k+1}^{r_\theta} R_{k+1,\theta}, \ldots, u_\ell^{r_\theta} R_{\ell,\theta}) \right\}_{\theta \in \mathsf{Path}(\mathsf{ID}|_k)}.$$

$\mathsf{KeyUp}(\mathsf{msk}, \mathsf{st}_0, RL_0, T)$: The state information st_0 contains the binary tree BT_0. Compute a set $\mathsf{KUNode}(\mathsf{BT}_0, RL_0, T)$. For each $\theta \in \mathsf{KUNode}(\mathsf{BT}_0, RL_0, T)$ recall $\mathsf{msk\text{-}shade}$ P_θ if it is defined. Otherwise, a new $\mathsf{msk\text{-}shade}$ at θ is chosen such that $P_\theta \xleftarrow{\$} \mathbb{G}_{p_1}$ using g, and store it in the corresponding node in the tree. Finally, the key update $\mathsf{ku}_{0,T}$ is generated as follows: Choose $r_\theta, t_\theta \xleftarrow{\$} \mathbb{Z}_n$ and $\bar{R}_{3,\theta}, \bar{R}'_{3,\theta}, \bar{R}''_{3,\theta}, \bar{R}_{1,\theta}, \ldots, \bar{R}_{\ell,\theta} \xleftarrow{\$} \mathbb{G}_{p_3}$ using X_3 for each $\theta \in \mathsf{KUNode}(\mathsf{BT}_0, RL_0, T)$ and compute $\mathsf{ku}_{0,T}$ as follows.

$$\left\{ (P_\theta^{-1} g^\alpha (u_0^{\mathsf{I}_0} h)^{r_\theta} (u'^T h')^{t_\theta} \bar{R}_{3,\theta}, g^{r_\theta} \bar{R}'_{3,\theta}, g^{t_\theta} \bar{R}''_{3,\theta}, u_1^{r_\theta} \bar{R}_{1,\theta}, \ldots, u_\ell^{r_\theta} \bar{R}_{\ell,\theta}) \right\}_\theta,$$

where the set is for every $\theta \in \mathsf{KUNode}(\mathsf{BT}_0, RL_0, T)$.

$\mathsf{DKGen}(\mathsf{sk}_{\mathsf{ID}|_k}, \mathsf{ku}_{\mathsf{ID}|_{k-1},T})$: Let $\mathsf{ID}|_k = (\mathsf{I}_0, \ldots, \mathsf{I}_k)$. Parse $\mathsf{ku}_{\mathsf{ID}|_{k-1},T} = \{\tilde{a}_{0,\theta}, \tilde{a}_{1,\theta}, \tilde{a}_{2,\theta}, \tilde{b}_{k,\theta}, \ldots, \tilde{b}_{\ell,\theta}\}_{\theta \in \mathsf{S}}$ and $\mathsf{sk}_{\mathsf{ID}|_k} = \{(a_{0,\theta}, a_{1,\theta}, b_{k+1,\theta}, \ldots, b_{\ell,\theta})\}_{\theta \in \mathsf{S}'}$, where S and S' are sets of nodes. If $(\mathsf{ID}|_k, \cdot, \cdot) \notin RL_{\mathsf{ID}|_{k-1}}$, then there should be at least one node θ in $\mathsf{Path}(\mathsf{ID}|_k) \cap \mathsf{S}$. For such θ, compute

$$(A_0, \ A_1, \ A_2, \ B_{k+1}, \ldots, B_\ell)$$

$$= (a_{0,\theta} \cdot \tilde{a}_{0,\theta} \cdot \tilde{b}_{k,\theta}^{\mathsf{I}_k}, \ a_{1,\theta} \cdot \tilde{a}_{1,\theta}, \ \tilde{a}_{2,\theta}, \ b_{k+1,\theta} \cdot \tilde{b}_{k+1,\theta}, \ldots, b_{\ell,\theta} \cdot \tilde{b}_{\ell,\theta}).$$

Finally, re-randomize the result and output it as $\mathsf{dk}_{\mathsf{ID}|_k,T}$. We explain how to re-randomize it later.

$\mathsf{KeyUp}(\mathsf{dk}_{\mathsf{ID}|_{k-1},T}, \mathsf{st}_{\mathsf{ID}|_{k-1}}, RL_{\mathsf{ID}|_{k-1}}, T)$:
For each $\theta \in \mathsf{KUNode}(\mathsf{BT}_{\mathsf{ID}|_{k-1}}, RL_{\mathsf{ID}|_{k-1}}, T)$, recall P_θ if it is stored. Otherwise, choose $P_\theta \xleftarrow{\$} \mathbb{G}_{p_1}$ using g and store it in the corresponding node in the tree. Let $\mathsf{dk}_{\mathsf{ID}|_{k-1},T}$ be $(A_0, A_1, A_2, B_k, \ldots, B_\ell)$. For each $\theta \in \mathsf{KUNode}(\mathsf{BT}_{\mathsf{ID}|_{k-1}}, RL_{\mathsf{ID}|_{k-1}}, T)$, re-randomize the decryption key with fresh randomness $r_\theta, t_\theta, s_{0,\theta}, s_{1,\theta}, s_{2,\theta}, s'_{k,\theta}, \ldots, s'_{\ell,\theta} \xleftarrow{\$} \mathbb{Z}_n$ so that obtain $(a_{0,\theta}, a_{1,\theta}, a_{2,\theta}, b_{k,\theta}, \ldots, b_{\ell,\theta})$. Finally, $\mathsf{ku}_{\mathsf{ID}|_{k-1},T}$ is generated as

$$\left\{ (P_\theta^{-1} a_{0,\theta}, \ a_{1,\theta}, \ a_{2,\theta}, \ b_{k,\theta}, \ldots, b_{\ell,\theta}) \right\}_{\theta \in \mathsf{KUNode}(\mathsf{BT}_{\mathsf{ID}|_{k-1}}, RL_{\mathsf{ID}|_{k-1}}, T)}.$$

$\mathsf{Enc}(M, \mathsf{ID}|_k, T)$: Let $\mathsf{ID}|_k = (\mathsf{I}_0, \ldots, \mathsf{I}_k)$. Choose an integer $s \xleftarrow{\$} \mathbb{Z}_n$ at random and compute

$$\mathsf{CT} = (M \cdot e(g,g)^{\alpha s}, \ g^s, \ (u_0^{\mathsf{I}_0} \cdots u_k^{\mathsf{I}_k} h)^s, \ (u'^T h')^s).$$

$\mathsf{Dec}(\mathsf{CT}, \mathsf{dk}_{\mathsf{ID}|_k, T})$: Let $\mathsf{ID}|_k = (\mathsf{I}_0, \dots, \mathsf{I}_k)$. Parse $\mathsf{CT} = (C', C_0, C_1, C_2)$ and $\mathsf{dk}_{\mathsf{ID}|_k, T} = (A_0, A_1, A_2, B_{k+1}, \dots, B_\ell)$. Output $C' \cdot \frac{e(A_1, C_1) \cdot e(A_2, C_2)}{e(A_0, C_0)}$.

$\mathsf{Revoke}(\mathsf{ID}|_k, T, RL_{\mathsf{ID}|_{k-1}}, \mathsf{st}_{\mathsf{ID}|_{k-1}})$: Find a leaf node $\zeta_{\mathsf{ID}|_k}$, associating with $\mathsf{ID}|_k$, in $\mathsf{BT}_{\mathsf{ID}|_{k-1}}$. Update $RL_{\mathsf{ID}|_{k-1}}$ by adding a triple $(\mathsf{ID}|_k, T, \zeta_{\mathsf{ID}|_k})$.

Correctness: From the shape of the ciphertext for $\mathsf{ID}|_k$ and T, we can easily see that if $\mathsf{dk}_{\mathsf{ID}|_k, T}$ is of the form $(g^\alpha (u_0^{\mathsf{I}_0} \cdots u_k^{\mathsf{I}_k} h)^r (u'^T h')^t R_3', g^r R_3, g^t R_3'', b_{k+1} R_{k+1}, \dots, b_\ell R_\ell)$, then the DKGen algorithm correctly outputs the plaintext with probability 1. We show that $\mathsf{dk}_{\mathsf{ID}|_k, T}$, which is obtained by k times repeatedly performing SKGen, KeyUp and DKGen, is of the desired form $(g^\alpha (u_0^{\mathsf{I}_0} \cdots u_k^{\mathsf{I}_k} h)^r (u'^T h')^t R_3', \ g^r R_3, \ g^t R_3'', \ u_{k+1}^r R_{k+1}, \dots, u_\ell^r R_\ell)$ for some r and t. Then, it is sufficient for the perfect correctness. To this end, we will use the mathematical induction; for the first level ancestor, $\mathsf{sk}_{\mathsf{ID}|_1}$ and $\mathsf{ku}_{0,T}$ are respectively defined as

$$\left\{ (P_\theta (u_0^{\mathsf{I}_0} u_1^{\mathsf{I}_1} h)^{r_\theta} R_{3,\theta}', g^{r_\theta} R_{3,\theta}, u_2^{r_\theta} R_{2,\theta}, \dots, u_\ell^{r_\theta} R_{\ell,\theta}) \right\}_{\theta \in \mathsf{Path}(\mathsf{ID}|_1)} \quad \text{and}$$

$$\left\{ (P_\theta^{-1} g^\alpha (u_0^{\mathsf{I}_0} h)^{r_\theta'} (u'^T h')^{t_\theta} \bar{R}_{3,\theta}, g^{r_\theta'} \bar{R}_{3,\theta}', g^{t_\theta} \bar{R}_{3,\theta}'', u_1^{r_\theta'} \bar{R}_{1,\theta}, \dots, u_\ell^{r_\theta'} \bar{R}_{\ell,\theta}) \right\}_\theta,$$

where $\theta \in \mathsf{KUNode}(\mathsf{BT}_0, RL_0, T)$. For $\theta \in \mathsf{Path}(\mathsf{ID}|_1) \cap \mathsf{KUNode}(\mathsf{BT}_0, RL_0, T)$, $\mathsf{dk}_{\mathsf{ID}|_1}$ is equal to

$$\Big(g^\alpha (u_0^{\mathsf{I}_0} u_1^{\mathsf{I}_1} h)^{r_\theta + r_\theta'} (u'^T h')^{t_\theta} (R_{3,\theta}' \bar{R}_{3,\theta} \bar{R}_{1,\theta}), \ g^{r_\theta + r_\theta'} R_{3,\theta}, \ g^{t_\theta} \bar{R}_{3,\theta},$$

$$u_2^{r_\theta + r_\theta'} (R_{2,\theta} \bar{R}_{2,\theta}), \dots, u_\ell^{r_\theta + r_\theta'} (R_{\ell,\theta} \bar{R}_{\ell,\theta}) \Big)$$

and later it is re-randomized. Thus, $\mathsf{dk}_{\mathsf{ID}|_1}$ is of the desired form. As in the similar confirmation, we can check the case of $k \geq 2$ under the assumption that $\mathsf{dk}_{\mathsf{ID}|_{k-1}}$ is of the above form.

Decryption Key Re-randomization: The decryption key for $\mathsf{ID}|_k = (\mathsf{I}_0, \dots, \mathsf{I}_k)$ at T has the form $(g^\alpha (u_0^{\mathsf{I}_0} \cdots u_k^{\mathsf{I}_k} h)^r (u'^T h')^t R_3', \ g^r R_3, g^t R_3'', u_{k+1}^r R_{k+1}, \dots, u_\ell^r R_\ell)$. Since u_i's, h, g, u', h' and X_3 are publicly available, anyone can re-randomize $\mathsf{dk}_{\mathsf{ID}|_k, T} := (a_0, a_1, a_2, b_{k+1}, \dots, b_\ell)$ by computing

$$\Big(a_0 (u_0^{\mathsf{I}_0} \cdots u_k^{\mathsf{I}_k} h)^{r'} (u'^T h')^{t'} X_3^{s_0}, \ a_1 g^{r'} X_3^{s_1}, \ a_2 g^{t'} X_3^{s_2},$$

$$b_{k+1} u_{k+1}^{r'} X_3^{s_{k+1}'}, \dots, b_\ell u_\ell^{r'} X_3^{s_\ell'} \Big)$$

with randomly chosen $r', t', s_0, s_1, s_2, s_{k+1}', \dots, s_\ell' \xleftarrow{\$} \mathbb{Z}_n$.

Theorem 1. *The proposed RHIBE scheme is IND-RID-CPA secure under Assumptions 1, 2, and 3.*

Before giving the proof, we explain our strategy. In particular, how to apply dual system encryption methodology to R(H)IBE. In dual system encryption, normal ciphertext can be decrypted by using either normal decryption key or semi-functional key, but semi-functional ciphertext cannot be decrypted by using semi-functional decryption key. In R(H)IBE, a decryption key $\mathsf{dk}_{\mathsf{ID},T}$ is computed from a secret key $\mathsf{sk}_{\mathsf{ID}}$ and key update information ku_T. So, in our strategy, we newly define normal key update information/secret key and semi-functional key update information/secret key.

- From a normal secret key and normal key update information, a (non-revoked) user can compute a normal decryption key.
- From a semi-functional secret key and normal key update information, a (non-revoked) user can compute a semi-functional decryption key.
- From a normal secret key and semi-functional key update information, a (non-revoked) user can compute a semi-functional decryption key.
- From a semi-functional secret key and semi-functional key update information, a (non-revoked) user can compute a semi-functional decryption.

Due to the functionality of RHIBE, we further need to consider the cases that key update information is generated by decryption keys.

- From a semi-functional decryption key, the KeyUp algorithm outputs semi-functional key update information.

Remark: It is particularly worth noting that, in the security proof of the Lewko-Waters HIBE scheme, the simulator $\mathcal{B}$ is compelled to be able to only create a "nominally" semi-functional key, which still works for a semi-functional ciphertext, in order to prevent that $\mathcal{B}$ attempts to test itself whether key is semi-functional by creating a semi-functional ciphertext and trying to decrypt. Since there are many routes to create decryption keys in the RHIBE context, we need to guarantee that still $\mathcal{B}$ can only create a nominally semi-functional key regardless of the decryption key generation route.

Semi-functional Ciphertext: Let g_2 be a generator of $\mathbb{G}_{p_2}$, and let (C', C_0', C_1', C_2') be a ciphertext generated by the encryption algorithm. Choose $x, z_c \xleftarrow{\$} \mathbb{Z}_n$, and set $(C'', C_0', C_1', C_2') = (C', C_0'g_2^x, C_1'g_2^{xz_c}, C_2')$.

Semi-functional Secret Key: Let $\{(a'_{0,\theta}, a'_{1,\theta}, b'_{i+1,\theta}, \ldots, b'_{\ell,\theta})\}_\theta$ be a secret key $\mathsf{sk}_{\mathsf{ID}|_{i-1}}$ by the SKGen algorithm. Choose $z_k \xleftarrow{\$} \mathbb{Z}_n$ and for each θ choose $\gamma_\theta, z_{i+1,\theta}, \ldots, z_{\ell,\theta} \xleftarrow{\$} \mathbb{Z}_n$, and set $\{(a_{0,\theta}, a_{1,\theta}, k_{i+1,\theta}, \ldots, k_{\ell,\theta})\}_\theta = \{(a_{0,\theta}g_2^{\gamma_\theta z_k}, a_{1,\theta}g_2^{\gamma_\theta}, k_{i+1,\theta}g_2^{\gamma_\theta z_{i+1,\theta}}, \ldots, k_{\ell,\theta}g_2^{\gamma_\theta z_{\ell,\theta}})\}_\theta$.

Semi-functional Key Update Information: Let $\{(\tilde{a}'_{0,\theta}, \tilde{a}'_{1,\theta}, \tilde{a}'_{2,\theta}, \tilde{k}'_{i+1,\theta}, \ldots, \tilde{k}'_{\ell,\theta})\}_\theta$ be key update information $\mathsf{ku}_{\mathsf{ID}|_{i-1},T}$ by the KeyUp algorithm. Choose $z_k \xleftarrow{\$} \mathbb{Z}_n$ and for each θ choose $\gamma_\theta, z_{i+1,\theta}, \ldots, z_{\ell,\theta} \xleftarrow{\$} \mathbb{Z}_n$, and set $\{\tilde{a}_{0,\theta}, \tilde{a}_{1,\theta}, \tilde{a}_{2,\theta}, \tilde{k}_{i+1,\theta}, \ldots, \tilde{k}_{\ell,\theta}\}_\theta = \{(\tilde{a}'_{0,\theta}g_2^{\gamma_\theta z_k}, \tilde{a}'_{1,\theta}g_2^{\gamma_\theta}, \tilde{a}'_{2,\theta}, \tilde{k}'_{i+1,\theta}g_2^{\gamma_\theta z_{i+1,\theta}}, \ldots, \tilde{k}'_{\ell,\theta}g_2^{\gamma_\theta z_{\ell,\theta}})\}_\theta$.

Semi-functional Decryption Key: Let $(A'_0, A'_1, A'_2, B'_{i+1}, \ldots, B'_\ell)$ be a decryption key generated by the DKGen algorithm. Choose $\gamma, z_k, z_{i+1}, \ldots, z_\ell \xleftarrow{\$} \mathbb{Z}_n$, and set $(A_0, A_1, A_2, B_{i+1}, \ldots, B_\ell) = (A'_0 g_2^{\gamma z_k}, A'_1 g_2^\gamma, A'_2, B'_{i+1} g_2^{\gamma z_{i+1}}, \ldots, B'_\ell g_2^{\gamma z_\ell})$.

Semi-functional Ciphertext×Normal Decryption Key: Let a semi-functional ciphertext be $\mathsf{CT} = (M \cdot e(g,g)^{\alpha s}, \ g^s g_2^x, \ (u_0^{l_0} \cdots u_i^{l_i} h)^s g_2^{x z_c}, (u'^T h')^s)$ and and a normal decryption key be $(g^\alpha (u_0^{l_0} \cdots u_i^{l_i} h)^r (u'^T h')^t R'_3, g^r R_3, g^t R''_3, b_{i+1} R_{i+1}, \ldots, b_\ell R_\ell)$. Then, the Dec algorithm computes

$$\frac{e(g^r R_3, (u_0^{l_0} \cdots u_i^{l_i} h)^s g_2^{x z_c}) e(g^t R''_3, (u'^T h')^s)}{e(g^\alpha (u_0^{l_0} \cdots u_i^{l_i} h)^r (u'^T h')^t R'_3, g^s g_2^x)} = \frac{1}{e(g,g)^{\alpha s}}$$

Therefore, the Dec algorithm normally works.

Semi-functional Ciphertext×Semi-functional Decryption Key: Let a semi-functional ciphertext be $\mathsf{CT} = (M \cdot e(g,g)^{\alpha s}, \ g^s g_2^x, \ (u_0^{l_0} \cdots u_i^{l_i} h)^s g_2^{x z_c}, (u'^T h')^s)$ and a semi-functional decryption key be $(g^\alpha (u_0^{l_0} \cdots u_i^{l_i} h)^r (u'^T h')^t R'_3 g_2^{\gamma z_k}, g^r R_3 g_2^\gamma, g^t R''_3, b_{i+1} R_{i+1} g_2^{\gamma z_{i+1}}, \ldots, b_\ell R_\ell g_2^{\gamma z_\ell})$. Then, the Dec algorithm computes

$$\frac{e(g^r R_3 g_2^\gamma, (u_0^{l_0} \cdots u_i^{l_i} h)^s g_2^{x z_c}) e(g^t R''_3, (u'^T h')^s)}{e(g^\alpha (u_0^{l_0} \cdots u_i^{l_i} h)^r (u'^T h')^t R'_3 g_2^{\gamma z_k}, g^s g_2^x)} = \frac{e(g_2, g_2)^{x\gamma(z_c - z_k)}}{e(g,g)^{\alpha s}}$$

That is, as in the Lewko-Waters HIBE scheme, when a semi-functional key is used to decrypt a semi-functional ciphertext, the Dec algorithm will compute the blinding factor multiplied by the additional term $e(g_2, g_2)^{x\gamma(z_c - z_k)}$. If $z_c = z_k$, then the decryption still works and we say that the decryption key is nominally semi-functional.

Organization of the Proof: In the following security proof, first the challenge ciphertext is changed to be semi-functional. Next, a decryption key is changed to be semi-functional one by one. We remark that key update information generated by a semi-functional decryption key is also semi-functional. That is, we need to consider the following case: $\mathcal{A}$ issues a decryption key query for $\mathsf{ID}|_i$ and the simulator $\mathcal{B}$ sends a normal decryption key. Later, $\mathcal{A}$ issues a decryption key query of its ancestor's identity, say $\mathsf{ID}|_{i-1}$, and $\mathcal{B}$ sends a semi-functional decryption key. Moreover, $\mathcal{A}$ issues a secret key query of this identity $\mathsf{ID}|_{i-1}$ and $\mathcal{B}$ sends a normal secret key. Then, $\mathcal{A}$ can compute key update information from the semi-functional decryption key, and therefore can compute a decryption key of $\mathsf{ID}|_i$ by myself. Then, a decryption key is semi-functional. So, we need to guarantee that all decryption keys appeared in the games are normal or nominally semi-functional regardless of its generation routes.

Next, key update information is changed to be semi-functional one by one. Though each $\mathsf{ku}_{\mathsf{ID},T}$ contains $O(R \log(N/R))$-size subkeys, we replace all subkeys simultaneously since the differences of subkeys are randomness part only. As in the previous games, we need to guarantee that all decryption keys appeared in the games are normal or nominally semi-functional regardless of its generation routes.

Next, a secret key is changed to be semi-functional one by one. Though each $\mathsf{sk}_{\mathsf{ID}}$ contains $O(\log N)$-size subkeys, we replace all subkeys simultaneously since the differences of subkeys are randomness part only. Finally, the plaintext of the challenge ciphertext is replaced as a random value.

As the most different point between HIBE and RHIBE, an adversary is allowed to obtain the secret key of the challenge identity (or its ancestors). In the scheme, for history-free update, no ancestor's secret value is contained in secret keys. Moreover, the simulator can choose state information P_θ directly (which helps us to prove insider security). Therefore the simulator can generate the secret key of the challenge identity (or its ancestors) directly. So, in the following all games, $\mathcal{B}$ chooses P_θ and stores it as state information.

Let Game_{real} be the real security game. The next game, $\mathrm{Game}_{restricted}$ is the same as Game_{real} except that the adversary is not allowed to issue key generation queries for identities which are prefixes of the challenge identity modulo p_2.

Lemma 1 ([6]). *Suppose there exists an adversary $\mathcal{A}$ such that $\mathrm{Game}_{real}\mathbf{Adv}_{\mathcal{A}} - \mathrm{Game}_{restricted}\mathbf{Adv}_{\mathcal{A}} = \epsilon$ Then, we can build an algorithm $\mathcal{B}$ with advantage $\geq \epsilon/2$ in breaking Assumption 2.*

The next game, Game_0 is the same as $\mathrm{Game}_{restricted}$ except that the challenge ciphertext is semi-functional.

Lemma 2. *Suppose there exists an adversary $\mathcal{A}$ such that $\mathrm{Game}_{restricted}\mathbf{Adv}_{\mathcal{A}} - \mathrm{Game}_0\mathbf{Adv}_{\mathcal{A}} = \epsilon$ Then, we can build an algorithm $\mathcal{B}$ with advantage ϵ in breaking Assumption 1.*

Proof. First, $\mathcal{B}$ is given $(g, X_3, \bar{T})$. $\mathcal{B}$ chooses $\alpha, a_0, \ldots, a_\ell, b \xleftarrow{\$} \mathbb{Z}_n$ and sets $g = g$, $u_i = g^{a_i}$ $(i \in [0, \ell])$, and $h = g^b$. Moreover, $\mathcal{B}$ guesses T^* (with success probability at least $1/|\mathcal{T}|$), and chooses $c, d \xleftarrow{\$} \mathbb{Z}_n$, and sets $u' = g^c$ and $h' = u'^{-T^*} g^d$. $\mathcal{B}$ sends public parameters $\mathsf{mpk}^{\mathsf{mLW}} = \{n, g, h, u_0, \ldots, u_\ell, e(g, g)^\alpha, u', h', X_3\}$ to $\mathcal{A}$.

For a key query $\mathsf{ID}|_j = (\mathsf{I}_0, \ldots, \mathsf{I}_j)$, $\mathcal{B}$ generates a normal secret key key as follows. $\mathcal{B}$ chooses $r_\theta, t_\theta, w_\theta, v_{j_1,\theta}, \ldots, v_\ell \xleftarrow{\$} \mathbb{Z}_n$ for each θ, and computes

$$\left\{ (P_\theta(u_0^{\mathsf{I}_0} \cdots u_j^{\mathsf{I}_j} h)^{r_\theta} X_3^{w_\theta}, g^{r_\theta} X_3^{t_\theta}, u_{j+1}^{r_\theta} X_3^{v_{j+1,\theta}}, \ldots, u_\ell^{r_\theta} X_3^{v_{\ell,\theta}} \right\}_\theta.$$

Since $\mathcal{B}$ knows α, $\mathcal{B}$ can answer all key update information, secret key, and decryption key queries. $\mathcal{B}$ generates normal values for these queries.

$\mathcal{A}$ sends two messages M_0 and M_1 and the challenge identity $\mathsf{ID}|_{k^*}^* = (\mathsf{I}_0^*, \ldots, \mathsf{I}_{k^*}^*)$ to $\mathcal{B}$. $\mathcal{B}$ chooses $b \xleftarrow{\$} \{0, 1\}$, and computes $C' = M_b \cdot e(\bar{T}, g)^\alpha$, $C_0 = \bar{T}^{a_0 \mathsf{I}_0^* + \cdots + a_{k^*} \mathsf{I}_{k^*}^* + b}$, $C_1 = \bar{T}$, and $C_2 = \bar{T}^d$. Here, the $\mathbb{G}_{p_1}$ part of $\bar{T}$ is implicitly set g^s and the $\mathbb{G}_{p_1}$ part of $\bar{T}^d$ is implicitly set $(u'^{T^*} h')^s$ if $\mathcal{B}$'s guessing is correct. If $T \in \mathbb{G}_{p_1 p_2}$, then this ciphertext is semi-functional with

$z_c = a_1 \mathsf{l}_1^* + \cdots + a_{k^*} \mathsf{l}_{k^*}^* + b$. If $T \in \mathbb{G}_{p_1}$, then this ciphertext is normal. So, $\mathcal{B}$ uses the output of $\mathcal{A}$ for breaking Assumption 1. □

For the number of decryption key queries q_{dk}, $\mathsf{Game}_k^{\mathsf{dk}}$ ($k \in [1, q_{\mathsf{dk}}]$) is the same as that of $\mathsf{Game}_{k-1}^{\mathsf{dk}}$ except that the first k decryption keys are semi-functional and the rest of the keys are normal. We remark that $\mathsf{Game}_0^{\mathsf{dk}} = \mathsf{Game}_0$.

Lemma 3. *Suppose there exists an adversary $\mathcal{A}$ such that $\mathsf{Game}_{k-1}^{\mathsf{dk}} \mathbf{Adv}_{\mathcal{A}} - \mathsf{Game}_k^{\mathsf{dk}} \mathbf{Adv}_{\mathcal{A}} = \epsilon$ Then, we can build an algorithm $\mathcal{B}$ with advantage ϵ in breaking Assumption 2.*

Proof. First, $\mathcal{B}$ is given $(g, X_1 X_2, X_3, Y_2 Y_3, \bar{T})$. $\mathcal{B}$ chooses $\alpha, a_0, \ldots, a_\ell, b \xleftarrow{\$} \mathbb{Z}_n$ and sets $g = g$, $u_i = g^{a_i}$ ($i \in [0, \ell]$), and $h = g^b$. Moreover, $\mathcal{B}$ guesses T^* (with success probability at least $1/|\mathcal{T}|$), chooses $c, d \xleftarrow{\$} \mathbb{Z}_n$, and sets $u' = g^c$ and $h' = u'^{-T^*} g^d$. $\mathcal{B}$ sends public parameters $\mathsf{mpk}^{\mathsf{mLW}} = \{n, g, h, u_0, \ldots, u_\ell, e(g, g)^\alpha, u', h', X_3\}$ to $\mathcal{A}$. Since $\mathcal{B}$ knows α, $\mathcal{B}$ can answer all key update information and secret key queries. $\mathcal{B}$ generates normal values for these queries.

For the i-th decryption key query $(\mathsf{ID}|_j, T)$ where $\mathsf{ID}|_j = (\mathsf{l}_0, \ldots, \mathsf{l}_j)$ and $i < k$, $\mathcal{B}$ generates a semi-functional decryption key. $\mathcal{B}$ chooses $r, z, t, w, w', z_{j+1}, \ldots, z_\ell \xleftarrow{\$} \mathbb{Z}_n$, and computes

$$(g^\alpha (u_0^{\mathsf{l}_0} \cdots u_j^{\mathsf{l}_j} h)^r (u'^T h')^t (Y_2 Y_3)^z, \ g^r (Y_2 Y_3)^w, \ g^t X_3^{w'},$$

$$u_{j+1}^r (Y_2 Y_3)^{z_{j+1}}, \ldots, u_\ell^r (Y_2 Y_3)^{z_\ell}).$$

For $i > k$, $\mathcal{B}$ generates a normal decryption key by using α. For $i = k$, $\mathcal{B}$ sets $z_k = a_0 \mathsf{l}_0 + \cdots + a_j \mathsf{l}_j + b$, chooses $t, w, w', w_{j+1}, \ldots, w_\ell \xleftarrow{\$} \mathbb{Z}_n$, and computes $\mathsf{dk}_{\mathsf{ID}|_j, T} = (g^\alpha \bar{T}^{z_k} (u'^T h')^t X_3^w, \bar{T}, g^t X_3^{w'}, \bar{T}^{a_{j+1}} X_3^{w_{j+1}}, \ldots, \bar{T}^{a_\ell} X_3^{w_\ell})$. If $\bar{T} \in \mathbb{G}_{p_1 p_3}$, then this key is normal where the $\mathbb{G}_{p_1}$ part of $\bar{T}$ is set as g^r. If $\bar{T} \in \mathbb{G}$, this key is semi-functional.

$\mathcal{A}$ sends two messages M_0 and M_1 and the challenge identity $\mathsf{ID}|_{k^*}^* = (\mathsf{l}_0^*, \ldots, \mathsf{l}_{k^*}^*)$ to $\mathcal{B}$. $\mathcal{B}$ chooses $b \xleftarrow{\$} \{0, 1\}$, and computes $C' = M_b \cdot e(X_1 X_2, g)^\alpha$, $C_0 = (X_1 X_2)^{a_0 \mathsf{l}_0^* + \cdots + a_{k^*} \mathsf{l}_{k^*}^* + b}$, $C_1 = X_1 X_2$, and $C_2 = X_1^d$. Here, X_1 is implicitly set g^s and $z_c = a_0 \mathsf{l}_0^* + \cdots + a_{k^*} \mathsf{l}_{k^*}^* + b$. If $\bar{T} \in \mathbb{G}_{p_1 p_3}$, then $\mathcal{B}$ has properly simulated $\mathsf{Game}_{k-1}^{\mathsf{dk}}$. If $\bar{T} \in \mathbb{G}$, then $\mathcal{B}$ has properly simulated $\mathsf{Game}_k^{\mathsf{dk}}$.

If $\mathcal{B}$ tries to test whether the answer of k-th decryption key query $\mathsf{dk}_{\mathsf{ID}|_j, T}$ is semi-functional or not by computing a semi-functional ciphertext, $\mathcal{B}$ can only create a nominally semi-functional decryption key with $z_k = z_c$. The answer of k-th decryption key query $\mathsf{dk}_{\mathsf{ID}|_j, T}$ is nominally semi-functional since $z_k = z_c$. So, we have to consider the case that key update information $\mathsf{ku}_{\mathsf{ID}|_j, T}$ is generated by this $\mathsf{dk}_{\mathsf{ID}|_j, T}$ (then $\mathsf{ku}_{\mathsf{ID}|_j, T}$ is semi-functional), and $\mathsf{dk}_{\mathsf{ID}|_{j+1}, T}$ is computed by $\mathsf{ku}_{\mathsf{ID}|_j, T}$ and a normal secret key $\mathsf{sk}_{\mathsf{ID}|_{j+1}}$. $\mathsf{ku}_{\mathsf{ID}|_j, T}$ is represented as

$$\left\{ (P_\theta^{-1} g^\alpha \bar{T}^{z_k} (u_0^{\mathsf{l}_0} \cdots u_j^{\mathsf{l}_j} h)^{r_\theta} (u'^T h')^{t+t_\theta} X_3^{w+s_{0,\theta}}, \ \bar{T} g^{r_\theta} X_3^{s_{1,\theta}}, \ g^{t+t_\theta} X_3^{w'+s_{2,\theta}}, \right.$$

$$\bar{T}^{a_{j+1}} u_{j+1}^{r_\theta} X_3^{w_{j+1}+s'_{j+1,\theta}}, \dots, \bar{T}^{a_\ell} u_\ell^{r_\theta} X_3^{w_\ell+s'_{\ell,\theta}})\Big\}_\theta$$

where $r_\theta, t_\theta, s_{0,\theta}, s_{1,\theta}, s_{2,\theta}, s'_{j+1,\theta}, \dots, s'_{\ell,\theta} \xleftarrow{\$} \mathbb{Z}_n$ are chosen for each θ. Then since a normal secret key $\mathsf{sk}_{\mathsf{ID}|_{j+1}}$ is represented as

$$\Big\{ (P_\theta(u_0^{l_0} \cdots u_{j+1}^{l_{j+1}} h)^{r'_\theta} R'_{3,\theta}, g^{r'_\theta} R_{3,\theta}, u_{j+1}^{r'_\theta} R_{j+1,\theta}, \dots, u_\ell^{r'_\theta} R_{\ell,\theta}) \Big\}_\theta,$$

the first component of $\mathsf{dk}_{\mathsf{ID}|_{j+1},T}$ is represented as

$$g^{\alpha} \bar{T}^{z_k + a_{j+1}l_{j+1}} (u_0^{l_0} \cdots u_{j+1}^{l_{j+1}} h)^{r_\theta + r'_\theta + r'} (u'^T h')^{t+t_\theta+t'} (X_3^{s_{0,\theta}+(w_{j+1}+s'_{j+1,\theta})l_{j+1}+s_0} R'_{3,\theta})$$

where $r', t', s_0 \in \mathbb{Z}_n$ are for re-randomization. As we can see, $\mathsf{dk}_{\mathsf{ID}|_{j+1},T}$ preserves the form with $z'_k := z_k + a_{j+1}l_{j+1}$ and therefore, $\mathsf{dk}_{\mathsf{ID}|_{j+1},T}$ is nominally semi-functional.

From the above estimations, we confirmed that even if $\mathcal{B}$ tries to test, $\mathcal{B}$ can only create a nominally semi-functional decryption key. So, $\mathcal{B}$ uses the output of $\mathcal{A}$ for breaking Assumption 2. $\square$

For the number of key update information queries q_{ku}, $\mathsf{Game}_k^{\mathsf{ku}}$ ($k \in [1, q_{\mathsf{ku}}]$) is the same as that of $\mathsf{Game}_{k-1}^{\mathsf{ku}}$ except that the first k key update information are semi-functional and the rest of these are normal. We remark that $\mathsf{Game}_0^{\mathsf{ku}} = \mathsf{Game}_{q_{\mathsf{dk}}}^{\mathsf{dk}}$.

Lemma 4. *Suppose there exists an adversary $\mathcal{A}$ such that $\mathsf{Game}_{k-1}^{\mathsf{ku}} \mathbf{Adv}_{\mathcal{A}} - \mathsf{Game}_k^{\mathsf{ku}} \mathbf{Adv}_{\mathcal{A}} = \epsilon$ Then, we can build an algorithm $\mathcal{B}$ with advantage ϵ in breaking Assumption 2.*

Proof. First, $\mathcal{B}$ is given $(g, X_1X_2, X_3, Y_2Y_3, \bar{T})$. $\mathcal{B}$ chooses $\alpha, a_0, \dots, a_\ell, b \xleftarrow{\$} \mathbb{Z}_n$ and sets $g = g$, $u_i = g^{a_i}$ ($i \in [0, \ell]$), and $h = g^b$. Moreover, $\mathcal{B}$ guesses T^* (with success probability at least $1/|\mathcal{T}|$), chooses $c, d \xleftarrow{\$} \mathbb{Z}_n$, and sets $u' = g^c$ and $h' = u'^{-T^*} g^d$. $\mathcal{B}$ sends public parameters $\mathsf{mpk}^{\mathsf{mLW}} = \{n, g, h, u_0, \dots, u_\ell, e(g,g)^\alpha, u', h', X_3\}$ to $\mathcal{A}$. Since $\mathcal{B}$ knows α, $\mathcal{B}$ can answer all secret key and decryption key queries. $\mathcal{B}$ generates normal secret keys. Moreover, $\mathcal{B}$ computes semi-functional decryption keys of $(\mathsf{ID}|_j, T)$ where $\mathsf{ID}|_j = (l_0, \dots, l_j)$ as folows. $\mathcal{B}$ chooses $r, z, t, w, w', z_{j+1,\theta}, \dots, z_{\ell,\theta} \xleftarrow{\$} \mathbb{Z}_n$, and computes

$$(g^\alpha (u_0^{l_0} \cdots u_j^{l_j} h)^r (u'^T h')^t (Y_2Y_3)^z, \; g^r(Y_2Y_3)^w, \; g^t X_3^{w'},$$

$$u_{j+1}^r (Y_2Y_3)^{z_{j+1}}, \dots, u_\ell^r (Y_2Y_3)^{z_\ell}).$$

For the i-th key update information query $(\mathsf{ID}|_j, T)$ where $\mathsf{ID}|_j = (l_0, \dots, l_j)$ and $i < k$, $\mathcal{B}$ generates semi-functional key update information $\mathsf{ku}_{\mathsf{ID}|_j,T}$ as follows. $\mathcal{B}$ chooses $r_\theta, z_\theta, t_\theta, w_\theta, w'_\theta, z_{j+1,\theta}, \dots, z_{\ell,\theta} \xleftarrow{\$} \mathbb{Z}_n$ for each θ, and computes

$$\Big\{ (P_\theta^{-1} g^\alpha (u_0^{l_0} \cdots u_j^{l_j} h)^{r_\theta} (u'^T h')^{t_\theta} (Y_2Y_3)^{z_\theta}, \; g^{r_\theta}(Y_2Y_3)^{w_\theta}, \; g^{t_\theta} X_3^{w'_\theta},$$

$$u_{j+1}^{r_\theta}(Y_2Y_3)^{z_{j+1,\theta}}, \ldots, u_\ell^{r_\theta}(Y_2Y_3)^{z_{\ell,\theta}}\Big)\Big\}_\theta.$$

For $i > k$, $\mathcal{B}$ generates normal key update information by using α. For $i = k$, $\mathcal{B}$ sets $z_k = a_0 l_0 + \cdots + a_j l_j + b$, chooses $\gamma_\theta, t_\theta, w_\theta, w'_\theta, w_{j+1,\theta}, \ldots, w_{\ell,\theta} \xleftarrow{\$} \mathbb{Z}_n$, and computes $\mathsf{ku}_{\mathsf{ID}|_j, T}$ as

$$\Big\{(P_\theta^{-1} g^\alpha \bar{T}^{\gamma_\theta z_k}(u'^T h')^{t_\theta} X_3^{w_\theta}, \ \bar{T}^{\gamma_\theta}, \ g^{t_\theta} X_3^{w'_\theta},$$

$$\bar{T}^{\gamma_\theta a_{j+1}} X_3^{w_{j+1,\theta}}, \ldots, \bar{T}^{\gamma_\theta a_\ell} X_3^{w_{\ell,\theta}})\Big\}_\theta.$$

If $\bar{T} \in \mathbb{G}_{p_1 p_3}$, then this key is normal where the $\mathbb{G}_{p_1}$ part of $\bar{T}^{\gamma_\theta}$ is set as g^{r_θ}. If $\bar{T} \in \mathbb{G}$, this key is semi-functional.

$\mathcal{A}$ sends two messages M_0 and M_1 and the challenge identity $\mathsf{ID}|_{k^*}^* = (l_0^*, \ldots, l_{k^*}^*)$ to $\mathcal{B}$. $\mathcal{B}$ chooses $b \xleftarrow{\$} \{0, 1\}$, and computes $C' = M_b \cdot e(X_1 X_2, g)^\alpha$, $C_0 = (X_1 X_2)^{a_0 l_0^* + \cdots + a_{k^*} l_{k^*}^* + b}$, $C_1 = X_1 X_2$, and $C_2 = X_1^d$. Here, X_1 is implicitly set g^s and $z_c = a_0 l_0^* + \cdots + a_{k^*} l_{k^*}^* + b$. If $\bar{T} \in \mathbb{G}_{p_1 p_3}$, then $\mathcal{B}$ has properly simulated $\mathsf{Game}_{k-1}^{\mathsf{ku}}$. If $\bar{T} \in \mathbb{G}$, then $\mathcal{B}$ has properly simulated $\mathsf{Game}_k^{\mathsf{ku}}$.

As in $\mathsf{Game}_k^{\mathsf{ku}}$, we can easily check $\mathsf{dk}_{\mathsf{ID}|_{j+1}, T}$ is a nominally semi-functional decryption key in the case that $\mathsf{dk}_{\mathsf{ID}|_{j+1}, T}$ is computed by $\mathsf{ku}_{\mathsf{ID}|_j, T}$ and a normal secret key $\mathsf{sk}_{\mathsf{ID}|_{j+1}}$. So, $\mathcal{B}$ uses the output of $\mathcal{A}$ for breaking Assumption 2.   □

For the number of secret key queries q_{sk}, $\mathsf{Game}_k^{\mathsf{sk}}$ ($k \in [1, q_{\mathsf{sk}}]$) is the same as that of $\mathsf{Game}_{k-1}^{\mathsf{sk}}$ except that the first k secret keys are semi-functional and the rest of keys are normal. We remark that $\mathsf{Game}_0^{\mathsf{sk}} = \mathsf{Game}_{q_{\mathsf{ku}}}^{\mathsf{ku}}$.

Lemma 5. *Suppose there exists an adversary $\mathcal{A}$ such that $\mathsf{Game}_{k-1}^{\mathsf{sk}} \mathbf{Adv}_{\mathcal{A}} - \mathsf{Game}_k^{\mathsf{sk}} \mathbf{Adv}_{\mathcal{A}} = \epsilon$ Then, we can build an algorithm $\mathcal{B}$ with advantage ϵ in breaking Assumption 2.*

Proof. First, $\mathcal{B}$ is given $(g, X_1 X_2, X_3, Y_2 Y_3, \bar{T})$. $\mathcal{B}$ chooses $\alpha, a_0, \ldots, a_\ell, b \xleftarrow{\$} \mathbb{Z}_n$ and sets $g = g$, $u_i = g^{a_i}$ ($i \in [0, \ell]$), and $h = g^b$. Moreover, $\mathcal{B}$ guesses T^* (with success probability at least $1/|T|$), chooses $c, d \xleftarrow{\$} \mathbb{Z}_n$, and sets $u' = g^c$ and $h' = u'^{-T^*} g^d$. $\mathcal{B}$ sends public parameters $\mathsf{mpk}^{\mathsf{mLW}} = \{n, g, h, u_0, \ldots, u_\ell, e(g, g)^\alpha, u', h', X_3\}$ to $\mathcal{A}$.

For a decryption key query $(\mathsf{ID}|_j, T)$ where $\mathsf{ID}|_j = (l_0, \ldots, l_j)$, $\mathcal{B}$ computes a semi-functional decryption key $\mathsf{dk}_{\mathsf{ID}|_j, T}$ as follows. $\mathcal{B}$ sets $z_k = a_0 l_0 + \cdots + a_j l_j + b$, and chooses $r, z, t, w, w', z_{j+1,\theta}, \ldots, z_{\ell,\theta} \xleftarrow{\$} \mathbb{Z}_n$, and computes

$$(g^\alpha(u_0^{l_0} \cdots u_j^{l_j} h)^r (u'^T h')^t (Y_2 Y_3)^{z z_k}, \ g^r (Y_2 Y_3)^{w z_k}, \ g^t X_3^{w'},$$

$$u_{j+1}^r (Y_2 Y_3)^{z_{j+1} z_k}, \ldots, u_\ell^r (Y_2 Y_3)^{z_\ell z_k}).$$

Similarly, for a key update information query $(\mathsf{ID}|_j, T)$ where $\mathsf{ID}|_j = (l_0, \ldots, l_j)$, $\mathcal{B}$ computes semi-functional key update information $\mathsf{ku}_{\mathsf{ID}|_j, T}$ as follows. $\mathcal{B}$ sets $z_k = a_0 l_0 + \cdots + a_j l_j + b$, and chooses $r_\theta, z_\theta, t_\theta, w_\theta,$ $w'_\theta, z_{j+1,\theta}, \ldots, z_{\ell,\theta} \xleftarrow{\$} \mathbb{Z}_n$ for each θ, and computes

$$\left\{ (P_\theta^{-1} g^\alpha (u_0^{l_0} \cdots u_j^{l_j} h)^{r_\theta} (u'^T h')^{t_\theta} (Y_2 Y_3)^{z_\theta z_k}, \; g^{r_\theta} (Y_2 Y_3)^{w_\theta z_k}, \; g^{t_\theta} X_3^{w'_\theta}, \right.$$

$$\left. u_{j+1}^{r_\theta} (Y_2 Y_3)^{z_{j+1,\theta} z_k}, \ldots, u_\ell^{r_\theta} (Y_2 Y_3)^{z_{\ell,\theta} z_k}) \right\}_\theta.$$

For the i-th secret key query $\mathsf{ID}|_j$ where $\mathsf{ID}|_j = (l_0, \ldots, l_j)$ and $i < k$, $\mathcal{B}$ generates a semi-functional secret key $\mathsf{sk}_{\mathsf{ID}|_j}$ as follows. $\mathcal{B}$ sets $z_k = a_0 l_0 + \cdots + a_j l_j + b$, chooses $r_\theta, z_\theta, t_\theta, w_\theta, w'_\theta, z_{j+1,\theta}, \ldots, z_{\ell,\theta} \xleftarrow{\$} \mathbb{Z}_n$ for each θ, and computes

$$\left\{ (P_\theta (u_0^{l_0} \cdots u_j^{l_j} h)^{r_\theta} (u'^T h')^{t_\theta} (Y_2 Y_3)^{z_\theta z_k}, \; g^{r_\theta} (Y_2 Y_3)^{w_\theta z_k}, \; g^{t_\theta} X_3^{w'_\theta}, \right.$$

$$\left. u_{j+1}^{r_\theta} (Y_2 Y_3)^{z_{j+1,\theta} z_k}, \ldots, u_\ell^{r_\theta} (Y_2 Y_3)^{z_{\ell,\theta} z_k}) \right\}_\theta.$$

For $i > k$, $\mathcal{B}$ generates normal secret keys. For $i = k$, $\mathcal{B}$ sets $z_k = a_0 l_0 + \cdots + a_j l_j + b$, chooses $\gamma_\theta, t_\theta, w_\theta, w'_\theta, w_{j+1,\theta}, \ldots, w_{\ell,\theta} \xleftarrow{\$} \mathbb{Z}_n$, and computes $\mathsf{sk}_{\mathsf{ID}|_j}$ as

$$\left\{ (P_\theta \bar{T}^{\gamma_\theta z_k} (u'^T h')^{t_\theta} X_3^{w_\theta}, \bar{T}^{\gamma_\theta}, g^{t_\theta} X_3^{w'_\theta}, \bar{T}^{\gamma_\theta a_{j+1}} X_3^{w_{j+1,\theta}}, \ldots, \bar{T}^{\gamma_\theta a_\ell} X_3^{w_{\ell,\theta}}) \right\}_\theta.$$

If $\bar{T} \in \mathbb{G}_{p_1 p_3}$, then this key is normal where the $\mathbb{G}_{p_1}$ part of $\bar{T}^{\gamma_\theta}$ is set as g^{r_θ}. If $\bar{T} \in \mathbb{G}$, this key is semi-functional.

$\mathcal{A}$ sends two messages M_0 and M_1 and the challenge identity $\mathsf{ID}|_{k^*}^* = (l_0^*, \ldots, l_{k^*}^*)$ to $\mathcal{B}$. $\mathcal{B}$ chooses $b \xleftarrow{\$} \{0,1\}$, and computes $C' = M_b \cdot e(X_1 X_2, g)^\alpha$, $C_0 = (X_1 X_2)^{a_0 l_0^* + \cdots + a_{k^*} l_{k^*}^* + b}$, $C_1 = X_1 X_2$, and $C_2 = X_1^d$. Here, X_1 is implicitly set g^s and $z_c = a_0 l_0^* + \cdots + a_{k^*} l_{k^*}^* + b$. If $\bar{T} \in \mathbb{G}_{p_1 p_3}$, then $\mathcal{B}$ has properly simulated $\mathsf{Game}_{k-1}^{\mathsf{sk}}$. If $\bar{T} \in \mathbb{G}$, then $\mathcal{B}$ has properly simulated $\mathsf{Game}_k^{\mathsf{sk}}$.

If $\mathcal{B}$ tries to test whether the answer of k-th secret key query $\mathsf{sk}_{\mathsf{ID}|_j}$ is semi-functional or not by computing a semi-functional ciphertext, $\mathcal{B}$ can only create a nominally semi-functional decryption key with $z_k = z_c$. Let us consider the case that a decryption key $\mathsf{dk}_{\mathsf{ID}|_j,T}$ is computed by semi-functional key update information $\mathsf{ku}_{\mathsf{ID}|_{j-1},T}$ and $\mathsf{sk}_{\mathsf{ID}|_j}$, where $\mathsf{ku}_{\mathsf{ID}|_{j-1},T}$ is defined as

$$\left\{ (P_\theta^{-1} g^\alpha (u_0^{l_0} \cdots u_{j-1}^{l_{j-1}} h)^{r'_\theta} (u'^T h')^{t'_\theta} (Y_2 Y_3)^{z'_\theta z_k}, \; g^{t_\theta} X_3^{w'_\theta}, \right.$$

$$\left. u_j^{r'_\theta} (Y_2 Y_3)^{z_{j,\theta} z_k}, \ldots, u_\ell^{r'_\theta} (Y_2 Y_3)^{z_{\ell,\theta} z_k}) \right\}.$$

For some θ, the first component of $\mathsf{dk}_{\mathsf{ID}|_j,T}$ output by the DKGen algorithm is represented as

$$g^\alpha \bar{T}^{\gamma_\theta z_k} (u_0^{l_0} \cdots u_j^{l_j} h)^{r'_\theta + r_\theta} (u'^T h')^{t'_\theta + t_\theta} (Y_2 Y_3)^{z_k (z'_\theta + z_{j,\theta} l_j)}$$

That is, the $\mathbb{G}_{p_2}$ part of this value can be represented as $g_2^{\gamma z_k}$ for some $\gamma \in \mathbb{Z}_n$. Therefore, $\mathsf{dk}_{\mathsf{ID}|_j,T}$ is a nominally semi-functional decryption key.

Similarly, we have to care about the case that $\mathsf{ku}_{\mathsf{ID}|_{j-1},T}$ is computed a semi-functional decryption key $\mathsf{dk}_{\mathsf{ID}|_{j-1},T}$, and $\mathsf{dk}_{\mathsf{ID}|_j,T}$ is computed by $\mathsf{ku}_{\mathsf{ID}|_{j-1},T}$ and $\mathsf{sk}_{\mathsf{ID}|_j}$. Since $\mathsf{ku}_{\mathsf{ID}|_{j-1},T}$ is computed by multiplicating P_θ^{-1} to $\mathsf{dk}_{\mathsf{ID}|_j,T}$ and the re-randomization process is independent of the $\mathbb{G}_{p_2}$ part, $\mathsf{dk}_{\mathsf{ID}|_j,T}$ is a nominally semi-functional decryption key.

From the above estimations, we confirmed that even if $\mathcal{B}$ tries to test, $\mathcal{B}$ can only create a nominally semi-functional decryption key. So, $\mathcal{B}$ uses the output of $\mathcal{A}$ for breaking Assumption 2.   □

The next game, Game_{final} is the same as $\mathrm{Game}^{\mathsf{sk}}_{q_{\mathsf{sk}}}$ except that the challenge ciphertext is semi-functional of a random message (not one of the challenge messages).

Lemma 6. *Suppose there exists an adversary $\mathcal{A}$ such that $\mathrm{Game}^{\mathsf{sk}}_{q_{\mathsf{sk}}}\mathbf{Adv}_{\mathcal{A}} - \mathrm{Game}_{final}\mathbf{Adv}_{\mathcal{A}} = \epsilon$ Then, we can build an algorithm $\mathcal{B}$ with advantage ϵ in breaking Assumption 3.*

Proof. First, $\mathcal{B}$ is given $(g, g^\alpha X_2, X_3, g^s Y_2, Z_2, \bar{T})$. $\mathcal{B}$ chooses $a_0, \ldots, a_\ell, b \xleftarrow{\$} \mathbb{Z}_n$ and sets $g = g$, $u_i = g^{a_i}$ ($i \in [0, \ell]$), $h = g^b$, and $e(g, g)^\alpha = e(g^\alpha X_2, g)$. Moreover, $\mathcal{B}$ guesses T^* (with success probability at least $1/|\mathcal{T}|$), chooses $c, d \xleftarrow{\$} \mathbb{Z}_n$, and sets $u' = g^c$ and $h' = u'^{-T^*} g^d$. $\mathcal{B}$ sends public parameters $\mathsf{mpk}^{\mathsf{mLW}} = \{n, g, h, u_0, \ldots, u_\ell, e(g, g)^\alpha, u', h', X_3\}$ to $\mathcal{A}$.

For a secret key query $\mathsf{ID}|_j = (\mathsf{I}_0, \ldots, \mathsf{I}_j)$, $\mathcal{B}$ generates a semi-functional secret key as follows. $\mathcal{B}$ chooses $c_\theta, r_\theta, t_\theta, w_\theta, z_\theta, z_{j+1,\theta}, \ldots, z_{\ell,\theta}, w_{j+1,\theta}, \ldots, w_{\ell,\theta} \xleftarrow{\$} \mathbb{Z}_n$ for each θ, and computes $\mathsf{sk}_{\mathsf{ID}|_j}$ as

$$\left\{ (P_\theta Z_2^{c_\theta}(u_0^{\mathsf{I}_0} \cdots u_j^{\mathsf{I}_j} h)^{r_\theta} X_3^{w_\theta}, \; g^{r_\theta} Z_2^{z_\theta} X_3^{t_\theta}, \right.$$

$$\left. u_{j+1}^{r_\theta} Z_2^{z_{j+1,\theta}} X_3^{w_{j+1,\theta}}, \ldots, u_\ell^{r_\theta} Z_2^{z_{\ell,\theta}} X_3^{w_{\ell,\theta}}) \right\}_\theta.$$

For a key update information query $(\mathsf{ID}|_j, T)$ where $\mathsf{ID}|_j = (\mathsf{I}_0, \ldots, \mathsf{I}_j)$, $\mathcal{B}$ generates semi-functional key update information as follows. $\mathcal{B}$ chooses $c_\theta, r_\theta, t_\theta, t'_\theta, w_\theta, w'_\theta, z_\theta, z_{j+1,\theta}, \ldots, z_{\ell,\theta}, w_{j+1,\theta}, \ldots, w_{\ell,\theta} \xleftarrow{\$} \mathbb{Z}_n$ for each θ, and computes $\mathsf{ku}_{\mathsf{ID}|_j,T}$ as

$$\left\{ (P_\theta^{-1} g^\alpha X_2 Z_2^{c_\theta}(u_0^{\mathsf{I}_0} \cdots u_j^{\mathsf{I}_j} h)^{r_\theta} (u'^T h')^{t_\theta} X_3^{w_\theta}, \; g^{r_\theta} Z_2^{z_\theta} X_3^{t'_\theta}, \; g^{t_\theta} X_3^{w'_\theta}, \right.$$

$$\left. u_{j+1}^{r_\theta} Z_2^{z_{j+1,\theta}} X_3^{w_{j+1,\theta}}, \ldots, u_\ell^{r_\theta} Z_2^{z_{\ell,\theta}} X_3^{w_{\ell,\theta}}) \right\}_\theta.$$

For a decryption key query $(\mathsf{ID}|_j, T)$ where $\mathsf{ID}|_j = (\mathsf{I}_0, \ldots, \mathsf{I}_j)$, $\mathcal{B}$ generates semi-functional decryption key as follows. $\mathcal{B}$ chooses $c, r, t, t', w, w', z, z_{j+1}, \ldots, z_\ell, w_{j+1}, \ldots, w_\ell \xleftarrow{\$} \mathbb{Z}_n$, and computes $\mathsf{dk}_{\mathsf{ID}|_j,T}$ as

$$(g^\alpha X_2 Z_2^c(u_0^{\mathsf{I}_0} \cdots u_j^{\mathsf{I}_j} h)^r (u'^T h')^t X_3^w, \; g^r Z_2^z X_3^{t'}, \; g^t X_3^{w'},$$

$$u_{j+1}^r Z_2^{z_{j+1}} X_3^{w_{j+1}}, \ldots, u_\ell^{r_\theta} Z_2^{z_\ell} X_3^{w_\ell}).$$

$\mathcal{A}$ sends two messages M_0 and M_1 and the challenge identity $\mathsf{ID}|_{k^*}^* = (\mathsf{I}_0^*, \ldots, \mathsf{I}_{k^*}^*)$ to $\mathcal{B}$. $\mathcal{B}$ chooses $b \xleftarrow{\$} \{0,1\}$, and computes $C' = M_b \cdot \bar{T}$, $C_0 = (g^s Y_2)^{a_0 \mathsf{I}_0^* + \cdots + a_{k^*} \mathsf{I}_{k^*}^* + b}$, $C_1 = g^s Y_2$, and $C_2 = (g^s Y_2)^d$. If $\bar{T} = e(g,g)^{\alpha s}$, then $C' = M_b e(g,g)^{\alpha s}$. Therefore, this is a semi-functional ciphertext of M_b. If $\bar{T}$ is a random element of $\mathbb{G}_T$, then this is a semi-functional ciphertext of a random message. So, $\mathcal{B}$ uses the output of $\mathcal{A}$ for breaking Assumption 3. $\qquad\square$

5   Conclusion

In this paper, we propose the first adaptive-ID secure and scalable RHIBE scheme. Moreover, our construction has the advantages of the Seo-Emura RHIBE scheme [13]; that is, it has the history-free update, security against insiders, short ciphertexts, and decryption key exposure resistance. Since our scheme is constructed over composite order bilinear groups, it is a natural open problem to construct adaptive-ID secure RHIBE scheme over prime order settings. Lewko's technique for prime-order construction [5] might be useful for this problem.

References

1. Boldyreva, A., Goyal, V., Kumar, V.: Identity-based encryption with efficient revocation. In: ACM CCS, pp. 417–426 (2008)
2. Boneh, D., Boyen, X., Goh, E.-J.: Hierarchical identity based encryption with constant size ciphertext. In: Cramer, R. (ed.) EUROCRYPT 2005. LNCS, vol. 3494, pp. 440–456. Springer, Heidelberg (2005)
3. Boneh, D., Franklin, M.: Identity-based encryption from the weil pairing. In: Kilian, J. (ed.) CRYPTO 2001. LNCS, vol. 2139, pp. 213–229. Springer, Heidelberg (2001)
4. Boneh, D., Franklin, M.K.: Identity-based encryption from the weil pairing. SIAM J. Comput. **32**(3), 586–615 (2003)
5. Lewko, A.: Tools for simulating features of composite order bilinear groups in the prime order setting. In: Pointcheval, D., Johansson, T. (eds.) EUROCRYPT 2012. LNCS, vol. 7237, pp. 318–335. Springer, Heidelberg (2012)
6. Lewko, A., Waters, B.: New techniques for dual system encryption and fully secure HIBE with short ciphertexts. In: Micciancio, D. (ed.) TCC 2010. LNCS, vol. 5978, pp. 455–479. Springer, Heidelberg (2010)
7. Libert, B., Vergnaud, D.: Adaptive-ID secure revocable identity-based encryption. In: Fischlin, M. (ed.) CT-RSA 2009. LNCS, vol. 5473, pp. 1–15. Springer, Heidelberg (2009)
8. Naor, D., Naor, M., Lotspiech, J.: Revocation and tracing schemes for stateless receivers. In: Kilian, J. (ed.) CRYPTO 2001. LNCS, vol. 2139, pp. 41–62. Springer, Heidelberg (2001)
9. Seo, J.H., Emura, K.: Efficient delegation of key generation and revocation functionalities in identity-based encryption. In: Dawson, E. (ed.) CT-RSA 2013. LNCS, vol. 7779, pp. 343–358. Springer, Heidelberg (2013)
10. Seo, J.H., Emura, K.: Revocable identity-based encryption revisited: security model and construction. In: Kurosawa, K., Hanaoka, G. (eds.) PKC 2013. LNCS, vol. 7778, pp. 216–234. Springer, Heidelberg (2013)

11. Seo, J.H., Emura, K.: Revocable hierarchical identity-based encryption. Theor. Comput. Sci. **542**, 44–62 (2014)
12. Seo, J.H., Emura, K.: Revocable identity-based cryptosystem revisited: Security models and constructions. IEEE Trans. Inf. Forensics Secur. **9**(7), 1193–1205 (2014)
13. Seo, J.H., Emura, K.: Revocable hierarchical identity-based encryption: history-free update, security against insiders, and short ciphertexts. In: Nyberg, K. (ed.) CT-RSA 2015. LNCS, vol. 9048, pp. 106–123. Springer, Heidelberg (2015)
14. Tsai, T., Tseng, Y., Wu, T.: RHIBE: constructing revocable hierarchical ID-based encryption from HIBE. Informatica Lith. Acad. Sci. **25**(2), 299–326 (2014)
15. Waters, B.: Dual system encryption: realizing fully secure IBE and HIBE under simple assumptions. In: Halevi, S. (ed.) CRYPTO 2009. LNCS, vol. 5677, pp. 619–636. Springer, Heidelberg (2009)

Elliptic Curve Cryptography

Invalid Curve Attacks in a GLS Setting

Taechan Kim[✉] and Mehdi Tibouchi

NTT Secure Platform Laboratories, Tokyo, Japan
{taechan.kim,tibouchi.mehdi}@lab.ntt.co.jp

Abstract. In recent years, most speed records for implementations of elliptic curve cryptosystems have been achieved on curves endowed with nontrivial fast endomorphisms, particularly based on the technique introduced by Galbraith, Lin and Scott (GLS). Therefore, studying the security of those curves is of prime importance.

In this paper, we examine the applicability of the class of attacks introduced by Biehl et al., known as *invalid curve attacks*, to cryptographic implementations based on GLS curves. In invalid curve attacks, a cryptographic device that computes a secret scalar multiplication $P \mapsto kP$ on a certain elliptic curve $E/\mathbb{F}_q$ receives as input an arbitrary "invalid" point $\widetilde{P} \notin E(\mathbb{F}_q)$. Biehl et al. observed that the device then computes the scalar multiplication by k on a different elliptic curve $\widetilde{E}/\mathbb{F}_q$, and if that curve is weaker than E, the attacker can use the result to recover information about the secret k.

The attack doesn't readily adapt to the GLS setting, since the device computes the scalar multiplication as $P \mapsto k_1 P + k_2 \psi(P)$ where ψ is the efficient endomorphism of the GLS curve E, and if it receives an arbitrary invalid point $\widetilde{P}$ on a curve $\widetilde{E} \neq E$, the computation of the map ψ yields a point on a completely different curve again, and the scalar multiplication outputs gibberish. We show, however, that a large family of invalid points $\widetilde{P}$ lie on curve stable under ψ, and using that observation we can modify the attack of Biehl et al. to effectively recover the secrets k_1 and k_2, although the result of the computation on an invalid point doesn't have the "correct" discrete logarithm.

Keywords: Elliptic curve cryptography · GLS method · Cryptanalysis

1 Introduction

Elliptic Curve Cryptography. Since its invention by Miller and Koblitz in the mid-1980s [24,27], elliptic curve cryptography has gained broad adoption in the cryptographic community. Compared with traditional settings like RSA and finite field discrete logarithms, cryptosystems based on elliptic curves offer numerous advantages, achieving higher efficiency with a much smaller key size. This makes them particularly well-suited for efficient implementations.

A number of elliptic curve-based primitives, particularly for key exchange and signatures, have been standardized starting from the early 2000s [1,13,20] and are now widely deployed in embedded applications and on the Internet.

© Springer International Publishing Switzerland 2015
K. Tanaka and Y. Suga (Eds.): IWSEC 2015, LNCS 9241, pp. 41–55, 2015.
DOI: 10.1007/978-3-319-22425-1_3

Invalid Curve Attacks. The security of most elliptic curve cryptosystems is based on the hardness of the elliptic curve discrete logarithm problem (ECDLP) and of related algorithmic problems. ECDLP is the problem of finding the secret scalar k given an elliptic curve point P and its multiple kP in the group of points of the curve. When P is a point of large prime order (or when its order is divisible by a large prime), the problem is believed to be hard (on almost all elliptic curves). However, it is of course easy when P has very small (or very smooth) order, and hence the corresponding problem instances must be avoided.

However, even in a system which is properly designed to use points of large order, a dishonest party might try to adversarially introduce such weak group elements and use them to extract information about the secret key. For instance, in the Diffie-Hellman protocol, choosing one's public key P as an element of a small subgroup allows to recover information about the secret key of the other party, as discussed e.g. in [2,5,26].

In 2000, Biehl, Meyer and Müller [7] carried out a thorough formal analysis of such attacks on ECDLP, which later came to be known as *invalid curve attack*.

Their key observation was that, on an elliptic curve in Weierstrass form, at least one of the curve parameters does not intervene in the expression of the addition and doubling formulas; more precisely, in commonly used coordinate systems, the addition and doubling formulas on:

$$E : y^2 = x^3 + ax + b$$

do not involve the parameter b at all. As a result, if one perturbs the computation of a scalar multiplication kP by modifying the coordinates of P into those of a different point $\widetilde{P}$ not on the curve E, the computation carried out by the device is still an elliptic curve scalar multiplication $k\widetilde{P}$, but on a different curve:

$$\widetilde{E} : y^2 = x^3 + ax + \tilde{b} \quad \text{where} \quad \tilde{b} = y_{\widetilde{P}}^2 - x_{\widetilde{P}}^3 - a x_{\widetilde{P}}.$$

If the discrete logarithm problem happens to be easy on $\widetilde{E}$ (e.g. when the group order of $\widetilde{E}$ is a smooth number, or when $\widetilde{P}$ is a point of small order on $\widetilde{E}$), information about k can be deduced from the result $k\widetilde{P}$ of the faulty scalar multiplication. More recently, Wang and Zhan [30] established that Biehl et al.'s attack has subexponential time complexity under reasonable heuristics.

A number of variants of that invalid curve attacks have since been considered in the literature, such as the twist attack of Fouque et al. [14], which assumes that the input point is in compressed form, and perturbs it into a point on the twisted curve of E, which is often weaker; Karabina and Ustaoğlu [22] discussed invalid curve attacks on hyperelliptic curve cryptosystems; Ciet and Joye [8] extended the work of Biehl et al. to attacks that directly tamper with certain curve parameters, and that work was recently revisited by Kim and Tibouchi [23].

Our Contributions. In this paper, we are particularly interested in invalid curve attacks in a tamper-proof device that employs non-generic scalar multiplication methods. Since multiplication of an elliptic curve point by a scalar is the fundamental operation in ECC, there have been several efforts to improve this operation using efficiently computable endomorphisms on base curves [12,15,17,25].

The basic idea of using an efficiently computable endomorphism for scalar multiplication is first introduced by Gallant, Lambert, and Vanstone [17], but their method is only applicable to very special elliptic curves. The method proposed later by Galbraith, Lin, and Scott [15,16] uses curves over extension fields to provide a much larger family of elliptic curves with endomorphisms, which gives a lot more leeway in selecting efficient and secure parameters, and lets practitioners avoid overly special curves.

Consider a GLV/GLS curve $E/\mathbb{F}_q$ and its efficient endomorphism ψ. The endomorphism ψ acts on any subgroup of $E(\mathbb{F}_q)$ of prime order by multiplication by a certain constant λ: for an element P of that subgroup, $\psi(P) = \lambda P$, and the computation of a scalar multiplication kP is thus carried out by first applying ψ to P and then computing the multiexponentiation $k_1 P + k_2 \psi(P)$ with small $k_1, k_2 \approx k^{1/2}$ satisfying $k \equiv k_1 + k_2 \lambda \pmod{r}$, where r is the order of P.

In this paper, we mainly focus on extending Biehl et al.'s invalid curve attack to the GLS setting. The attacker sends an invalid point (outside $E(\mathbb{F}_q)$) as input to a cryptographic device implementing a fast scalar multiplication on a GLS curve, and tries to recover informations on the secret scalar from the corresponding output.

Adapting Biehl et al.'s attack to this setting is not straightforward, because for an arbitrary invalid input point $\widetilde{P}$ on an invalid curve $\widetilde{E}$, what the device will compute as the image of $\widetilde{P}$ under ψ is overwhelmingly likely to be outside of either E or $\widetilde{E}$, and hence making sense of the invalid computation of the scalar multiple is largely hopeless. We show, however, that a careful choice of invalid curves lets us overcome that difficulty and carry out an attack that largely resembles Biehl et al.'s; there are further differences, however, related to the fact that, even with carefully chosen input points, the discrete logarithms of the corresponding outputs are not equal to the original secret k.

We note that the attack, like Biehl et al.'s, is applicable to arbitrary elliptic curve implementations in which addition and doubling formulas are independent of at least one curve parameter; this includes not just curves in Weierstrass form but also, for example, (twisted) Edwards curve using extended coordinates [19].

2   Preliminaries

Elliptic Curves. An elliptic curve over a finite prime field $\mathbb{F}_p$, $p > 3$, in Weierstrass form, can be described as the set of points (x, y) on the affine plane curve

$$E : y^2 = x^3 + ax + b \tag{1}$$

for some coefficients $a, b \in \mathbb{F}_p$ with $4a^3 + 27b^2 \neq 0$, together with the point at infinity on the projective closure of the affine curve. This set $E(\mathbb{F}_p)$ is endowed with a natural abelian group law which can be defined by a chord-and-tangent process.

The formulas in affine coordinates do not involve the curve parameter b, and it has been a part of the key observation in invalid curve attacks. Such attacks can in principle be applied in any other coordinate system in which formulas for addition and doubling are independent of at least one curve parameter. This includes projective/Jacobian coordinates in Weierstrass form, as well as projective coordinates on Hessian and generalized Hessian curves [11], Huff and twisted Huff curves [21], etc. For Edwards and twisted Edwards curves [4,6], some formulas depend on all curve parameters, but the extended coordinates of Hisil et al. [19], which are some of the most efficient available, are independent of one curve parameter.

ECDLP. Consider again an elliptic curve E over $\mathbb{F}_p$, and a point $P \in E(\mathbb{F}_p)$ of order N in the group (usually a generator of either $E(\mathbb{F}_p)$ itself or a subgroup of small index). The elliptic curve discrete logarithm problem (ECDLP) in (the subgroup generated by P in) E is the computational problem of finding $k \in \{0, \ldots, N-1\}$ given P and the scalar multiple kP. For almost all isomorphism classes of elliptic curves, this problem is considered hard, and the best known attack has a complexity of $O(\sqrt{N})$.

GLS Scalar Multiplications. In this section, we briefly summarize the GLS scalar multiplication method proposed by Galbraith, Lin and Scott [15,16]. For an elliptic curve E over $\mathbb{F}_{p^2}$ and a point $P \in E$, they observed that there exists an endomorphism ψ on E satisfying $\psi^2(P) = -P$. In particular, $\psi(P)$ happens to be a multiple of P as soon as the order of P is sufficiently large, so one can apply the GLV method to obtain efficient scalar multiplication by computing $kP = k_1 P + k_2 \psi(P)$. The endomorphism ψ is constructed by the composition $\phi \circ (\mathrm{Frob}_p) \circ \phi^{-1}$, where Frob_p is the p-th Frobenius map and ϕ is the twisting isomorphism over $\mathbb{F}_{p^4}$.

Lemma 1 ([15]). *Let $p > 3$ be a prime and let E' be an elliptic curve over $\mathbb{F}_p$ with $p + 1 - t$ points. Let E over $\mathbb{F}_{p^2}$ be the quadratic twist of $E'(\mathbb{F}_{p^2})$. Let $\phi : E' \to E$ be the twisting isomorphism defined over $\mathbb{F}_{p^4}$, i.e. $\phi(x, y) = (ux, u^{3/2}y)$ for a non-square $u \in \mathbb{F}_{p^2}$. Let $\psi = \phi \circ (\mathrm{Frob}_p) \circ \phi^{-1}$. Let $P \in E(\mathbb{F}_{p^2})$ be a point of prime order $r > 2p$, then $\psi(P) = \lambda P$ where $\lambda = t^{-1}(p - 1) \pmod{r}$.*

In particular, in the case of $p \equiv 5 \bmod 8$ we have

$$\psi(x, y) = (-x^p, iy^p), \tag{2}$$

for $i \in \mathbb{F}_p$ satisfying $i^2 = -1$.

3   Invalid Curve Attack on GLS Scalar Multiplication

We consider a cryptographic device that computes the scalar multiplication on a base point by a secret scalar k in a GLS setting. We then try to recover the secret k from the outputs obtained by inputting modified base points to the device.

Description of Attacks. Let E be an elliptic curve defined over $\mathbb{F}_{p^2}$ with efficiently computable endomorphism ψ as described in Lemma 1. Assume that the device computes kP on $P \in E(\mathbb{F}_{p^2})$ using the GLS method, in a coordinate system whose addition and doubling formulas are independent of one of the curve parameters (we assume Weierstrass coordinates in our exposition below, but any such coordinate system has the same vulnerability).

Let us give an overview of how to generalize the *invalid curve attack* of Biehl et al. [7] to this GLS setting. To fix ideas, we suppose that the endomorphism ψ is given by the formula of (2).

Recall that in the invalid curve attack, the attack sends to the scalar multiplication device a point $\widetilde{P} = (x_{\widetilde{P}}, y_{\widetilde{P}})$ that deliberately does not lie on the curve E. Instead, it lies on a different curve:

$$\widetilde{E} : y^2 = x^3 + ax + \widetilde{b} \quad \text{for} \quad \widetilde{b} = y_{\widetilde{P}}^2 - x_{\widetilde{P}}^3 - ax_{\widetilde{P}},$$

and since the addition and doubling formulas do not depend on the parameter b, the scalar multiplication involving $\widetilde{P}$ will really be a scalar multiplication on $\widetilde{E}$.

In the GLS setting, however, the multiplication $k\widetilde{P}$ is in fact computed as $k_1\widetilde{P} + k_2\psi(\widetilde{P})$, where k_1, k_2 are half-length exponents and $\psi(\widetilde{P})$ is what the device computes when trying to apply the endomorphism ψ of E to the invalid point $\widetilde{P}$, which does not lie on E. In our case, the result of that computation is $\widetilde{P}' = (-x_{\widetilde{P}}^p, i \cdot y_{\widetilde{P}}^p)$, which typically does *not* lie on the same curve $\widetilde{E}$ as $\widetilde{P}$ anymore. And this is where the invalid curve attack normally breaks down in the GLS setting: the GLS scalar multiplication ends up adding together two points that lie on different elliptic curves and outputting a result which is most likely gibberish.

Nevertheless, the attack can be salvaged with a careful choice of the invalid input point $\widetilde{P}$: we show that there is an exponentially large family of curves $\widetilde{E}$ whose groups of points are *stable* under the map $(x, y) \mapsto (-x^p, i \cdot y^p)$, so that $\widetilde{P}'$ remains on the same curve as $\widetilde{P}$. When $\widetilde{P}$ is chosen on such a curve $\widetilde{E}$, the device ultimately outputs the point $\widetilde{Q} := k_1\widetilde{P} + k_2\widetilde{P}'$, where the two scalar multiplications and the final point addition are all computed in the group $\widetilde{E}(\mathbb{F}_{p^2})$. Moreover, we show that $\widetilde{Q}$ belongs to the cyclic subgroup $\langle\widetilde{P}\rangle$ generated by $\widetilde{P}$. If $\widetilde{P}$ is chosen in such a way that the discrete logarithm problem in this group is easy, computing the discrete logarithm yields a modular relation of the form $\log_{\widetilde{P}}(\widetilde{Q}) \equiv k_1 + k_2 \cdot \log_{\widetilde{P}}(\widetilde{P}') \bmod s$, where s is the order of $\widetilde{P}$ in $\widetilde{E}(\mathbb{F}_{p^2})$. Note that the discrete logarithm of $\widetilde{P}'$ in that relation is *not* the same value λ as for valid points, but we show that it is easy to find nonetheless.

More generally, even if the order s of $\widetilde{P}$ has large prime factors making the DLP too hard to solve completely, one can easily compute the discrete logarithm of $\widetilde{Q}$ modulo any small prime factor of s, and thus recover partial information about k_1 and k_2. Repeating this step with several base points (possibly on different curves) and combining the results using Chinese remainder theorem (CRT), one can obtain a bivariate modular equation with unknowns k_1 and k_2, which can be solved efficiently using the LLL algorithm if the modulus is large enough.

In the subsequent paragraphs, we provide full details about this new attack. In particular, we show in Lemma 2 that:

1. there is a large choice of invalid points $\widetilde{P}$ such that $\widetilde{P}'$ lies on the same curve $\widetilde{E}$: precisely, we propose a way to generate curves $\widetilde{E}$ stable under the map ψ, and show that there are exponentially many such curves;
2. for any such curve $\widetilde{E}$, we show that $\widetilde{P}' \in \langle \widetilde{P} \rangle$, and that the discrete logarithm $\log_{\widetilde{P}}(\widetilde{P}')$ can easily be found without resorting to generic DLP algorithms such as Pollard rho.

Choice of the Invalid Curve $\widetilde{E}$. As mentioned above, Biehl et al.'s attack cannot be applied directly in a GLS setting, due to the fact that the endomorphism ψ is not well-defined for arbitrary chosen base points. In this paragraph, however, we show how to generate a curve $\widetilde{E}$ so that it is also stable under the same endomorphism ψ.

Denote by $E_{a,b}$ the elliptic curve defined by Weierstrass [1] equation $y^2 = x^3 + ax + b$, and let $E := E_{a,b}$ over $\mathbb{F}_{p^2}$ be the GLS curve to be attacked. Recall that the curve E is the quadratic twist over $\mathbb{F}_{p^2}$ of an elliptic curve $E_{\alpha,\beta}$ defined over $\mathbb{F}_p$. In other words, $a = \alpha u^2$ and $b = \beta u^3$ for $u \in \mathbb{F}_{p^2}$ some non quadratic residue. For the GLS curve E be secure, the parameters α and β are chosen such that the curve order $\#E(\mathbb{F}_{p^2}) = (p-1)^2 + t^2$ is almost prime, where t is the trace of Frobenius acting on the curve $E_{\alpha,\beta}$.

To generate an invalid curve $\widetilde{E}$, we need to keep the curve parameter a unchanged compared with the original curve E, so that the addition and doubling formulas in the device also compute the arithmetic operations on $\widetilde{E}$: thus, we let $\widetilde{E} = E_{a,\widetilde{b}}$ for some $\widetilde{b} \in \mathbb{F}_{p^2}$. We choose $\widetilde{b}$ in the following way: First, one chooses an element $\widetilde{\beta}$ randomly from $\mathbb{F}_p$, and sets an elliptic curve $E_{\alpha,\widetilde{\beta}}$ over $\mathbb{F}_p$, then sets $\widetilde{b} := \widetilde{\beta} u^3$. In this case, the curve $\widetilde{E}$ is no other than the quadratic twist of $E_{\alpha,\widetilde{\beta}}$, so we can compute the order of the curve efficiently: $\#\widetilde{E}(\mathbb{F}_{p^2}) = (p-1)^2 + \widetilde{t}^2$, where $\widetilde{t}$ is the trace of the Frobenius on $E_{\alpha,\widetilde{\beta}}$. With high probability, the order $\#\widetilde{E}$ is no longer almost prime, and thus is likely to have relatively large subgroups in which the discrete logarithm is tractable. Also we have many possible choices for $\widetilde{E}$ (roughly, there are about p choices by varying $\widetilde{\beta} \in \mathbb{F}_p$).

Endomorphism on $\widetilde{E}$. Consider the endomorphism $\widetilde{\psi} \in \mathrm{End}(\widetilde{E})$ defined by $\widetilde{\psi} = \widetilde{\phi} \circ (\mathrm{Frob}_p) \circ \widetilde{\phi}^{-1}$, where $\widetilde{\phi} : E_{\alpha,\widetilde{\beta}} \to \widetilde{E}$ is the twisting isomorphism. Remark that the twisting isomorphisms $\phi : E_{\alpha,\beta} \to E$ and $\widetilde{\phi}$ are given by the same formula as $(x,y) \mapsto (ux, \sqrt{u}^3 y)$. Thus the endomorphism $\psi = \phi \circ (\mathrm{Frob}_p) \circ \phi^{-1}$ and $\widetilde{\psi}$ on each curve are also of the same form.

---

[1] We also remark that the same results hold analogously for the Edwards curve of equation $ax^2 + y^2 = 1 + dx^2y^2$, by letting the curve parameter d vary while the curve parameter a is fixed. The addition/doubling formulas are independent of d when extended coordinates [19] are used.

As a result, the endomorphism computed by the device when evaluating the formula for ψ at $\widetilde{P}$ does indeed correspond to the computation of the endomorphism $\widetilde{\psi}$ on the curve $\widetilde{E}$.

However, the output of the perturbed computation is different from $k\widetilde{P}$, since, in a modified setting, $\widetilde{\psi}(\widetilde{P})$ is not identical to $\lambda\widetilde{P}$ at all. Instead, we observe that $\widetilde{\psi}(\widetilde{P}) = \widetilde{\lambda}\widetilde{P}$ for another constant $\widetilde{\lambda}$. It simply comes from the result of [15], recall that $\widetilde{E}$ is also another GLS curve.

Lemma 2. *Let $u \in \mathbb{F}_{p^2}$ be a non-square in $\mathbb{F}_{p^2}$ and $E_{\alpha,\beta}$ be an elliptic curve defined over $\mathbb{F}_p$ with the trace t. Let $E = E_{a,b} = E_{\alpha u^2, \beta u^3}$ be the quadratic twist of $E_{\alpha,\beta}(\mathbb{F}_{p^2})$. For $\widetilde{\beta} \neq \beta \in \mathbb{F}_p$, define new elliptic curve by $\widetilde{E} = E_{a,\widetilde{b}} = E_{\alpha u^2, \widetilde{\beta}u^3}$ so that it is the quadratic twist of $E_{\alpha,\widetilde{\beta}}$. Let $s \mid \#\widetilde{E}(\mathbb{F}_{p^2})$ be a prime factor satisfying $s^2 \nmid \#\widetilde{E}(\mathbb{F}_{p^4})$. If a point $\widetilde{P} \in \widetilde{E}$ is of order s, then $\widetilde{\psi}(\widetilde{P}) = \widetilde{\lambda}\widetilde{P}$ where $\widetilde{\lambda} = \widetilde{t}^{-1}(p-1) \pmod{s}$, where $\widetilde{t}$ is the trace of $E_{\alpha,\widetilde{\beta}}$.*

Proof. This is a direct consequence of [15, Lemma 2]. □

Remark 1. With the notations of the previous lemma, the statement of [15, Lemma 2] actually makes the stronger assumption that $s > 2p$, but one can readily observe from its proof that the much weaker condition that $s^2 \nmid \#\widetilde{E}(\mathbb{F}_{p^4})$ is sufficient.

We also note that, under the heuristic that orders of random elliptic curves behave like random integers of the same size with respect to the distribution of their (not extremely small) prime factors (this is well supported by the theorems of Gekeler [18]; see also the discussion in [23, Sect. 3.3]), s^2 divides $\#\widetilde{E}(\mathbb{F}_{p^4})$ with probability about $1/s$, which is quite small for any moderately large prime factor s of the order of $\widetilde{E}(\mathbb{F}_{p^2})$.

Solving the ECDLP. We now discuss how combining the results from a few invalid curves allows to recover the secret k.

Choose an elliptic curve $\widetilde{E}$ as described above. Suppose that the order $\#\widetilde{E}(\mathbb{F}_{p^2})$ has a prime factor s satisfying the conditions of Lemma 2, and small enough that the ECDLP in a group of order s is tractable. Pick any point $\widetilde{P} \in \widetilde{E}(\mathbb{F}_{p^2})$ of order s; the constant $\widetilde{\lambda}$ such that $\widetilde{\psi}(\widetilde{P}) = \widetilde{\lambda}\widetilde{P}$ is given by the formula of Lemma 2. Thus, the point $\widetilde{Q}$ outputted by the device satisfies $\widetilde{Q} = (k_1 + k_2\widetilde{\lambda})\widetilde{P}$. By solving the discrete logarithm of $\widetilde{Q}$ with respect to $\widetilde{P}$, one obtains a modular equation $\log_{\widetilde{P}}(\widetilde{Q}) \equiv k_1 + k_2 \cdot \widetilde{\lambda} \bmod s$, where $\widetilde{\lambda} = \widetilde{t}^{-1}(p-1) \pmod{s}$.

Repeating this step with several points $\widetilde{P}$ on possibly different curves $\widetilde{E}$ and combining the results with the CRT, one obtains an equation of the form

$$\mathbf{k} \equiv k_1 + k_2\Lambda \bmod \mathbf{s}, \tag{3}$$

where $\mathbf{s}$ is the product of the orders of all points $\widetilde{P}$, Λ is the CRT value of all the $\widetilde{\lambda}$'s, and $\mathbf{k}$ is the CRT value of all the discrete logarithms $\log_{\widetilde{P}}(\widetilde{Q})$.

Hence the problem reduces to solving the bivariate modular equation (3) in k_1, k_2, and this can be done very easily using the LLL algorithm in a 4-dimensional lattice (or even with continued fraction computations) as long as

$$|k_1|, |k_2| \lesssim s^{1/2} \tag{4}$$

since the modular equation is an instance of the $(1, 2)$-ME problem in the terminology of Takayasu and Kunihiro [29] (see also [9]).

One can use as many invalid points as necessary until the desired bound (4) is satisfied. In practice, $k_1, k_2 \approx \log_2 p$, so the bound is satisfied whenever $\log_2 s > 2 \log_2 p = \log_2 r$.

Algorithm 1. Invalid curves attack in GLS setting

Input: A device computing GLS multiplication $k_1 P + k_2 \psi(P)$ on a curve $E = E_{a,b} = E_{\alpha u^2, \beta u^3}/\mathbb{F}_{p^2}$ with the endomorphism ψ and fixed secret scalars $k_1, k_2 \approx \log_2 p$, a positive constant B

Output: k_1, k_2

1: Set $\Lambda = 0 \pmod 1$, $\mathbf{k} = 0 \pmod 1$ and $\mathbf{s} = 1$
2: Set $S = \{\}$
3: **repeat**
4:     Set $\widetilde{s} = 1$
5:     Choose a random $\widetilde{\beta} \in \mathbb{F}_p$ and compute $\widetilde{t} = p + 1 - \#E_{\alpha, \widetilde{\beta}}(\mathbb{F}_p)$
6:     Set $\widetilde{E} = E_{\alpha u^2, \widetilde{\beta} u^3}$ then $\#\widetilde{E}(\mathbb{F}_{p^2}) = (p-1)^2 + \widetilde{t}^2$
7:     Choose a random point $\widetilde{P} \in \widetilde{E}(\mathbb{F}_{p^2})$
8:     Call the device with input $\widetilde{P}$ to obtain $\widetilde{Q}$
9:     Compute the list $\widetilde{S}$ of prime factors of $\#\widetilde{E}(\mathbb{F}_{p^2})$ up to B
10:    Remove from $\widetilde{S}$ the primes already in S and those whose square divides $\#\widetilde{E}(\mathbb{F}_{p^4})$

11:    **for** $s \in \widetilde{S}$ **do**
12:        Compute $P_s := (\#\widetilde{E}/s)\widetilde{P}$ and $Q_s := (\#\widetilde{E}/s)\widetilde{Q}$
13:        Compute the discrete logarithm $\log_{P_s}(Q_s)$ using Pollard's rho algorithm
14:        $\widetilde{s} \leftarrow \widetilde{s} \times s$, $\mathbf{k} \leftarrow \mathrm{CRT}(\mathbf{k}, \log_{P_s}(Q_s))$
15:        Append s to S
16:    **end for**
17:    Set $\widetilde{\lambda} = \widetilde{t}^{-1}(p-1) \pmod{\widetilde{s}}$
18:    $\Lambda \leftarrow \mathrm{CRT}(\Lambda, \widetilde{\lambda})$ and $\mathbf{s} \leftarrow \mathbf{s} \times \widetilde{s}$
19: **until** $\log_2 \mathbf{s} > 2 \log_2 p$
20: Solve $\mathbf{k} \equiv k_1 + k_2 \Lambda \bmod \mathbf{s}$ using the LLL algorithm
21: **return** k_1, k_2

---

4   Complexity Analysis

In this section, we discuss the expected number of invalid curves required for the attack to succeed and retrieve the full secret exponent k the attack, i.e.

the number of iterations in Algorithm 1, and its consequences on the overall complexity of the attack.

Clearly, the crucial parameter that influences the expected number of iterations is the bound B chosen on prime orders s with respect to which one chooses to carry out the generic discrete logarithm computations (steps 11–16 of Algorithm 1). For the attack to succeed, we need to obtain invalid curves until the size of the product of the B-smooth parts of the orders of all curves becomes larger than $2 \log_2 p$.

As a result, under the aforementioned reasonable heuristic assumption that the order of a random elliptic curve behaves as a random integer of the same size in terms of the distribution of the sizes of moderately large prime factors (see again [18]), the problem essentially reduces to estimating the average size of the B-smooth part of a random integer, which we do below.

We find, for example, that the 2^{40}-smooth part of a 256-bit integer is about 34 to 35 bits long on average, so that, when attacking a 256-bit GLS curve with $B = 2^{40}$, we expect to use around 7 or 8 invalid curves on average to recover the whole secret k, and hence solve what amounts to 7 to 8 instances of the Pohlig–Hellman algorithm in elliptic curve subgroups of at most 40 bits inside 256-bit curves, which only takes a few minutes at most on a recent consumer-grade computer. This is well verified in our experiments, as discussed in the next section.

More generally, for any constant u, we would like to estimate the relative size of the B-smooth part of n for $B = n^{1/u}$, on average over all integers n less than some large bound x. Denoting by $S_B(n)$ be the B-smooth part of an integer n, what we want to estimate is the average $\xi(u, x)$ of $\frac{\log S_B(n)}{\log n}$ over integers $1 < n \leq x$, or more realistically, its limit $\xi(u)$ as $x \to \infty$.

We can show that $\xi(u)$ exists and obtain a recurrence formula for it using the random bisection method introduced by Bach and Peralta [3]. Let n_1 be the largest prime factor of n. Let $F(t)$ be the asymptotic probability that $n_1 \leq n^t$, then it is well-known that $F(1/t) = \rho(t)$, where ρ is Dickman's function [10].

As done in [3], we condition on the largest length λ produced by random bisection of $[0, 1]$. If the largest prime factor $n_1 = n^\lambda$ is less than $B = n^{1/u}$, then the relative size of B-smooth part of n is 1. On the other hand, if $\lambda > 1/u$, then the relative length of B-smooth part is given by $(1 - \lambda) \cdot \xi(u(1 - \lambda))$ (the B-smooth part of n is the same with that of n/n_1 and the expected relative size of B-smooth part of n/n_1 is $\xi(u(1-\lambda))$). Therefore, we have the relation (expected value is computed with the cumulative distribution function $F(\lambda)$):

$$\xi(u) = \int_0^{1/u} dF(\lambda) + \int_{1/u}^1 (1 - \lambda) \cdot \xi(u(1 - \lambda)) dF(\lambda). \tag{5}$$

Together with $dF(\lambda) = F(\frac{\lambda}{1-\lambda}) \frac{d\lambda}{\lambda}$, we obtain the following theorem.

Theorem 1. *Let $\xi(u)$ be the asymptotic expected value of the relative size of $n^{1/u}$-smooth part of integers $n \leq x$ as $x \to \infty$. Then we have $\xi(u) = 1$ for*

$0 \leq u \leq 1$, *and*

$$\xi(u) = \rho(u) + \frac{1}{u} \int_0^{u-1} \frac{t}{u-t} \cdot \rho\left(\frac{t}{u-t}\right) \xi(t)dt \quad for \quad u \geq 1.$$

Proof. By the definition, $\xi(u) = 1$ for $0 \leq u \leq 1$. For $u \geq 1$, substituting by $u(1 - \lambda) = t$ and $F(1/u) = \rho(u)$ in Equation (5), we have the desired relation.   □

One can then recursively solve the integral equation of Theorem 1 on successive integer intervals to find an analytic expression for $\xi(u)$, $u \in [m - 1, m]$, for any given positive integer m. The computation is carried out in Appendix A for the first few values of m.

Alternatively, Theorem 1 also lets us evaluate $\xi(u)$ numerically for arbitrary real values; we have carried out this computation for $u \in [0, 25]$, which is sufficient to estimate the required number of iterations in Algorithm 1 for reasonable curve sizes and choices of B. For reference, values of ξ at integers up to 25 are provided in Table 1

As a side note, our computations indicate that $0 \leq 1 - u\xi(u) \leq \frac{1}{\log u}$, which implies in particular that $\xi(u) \sim \frac{1}{u}$. That estimate, which is certainly consistent with what one would expect heuristically, has interesting consequences on the asymptotic complexity of the attack.

Theorem 1 says that the size of $n^{1/u}$-smooth part of an integer n is expected to be $\xi(u) \log n$. Thus, if $B = n^{1/u}$, we need $1/\xi(u)$ invalid curves to satisfy the bound (4) and complete the attack. This holds for any constant u, but if, as is likely, the estimate $\xi(x, u) = \xi(u) \cdot (1 + o(1))$ holds in a larger range for u,

Table 1. The value of $\xi(u)$ on $[2, 25]$ computed using trapezoid method with step size 2^9.

u	$\xi(u)$	u	$\xi(u)$
2	0.500000000000000	14	0.0613410089780015
3	0.302715664266197	15	0.0572426215494454
4	0.217600344410167	16	0.0536583200347903
5	0.173287653074599	17	0.0504970117898520
6	0.144042277957326	18	0.0476879169200047
7	0.123179768322826	19	0.0451752206823007
8	0.107625187526379	20	0.0429143172780390
9	0.0955848422867871	21	0.0408691341252017
10	0.0859784800492326	22	0.0390101804315816
11	0.0781304098434759	23	0.0373131125968829
12	0.0715967896346687	24	0.0357576549323112
13	0.0660727077964555	25	0.0343267849366367

such as $u = O\big((\log x)^{1-\varepsilon}\big)$, $\varepsilon > 0$ (as is conjectured for ρ), one can derive a quasipolynomial bound on the complexity of the overall attack.

Indeed, suppose that the target elliptic curve E is defined over $\mathbb{F}_q = \mathbb{F}_{p^2}$, and let $u = (\log q)^{1-\varepsilon}$ and $B = q^{1/u} = q^\varepsilon$. The expected number of iterations of the entire algorithm satisfies is given by $1/\xi(q, u) \sim 1/\xi(u) \sim u$, and each iteration takes time $O(B)$, because finding all prime factors of $\#\widetilde{E}(\mathbb{F}_q)$ up to be can be done in time $O(B)$ by trial division, the Pohlig–Hellman algorithm itself has a complexity bound of $O(B^{1/2})$, and all other steps including LLL are polynomial in $\log q$. Overall, it follows that the attack completes in expected time $O(uB) = O(q^\varepsilon)$, and is thus heuristically quasipolynomial.

If $\xi(x, u) = \xi(u) \cdot (1 + o(1))$ held even up to $u = O(\log x / \log \log x)$, one could even say that the attack is polynomial, but such a wide range is unlikely as we don't even expect the counterpart for ρ to hold that far.

5   Implementations

We have implemented the attacks described in the previous section using the Sage computer algebra system [28], and carried out experiments on a single core of a laptop computer with 1.7 GHz Intel Core i7 CPU and 8GB RAM. We experimented with the following curve parameters introduced by Galbraith et al. [15]:

- GLS curve in Weierstrass form $E_1/\mathbb{F}_{p_1^2} = E_{-3,1028u}$, where $p_1 = 2^{127} + 29$ and $\#E_1(\mathbb{F}_{p_1^2}) = r$ for 254-bit prime r. We use $\mathbb{F}_{p_1^2} = \mathbb{F}_{p_1}[u]/(u^2 - 2)$. It is the quadratic twist of $E/\mathbb{F}_{p_1} = E_{-3/2,514}$ with the endomorphism $\psi(x, y) = (-x^{p_1}, iy^{p_1})$ for $i \in \mathbb{F}_{p_1}$ satisfying $i^2 = -1$.
- GLS Edwards curve $E_2/\mathbb{F}_{p_2^2} = E_{Edw,(u,42u)}$, where $p_2 = 2^{127} - 1$. We use $\mathbb{F}_{p_2^2} = \mathbb{F}_{p_2}[i]/(i^2 + 1)$ and $u = i + 2$. It is the quadratic twist of $E/\mathbb{F}_{p_2} = E_{Edw,(1,42)}$ with the endomorphism $\psi(x, y) = (u^{(1-p_2)/2}x^{p_2}, y^{p_2})$, where the equation of the Edwards curve is given by $E_{Edw,(a,d)} : ax^2 + y^2 = 1 + dx^2 y^2$.

For the Edwards curve case, we assume that the (non-unified) extended coordinates of [19] are used (their addition and doubling formulas are independent of the curve parameter d).

As described in the previous section, the complexity of Algorithm 1 mainly depends on the smoothness bound B. In this section, we report the experimental results for various B.

Consider the B-smooth part of the order of a 254-bit curve. For $B = 2^{40}$, the bit size of the B-smooth part is expected to be $\xi(254/40) \cdot 254 \approx 34.54$ and the expected number of invalid curves is $1/\xi(254/40) \approx 7.35$. Indeed, we found in our experiments that 4–12 invalid curves were required for the attack to succeed against the curve E_1, and 5–11 against the curve E_2, which is quite consistent with the expected amount.

The most CPU-demanding part of the attack is the Pohlig–Hellman algorithm (line 11 to line 16, Algorithm 1). Our implementation uses Sage's (not particularly optimized) generic group implementation of Pohlig–Hellman, and

Table 2. Timing results in various settings.

(a) Timing results for E_1

	E_1 (Weierstrass form)		
	$B = 2^{40}$	$B = 2^{30}$	$B = 2^{20}$
# of invalid curves (expected)	7.35	9.84	14.78
# of invalid curves (actual)	4–12	7–21	16–36
DL recovery time (s)	14.51–110.82	2.42–9.72	0.38–0.92
LLL recovery time (ms)	1.19–1.56	1.43–15.28	1.15–1.38

(b) Timing results for E_2

	E_2 (Edwards curve)		
	$B = 2^{40}$	$B = 2^{30}$	$B = 2^{20}$
# of invalid curves (expected)	7.35	9.84	14.78
# of invalid curves (actual)	5–11	9–17	15–25
DL recovery time (s)	25.06–204.57	1.38–8.90	0.58–0.83
LLL recovery time (ms)	0.95–1.05	1.00–7.6	1.13–1.51

even so, the required computation time for the whole attack never exceeds a few minutes with $B = 2^{40}$, and is even faster with lower values of B. The LLL algorithm to solve the modular bivariate equation (line 20) takes on the order of milliseconds.

We recorded the expected/actual number of invalid curves and timing results for each curve E_i with $B = 2^{40}, 2^{30}$ and 2^{20} in Table 2. The result shows that the smaller values of B yield faster attacks, but they of course require a higher number of invalid curves, so the right choice of B may depend on how practical it is to feed invalid input points to the cryptographic device.

6   Conclusion

We have extended the invalid curve attack of Biehl et al. to the GLS setting and showed that it works well in practice when invalid curves are chosen carefully to be stable under the fixed GLS endomorphism. This makes it possible to break the ECDLP efficiently with only a few invalid input points.

The most natural countermeasure against this attack is the same as in the original setting: simply checking that input points lie on the expected curves. This is simple enough, but our attack shows that this check cannot be dispensed with even in the presence of efficient endomorphisms.

An interesting open problem would be to extend this attack further to GLV curves, or to multidimensional GLV–GLS curves, which are also popular in record implementations of elliptic and hyperelliptic cryptography. Unlike GLS curves, GLV curves tend to be very sparse, which makes it difficult to construct invalid curves that still admit an efficient endomorphism computed with the same formula. As a result, the invalid computation is likely to involve points on two different curves and to output a result that is difficult to exploit. Nevertheless, one should expect this result to leak some information about the secret scalar, and it would be interesting to find a way of extracting it.

A   Computation of $\xi(u)$ on $u \in [1,3]$

First, we compute $\xi(u)$ on $1 \le u \le 2$,

$$
\xi(u) = \rho(u) + \frac{1}{u} \int_0^{u-1} \frac{t}{u-t} \cdot \rho\left(\frac{t}{u-t}\right) \xi(t) dt
$$

$$
= \rho(u) + \frac{1}{u} \int_0^{u-1} \frac{t}{u-t} dt = \rho(u) + \frac{1}{u}\left[-t - u\log(t-u)\right]_0^{u-1}
$$

$$
= \rho(u) + \frac{1}{u}\left(-(u-1) + u\log u\right) = \frac{1}{u},
$$

where the second equality comes from that $\xi(t) = 1$ and $\rho\left(\frac{t}{u-t}\right) = 1$ for $0 \le t \le u-1$, and the last equality comes from that $\rho(u) = 1 - \log u$ for $1 \le u \le 2$.
For $2 \le u \le 3$, we have

$$
\xi(u) = \rho(u) + \frac{1}{u} \int_0^{u-1} \frac{t}{u-t} \cdot \rho\left(\frac{t}{u-t}\right) \xi(t) dt
$$

$$
= \rho(u) + \frac{1}{u}\left(\int_0^1 \frac{t}{u-t} dt + \int_1^{u-1} \frac{t}{u-t} \cdot \rho\left(\frac{t}{u-t}\right) \frac{dt}{t}\right)
$$

$$
= \rho(u) + \frac{1}{u}\left(-1 + u\log\left(\frac{u}{u-1}\right) + \rho\left(\frac{u}{u-1}\right) - \rho(u)\right)
$$

$$
= \frac{u-1}{u}\left(\rho(u) + \log\left(\frac{u}{u-1}\right)\right)
$$

substituting by $\frac{t}{u-t} = s$ and using $\rho'(s) = -\rho(s-1)/s$ for the later integral in the second line. In the last equality, we used that $\rho\left(\frac{u}{u-1}\right) = 1 - \log\left(\frac{u}{u-1}\right)$ since $1 \le \frac{u}{u-1} \le 2$ for $u \ge 2$.

References

1. ANSI X9.63: Public Key Cryptography for the Financial Services Industry, Key Agreement and Key Transport Using Elliptic Curve Cryptography. ANSI, Washington DC (2001)
2. Antipa, A., Brown, D.R.L., Menezes, A., Struik, R., Vanstone, S.: Validation of elliptic curve public keys. In: Desmedt, Y.G. (ed.) PKC 2003. LNCS, vol. 2567, pp. 211–223. Springer, Heidelberg (2003)
3. Bach, E., Peralta, R.: Asymptotic semismoothness probabilities. Math. Comput. **65**(216), 1701–1715 (1996)
4. Bernstein, D.J., Birkner, P., Joye, M., Lange, T., Peters, C.: Twisted Edwards curves. In: Vaudenay, S. (ed.) AFRICACRYPT 2008. LNCS, vol. 5023, pp. 389–405. Springer, Heidelberg (2008)

5. Bernstein, D.J., Lange, T.: SafeCurves: choosing safe curves for elliptic-curve cryptography. http://safecurves.cr.yp.to

6. Bernstein, D.J., Lange, T.: Faster addition and doubling on elliptic curves. In: Kurosawa, K. (ed.) ASIACRYPT 2007. LNCS, vol. 4833, pp. 29–50. Springer, Heidelberg (2007)

7. Biehl, I., Meyer, B., Müller, V.: Differential fault attacks on elliptic curve cryptosystems. In: Bellare, M. (ed.) CRYPTO 2000. LNCS, vol. 1880, pp. 131–146. Springer, Heidelberg (2000)

8. Ciet, M., Joye, M.: Elliptic curve cryptosystems in the presence of permanent and transient faults. Des. Codes Crypt. **36**(1), 33–43 (2005)

9. Coron, J.-S., Naccache, D., Tibouchi, M.: Fault attacks against EMV signatures. In: Pieprzyk, J. (ed.) CT-RSA 2010. LNCS, vol. 5985, pp. 208–220. Springer, Heidelberg (2010)

10. Dickman, K.: On the frequency of numbers containing prime factors of a certain relative magnitude. Arkiv för Matematik, Astronomi och Fysik **22A**(10), 1–14 (1930)

11. Farashahi, R.R., Joye, M.: Efficient arithmetic on Hessian curves. In: Nguyen, P.Q., Pointcheval, D. (eds.) PKC 2010. LNCS, vol. 6056, pp. 243–260. Springer, Heidelberg (2010)

12. Faz-Hernández, A., Longa, P., Sánchez, A.H.: Efficient and secure algorithms for GLV-based scalar multiplication and their implementation on GLV-GLS curves. In: Benaloh, J. (ed.) CT-RSA 2014. LNCS, vol. 8366, pp. 1–27. Springer, Heidelberg (2014)

13. FIPS PUB 186-3: Digital Signature Standard (DSS). NIST (2009)

14. Fouque, P.-A., Lercier, R., Réal, D., Valette, F.: Fault attack on elliptic curve montgomery ladder implementation. In: Breveglieri, L., Gueron, S., Koren, I., Naccache, D., Seifert, J.-P., (eds) FDTC, pp. 92–98 (2008)

15. Galbraith, S.D., Lin, X., Scott, M.: Endomorphisms for faster elliptic curve cryptography on a large class of curves. In: Joux, A. (ed.) EUROCRYPT 2009. LNCS, vol. 5479, pp. 518–535. Springer, Heidelberg (2009)

16. Galbraith, S.D., Lin, X., Scott, M.: Endomorphisms for faster elliptic curve cryptography on a large class of curves. J. Crypt. **24**(3), 446–469 (2011)

17. Gallant, R.P., Lambert, R.J., Vanstone, S.A.: Faster point multiplication on elliptic curves with efficient endomorphisms. In: Kilian, J. (ed.) CRYPTO 2001. LNCS, vol. 2139, pp. 190–200. Springer, Heidelberg (2001)

18. Gekeler, E.-U.: The distribution of group structures on elliptic curves over finite prime fields. Documenta Mathematica **11**, 119–142 (2006)

19. Hisil, H., Wong, K.K.-H., Carter, G., Dawson, E.: Twisted Edwards curves revisited. In: Pieprzyk, J. (ed.) ASIACRYPT 2008. LNCS, vol. 5350, pp. 326–343. Springer, Heidelberg (2008)

20. ISO/IEC 18033-2: Information technology - Security techniques - Encryption algorithms - Part 2: Asymmetric ciphers. ISO, Geneva (2006)

21. Joye, M., Tibouchi, M., Vergnaud, D.: Huff's model for elliptic curves. In: Hanrot, G., Morain, F., Thomé, E. (eds.) ANTS-IX. LNCS, vol. 6197, pp. 234–250. Springer, Heidelberg (2010)

22. Karabina, K., Ustaoğlu, B.: Invalid-curve attacks on (hyper)elliptic curve cryptosystems. Adv. in Math. of Comm. **4**(3), 307–321 (2010)

23. Kim, T., Tibouchi, M.: Bit-flip faults on elliptic curve base fields, revisited. In: Boureanu, I., Owesarski, P., Vaudenay, S. (eds.) ACNS 2014. LNCS, vol. 8479, pp. 163–180. Springer, Heidelberg (2014)

24. Koblitz, N.: Elliptic curve cryptosystems. Math. Comp. **48**, 203–209 (1987)
25. Longa, P., Sica, F.: Four-dimensional Gallant-Lambert-Vanstone scalar multipli-
    cation. In: Wang, X., Sako, K. (eds.) ASIACRYPT 2012. LNCS, vol. 7658, pp.
    718–739. Springer, Heidelberg (2012)
26. Menezes, A., Ustaoglu, B.: On the importance of public-key validation in the MQV
    and HMQV Key agreement protocols. In: Barua, R., Lange, T. (eds.) INDOCRYPT
    2006. LNCS, vol. 4329, pp. 133–147. Springer, Heidelberg (2006)
27. Miller, V.S.: Use of elliptic curves in cryptography. In: Williams, H.C. (ed.)
    CRYPTO 1985. LNCS, vol. 218, pp. 417–426. Springer, Heidelberg (1986)
28. Stein, W., et al.: Sage Mathematics Software (Version 5.11). The Sage Development
    Team, 2013. http://www.sagemath.org
29. Takayasu, A., Kunihiro, N.: Better lattice constructions for solving multivariate
    linear equations modulo unknown divisors. In: Boyd, C., Simpson, L. (eds.) ACISP.
    LNCS, vol. 7959, pp. 118–135. Springer, Heidelberg (2013)
30. Wang, M., Zhan, T.: Analysis of the fault attack ECDLP over prime field. J. Appl.
    Math. **2011**, 1–11 (2011)

New Fast Algorithms for Elliptic Curve Arithmetic in Affine Coordinates

Wei Yu[1,2], Kwang Ho Kim[3]([✉]), and Myong Song Jo[4]

[1] Institute of Information Engineering, Chinese Academy of Sciences,
Beijing 100093, China
yuwei_1_yw@163.com
[2] Data Assurance and Communication Security Research Center,
Chinese Academy of Sciences, Beijing 100093, China
[3] Department of Algebra and Geometry, Institute of Mathematics,
National Academy of Sciences, Pyongyang, D.P.R. of Korea
math.inst@star-co.net.kp
[4] KumSong School, Pyongyang, D.P.R. of Korea

Abstract. We present new algorithms computing $3P$ and $2P + Q$ by removing the same part of numerators and denominators of their formulas, given two points P and Q on elliptic curves defined over prime fields and binary fields in affine coordinates. Our algorithms save one or two field multiplications compared with ones presented by Ciet, Joye, Lauter, and Montgomery. Since $2P + Q$ takes $\frac{1}{3}$ proportion, 28.5 % proportion, and 25.8 % proportion of all point operations by non-adjacent form, binary/ternary approach and tree approach to compute scalar multiplications respectively, $3P$ occupies 42.9 % proportion and 33.4 % proportion of all point operations by binary/ternary approach and tree approach to compute scalar multiplications respectively, utilizing our new formulas of $2P + Q$ and $3P$, scalar multiplications by using non-adjacent form, binary/ternary approach and tree approach are improved.

Keywords: Elliptic curve · Double-base number system · Elliptic curve arithmetic · Scalar multiplication

1 Introduction

Since Koblitz [1] and Miller [2] devised to use elliptic curves in cryptography in 1985 independently, the research for speeding up the elliptic curve cryptosystems (ECC) has been done very actively in the information security and cryptography field. Compared to other public-key cryptosystems, e.g. RSA, ECC can provide equivalent security with much shorter key length. So it is more suitable for

This research is supported in part by National Research Foundation of China under Grant No. 61379137, No. 61272040, and in part by National Basic Research Program of China(973) under Grant No.2013CB338001.

© Springer International Publishing Switzerland 2015
K. Tanaka and Y. Suga (Eds.): IWSEC 2015, LNCS 9241, pp. 56–64, 2015.
DOI: 10.1007/978-3-319-22425-1_4

cryptographic applications on memory and power constrained embedded systems such as Smart Card or RFID. The bottleneck of such cryptographic applications is the time and space consumption of the cryptographic scheme. Therefore, speed and memory optimization of elliptic curve arithmetic has been continued to be explored as one of the most attractive themes.

There are a great development of Point operations in projective coordinates and Jacobian coordinates [3–6]. For the projective coordinates and Jacobian coordinates take more space, we focus on affine coordinates. There are some algorithms to compute scalar multiplication efficiently. Non-adjacent form (NAF) [7] is the most efficient representation among all binary representations calculating scalar multiplications which need compute $2P, 2P + Q/P + Q$ for given two points P and Q on elliptic curves. Double-base number system is extensively applied to speed up scalar multiplications [8–15] which need the computation of $2P, 2P + Q/P + Q, 3P$ for given two points P and Q on elliptic curves in the last ten years.

This work focuses on the speedup of the computation of $2P+Q, 3P$ on elliptic curves over prime fields and binary fields in affine coordinates. Our main idea is removing the same part of numerators and denominators of their formulas. We improve the formulas of $2P + Q$ and $3P$ on elliptic curves defined over prime fields and binary fields. However, this method can not speed up the operations $P+Q, 2P$. Let $P(X_1, y_1)$, $Q(x_2, y_2)$, $d = (x_2 - x_1)^2 \cdot (2x_1 + x_2) - (y_2 - y_1)^2$, in [16] to compute $2P + Q$, the inversion is $\frac{1}{d \cdot (x_2 - x_1)}$. Considering $(x_3, y_3) = 2P + Q$, the inversion only is $\frac{1}{d}$, which deduces to save 2 field multiplications over prime fields. Using this trick to $2P + Q$ over binary fields, $3P$ over prime fields, and $3P$ over binary fields, the formulas save 1 field multiplication, 1 field squaring, and 2 field multiplications plus 2 field squarings respectively.

$2P + Q$ take $\frac{1}{3}$ of all point operations in NAF [7] and $3P$ take 33.3 % in binary/ternary [16] and 26.6 % in tree approach [10]. In one word, the improvements of $2P + Q$ and $3P$ lead to the speedup of scalar multiplication.

2   Preliminary

In the context of elliptic curve cryptography, two types of curves are mainly used. One type is the non-supersingular elliptic curves over binary fields:

$$E_2 : y^2 + xy = x^3 + ax^2 + b, \tag{1}$$

where $a, b \in \mathbb{F}_{2^n}, b \neq 0$. Another is the elliptic curves over large characteristic fields(of characteristic greater than three, in the remainder of this work, denoted by $\mathbb{F}_p$):

$$E_p : y^2 = x^3 + ax + b, \tag{2}$$

where $a, b \in \mathbb{F}_p, \Delta = 4a^3 + 27b^2 \neq 0$.

As well known, all rational points on an elliptic curve form an abelian group with addition law defined by the chord-tangent rule. For two points $P = (x_1, y_1), Q = (x_2, y_2)(x_1 \neq x_2)$ on elliptic curve (1) and (2), the primitive formulas for computing [2] $P = P + P$ and $P + Q$ are followings

(See [6,17,18]). These operations are usually called as point doubling and point addition, respectively. The outputs of these operations are denoted as (x_3, y_3).

For the curve (1) over $\mathbb{F}_{2^n}$,

Point Doubling

$$\lambda = \frac{x_1^2 + y_1}{x_1}$$
$$x_3 = \lambda^2 + \lambda + a \tag{3}$$
$$y_3 = (x_1 + x_3)\lambda + x_3 + y_1$$

Point Addition

$$\lambda = \frac{y_1 + y_2}{x_1 + x_2}$$
$$x_3 = \lambda^2 + \lambda + a + x_1 + x_2 \tag{4}$$
$$y_3 = (x_1 + x_3)\lambda + x_3 + y_1.$$

For the curve (2) over $\mathbb{F}_p$,

Point Doubling

$$\lambda = \frac{3x_1^2 + a}{2y_1}$$
$$x_3 = \lambda^2 - 2x_1 \tag{5}$$
$$y_3 = \lambda(x_1 - x_3) - y_1$$

Point Addition

$$\lambda = \frac{y_2 - y_1}{x_2 - x_1}$$
$$x_3 = \lambda^2 - x_1 - x_2 \tag{6}$$
$$y_3 = \lambda(x_1 - x_3) - y_1.$$

Obviously, computation of $2P + Q$ is possible in two ways. One is to compute $2P$ by using (3) or (5) and then add Q to the result using (4) or (6). Another is to compute $2P + Q$ as $(P + Q) + P$ by two successive point additions.

Of course, the results are equal, but computation for checking it is non-trivial. So, in works for speeding up ECC, it is usual to consider the two ways to be different. In [19], it was first shown that the $(P + Q) + P$ way is more efficient than the $2P + Q$ way.

Their discovery is as follows. For the curves (2) over $\mathbb{F}_p$ for instance, $(x_5, y_5) = ((x_1, y_1) + (x_2, y_2)) + (x_1, y_1)$ is computed by the $(P + Q) + P$ way, i.e.,

$$\lambda = \frac{y_2 - y_1}{x_2 - x_1}$$
$$x_3 = \lambda^2 - x_1 - x_2$$
$$y_3 = \lambda(x_1 - x_3) - y_1$$
$$\lambda_1 = \frac{y_3 - y_1}{x_3 - x_1} \tag{7}$$
$$x_5 = \lambda_1^2 - x_1 - x_3$$
$$y_5 = \lambda_1(x_1 - x_5) - y_1.$$

Then, since $\lambda_1 = \frac{y_3 - y_1}{x_3 - x_1} = -\lambda + \frac{2y_1}{x_1 - x_3}$, y_3 is needless to get (x_5, y_5) so a field multiplication in the y_3 computation can be saved.

In this work, we will follow the general approach of expressing the cost of point operations in terms of field inversions (I), multiplications (M) and squarings (S), disregarding the cost of field addition/subtractions and multiplications by small constants for simplification purposes.

In [16], it was shown that not only y_3, but also x_3 can be ignored in the computation of (x_5, y_5) and there exists a more efficient algorithm for $2P + Q$, when the cost of one field inversion is not cheaper than that of six field multiplications. Here is the algorithm presented in [16] for $2P + Q$ over $\mathbb{F}_p$ in affine coordinates (Henceforth it will be abbreviated as A):

[**Algorithm 1**] ($2P + Q$ over $\mathbb{F}_p$ [16])
Input: $P = (x_1, y_1), Q = (x_2, y_2) \in E_p$
Output: $(x_3, y_3) = 2P + Q$

$$d = (x_2 - x_1)^2 \cdot (2x_1 + x_2) - (y_2 - y_1)^2, D = d \cdot (x_2 - x_1), I = D^{-1},$$
$$\lambda = (y_2 - y_1) \cdot dI, \lambda_1 = -\lambda + 2y_1 \cdot (x_2 - x_1)^3 I,$$
$$x_3 = (\lambda_1 - \lambda) \cdot (\lambda_1 + \lambda) + x_2, y_3 = (x_1 - x_3) \cdot \lambda_1 - y_1.$$

As shown in [16,20,21], $1I \geq 8M$ and thus Algorithm 1 is superior to the algorithm presented in [19].

The same idea as in algorithm 1 also applies to curve (1) over binary fields, reducing $2P + Q$ cost to $1I + 9M + 2S$ (see [16]).

Now, we will further improve these reduced costs.

3 Speeding up Some Elliptic Curve Operations in Affine Coordinates

Based on the formulas of $2P + Q, 3P$ in [16], we give our new formulas.

Lemma 1. *There exists an algorithm which computes $2P + Q = (x_3, y_3)$ at the cost of $1I + 7M + 2S$, given points $P = (x_1, y_1)$ and $Q = (x_2, y_2)$ on elliptic curve $y^2 = x^3 + ax + b$ over large characteristic field $\mathbb{F}_p$.*

Proof. Let $d = (x_2 - x_1)^2 \cdot (2x_1 + x_2) - (y_2 - y_1)^2, \lambda = (y_2 - y_1) \cdot (x_2 - x_1)^{-1}$. In Algorithm 1, we notice that

$$x_3 -2(-\lambda + y_1 \cdot (x_2 - x_1)^2 d^{-1}) \cdot (2y_1 \cdot (x_2 - x_1)^2 \cdot d^{-1}) + x_2$$
$$=2(y_1 - y_2 + y_1 \cdot (x_2 - x_1)^3 d^{-1}) \cdot (2y_1 \cdot (x_2 - x_1) \cdot d^{-1}) + x_2.$$

And y_3 can be computed as

$$y_3 =(x_1 - x_3) \cdot (-\lambda + 2y_1 \cdot (x_2 - x_1)^2 d^{-1}) - y_1$$
$$=[-2(y_1 - y_2 + y_1 \cdot (x_2 - x_1)^3 d^{-1}) \cdot (2y_1 \cdot (x_2 - x_1) \cdot d^{-1}) + (x_1 - x_2)]$$
$$\cdot (-\lambda + 2y_1 \cdot (x_2 - x_1)^2 d^{-1}) - y_1$$

$$= [-2(y_1 - y_2 + y_1 \cdot (x_2 - x_1)^3 d^{-1}) \cdot (2y_1 \cdot d^{-1}) - 1]$$
$$\cdot (x_2 - x_1)(-\lambda + 2y_1 \cdot (x_2 - x_1)^2 d^{-1}) - y_1$$
$$= [-2(y_1 - y_2 + y_1 \cdot (x_2 - x_1)^3 d^{-1}) \cdot (2y_1 \cdot d^{-1}) - 1]$$
$$\cdot (-(y_2 - y_1) + 2y_1 \cdot (x_2 - x_1)^3 d^{-1}) - y_1.$$

The inversion only need $\frac{1}{d}$. Then $2P + Q = (x_3, y_3)$ are shown in Algorithm 2 at the cost of $1I + 7M + 2S$.

[Algorithm 2] $2P + Q$ over $\mathbb{F}_p$
Input: $P = (x_1, y_1), Q = (x_2, y_2) \in E_p$
Output: $(x_3, y_3) = 2P + Q$

$$d = (x_2 - x_1)^2 \cdot (2x_1 + x_2) - (y_2 - y_1)^2, I = d^{-1}, A = y_1 \cdot I,$$
$$B = A \cdot (x_2 - x_1), C = B \cdot (x_2 - x_1)^2, \lambda_2 = C - (y_2 - y_1),$$
$$x_3 = x_2 + 4B \cdot \lambda_2, y_3 = -y_1 + (-1 - 4A \cdot \lambda_2) \cdot (\lambda_2 + C).$$

Let $d = (x_2 - x_1)^2 \cdot (2x_1 + x_2) - (y_2 - y_1)^2$. In the computing of $2P + Q$ over prime fields, we obtained (x_3, y_3) without computing by $\frac{1}{d}$ instead of $\frac{1}{d(x_2 - x_1)}$. This trick also can be used in computing $2P + Q$ over binary field.

Lemma 2. *There exists an algorithm that computes $2P + Q = (x_3, y_3)$ at the cost of $1I + 8M + 2S$, given points $P = (x_1, y_1)$ and $Q = (x_2, y_2)$ on non-supersingular elliptic curve $y^2 + xy = x^3 + ax^2 + b$ over binary field $\mathbb{F}_{2^n}$.*

Proof. It can be directly checked that following algorithm computes (x_3, y_3) at the cost of $1I + 8M + 2S$.

[Algorithm 3] $(2P + Q$ over $\mathbb{F}_{2^n})$
Input: $P = (x_1, y_1), Q = (x_2, y_2) \in E_p$
Output: $(x_3, y_3) = 2P + Q$

$$d = x_1 \cdot (y_2 + x_1^2) + x_2 \cdot (y_1 + x_1^2), I = d^{-1},$$
$$N = I \cdot (x_1^2 + x_1 \cdot x_2), M = (x_1 + x_2) \cdot N, x_3 = M^2 + M + x_2,$$
$$y_3 = [1 + N \cdot (1 + M)] \cdot [y_1 + y_2 + (x_1 + x_2) \cdot (1 + M)] + x_3 + y_1.$$

This trick can be also applied to the case of $P = Q$, namely $3P$. To the best of our knowledge, till now the affine speed record of $3P$ computation was $1I + 7M + 4S$ in [16] on both of large characteristic and binary field elliptic curves.

Lemma 3. *There exists an algorithm that computes $3P = (x_3, y_3)$ at the cost of $1I + 7M + 3S$, given a point $P = (x, y)$ on elliptic curve $y^2 = x^3 + ax + b$ over large characteristic field $\mathbb{F}_p$.*

Proof. As above, we will give a constructive proof. The following algorithm computing $3P$ costs $1I + 7M + 3S$.

[**Algorithm 4**] ($3P$ over $\mathbb{F}_p$)
Input: $P = (x, y) \in E_p$
Output: $(x_3, y_3) = 3P$

$$d = 12x \cdot y^2 - (3x^2 + a)^2, I = d^{-1}, A = 4y \cdot I,$$
$$B = 8y^2 \cdot I, C = y^2 \cdot B, x_3 = B \cdot (C - 3x^2 - a) + x,$$
$$y_3 = A \cdot (3x^2 + a - C) \cdot (2C - 3x^2 - a) - y.$$

The trick applied to the binary field $3P$ computation provides with more saving.

Lemma 4. *There exists an algorithm that computes $3P = (x_3, y_3)$ at the cost of $1I + 5M + 2S$, given a point $P = (x_1, y_1)$ on binary field elliptic curve $y^2 + xy = x^3 + ax^2 + b$.*

Proof. Following algorithm validates the lemma.

[**Algorithm 5**] ($3P$ over $\mathbb{F}_{2^n}$)
Input: $P = (x_1, y_1) \in E_2$
Output: $(x_3, y_3) = 3P$

$$d = x_1^2 \cdot (x_1^2 + x_1) + b, I = d^{-1}, M = I \cdot x_1^2,$$
$$N = x_1 \cdot M, x_3 = N^2 + N + x_1,$$
$$y_3 = (1 + N) \cdot [N^2 + M \cdot (x_1^2 + y_1)] + x_1 + y_1.$$

4   Comparisons

To compare with other works, we focused on the cost of point operations and their use in scalar multiplications.

4.1   Point Operations

Algorithm 2–5 presented in this work improve the speed records of $2P + Q$ and $3P$ in affine coordinates both on prime fields and binary fields. Table 1 shows comparison between the costs of algorithms presented in this work and the previous speed records.

The formulas of $2P + Q$ save $2M$ compared with [16] over prime fields and $1M$ over binary fields. And the formulas of $3P$ save $1S$ compared with [16] over prime fields and $2M + 2S$ over binary fields.

Next, we will analyze how the improvements of point operations affect scalar multiplications.

Table 1. Cost comparison of various arithmetic algorithms on elliptic curves over $\mathbb{F}_p$

	Ciet-Joye-Lauter-Montgomery [16]	This work	Saved cost
$2P + Q$	$1I + 9M + 2S$	$1I + 7M + 2S$	$2M$
$3P$	$1I + 7M + 4S$	$1I + 7M + 3S$	$1S$

Table 2. Cost comparison of various arithmetic algorithms on elliptic curves over $\mathbb{F}_{2^n}$

	Ciet-Joye-Lauter- Montgomery [16]	This work	Saved cost
$2P+Q$	$1I + 9M + 2S$	$1I + 8M + 2S$	$1M$
$3P$	$1I + 7M + 4S$	$1I + 5M + 2S$	$2M + 2S$

Table 3. Ratio of point operations by different methods

	NAF [7]	Binary/ternary [16]	Tree approach [10]
$2P$	2/3	0.2285/0.7996=0.286	0.3415/0.8364=0.408
$2P+Q$	1/3	0.2284/0.7996=0.285	0.2154/0.8364=0.258
$3P$	-	0.3427/0.7996=0.429	0.2795/0.8364=0.334

Table 4. Cost of scalar multiplication on elliptic curve over prime fields in affine coordinates per bit

	NAF [7]	Binary/ternary [16]	Tree approach [10]
[6, 16]	$I + \frac{13}{3}M + 2S$	$0.7996I + 4.9115M + 2.2846S$	$0.8364I + 4.5981M + 2.2318S$
this work	$I + \frac{11}{3}M + 2S$	$0.7996I + 4.4547M + 1.9419S$	$0.8364I + 4.1473M + 1.9523S$

Table 5. Cost of scalar multiplication on elliptic curve over binary fields in affine coordinates per bit

	NAF [7]	Binary/ternary [16]	Tree approach [10]
[6, 16]	$I + \frac{13}{3}M + 2S$	$0.7996I + 4.9115M + 2.2846S$	$0.8364I + 4.5981M + 2.2318S$
this work	$I + 4M + 2S$	$0.7996I + 3.9977M + 1.5992S$	$0.8364I + 3.8037M + 1.6728S$

4.2   Scalar Multiplications

In [6], the cost of $P + Q$ and $2P$ are $I + 2M + S$ and $I + 2M + 2S$ respectively both on prime fields and binary fields.

The ratio of point operations of $P + Q, 2P, 2P + Q$, and $3P$ used in different scalar multiplication algorithms such as NAF, binary/ternary approach, and tree approach are shown in Table 3 . $2P + Q$ takes $\frac{1}{3}$ proportion, 28.5 % proportion, and 25.8 % proportion of all point operations by non-adjacent form, binary/ternary approach and tree approach to compute scalar multiplications respectively. $3P$ occupies 42.9 % proportion and 33.4 % proportion of all point operations in binary/ternary approach and tree approach which return double-base representations to compute scalar multiplications respectively.

The cost of scalar multiplications per bit using NAF, binary/ternary approach, and tree approach over prime fields and binary fields are summarized in Tables 4 and 5 respectively which are calculated from Tables 1, 2 and 3 immediately. The results in Tables 4 and 5 show that scalar multiplication using the formulas proposed in this work are obviously faster than those using [16].

5   Conclusions

We proposed new formulas for point operations $2P + Q$ and $3P$ in affine coordinates both on prime fields and binary fields which are faster than the responding formulas in [16]. Although the point operations in affine coordinates are usually slower than those in projective coordinates when the cost of one field inversion > the cost of eight field multiplications, they usually take smaller space. The point operations $2P + Q$ and $3P$ proposed in this work are faster than those in projective coordinates in [3,4] when the cost of one field inversion < the cost of four field multiplications.

References

1. Koblitz, N.: Elliptic curve cryptosystems. Math. Comput. **48**, 203–209 (1987)
2. Miller, V.S.: Use of elliptic curves in cryptography. In: Williams, H.C. (ed.) CRYPTO 1985. LNCS, vol. 218, pp. 417–426. Springer, Heidelberg (1986)
3. Longa, P., Gebotys, C.: Fast multibase methods and other several optimizations for elliptic curve scalar multiplication. In: Jarecki, S., Tsudik, G. (eds.) PKC 2009. LNCS, vol. 5443, pp. 443–462. Springer, Heidelberg (2009)
4. Longa, P., Miri, A.: Fast and flexible elliptic curve point arithmetic over prime fields. IEEE Trans. Comput. **57**(3), 289–302 (2008)
5. Le, D.P., Nguyen, B.Pb.: Fast point quadupling on elliptic curve. In: SoICT 2012, pp. 218–222. ACM (2012)
6. Bernstein, D.J., Lange, T.: http://www.hyperelliptic.org/EFD/ (2015)
7. Reitwiesner, G.W.: Binary arithmetic. Adv. Comput. **1**, 231–308 (1960)
8. Dimitrov, V.S., Imbert, L., Mishra, P.K.: Efficient and secure elliptic curve point multiplication using double-base chains. In: Roy, B. (ed.) ASIACRYPT 2005. LNCS, vol. 3788, pp. 59–78. Springer, Heidelberg (2005)
9. Dimitrov, V.S., Imbert, L., Mishra, P.K.: The double-base number system and its application to elliptic curve cryptography. Math. Comp. **77**(262), 1075–1104 (2008)
10. Doche, C., Habsieger, L.: A tree-based approach for computing double-base chains. In: Mu, Y., Susilo, W., Seberry, J. (eds.) ACISP 2008. LNCS, vol. 5107, pp. 433–446. Springer, Heidelberg (2008)
11. Méloni, N., Hasan, M.A.: Elliptic curve scalar multiplication combining Yao's algorithm and double bases. In: Clavier, C., Gaj, K. (eds.) CHES 2009. LNCS, vol. 5747, pp. 304–316. Springer, Heidelberg (2009)
12. Méloni, N., Hasan, M.A.: Efficient double bases for scalar multiplication. IEEE Trans. Comput. **PP**(99), 1 (2015)
13. Doche, C.: On the enumeration of double-base chains with applications to elliptic curve cryptography. In: Sarkar, P., Iwata, T. (eds.) ASIACRYPT 2014. LNCS, vol. 8873, pp. 297–316. Springer, Heidelberg (2014)
14. Adikari, J., Dimitrov, V.S., Imbert, L.: Hybrid binary ternary number system for elliptic curve cryptosystems. IEEE Trans. Comput. **60**, 254–265 (2011)
15. Doche, C., Sutantyo, D.: New and improved methods to analyze and compute double-scalar multiplications. IEEE Trans. Comput. **63**(1), 230–242 (2014)

16. Ciet, M., Joye, M., Lauter, K., Montgomery, P.L.: Trading inversions for multiplications in elliptic curve cryptography. Des. Codes Crypt. **39**(2), 189–206 (2006)
17. Blake, I.F., Seroussi, G., Smart, N.P.: Elliptic Curves in Cryptography. Cambridge University Press, Cambridge (1999)
18. Avanzi, R.M., Cohen, H., Doche, C., Frey, G., Lange, T., Nguyen, K., Vercauteren, F.: Handbook of Elliptic and Hyperelliptic Curve Cryptography. CRC Press, Boca Raton (2005)
19. Eisenträger, K., Lauter, K., Montgomery, P.L.: Fast elliptic curve arithmetic and improved weil pairing evaluation. In: Joye, M. (ed.) CT-RSA 2003. LNCS, vol. 2612, pp. 343–354. Springer, Heidelberg (2003)
20. Brown, M., Hankerson, D., López, J., Menezes, A.: Software implementation of the NIST elliptic curves over prime fields. In: Naccache, D. (ed.) CT-RSA 2001. LNCS, vol. 2020, pp. 250–265. Springer, Heidelberg (2001)
21. Dahmen, E., Okeya, K., Schepers, D.: Affine precomputation with sole inversion in elliptic curve cryptography. In: Pieprzyk, J., Ghodosi, H., Dawson, E. (eds.) ACISP 2007. LNCS, vol. 4586, pp. 245–258. Springer, Heidelberg (2007)

Factoring

Implicit Factorization of RSA Moduli Revisited (Short Paper)

Liqiang Peng[1,2,3], Lei Hu[1,2(✉)], Yao Lu[1,4], Zhangjie Huang[1,2], and Jun Xu[1,2]

[1] State Key Laboratory of Information Security, Institute of Information
Engineering, Chinese Academy of Sciences, Beijing, China
{pengliqiang,hulei,huangzhangjie,xujun}@iie.ac.cn
[2] Data Assurance and Communication Security Research Center,
Chinese Academy of Sciences, Beijing, China
[3] University of Chinese Academy of Sciences, Beijing, China
[4] The University of Tokyo, Tokyo, Japan

Abstract. In this paper, we revisit the problem of factoring RSA moduli with implicit hint, where primes of two RSA moduli share some number of middle bits. Suppose that for two n-bit RSA moduli $N_1 = p_1q_1$ and $N_2 = p_2q_2$, q_1 and q_2 are (αn)-bit primes, p_1 and p_2 share tn bits at positions from t_1n to $t_2n = (t_1 + t)n$. Faugère et al. (PKC 2010) showed that when $t \geq 4\alpha$, one can factor N_1 and N_2 in polynomial time. In this paper, we improve this bound to $t > 4\alpha - 3\alpha^2$ by presenting a new method of solving a homogeneous linear equation modulo unknown divisors. Our method is verified by experiments.

Keywords: RSA modulus · Factorization with implicit hint · Coppersmith's technique · Middle bit

1 Introduction

How to efficiently factor integers which are composed of large primes is one of the most concern problems in algorithmic number theory. However, for now it does not exist any polynomial time algorithm. Therefore, many cryptosystems based on the difficulty of factorization problem are designed. Since its invention [18], the RSA public key cryptosystem is the most studied scheme in cryptology and has been widely used in practical applications due to its effective encryption and decryption. From the work of Coron and May [7], it has been proved that recovering the private key of the RSA cryptosystem and factoring the moduli are determinately equivalent in polynomial time.

However, there still exist many weaknesses in the RSA cryptosystem. For example, to achieve high efficiency in the decryption phase, small decryption exponents are often adopted and the security of such an RSA cryptosystem may be threatened by cryptanalysis such as small private exponent attack [4,20], small CRT-exponent attack [11] and so on. Moreover, the pseudo random number

© Springer International Publishing Switzerland 2015
K. Tanaka and Y. Suga (Eds.): IWSEC 2015, LNCS 9241, pp. 67–76, 2015.
DOI: 10.1007/978-3-319-22425-1_5

generators which are used in the key generation algorithm in the RSA cryptosystem may also threaten the security. Recently, Lenstra et al. [13] and Bernstein et al. [3] discovered this weakness and successfully factor some RSA moduli which are used in the real world. Hence, along this direction many researchers have paid many attentions to factoring RSA moduli with some specific hints.

Implicit Factorization. For the convenience of describing the problem of implicit factorization, we begin with a simple example. Assume that there are two n-bit RSA moduli $N_1 = p_1 q_1$ and $N_2 = p_2 q_2$, where q_1, q_2 are (αn)-bit prime integers.

In PKC 2009, May and Ritzenhofen [16] firstly proposed an efficient method to factor the RSA moduli if p_1 and p_2 share a large number of the least significant bits (LSBs). It has been rigorously proved in [16], if $tn \geq \alpha n + 3$, then (q_1, q_2) is the shortest vector in a related two-dimensional lattice. Once (q_1, q_2) is found by some lattice basis reduction method, the two RSA moduli are factored. May and Ritzenhofen also heuristically generalize their method to deal with implicit factorization of multiple RSA moduli.

Shortly later, Faugère et al. [8] analyzed the problem of implicit factorization where the primes share most significant bits (MSBs) or bits in the middle. According to Faugère et al.'s work, when p_1 and p_2 share $tn \geq 2\alpha n + 3$ MSBs, (q_1, q_2) can be found from a two-dimensional lattice. In the case of tn bits shared in the middle of the binary expressions of p_1 and p_2, they gave a heuristic bound that for the case of $tn \geq 4\alpha n + 7$, and q_1 and q_2 can be recovered from a three-dimensional lattice.

Related Works. Since the problem of implicit factorization has been proposed, it attracts a lot of attentions. Sarkar and Maitra [19] combined the implicit factorization and approximate integer common divisor problem, and by solving modular equations, they obtained the same bound of [8,16] for both LSBs case and MSBs case. Then Kurosawa and Ueda [12] reconsidered the method of [16] and gave a more tighter bound on the numbers of shared LSBs. In 2014, Peng et al. [17] and Lu et al. [15] used two different methods to improve the bound for both LSBs case and MSBs case. The intriguing point is that these two completely different methods obtained the same bounds on the numbers of shared LSBs or MSBs and it is worth to do further investigation to find the internal relations. However, all the above mentioned methods do not work for the case that the primes share middle bits.

Our Contribution. Recall the work of [17], Peng et al. firstly used a low dimensional lattice which is exactly considered in [8,16] to obtain a reduced basis, then they represented the desired vector as a linear equation of the reduced basis, they solved out the linear equation by using Coppersmith's technique, and finally obtain an improved bound.

In this paper, inspired by the idea of [17], for the first time we optimize the bound on the number of shared bits in the middle position. As it has been shown

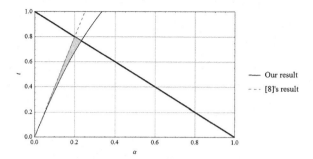

Fig. 1. Comparison with previous ranges on t with respect to α. Since $t \leq 1 - \alpha$, any valid range is under the thick solid diagonal line. Here the dotted line denotes the lower bound on t in [8] and the thin solid line denotes that in this paper. The grey shaded area is a new improvement presented in this paper.

in [8], if there are enough shared middle bits, the desired factorization can be directly obtained from the L^3 lattice basis reduction algorithm. We present a method to deal with the case where the shared middle bits are not enough to ensure that the desired factorization is included in the output of the L^3 algorithm. The starting point is that we represent the vector which we desire to find out as an integer linear combination of the reduced basis vectors of the lattice and obtain a modular equation system with three modular equations and three unknown variables. Then we transform the first two modular equations of the system to a modular equation by applying the Chinese remainder theorem and reduce one of the unknown variables by elimination with the last equation. Finally, we can obtain a homogeneous linear equation with two unknown variables modulo an unknown divisor of a known composite large integer. Once the small root of the modular equation has been solved out, the desired vectors can be recovered, which means the bound on the number of shared middle bits can be improved. Ignoring the small constant which is dependent on the bitlength n, the previous bound $t \geq 4\alpha$ can be improved to $t > 4\alpha - 3\alpha^2$. To the best of our knowledge, our lower bound on the number of the shared middle bits is the first improvement on the implicit factorization problem of middle bits case and experimental results also show this improvement.

An explicit description on our improvement is illustrated in Fig. 1.

The rest of this paper is organized as follows. Preliminaries on lattices are given in Sect. 2. In Sect. 3, we give a brief description of previous work of implicit factorization for middle bits case. Section 4 is our improvement and the experimental results. Finally, Sect. 5 is the conclusion.

2 Preliminaries

Consider the linear independent vectors $w_1, w_2, \cdots, w_k \in \mathbb{R}^n$. Then the lattice L spanned by $w_1, \cdots, w_k$ is the set of all integer linear combinations of $w_1, \cdots, w_k$. The number of vectors, namely k, is the dimension of L and the vectors $w_1, \cdots, w_k$ is a basis of L. Any lattice of dimension larger than 1 has infinitely many bases.

For a lattice, calculating its shortest vector is known to be NP-hard problem under randomized reductions [2]. However, since the L^3 lattice basis reduction algorithm which can output an approximation of shortest vector in polynomial time has been introduced in [14], lattice becomes a fundamental tool to analyze the security of public key cryptosystem.

Lemma 1. (L^3, [14]) Let L be a lattice of dimension k. Applying the L^3 algorithm to L, the output reduced basis vectors $v_1, \cdots, v_k$ satisfy that

$$\|v_1\| \leq \|v_2\| \leq \cdots \leq \|v_i\| \leq 2^{\frac{k(k-i)}{4(k+1-i)}} \det(L)^{\frac{1}{k+1-i}}, \text{for any } 1 \leq i \leq k.$$

In [6], a strategy which is usually called Coppersmith's technique has been discussed. It used lattice-based method to find small integer roots of modular equation with one variable, and of integer equation with two variables. In [10], Jochemsz and May extended the results and gave a general method to find roots of multivariate polynomials.

Given a polynomial $g(x_1, \cdots, x_k) = \sum\limits_{i_1, \cdots, i_k} a_{i_1, \cdots, i_k} x_1^{i_1} \cdots x_k^{i_k}$, we define

$$\|g(x_1, \cdots, x_k)\|^2 = \sum\limits_{i_1, \cdots, i_k} a_{i_1, \cdots, i_k}^2$$

The following lemma due to Howgrave-Graham's result [9] gives a sufficient condition under which modular roots are still satisfied for integer equations.

Lemma 2. (Howgrave-Graham, [9]) Let $g(x_1, \cdots, x_k) \in \mathbb{Z}[x_1, \cdots, x_k]$ be an integer polynomial with at most w monomials. Suppose that

1. $g(y_1, \cdots, y_k) \equiv 0 \pmod{p^m}$ for $|y_1| \leq X_1, \cdots, |y_k| \leq X_k$, and
2. $\|g(x_1 X_1, \cdots, x_k X_k)\| < \dfrac{p^m}{\sqrt{w}}$

Then $g(y_1, \cdots, y_k) = 0$ holds over the integers.

Lattice based approaches of solving small roots of a modular or integer equation are first to construct a lattice from the polynomial of the equation, then by lattice basis reduction algorithm obtain new short lattice vectors which correspond to new polynomials with small norms and with the same roots as the original polynomial. These approaches usually rely on the following heuristic assumption.

Assumption 1. Lattice based constructions always yield algebraically independent polynomials, and the common roots of these polynomials can be efficiently computed by using numerical or symbolic methods.

Gaussian Heuristic. In [1] a claim states that with overwhelming probability, the minima $\lambda_i(\mathcal{L})$ of a random n-dimensional lattice $\mathcal{L}$ are all asymptotically close to the Gaussian heuristic, that is, for all $1 \leq i \leq n$,

$$\frac{\lambda_i(\mathcal{L})}{\det(\mathcal{L})^{\frac{1}{n}}} \approx \sqrt{\frac{n}{2\pi e}},$$

where the minima $\lambda_i(\mathcal{L})$ denotes the i-th minimum of lattice $\mathcal{L}$, which means it is the radius of the smallest zero-centered ball containing at least i linearly independent lattice vectors.

Note that for our attack, the low-dimensional lattice we constructed is not a random lattice, however, according to our practical experiments, the lengths of the vectors of the lattice basis outputted from the L^3 algorithm to that specific lattice are indeed asymptotically close to the Gaussian heuristic. Based on this observation on the lengths of our reduced basis vectors, we give the following attacks and we also do experiments to verify our attacks.

3 Previous Method of Factoring Two RSA Moduli with Implicitly Common Middle Bits

Let $N_1 = p_1 q_1$ and $N_2 = p_2 q_2$ be two given RSA moduli of n bits, where q_1 and q_2 are (αn)-bit primes and p_1 and p_2 are primes that share tn bits at position from $t_1 n$ to $t_2 n = (t_1 + t)n$. For convenience, we write N_1 and N_2 as follows:

$$N_1 = p_1 q_1 = (p_{1_2} 2^{t_2 n} + p 2^{t_1 n} + p_{1_0}) q_1,$$
$$N_2 = p_2 q_2 = (p_{2_2} 2^{t_2 n} + p 2^{t_1 n} + p_{2_0}) q_2.$$

Then we reduce the equations by modulo $2^{t_2 n}$, one can obtain two equations with 5 unknown variables $p, p_{1_0}, p_{2_0}, q_1, q_2$:

$$N_1 \equiv (p 2^{t_1} + p_{1_0}) q_1 \bmod 2^{t_2 n}$$
$$N_2 \equiv (p 2^{t_1} + p_{2_0}) q_2 \bmod 2^{t_2 n}.$$

Faugère et al. transformed the problem of factoring N_1 and N_2 to finding short vectors of a three-dimensional lattice, more precisely, a lattice L defined by the row vectors of the following matrix

$$\begin{pmatrix} K & 0 & N_2 \\ 0 & K & -N_1 \\ 0 & 0 & 2^{t_2 n} \end{pmatrix},$$

where $K = 2^{(\alpha + t_1)n}$.

Clearly, the vector $v = (q_1 K, q_2 K, r)$ with r being the unique remainder in $(-2^{t_2 n-1}, 2^{t_2 n-1}]$ of $q_1 N_2 - q_2 N_1$ modulo $2^{t_2 n}$ is in L. Due to the work of [8], v is the shortest vector in L when

$$tn \geq 4\alpha n + 7.$$

Then one can obtain the primes (q_1, q_2) by a lattice basis reduction algorithm. Note that, for large n, we simplify the bound as $t \geq 4\alpha$ by ignoring the small constant $\frac{7}{n}$.

4   Our Improvement

In this section, we propose a method to deal with the failure case of $t < 4\alpha$ in the previous section and improve the lower bound on t.

Note that, when $t < 4\alpha$ the vector $(q_1 K, q_2 K, r)$ is not the shortest vector of L, which means $(q_1 K, q_2 K, r)$ is generally not included in the outputted basis of the L^3 algorithm.

To facilitate the description, we denote $\lambda_1 = (l_{11}, l_{12}, l_{13})$, $\lambda_2 = (l_{21}, l_{22}, l_{23})$ and $\lambda_3 = (l_{31}, l_{32}, l_{33})$ as the basis vectors of L_1 obtained from the L^3 algorithm. With overwhelming probability, the minima of a lattice are all asymptotically close to the Gaussian heuristic, hence we have that $\|\lambda_1\| \approx \|\lambda_2\| \approx \|\lambda_3\| \approx \det(L)^{\frac{1}{3}}$. Thus, the sizes of l_{ij} can be estimated as $\det(L)^{\frac{1}{3}} = 2^{\frac{2\alpha + 3t_1 + t}{3} n}$.

Write the vector $(q_1 K, q_2 K, r)$ as a linear combination of λ_1, λ_2 and λ_3 with integral coefficients x_0, y_0, z_0, namely $(q_1 K, q_2 K, r) = x_0 \lambda_1 + y_0 \lambda_2 + z_0 \lambda_3$. Moreover, the entry r is $q_1 N_2 - q_2 N_1 \bmod 2^{t_2 n} = q_1 q_2 (p_{2_0} - p_{1_0}) \bmod 2^{t_2 n}$ in $(-2^{t_2 n - 1}, 2^{t_2 n - 1}]$ and $|q_1 q_2 (p_{2_0} - p_{1_0})|$ is less than $2^{(2\alpha + t_1)n}$. Hence, when $t \geq 2\alpha$, we have that $r = q_1 q_2 (p_{2_0} - p_{1_0})$.

Then we get three modular equations modulo unknown prime numbers:

$$\begin{cases} x_0 l_{11} + y_0 l_{21} + z_0 l_{31} = q_1 K \equiv 0 \pmod{q_1}, \\ x_0 l_{12} + y_0 l_{22} + z_0 l_{32} = q_2 K \equiv 0 \pmod{q_2}, \\ x_0 l_{13} + y_0 l_{23} + z_0 l_{33} = q_1 q_2 (p_{2_0} - p_{1_0}) \equiv 0 \pmod{q_1 q_2} \end{cases} \tag{1}$$

Since $|l_{ij}| \approx 2^{\frac{2\alpha + 3t_1 + t}{3} n}$, the desired solutions of (1) can be estimated roughly by $x_0, y_0, z_0 \approx \frac{q_j K}{l_{ij}} \approx 2^{\frac{4\alpha - t}{3} n}$.

Using the Chinese remainder theorem, from the first two equations of (1) we get an equation with the form of

$$a x_0 + b y_0 + c z_0 \equiv 0 \pmod{q_1 q_2}, \tag{2}$$

where a is an integer satisfying $a \equiv l_{11} \pmod{N_1}$ and $a \equiv l_{12} \pmod{N_2}$, b is an integer satisfying $b \equiv l_{21} \pmod{N_1}$ and $b \equiv l_{22} \pmod{N_2}$ and c is an integer satisfying $c \equiv l_{31} \pmod{N_1}$ and $c \equiv l_{32} \pmod{N_2}$. Clearly, a, b and c can be calculated from $l_{11}, l_{12}, l_{21}, l_{22}, l_{31}, l_{32}, N_1$ and N_2 by the extended Euclidean algorithm.

Then we reduce the common variable z_0 from equation (2) and the third equation of (1) by elimination technique. Therefore, we can obtain a modular equation with the form of

$$a' x_0 + b' y_0 \equiv 0 \pmod{q_1 q_2}, \tag{3}$$

In order to recover the integral coefficients x_0 and y_0, we construct a modular equation

$$f(x, y) = a'x + b'y \equiv 0 \,(\mathrm{mod}\ q_1 q_2).$$

Since $\gcd(a', N_1 N_2)$ is 1, or else we have found a factor of $N_1 N_2$. Therefore, we can use $\widehat{f} = a'^{-1} f(x, y) \bmod N_1 N_2$.

Then we select polynomials as follows:

$$g_k(x, y) = y^{m-k} \widehat{f}^k(x, y)(N_1 N_2)^{\max\{s-k, 0\}}, \text{for } k = 0, 1, \cdots, m,$$

where m and s are integers which will be chosen later. Below we let $s \le m$ and $\sigma = \frac{s}{m} \in [0, 1]$.

Obviously, all the above polynomials have the same roots which are desired integral coefficients (x_0, y_0) modulo $(q_1 q_2)^s$ and the solutions can be roughly estimated by $|x_0| \simeq X(:= 2^{\frac{4\alpha-t}{3}n})$ and $|y_0| \simeq Y(:= 2^{\frac{4\alpha-t}{3}n})$, neglecting any small constant because N is relatively large.

Then we construct a matrix, whose row vectors are the coefficient vectors of $g_k(xX, yY)$ with respect to the monomials on x, y. It is easy to check that it is a triangular matrix, and its diagonal entries are

$$X^k Y^{m-k} (N_1 N_2)^{\max\{s-k, 0\}}, \text{ for } k = 0, \cdots, m$$

Let the row vectors of this matrix span a lattice L_1.

By construction, its determinant is easily determined as

$$\det(L_1) = X^{S_x} Y^{S_y} (N_1 N_2)^{S_N}$$

where the exponents S_x, S_y, S_N are calculated as follows:

$$S_x = \sum_{k=0}^{m} k = \frac{1}{2} m^2 + o(m^2),$$

$$S_y = \sum_{k=0}^{m} (m - k) = \frac{1}{2} m^2 + o(m^2),$$

$$S_N = \sum_{k=0}^{s-1} (s - k) = \frac{\sigma^2}{2} m^2 + o(m^2).$$

On the other hand, the dimension of L_1 is $\dim(L_1) = m + 1$. According to Lemmas 1 and 2, one can obtain polynomial equations which share the root (x_0, y_0) over integers if

$$\det(L_1)^{\frac{1}{\dim(L_1)}} < \gamma(q_1 q_2)^s,$$

where γ is a small constant. Now, for large N_1 and N_2, the required condition can be reduced as $\det(L_1)^{\frac{1}{\dim(L_1)}} < (q_1 q_2)^s$, namely,

$$X^{\frac{1}{2}m^2+o(m^2)} Y^{\frac{1}{2}m^2+o(m^2)} (N_1 N_2)^{\frac{\sigma^2}{2}m^2+o(m^2)} < (q_1 q_2)^{\sigma m^2+o(m^2)}$$

Table 1. For 1000-bit RSA moduli, theoretical and experimental results of middle bits problem

Bitsize of p_i, q_i $(1-\alpha)\log_2 N, \alpha\log_2 N$	Theo. of [8]	Theo. of ours	$\dim(L_1) = 21$		$\dim(L_1) = 41$	
			Expt	Time(sec)	Expt	Time(sec)
900,100	407	370	380	118.015	370	6732.652
850,150	607	533	560	196.863	540	10824.582
800,200	failed	680	710	294.561	690	15249.878

To obtain an asymptotic bound, we assume m goes to infinite and ignore the small terms $o(m^2)$. Putting the bounds X, Y into the above inequality, we obtain that

$$(\frac{4\alpha - t}{3}) \cdot \frac{1}{2} \cdot 2 + 2 \cdot \frac{\sigma^2}{2} < 2\alpha \cdot \sigma$$

For optimization, we let $\sigma = \alpha$, and finally we obtain the following bound on t:

$$t > 4\alpha - 3\alpha^2.$$

Then we can obtain several polynomial equations which share the root (x_0, y_0) over integers. Under Assumption 1, we can successfully collect the desired roots.

Experimental Results. We have implemented the experiment program in Magma 2.10 computer algebra system [5] on a PC with Intel(R) Core(TM) Duo CPU(2.53GHz, 1.9GB RAM Windows 7). In all experiments, we obtained several integer equations with desired roots (x_0, y_0) over $\mathbb{Z}$ and found that these equations had a common factor with the form of $ax + by$. In this situation, $ax_0 + by_0$ always equals to 0 and $\gcd(x_0, y_0)$ is small. Hence, the solution (x_0, y_0) can be solved out.

The following Table 1 lists some theoretical and experimental results on factoring two 1000-bit RSA moduli with shared middle bits.

Note that, in the case of $(800, 200)$, the theoretical bound of [8] is 807 bits larger than 800 bits, which means this case can not be found.

Extension to More RSA Mudoli. We heuristically generalize the above result from two RSA moduli to an arbitrary number of n-bit RSA moduli. By combining modulo equations and reducing common variables, we can similarly improve the previous bound of [8].

The key sketch of our method can be described as follows:

(1) For k RSA moduli, based on the work of [8], we firstly construct a $\frac{k(k+1)}{2}$-dimensional lattice. If the shared middle bits are not enough to ensure that the factorization is included in the output of the L^3 algorithm, we represent the desired vector as an integer linear combination of the reduced basis vectors of the lattice and obtain a modular equation system with $\frac{k(k+1)}{2}$ modular equations and $\frac{k(k+1)}{2}$ unknown variables.

(2) In this step, we reduce the unknown variables by elimination in order. At first, we can respectively obtain two homogeneous linear equations with $\frac{k(k+1)}{2}$ unknown variables modulo $q_i q_j$, for $1 \leq i, j \leq k$ and $i \neq j$ and reduce one of the unknown variables by elimination of these two equations.

Then we have an equation f_1 modulo $q_i q_j$. Note that, we can also obtain an equation f_2 modulo $q_i q_l$ and an equation f_3 modulo $q_j q_l$, where $l = 1, \cdots, k$ and $l \neq i, j$. By applying the Chinese remainder theorem, we can obtain an equation modulo $q_i q_j q_l$ from f_1 and f_2, similarly we can obtain another equation modulo $q_i q_j q_l$ from f_1 and f_3. Then we can reduce one unknown variable and obtain a homogeneous linear equation modulo $q_i q_j q_l$.

(3) Based on this order, we can finally obtain a homogeneous linear equation modulo $q_1 q_2 \cdots q_k$ and the number of unknown variables is

$$\frac{k(k+1)}{2} - 1 - 1 - 2 - \cdots - (k-2) = 2k - 2.$$

By solving this modular equation, we can obtain an improved bound.

5 Conclusion

In this paper, we revisited the problem of implicit factorization and we for the first time improved the bound of implicit factorization on the number of the middle bits that the primes share. Our method is to recover the coordinates of the expression of the desired vectors with respect to some reduced lattice basis. It is nice to see our theoretical bound and experimental results are both have an improvement on existing results.

Acknowledgements. The authors would like to thank anonymous reviewers for their helpful comments and suggestions. The work of this paper was supported by the National Key Basic Research Program of China (2013CB834203), the National Natural Science Foundation of China (Grants 61472417, 61402469, 61472416 and 61272478), the Strategic Priority Research Program of Chinese Academy of Sciences under Grant XDA06010702 and XDA06010703, and the State Key Laboratory of Information Security, Chinese Academy of Sciences.

References

1. Ajtai, M.: Generating random lattices according to the invariant distribution. Draft of March (2006)
2. Ajtai, M.: The shortest vector problem in L_2 is NP-hard for randomized reductions (extended abstract). In: Vitter, J.S. (ed.) STOC 1998. pp. 10–19. ACM (1998)
3. Bernstein, D.J., Chang, Y.-A., Cheng, C.-M., Chou, L.-P., Heninger, N., Lange, T., van Someren, N.: Factoring RSA keys from certified smart cards: Coppersmith in the wild. In: Sako, K., Sarkar, P. (eds.) ASIACRYPT 2013, Part II. LNCS, vol. 8270, pp. 341–360. Springer, Heidelberg (2013)
4. Boneh, D., Durfee, G.: Cryptanalysis of RSA with private key d less than $N^{0.292}$. IEEE Trans. Inf. Theor. **46**(4), 1339–1349 (2000)

5. Bosma, W., Cannon, J.J., Playoust, C.: The magma algebra system I: the user language. J. Symbolic Comput. **24**(3–4), 235–265 (1997)
6. Coppersmith, D.: Small solutions to polynomial equations, and low exponent RSA vulnerabilities. J. Crypt. **10**(4), 233–260 (1997)
7. Coron, J., May, A.: Deterministic polynomial-time equivalence of computing the RSA secret key and factoring. J. Crypt. **20**(1), 39–50 (2007)
8. Faugère, J.-C., Marinier, R., Renault, G.: Implicit factoring with shared most significant and middle bits. In: Nguyen, P.Q., Pointcheval, D. (eds.) PKC 2010. LNCS, vol. 6056, pp. 70–87. Springer, Heidelberg (2010)
9. Howgrave-Graham, N.: Finding small roots of univariate modular equations revisited. In: Darnell, M.J. (ed.) Cryptography and Coding 1997. LNCS, vol. 1355, pp. 131–142. Springer, Heidelberg (1997)
10. Jochemsz, E., May, A.: A strategy for finding roots of multivariate polynomials with new applications in attacking RSA variants. In: Lai, X., Chen, K. (eds.) ASIACRYPT 2006. LNCS, vol. 4284, pp. 267–282. Springer, Heidelberg (2006)
11. Jochemsz, E., May, A.: A polynomial time attack on RSA with private CRT-exponents smaller than $N^0.073$. In: Menezes, A. (ed.) CRYPTO 2007. LNCS, vol. 4622, pp. 395–411. Springer, Heidelberg (2007)
12. Kurosawa, K., Ueda, T.: How to factor N_1 and N_2 when $p_1 = p_2 \bmod 2^t$. In: Sakiyama, K., Terada, M. (eds.) IWSEC 2013. LNCS, vol. 8231, pp. 217–225. Springer, Heidelberg (2013)
13. Lenstra, A.K., Hughes, J.P., Augier, M., Bos, J.W., Kleinjung, T., Wachter, C.: Public keys. In: Safavi-Naini, R., Canetti, R. (eds.) CRYPTO 2012. LNCS, vol. 7417, pp. 626–642. Springer, Heidelberg (2012)
14. Lenstra, A.K., Lenstra, H.W., Lovász, L.: Factoring polynomials with rational coefficients. Math. Ann. **261**(4), 515–534 (1982)
15. Lu, Y., Peng, L., Zhang, R., Lin, D.: Towards optimal bounds for implicit factorization problem. IACR Crypt. ePrint Arch. **2014**, 825 (2014)
16. May, A., Ritzenhofen, M.: Implicit factoring: on polynomial time factoring given only an implicit hint. In: Jarecki, S., Tsudik, G. (eds.) PKC 2009. LNCS, vol. 5443, pp. 1–14. Springer, Heidelberg (2009)
17. Peng, L., Hu, L., Xu, J., Huang, Z., Xie, Y.: Further improvement of factoring RSA moduli with implicit hint. In: Pointcheval, D., Vergnaud, D. (eds.) AFRICACRYPT. LNCS, vol. 8469, pp. 165–177. Springer, Heidelberg (2014)
18. Rivest, R.L., Shamir, A., Adleman, L.M.: A method for obtaining digital signatures and public-key cryptosystems (reprint). Commun. ACM **26**(1), 96–99 (1983)
19. Sarkar, S., Maitra, S.: Approximate integer common divisor problem relates to implicit factorization. IEEE Trans. Inf. Theor. **57**(6), 4002–4013 (2011)
20. Wiener, M.J.: Cryptanalysis of short RSA secret exponents. IEEE Trans. Inf. Theor. **36**(3), 553–558 (1990)

Symmetric Cryptanalysis

Improved (Pseudo) Preimage Attacks on Reduced-Round GOST and Grøstl-256 and Studies on Several Truncation Patterns for AES-like Compression Functions

Bingke Ma[1,2,3]([✉]), Bao Li[1,2], Ronglin Hao[1,2,4], and Xiaoqian Li[1,2,3]

[1] State Key Laboratory of Information Security, Institute of Information Engineering, Chinese Academy of Sciences, Beijing 100093, China
[2] Data Assurance and Communication Security Research Center, Chinese Academy of Sciences, Beijing 100093, China
{bkma,lb,xqli}@is.ac.cn
[3] University of Chinese Academy of Sciences, Beijing, China
[4] Department of Electronic Engineering and Information Science, University of Science and Technology of China, Hefei 230027, China
haorl@mail.ustc.edu.cn

Abstract. In this paper, we present improved preimage attacks on the reduced-round GOST hash function family, which serves as the new Russian hash standard, with the aid of techniques such as the rebound attack, the Meet-in-the-Middle preimage attack and the multicollisions. Firstly, the preimage attack on 5-round GOST-256 is proposed which is the first preimage attack for GOST-256 at the hash function level. Then we extend the (previous) attacks on 5-round GOST-256 and 6-round GOST-512 to 6.5 and 7.5 rounds respectively by exploiting the involution property of the GOST transposition operation.

Secondly, inspired by the preimage attack on GOST-256, we also study the impacts of four representative truncation patterns on the resistance of the Meet-in-the-Middle preimage attack against AES-like compression functions, and propose two stronger truncation patterns which make it more difficult to launch this type of attack. Based on our investigations, we are able to slightly improve the previous pseudo preimage attacks on reduced-round Grøstl-256.

Keywords: Hash function · Cryptanalysis · Preimage · GOST · Grøstl-256 · The Meet-in-the-Middle preimage attack · Truncation patterns

This work was supported by the National High Technology Research and Development Program of China (863 Program, No.2013AA014002) and the National Natural Science Foundation of China (No.61379137).

K. Tanaka and Y. Suga (Eds.): IWSEC 2015, LNCS 9241, pp. 79–96, 2015.
DOI: 10.1007/978-3-319-22425-1_6

1   Introduction

Cryptographic hash function is one of the fundamental building blocks in modern cryptography. A hash function takes a message of arbitrary length as input and outputs a bit string of fixed length. For a secure hash function, three classical security properties are mainly concerned, namely, the collision resistance, the second preimage resistance, and the preimage resistance. Many state-of-the-art hash functions divide the input messages into short blocks and process each block with the compression function iteratively, such as the classical Merkle-Damgård [14,38] construction. Due to the generic attacks on the Merkle-Damgård construction [25,27,28], several new domain extension schemes are proposed to fix the inherent weaknesses of the Merkle-Damgård construction. One popular instance of these new constructions is the HAIFA framework proposed by Biham and Dunkelman in [9]. It adds a salt value and a counter which denotes the number of message bits hashed so far as extra input parameters to the compression function, thus makes each compression iteration distinct and is believed to resist certain generic attacks [27,28]. In practice, it has been adopted by several SHA-3 candidates including BLAKE [5], ECHO [7], SHAvite-3 [10], and also the new Russian hash standard GOST R 34.11-2012, to name but a few.

   The old GOST R 34.11-94 hash function [17] was theoretically broken in 2008 [35,36]. As a consequence, the new GOST R 34.11-2012 hash function [15,18,26] has replaced GOST R 34.11-94 as the new Russian national hash standard since January 1, 2013, and it is also included by IETF as RFC 6986 [16]. The new GOST hash function family consists of two members: GOST-256 and GOST-512, which correspond to two different output lengths. It adopts the HAIFA construction with a unique output transformation, and its compression function contains two parallel AES-like permutations in the Miyaguchi-Preneel mode, which is very similar to the Whirlpool hash function [6,23]. Several cryptanalytic results have already been presented since the announcement of the new GOST hash function. Interesting results on the GOST compression function are shown in [1,43], but they seem to have limited impacts on the GOST hash function. The first cryptanalytic result at the hash function level was given by Zou et al. at Inscrypt 2013 [45], which presented collision attacks on 5-round GOST-256/512 and a preimage attack on 6-round GOST-512. The recent preimage attack in [2] is similar to the one in [45]. Ma et al. proposed several improved attacks on GOST [33], including improved (memoryless) preimage attacks on 6-round GOST-512, collision attacks on 6.5/7.5-round GOST-256/512, and the limited-birthday distinguisher [24] on 9.5-round GOST-512. More recently, Guo et al. presented generic second preimage attacks on the full GOST-512 hash function [22] by exploiting the misuse of the counter in the HAIFA mode of GOST. However, their attacks cannot be extended to preimage attacks which allow to invert the hash function. Due to the truncation, their attacks do not work for GOST-256 either.

   To the best of our knowledge, there are no preimage attacks on the GOST-256 hash function except for the trivial brute-force attack. One of the crucial reasons that prevent the preimage attack on GOST-256 is the ChopMD-like mode adopted in the GOST-256 output transformation. The ChopMD mode [12], which truncates a fraction of the final output chaining value, is proven to be indifferentiable from a random oracle [11,12]. However, the specific truncation patterns in it have not been well studied yet. Many instances in practice just truncate the MSBs (or LSBs) of the chaining value as the digest, such as some of the most acknowledged hash functions SHA-3 [8], SHA-224/384 [39], GOST-256, Grøstl [20] and many other SHA-3 candidates. Although this type of truncation pattern is convenient to be implemented in both software and hardware, its concrete impacts on the security properties of the hash functions are less evaluated in literature. Hence, evaluating different truncation patterns and seeking for an optimal one is a very meaningful issue in cryptographic research.

Our Contributions. Firstly, we present improved preimage attacks on the reduced-round GOST hash function family. With the aid of the Meet-in-the-Middle (MitM) preimage attack [4] and the multicollisions [25] which are constructed with dedicated collisions, we overcome the obstacle of the ChopMD mode and present a preimage attack on 5-round GOST-256. Furthermore, by exploiting the weaknesses of the GOST transposition operation, if the MixRow operation of the last round is omitted, we are able to extend the 5-round preimage attack on GOST-256 and the previous preimage attack on 6-round GOST-512 [33] to 6.5 rounds and 7.5 rounds respectively. Although it is not natural to omit the MixRow operation in a single round for a hash function like GOST because it certainly disturbs the wide trail strategy [13], these attacks are theoretically meaningful since they show that any operations with the involution property (*i.e.*, the matrix transposition operation of GOST) would facilitate the attackers with more attacked rounds, thus further demonstrate that such operations with the involution property should be cautiously considered as candidates for achieving transposition in AES-like hash primitives. The results are summarized in Table 1. However, due to the limited number of rounds attacked, our results do not pose any threats to the practical use of the GOST hash function family.

Secondly, motivated by the preimage attack on GOST-256, we discuss the impacts of the truncation patterns on the resistance of the MitM preimage attack against AES-like compression functions with the ChopMD mode. We investigate four representative truncation patterns, and show that two of them certainly make it more difficult to launch this type of attack. Moreover, the last pattern studied even resists the 6-round MitM preimage attack for certain digest sizes. Based on the investigations, we are able to slightly improve the previous pseudo preimage attacks on the reduced-round SHA-3 finalist Grøstl-256.

Organization of the Paper. In Sect. 2, GOST hash function. Section 3 presents the improved preimage attacks on the reduced-round GOST hash function family. Section 4 concludes and summarizes the paper. Due to the space limit, the impacts of four representative truncation patterns on the resistance of the MitM preimage attack against AES-like compression functions and the applications to Grøstl-256 will be reported in the full version of this paper [34].

Table 1. Summary of previous and our (second) preimage attacks on GOST

Target	Rounds attacked	Time	Memory	Reference
GOST-256 (12 Rounds)	5	2^{192}	2^{64}	Section 3
		2^{208a}	2^{12}	
	6.5^a	2^{232}	2^{120}	
GOST-512 (12 Rounds)	6	2^{505}	2^{256}	[2]
		2^{505}	2^{64}	[45]
		2^{496}	2^{64}	[33]
		2^{504c}	2^{11}	
	7.5^a	2^{496}	2^{64}	Section 3
		2^{504c}	2^{11}	
	Full^b	$2x \cdot 2^{512-x}$, for $x < 178.67$	Not given	[22]
		2^{523-x}, for $x < 259$	Not given	

a : Require the omission of the last MixRow operation.
b : Only work as second preimage attacks.
c : The memory minimized attacks.

2   The GOST Hash Function Family

The GOST hash function takes any message up to 2^{512} bits as input, and outputs a 256- or 512-bit digest, *i.e.*, GOST-256 and GOST-512. These two variants are almost the same, except that they have different initial values, and GOST-256 only preserves the 256 MSBs of the final 512-bit chaining value as the digest. As depicted in Fig. 1, the GOST hash function family adopts the HAIFA construction with a unique output transformation. The hash computation contains three stages. Before we give specific descriptions of each stage, we define several notations.

$A||B$ The concatenation of two bit strings A and B.
M The input message, which is divided into 512-bit blocks.
M_i The i-th 512-bit message block of M.
$|M|$ The bit length of M.
Len The bit length of the last message block of M.
Σ The 512-bit checksum of all message blocks.
CF The compression function.
h_i The i-th 512-bit chaining variable.
CT_i The i-th 512-bit counter which denotes the total message bits processed before the i-th CF call.

In the initialization stage, M is padded into a multiple of 512 bits, *i.e.*, $M||1||0^*$ is the padded message, which is then divided into N 512-bit blocks $M_0||M_1||...||M_{N-1}$. h_0 is assigned to the predefined IV of GOST-256 or GOST-512. $|M|$, Σ and CT_0 are assigned to 0. In the compression stage, each block M_i is processed iteratively, *i.e.*, $h_{i+1} = CF(h_i, M_i, CT_i)$ for $i = 0, 1, ..., N-1$. After each computation of the compression function, $|M|$, Σ and CT_{i+1} are updated accordingly. In the finalization stage, the output chaining value of the

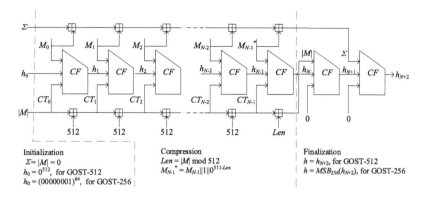

Fig. 1. Three stages of the GOST hash function

last message block h_N goes through the output transformation, *i.e.*, $h_{N+1} = CF(h_N, |M|, 0)$, $h_{N+2} = CF(h_{N+1}, \Sigma, 0)$. For GOST-512 (resp. GOST-256), h_{N+2} (resp. $MSB_{256}(h_{N+2})$) is the digest of M.

The compression function $CF(h_i, M_i, CT_i)$ can be seen as an AES-like block cipher E_K used in a Miyaguchi-Preneel-like mode, *i.e.*, $CF(h_i, M_i, CT_i) = E_{h_i \oplus CT_i}(M_i) \oplus M_i \oplus h_i$. As for the block cipher E_K, the 512-bit internal state is denoted as an 8×8 byte matrix. For the key schedule part, $h_i \oplus CT_i$ is assigned as the key K, then K_0 is computed from K as follows:

$$K_0 = L \circ P \circ S(K).$$

The round keys $K_1, K_2, ..., K_{12}$ are generated as follows:

$$K_{j+1} = L \circ P \circ S \circ XC(K_j) \ for \ j = 0, 1, 2, ...11,$$

where K_{12} is used as the post-whitening key:

- **AddRoundConstant(XC):** XOR a 512-bit constant predefined by the designers.
- **SubBytes(S):** process each byte of the state through the SBox layer.
- **Transposition(P):** transpose the k-th column to be the k-th row for $k = 0, 1, 2, ..., 7$, *i.e.*, transposition of the state matrix.
- **MixRows(L):** multiply each row of the state matrix by an MDS matrix.

For the data processing part, M_i is the plaintext, and is assigned to the initial state S_0. Then the state is updated 12 times with the round function as follows:

$$S_{j+1} = L \circ P \circ S \circ X(S_j), \ for \ j = 0, 1, 2, ...11,$$

where **AddRoundKey(X)** XOR the state with the round key K_j. Finally, the ciphertext $E_K(M_i)$ is computed with $S_{12} \oplus K_{12}$.

3   Improved Preimage Attacks on Reduced-Round GOST

This section illustrates improved preimage attacks on the reduced-round GOST hash function family. Firstly, we present the preimage attack on 5-round GOST-256. This attack overcomes the obstacles of the output transformation and the truncation operation of GOST-256 by combining the dedicated collision and preimage attacks on the compression function, and is the first preimage attack on GOST-256 in literature except for the trivial brute-force attack. Then by exploiting the weaknesses of the GOST transposition operation P, if the MixRow operation L in the last round is omitted, we are able to extend the 5-round preimage attack on GOST-256 and the previous 6-round preimage attack on GOST-512 to 6.5 rounds and 7.5 rounds respectively. Although such an omission seems unnatural for a construction like GOST since it certainly disturbs the wide trail strategy[1], the improved preimage attacks with more attacked rounds are of some theoretical interests since they further demonstrate that any operations with the involution property should be cautiously considered to achieve transposition in AES-like hash primitives from the aspect of preimage resistance, as similar deductions have already been presented from the aspect of collision-like attacks in [33].

3.1   Overview of the Preimage Attack on 5-Round GOST-256

Based on the preimage attacks on 6-round GOST-512 [33], a very straightforward strategy is to find a (pseudo) preimage[2] on the last compression function call of the output transformation, and convert it to a preimage attack on the GOST-256 hash function with the aid of tactfully constructed multicollisions. However, if the multicollisions are constructed with cascaded collisions which are generated with the birthday attack, the time complexity would be worse than the brute-force preimage attack on GOST-256. Moreover, the final truncation operation of GOST-256 also makes it more complex to launch the MitM preimage attack on the last compression function call of the output transformation.

To overcome the first obstacle, we could use dedicated methods to generate the collisions rather than the birthday attack, and construct the multicollisions with these dedicated cascaded collisions with unique structures. For the second problem, by a deeper look into the original MitM preimage attack framework on AES-like compression functions, we are able to find the optimal attack parameters for reduced-round GOST-256 although it performs the final truncation. Motivated

---

[1] Actually, similar observations have already been adopted [40], in which preimage attack on 7-round AES hashing modes was constructed by omitting the last Mix-Column operation, but it is natural for AES since there is no MixColumn operation in the last round of AES. However, the omission of MixColumn/MixRow in the last round is mainly adopted by block ciphers to achieve implementation advantage in the decryption algorithm, but such omission might be improper under the hash function setting since inverse computation is commonly not required for hash functions.

[2] A pseudo preimage is a preimage whose input chaining value is not equal to the given chaining value.

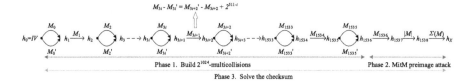

Fig. 2. Three phases of the preimage attack on reduced-round `GOST-256`

by the preimage attack on reduced-round `GOST-256`, more studies on the MitM preimage attack for `AES`-like compression functions with truncation are discussed in the full version of this paper [34].

As depicted in Fig. 2, the preimage attack on 5-round `GOST-256` consists of three phases. We briefly introduce the main thoughts of each phase which help to understand the whole attack.

Phase 1. 2^{1024}-multicollisions are constructed with 512 cascaded 4-multicollisions pairs, namely,
$(M_{3i}, M'_{3i})||M_{3i+1}||(M_{3i+2}, M'_{3i+2})$ for $i = 0, 1, ..., 511$. We choose a 3-block message pair from each of the 4-multicollisions, *i.e.*,
$(M_{3i}||M_{3i+1}||M_{3i+2}, M'_{3i}||M_{3i+1}||M'_{3i+2})$ for $i = 0, 1, ..., 511$, which satisfy that

$$M_{3i} + M_{3i+2} = M'_{3i} + M'_{3i+2} + 2^{511-i}. \tag{1}$$

Since the size of the message checksum is 512 bits, and 4-multicollisions are required for each $i = 0, 1, ..., 511$ to make Eq. (1) hold, thus we need to build 2^{1024}-multicollisions. Any possible value of the message checksum can be constructed with the 512 3-block message pairs for $i = 0, 1, ..., 511$ selected from the 2^{1024}-multicollisions due to the unique structures of the collision pairs. In order to build the cascaded collisions which form the multicollisions, we utilize the dedicated collision attacks on 5-round `GOST-256` compression function [45] which will be illustrated later.

Phase 2. With the output chaining value of the 2^{1024}-multicollisions h_{1536}, we randomly choose one more message block M_{1536} which satisfies padding and derive the message bit length $|M|$ as well. The value of h_{1538} can then be computed. Let h_X denote the given target, the last compression function call of the output transformation is then inverted with the MitM preimage attack, and the preimage $\Sigma(M)$ generated in this phase is the message checksum desired. Since the collision attack in phase 1 can only reach 5 rounds, we only focus on the preimage attack on 5-round `GOST-256` compression function. The exact procedures of this phase will be provided later as well.

Phase 3. After phase 1 and phase 2, we get hold of the 2^{1024}-multicollisions, and also have the message checksum desired. In order to produce the desired checksum, we make use of the methods of [19,33], which first write the desired checksum in binary form and then use the unique structure of the cascaded collision pairs to produce each item of the binary expression. We refer to [33] for details of this phase.

After all three phases succeed, we manage to generate a preimage for 5-round GOST-256.

3.2 Phase 1. Construct the Multicollisions

This phase needs to generate 2^{1024}-multicollisions which can be decomposed into 512 pairs of 4-multicollisions, namely, $(M_{3i}, M'_{3i})||M_{3i+1}||(M_{3i+2}, M'_{3i+2})$ for $i = 0, 1, ..., 511$ which satisfy Eq. (1). In order to do so, we need to utilize the collision attack on 5-round GOST-256 compression function in [45]. The truncated differential trail is depicted in Fig. 3, and the rebound attack [37] is used to derive the collisions. The layers of the inbound phase, which are covered with the SuperSBox technique [21,32], are denoted in red, while the layers of the outbound phase are denoted in blue. We refer to [45] for more descriptions, and only present the results: it requires 2^{120} time and 2^{64} memory to find a collision, and the expected number of collisions with a fixed input chaining value is 2^8.

Now we show how to generate the 4-multicollisions $(M_{3i}, M'_{3i})||M_{3i+1}||(M_{3i+2}, M'_{3i+2})$ for a specific i. As depicted in Fig. 2, the exact procedures are as follows:

1. From the chaining value h_{3i}, with the aid of the rebound attack and the SuperSBox technique, we find a collision pair (M_{3i}, M'_{3i}) on 5-round GOST-256 at the cost of 2^{120} time and 2^{64} memory. Notice that the only difference between M_{3i} and M'_{3i} lies in a single cell whose exact position is determined by the value of i, and the difference can take at most 2^8 possible values.
2. From h_{3i+1}, we randomly choose the value of M_{3i+1} and compute the output chaining value h_{3i+2}.
3. From h_{3i+2}, we find a collision pair (M_{3i+2}, M'_{3i+2}) on 5-round GOST-256. Then we check whether $M_{3i} - M'_{3i} = M'_{3i+2} - M_{3i+2} + 2^{511-i}$ holds. Note that the only difference between M_{3i+2} and M'_{3i+2} lies in the same active cell as well, thus this condition holds with probability 2^{-8}.
4. From a fixed h_{3i+2}, 2^8 collision pairs (M_{3i+2}, M'_{3i+2}) can be generated, thus we expect to find one such pair which makes Eq. (1) hold. However, if no desired pair is found for the fixed h_{3i+2}, we can utilize the two-block strategy, and go to step 2 with another value of M_{3i+1}, then redo step 3. As a result, step 3 will eventually succeed and the 4-multicollisions can be constructed.

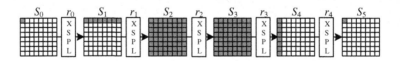

Fig. 3. Collision attack on 5-round GOST-256 compression function

Since the position of the active cell can be placed in any cell from the 64 possible positions, we can generate the 4-multicollisions for any item 2^{511-i} where

$i = 0, 1, ..., 511$. For instance, the differential trail in Fig. 3 is adopted to generate the 4-multicollisions for the item 2^{511-i} where $i = 0, 1, ..., 7$. Finally, after enumerating all 64 positions of the active cell and repeating the above procedures 512 times, the 2^{1024}-multicollisions are constructed with $512 \times (2^{120} + 2^{120+8}) \approx 2^{137}$ time and 2^{64} memory.

3.3   Phase 2. Invert the Output Transformation

As depicted in Fig. 2, given the target digest h_X and the chaining value h_{1538}, this phase inverts the last compression function call of the output transformation and derives the value of the message checksum $\Sigma(M)$. In order to do so, we utilize the MitM preimage attack on AES-like compression functions. Due to the close relevances, we give brief descriptions of the MitM preimage attack on AES-like compression functions. Then we discuss how the truncation pattern of GOST-256 influences our attack. Finally, we provide the optimal attack parameters derived.

The MitM Preimage Attack Framework on AES-like Compression Functions. The MitM preimage attack was first introduced by Aoki and Sasaki in [4]. The basic idea of this technique, which is known as *splice-and-cut*, aims to divide the target cipher into two sub-ciphers which can be computed independently. Several advanced techniques to further improve the basic attack are developed, such as *partial matching* [4], *initial structure* [41], *indirect partial matching* [3], *bicliques* [30] and *differential MitM attack* [31].

In [40], Sasaki proposed the first MitM preimage attack on AES-like compression functions. Two main techniques were presented, namely, *initial structure* in an AES-like compression function and *indirect partial matching* through an MixColumn layer. This work was later improved by Wu *et al.* in [44]. Thanks to the delicate descriptions of the MitM preimage attack framework on AES-like compression functions presented in [44], the chunk separations can be easily represented by introducing several essential integer parameters, and the best attack parameters can be easily derived through an exhaustive search. In [42], Sasaki *et al.* introduced the *guess-and-determine* approach to extend the basic attack by one more round. Based on these previous results, the basic MitM preimage attack framework on AES-like compression functions is achieved.

As shown in Fig. 4, without loss of generality, we use the chunk separation of 5-round GOST-256 in order to further illustrate the details of the attack framework. We first define several necessary notations which are used throughout this paper. These notations are also depicted in Fig. 4 for instance.

n Bit size of the digest.
N_c Bit size of the cell.
N_t Number of columns (or rows) in the state.
b Number of blue rows (or columns) in the initial structure.
r Number of red rows (or columns) in the initial structure.
c Number of constant cells in each row (or column) in the initial structure.

  g Number of guessed rows (or columns) in the backward (or forward) computation, which are denoted in purple. See Fig. 6 for instance.

D_b Freedom degrees of the blue chunk in bits.

D_r Freedom degrees of the red chunk in bits.

D_g Bit size of the guessed cells.

D_m Bit size of the match point.

TIME Time complexity of the preimage attack.

MEMORY Memory requirement of the preimage attack.

The attack procedures can be further divided into five steps which are illustrated as follows, and the interested readers are referred to [40, 44] for more details of the construction of the initial-structure and indirect-partial-matching through the MixRow layer, and [42] for the illustrations of the guess-and-determine strategy.

Step 1. Initial Structure. Choose random values for the constants which are used in the transformations between states #1 ↔ #2, and states #3 ↔ #4. Following the linear relations of the MixRow operation, compute the values for the forward chunk (in blue) which has D_b freedom degrees and the backward chunk (in red) which has D_r freedom degrees. After this step, the compression function is divided into two independent chunks thanks to the initial structure.

Step 2. Forward Computation. For all the blue and grey cells at state #4, the forward chunk can be computed forwards independently until state #5.

Step 3. Backward Computation. For all the red cells at #1, the backward chunk can be computed backwards independently until state #8. Although not shown in Fig. 4, the guess-and-determine strategy can be utilized at state #7 in order to extend the attack by one more round. Note that we do not consider the truncation operation currently when describing this general attack framework. The impacts brought by the truncation pattern through the feed-forward operation during the backward computation phase will be discussed later.

Step 4. Indirect-Partial-Matching through the MixRow Layer. The indirect-partial-matching is performed between states #5 ↔ #8 with partial known information of the red and blue cells from both directions by exploiting linear relations of the MixRow operation.

Step 5. Recheck. Check whether the guessed cells of the partial match derived in step 4 are guessed correctly. If so, check whether the partial match is also a full match. Repeat the above steps 1–5 until a preimage is found.

Deriving the Attack Parameters. As discussed in previous works, the attack parameters, e.g., the freedom degrees of the forward chunk D_b and the backward chunk D_r can be easily represented and enumerated with several predefined integer parameters (b, r, c) as shown in Fig. 4. However, the size of the match point D_m is rather ad-hoc, and needs to be treated carefully mainly due to impacts introduced by the truncation operation and the feed-forward operation.

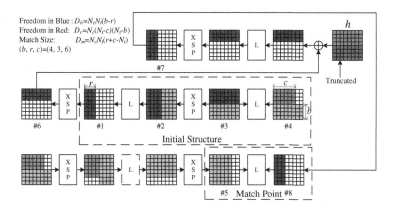

Freedom in Blue : $D_b = N_c N_t (b-r)$
Freedom in Red: $D_r = N_c (N_t c - c)(N_t - b)$
Match Size: $D_m = N_c N_t (r + c - N_t)$
$(b, r, c) = (4, 3, 6)$

Fig. 4. Preimage attack on 5-round `GOST-256` compression function

On the Match Point. For narrow pipe constructions, the state size equals with the digest size, and the indirect-partial-matching contributes to the match of the entire digest. As studied in previous works [40,42,44], the complexities can be denoted as follows (n denotes the digest size):

$$\texttt{TIME} = 2^n (2^{-D_r} + 2^{-D_b} + 2^{-D_m}),$$
$$\texttt{MEMORY} = min\{2^{D_r}, 2^{D_b}\}. \qquad (2)$$

However, since `GOST-256` is a wide pipe design which adopts the `ChopMD` mode, we have to ensure that the indirect-partial-matching operation between the forward blue chunk and the backward red chunk is only carried on the preserved grey cells at state #9, and is not related to the truncated green cells at state #9. As can be seen from Fig. 4, the red cells at state #7 are fully determined after the feed-forward operation of state #6 and state #9, and only the grey preserved cells at state #9 are involved in this feed-forward process, while the green truncated cells are totally irrelevant. Consequently, only the grey preserved cells of the digest at state #9 are matched through the indirect-partial-matching layer, and the attack complexities for narrow pipe constructions still hold for the `GOST-256` case. Hence, for the `GOST-256` case, the attack complexities can still be represented as Eq. (2).

Impacts of the Truncation. As depicted in Fig. 4, the truncation operation has direct influences on the matching part of the attack. More precisely, in the backward computation of the red chunks, the feed-forward operation can be conducted at state #6 or #7, and different selections will result in different attacks. However, note that the attack complexities still depend on the quartets, *i.e.*, (n, D_r, D_b, D_m), and thus the expressions of the complexities are the same as Eq. (2).

Case 1. Feed-forward at State #6. After the feed-forward operation with state #6, all information of the r red rows at #6 will be preserved due to the row-wise truncation as long as $r \leq 4$. Consequently, the backward computation would proceed by one more round and the original attack framework would still work as long as $r \leq 4$ is satisfied. We exhaustively search all possible attack parameters for case 1, and the optimal parameters offer the attack with time complexity 2^{192} which is optimal.

Case 2. Feed-forward at State #7. After the feed-forward operation with state #7, only the first four rows of the r red columns at #7 will be preserved due to the truncation. Although the number of attacked rounds remains unchanged, the size of the match point[3] is reduced by half which might have negative impacts on the overall complexity. Actually, we exhaustively search all possible attack parameters for case 2, and the optimal parameters offer the attack with time complexity 2^{208} which is not optimal.

Optimize Phase 2. Based on the previous discussions on the generic attack framework and the impacts introduced by the truncation, we choose state #6 as the position where the feed-forward operation is performed. Because (D_b, D_r, D_m) can be represented with (b, r, c), we can easily enumerate all possible values of (b, r, c) and search for the optimal attack parameters. The optimal chunk separation for 5-round GOST-256 is denoted in Fig. 4. Since the size of the target digest h_X is 256 bits, and we have 512 bits freedom degrees in $\Sigma(M)$, phase 2 will succeed with probability 1. Finally, according to the specific attack parameters in Fig. 4, we need 2^{192} time and 2^{64} memory to generate a preimage for 5-round GOST-256 compression function.

3.4   Phase 3. Generate the Preimage

We omit the details of this phase, since it follows almost the same procedures of the preimage attack in [33]. Note that this phase only needs several simple operations, thus the complexities are negligible.

3.5   Summarize the 5-Round Attack

Minimize the Memory Requirement. We can also minimize the memory requirement after a deeper look into the above attack. Since we have to store the 2^{1024}-multicollisions, the memory requirement is at least 2^{12} 512-bit blocks. We can launch the memoryless MitM preimage attack [29, 42] in phase 2, and generate a preimage with 2^{208} time and negligible memory using the parameters $(b, r, c) = (5, 4, 6)^{4}$. As for phase 1, the standard time/memory tradeoff can be utilized to reduce the memory requirement of the inbound phase as stated in [32, Appendix]. More precisely, referring to Fig. 3, the exact attack steps are as follows:

---

[3] Similar to the pseudo preimage attack on Grøstl [44], although the match point is cut to half by the feed-forward operation, the indirect-partial-matching is still only related to the preserved parts of the digest, and contains no information of the truncated parts.

[4] Note that $r = 4$ satisfy the above requirement $r \leq 4$ of Case 1.

1. We choose 2^s differences at S_4 and propagate the differences to $P^{-1} \circ L^{-1}(S_4)$, then save the 2^s differences in sorted lists for each SuperSBox.
2. We choose a random difference of $P^{-1} \circ L^{-1}(S_3)$, and compute the corresponding difference of $X(S_3)$.
3. For each specific SuperSBox, we enumerate all 2^{64} values according to the difference of $X(S_3)$, and compute the corresponding output differences. There are 2^s differences in the saved list, thus we expect to generate 2^s solutions (one solution average for each difference in the list) for this SuperSBox since we need to match a 64-bit difference. We can repeat step 3 for each SuperSBox independently.

Finally, we can generate 2^s solutions for the inbound phase with 2^{64} time and 2^s memory, or equivalently a single solution can be generated with 2^{64-s} time and 2^s memory. Combining the outbound phase, it takes $2^{64-s+120} = 2^{184-s}$ time and 2^s memory to derive a 5-round collision. The standard SuperSBox technique sets $s = 64$, and we could find a solution for the inbound phase with average time complexity one. But now we set $s = 12$, and we need 2^{64} time to find 2^{12} solutions for the inbound phase and 2^{172} time for a 5-round collision. Finally, in order to build and store the 2^{1024}-multicollisions, we need $512 \times (2^{172} + 2^{172+8}) \approx 2^{189}$ time and 2^{12} memory which is still not the bottleneck of the time complexity.

Complexity Analysis. Combining the complexities of the three phases, the preimage attack on 5-round GOST-256 requires 2^{192} time and 2^{64} memory. If we aim to minimize the memory requirement, the attack requires 2^{208} time and 2^{12} memory.

3.6   Extend the Preimage Attack to More Rounds

The transposition operation P of GOST is an involution, namely, $P(St) = P^{-1}(St)$ holds for any 512-bit state St. The collision and distinguishing attacks on GOST in [33] benefits from this fact with more attacked rounds. Now we show that the preimage attacks can also be extended by 1.5 more rounds by exploiting this property. More precisely, if we omit the MixRow operation L in the last round, due to the fact that the other round operations, namely, X, S, P are cell-independent operations, the backward computation can be further extended by 1.5 more rounds, and we are able to launch preimage attacks on 6.5-round GOST-256 and 7.5-round GOST-512. Although these improved attacks require the omission of the MixRow operation L in the last round which seems inappropriate for a primitive like GOST, they are worth mentioning since they further clarify that any operations with the involution property are not optimal candidates as the transposition operation of AES-like hash primitives. As a counter-example, we cannot extend the previous preimage attacks on 6-round Whirlpool [42] to 7.5 rounds by omitting the last MixRow operation, because the ShiftColumn operation of Whirlpool is not an involution.

Preimage Attack on 6.5-Round GOST-256. There are three phases in the preimage attack on 6.5-round GOST-256. The main ideas of each phase are identical to the 5-round attack, thus we only provide brief descriptions of the attack, and omit more specified details.

In phase 1, we generate 2^{1024}-multicollisions which can be decomposed into 512 pairs of 4-multicollisions similar to the 5-round preimage attack. In order to do so, we need to utilize the collision attack on 6.5-round GOST-256 compression function in [33]. The truncated differential trail is depicted in Fig. 5. By choosing different locations of the active column, we can build the 2^{1024}-multicollisions with $512 \times 2^{184} = 2^{193}$ time and 2^{64} memory by repeating the collision attack on the compression function 512 times.

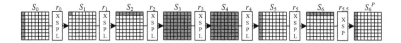

Fig. 5. Collision attack on 6.5-round GOST-256 compression function

In phase 2, we invert the last compression function call in the output transformation. We remove the MixRow operation L in the last round, and the optimal chunk separation for the 6.5-round MitM preimage attack is depicted in Fig. 6, and we can find a preimage $\Sigma(M)$ with complexities (TIME, MEMORY) = $(2^{232}, 2^{120})$.

Phase 3 is the same as the third phase of the 5-round preimage attack. Combining all 3 phases, it would require 2^{232} time and 2^{120} memory to generate a preimage for 6.5-round GOST-256. It is notable that although there exists a collision attack on 7.5-round GOST compression function [33] which can be utilized to build the 2^{1024}-multicollisions adopted in phase 1, we are not able to launch a preimage attack on 7.5-round GOST-256 due to its truncation operation.

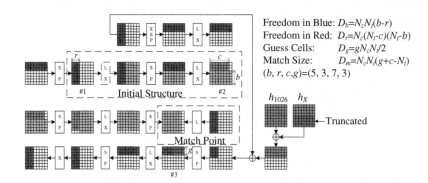

Fig. 6. Preimage attack on 6.5-round GOST-256 compression function

Preimage Attack on 7.5-Round GOST-256. The preimage attack on 7.5-round GOST-512 consists of three phases as well. The first and the last phases are the same as [33], while the improvement is carried out on the second phase. We remove the MixRow operation L in the last round, and the chunk separation for the second phase is depicted in Fig. 7. Finally, It requires 2^{496} time and 2^{64} memory to generate a preimage for 7.5-round GOST-512. We can also launch a memoryless variant of this attack following the methods in [33], and it requires 2^{504} time and 2^{11} memory to generate a preimage for 7.5-round GOST-512.

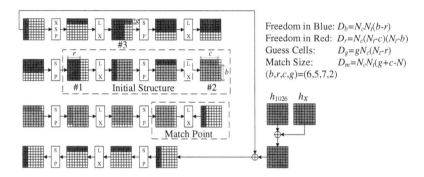

Freedom in Blue: $D_b = N_c N_t(b-r)$
Freedom in Red: $D_r = N_c(N_t-c)(N_t-b)$
Guess Cells: $D_g = g N_c(N_t-r)$
Match Size: $D_m = N_c N_t(g+c-N)$
$(b,r,c,g)=(6,5,7,2)$

Fig. 7. Preimage attack on 7.5-round GOST-512 compression function

4   Conclusion and Open Problems

In this paper, we present improved preimage attacks on the reduced-round GOST hash function family by combining the dedicated collision attack and the MitM preimage attack on the GOST compression function. As far as we know, our result is the first preimage attack on GOST-256 at the hash function level. We also investigate the impacts of four representative truncation patterns on the resistance of the MitM preimage attack against AES-like compression functions, and propose two strengthened truncation patterns, which make it more difficult to launch the MitM preimage attack.

However, due to the large search space, we are not able to study the impacts of all possible truncation patterns at the moment. An interesting and open problem is to seek for advanced approaches to efficiently enumerate all possible truncation patterns, and investigate their impacts on various security properties of many AES-based hash primitives.

References

1. AlTawy, R., Kircanski, A., Youssef, A.M.: Rebound attacks on stribog. In: Lee, H.-S., Han, D.-G. (eds.) ICISC 2013. LNCS, vol. 8565, pp. 175–188. Springer, Heidelberg (2014)

2. AlTawy, R., Youssef, A.M.: Preimage attacks on reduced-round stribog. In: Pointcheval, D., Vergnaud, D. (eds.) AFRICACRYPT. LNCS, vol. 8469, pp. 109–125. Springer, Heidelberg (2014)

3. Aoki, K., Guo, J., Matusiewicz, K., Sasaki, Y., Wang, L.: Preimages for step-reduced SHA-2. In: Matsui, M. (ed.) ASIACRYPT 2009. LNCS, vol. 5912, pp. 578–597. Springer, Heidelberg (2009)

4. Aoki, K., Sasaki, Y.: Preimage attacks on one-block MD4, 63-step MD5 and more. In: Avanzi, R.M., Keliher, L., Sica, F. (eds.) SAC 2008. LNCS, vol. 5381, pp. 103–119. Springer, Heidelberg (2009)

5. Aumasson, J.P., Henzen, L., Meier, W., Phan, R.C.W.: SHA-3 proposal BLAKE. Submission to NIST (Round 3) (2010). http://131002.net/blake/

6. Barreto, P., Rijmen, V.: The whirlpool hashing function. Submitted to NESSIE, September 2000. http://www.larc.usp.br/pbarreto/WhirlpoolPage.html

7. Benadjila, R., Billet, O., Gilbert, H., Macario-Rat, G., Peyrin, T., Robshaw, M., Seurin, Y.: SHA-3 proposal: ECHO. Submission to NIST (updated) (2009). http://crypto.rd.francetelecom.com/ECHO/

8. Bertoni, G., Daemen, J., Peeters, M., Van Assche, G.: The keccak reference. Submission to NIST (Round 3) (2011). http://keccak.noekeon.org/Keccak-reference-3.0.pdf

9. Biham, E., Dunkelman, O.: A framework for iterative hash functions - HAIFA. Cryptology ePrint Archive, Report 2007/278 (2007). http://eprint.iacr.org/2007/278

10. Biham, E., Dunkelman, O.: The SHAvite-3 hash function. Submission to NIST (Round 2) (2009), http://www.cs.technion.ac.il/~orrd/SHAvite-3/

11. Chang, D., Nandi, M.: Improved indifferentiability security analysis of chopMD hash function. In: Nyberg, K. (ed.) FSE 2008. LNCS, vol. 5086, pp. 429–443. Springer, Heidelberg (2008)

12. Coron, J.-S., Dodis, Y., Malinaud, C., Puniya, P.: Merkle-Damgård revisited: how to construct a hash function. In: Shoup, V. (ed.) CRYPTO 2005. LNCS, vol. 3621, pp. 430–448. Springer, Heidelberg (2005)

13. Daemen, J., Rijmen, V.: The wide trail design strategy. In: Honary, B. (ed.) Cryptography and Coding 2001. LNCS, vol. 2260, pp. 222–238. Springer, Heidelberg (2001)

14. Damgård, I.B.: A design principle for hash functions. In: Brassard, G. (ed.) CRYPTO 1989. LNCS, vol. 435, pp. 416–427. Springer, Heidelberg (1990)

15. Dolmatov, V., Degtyarev, A.: GOST R 34.11-2012: hash function (2013)

16. Dolmatov, V., Degtyarev, A.: Request for comments 6986: GOST R 34.11-2012: hash function. Internet Engineering Task Force (IETF) (2013). http://www.ietf.org/rfc/rfc6986.txt

17. Information protection and special communications of the federal security service of the Russian federation: GOST R 34.11-94, information technology cryptographic data security hashing function (1994). (In Russian)

18. Information protection and special communications of the federal security service of the Russian federation: GOST R 34.11-2012, information technology cryptographic data security hashing function (2012). http://www.tc26.ru/en/GOSTR3411-2012/GOST_R_34_11-2012_eng.pdf

19. Gauravaram, P., Kelsey, J.: Linear-XOR and additive checksums don't protect Damgård-Merkle hashes from generic attacks. In: Malkin, T. (ed.) CT-RSA 2008. LNCS, vol. 4964, pp. 36–51. Springer, Heidelberg (2008)

20. Gauravaram, P., Knudsen, L.R., Matusiewicz, K., Mendel, F., Rechberger, C., Schläffer, M., Thomsen, S.S.: Grøstl-a SHA-3 candidate. Submission to NIST (Round 3) (2011). http://www.groestl.info/Groestl.pdf
21. Gilbert, H., Peyrin, T.: Super-Sbox cryptanalysis: improved attacks for AES-like permutations. In: Hong, S., Iwata, T. (eds.) FSE 2010. LNCS, vol. 6147, pp. 365–383. Springer, Heidelberg (2010)
22. Guo, J., Jean, J., Leurent, G., Peyrin, T., Wang, L.: The usage of counter revisited: second-preimage attack on new Russian standardized hash function. In: Joux, A., Youssef, A. (eds.) SAC 2014. LNCS, vol. 8781, pp. 195–211. Springer, Heidelberg (2014)
23. International Organization for Standardization: ISO/IEC 10118–3:2004: information technology - security techniques - hash-functions - part 3: dedicated hash-functions (2004)
24. Iwamoto, M., Peyrin, T., Sasaki, Y.: Limited-birthday distinguishers for hash functions. In: Sako, K., Sarkar, P. (eds.) ASIACRYPT 2013, Part II. LNCS, vol. 8270, pp. 504–523. Springer, Heidelberg (2013)
25. Joux, A.: Multicollisions in iterated hash functions. Application to cascaded constructions. In: Franklin, M. (ed.) CRYPTO 2004. LNCS, vol. 3152, pp. 306–316. Springer, Heidelberg (2004)
26. Kazymyrov, O., Kazymyrova, V.: Algebraic aspects of the Russian hash standard GOST R 34.11-2012. Cryptology ePrint Archive, Report 2013/556 (2013). http://eprint.iacr.org/2013/556
27. Kelsey, J., Kohno, T.: Herding hash functions and the nostradamus attack. In: Vaudenay, S. (ed.) EUROCRYPT 2006. LNCS, vol. 4004, pp. 183–200. Springer, Heidelberg (2006)
28. Kelsey, J., Schneier, B.: Second preimages on n-bit hash functions for much less than 2^n work. In: Cramer, R. (ed.) EUROCRYPT 2005. LNCS, vol. 3494, pp. 474–490. Springer, Heidelberg (2005)
29. Khovratovich, D., Nikolić, I., Weinmann, R.-P.: Meet-in-the-middle attacks on SHA-3 candidates. In: Dunkelman, O. (ed.) FSE 2009. LNCS, vol. 5665, pp. 228–245. Springer, Heidelberg (2009)
30. Khovratovich, D., Rechberger, C., Savelieva, A.: Bicliques for preimages: attacks on Skein-512 and the SHA-2 family. In: Canteaut, A. (ed.) FSE 2012. LNCS, vol. 7549, pp. 244–263. Springer, Heidelberg (2012)
31. Knellwolf, S., Khovratovich, D.: New preimage attacks against reduced SHA-1. In: Safavi-Naini, R., Canetti, R. (eds.) CRYPTO 2012. LNCS, vol. 7417, pp. 367–383. Springer, Heidelberg (2012)
32. Lamberger, M., Mendel, F., Rechberger, C., Rijmen, V., Schläffer, M.: Rebound distinguishers: results on the full whirlpool compression function. In: Matsui, M. (ed.) ASIACRYPT 2009. LNCS, vol. 5912, pp. 126–143. Springer, Heidelberg (2009)
33. Ma, B., Li, B., Hao, R., Li, X.: Improved cryptanalysis on reduced-round GOST and whirlpool hash function. In: Boureanu, I., Owesarski, P., Vaudenay, S. (eds.) ACNS 2014. LNCS, vol. 8479, pp. 289–307. Springer, Heidelberg (2014)
34. Ma, B., Li, B., Hao, R., Li, X.: Improved (Pseudo) preimage attacks on reduced-round GOST and Grøstl-256 and studies on several truncation patterns for AES-like compression functions (Full version). Cryptology ePrint Archive (2015)
35. Mendel, F., Pramstaller, N., Rechberger, C.: A (Second) preimage attack on the GOST hash function. In: Nyberg, K. (ed.) FSE 2008. LNCS, vol. 5086, pp. 224–234. Springer, Heidelberg (2008)

36. Mendel, F., Pramstaller, N., Rechberger, C., Kontak, M., Szmidt, J.: Cryptanalysis of the GOST hash function. In: Wagner, D. (ed.) CRYPTO 2008. LNCS, vol. 5157, pp. 162–178. Springer, Heidelberg (2008)
37. Mendel, F., Rechberger, C., Schläffer, M., Thomsen, S.S.: The rebound attack: cryptanalysis of reduced whirlpool and Grøstl. In: Dunkelman, O. (ed.) FSE 2009. LNCS, vol. 5665, pp. 260–276. Springer, Heidelberg (2009)
38. Merkle, R.C.: One way hash functions and DES. In: Brassard, G. (ed.) CRYPTO 1989. LNCS, vol. 435, pp. 428–446. Springer, Heidelberg (1990)
39. National Institute of Standards and Technology (NIST): FIPS PUB 180–3: secure hash standard. Federal Information Processing Standards Publication 180–3, U.S. Department of Commerce, October 2008. http://csrc.nist.gov/publications/fips/fips180-3/fips180-3_final.pdf
40. Sasaki, Y.: Meet-in-the-middle preimage attacks on AES hashing modes and an application to whirlpool. In: Joux, A. (ed.) FSE 2011. LNCS, vol. 6733, pp. 378–396. Springer, Heidelberg (2011)
41. Sasaki, Y., Aoki, K.: Finding preimages in full MD5 faster than exhaustive search. In: Joux, A. (ed.) EUROCRYPT 2009. LNCS, vol. 5479, pp. 134–152. Springer, Heidelberg (2009)
42. Sasaki, Y., Wang, L., Wu, S., Wu, W.: Investigating fundamental security requirements on whirlpool: improved preimage and collision attacks. In: Wang, X., Sako, K. (eds.) ASIACRYPT 2012. LNCS, vol. 7658, pp. 562–579. Springer, Heidelberg (2012)
43. Wang, Z., Yu, H., Wang, X.: Cryptanalysis of GOST R hash function. Cryptology ePrint Archive, Report 2013/584 (2013). http://eprint.iacr.org/2013/584
44. Wu, S., Feng, D., Wu, W., Guo, J., Dong, L., Zou, J.: (Pseudo)Preimage attack on round-reduced Grøstl hash function and others. In: Canteaut, A. (ed.) FSE 2012. LNCS, vol. 7549, pp. 127–145. Springer, Heidelberg (2012)
45. Zou, J., Wu, W., Wu, S.: Cryptanalysis of the round-reduced GOST hash function. In: Lin, D., Xu, S., Yung, M. (eds.) Inscrypt 2013. LNCS, vol. 8567, pp. 309–322. Springer, Heidelberg (2014)

Improvement on the Method for Automatic Differential Analysis and Its Application to Two Lightweight Block Ciphers DESL and LBlock-s

Siwei Sun[1,2], Lei Hu[1,2]([⊠]), Kexin Qiao[1,2], Xiaoshuang Ma[1,2],
Jinyong Shan[1,2], and Ling Song[1,2]

[1] State Key Laboratory of Information Security, Institute of Information
Engineering, Chinese Academy of Sciences, Beijing 100093, China
[2] Data Assurance and Communication Security Research Center,
Chinese Academy of Sciences, Beijing 100093, China
{sunsiwei,hulei,qiaokexin,maxiaoshuang,shanjinyong,songling}@iie.ac.cn

Abstract. With the development of the ubiquitous computing and large-scale information processing systems, the demand for lightweight block ciphers which is suitable for resource constrained computing devices is increasing. Hence, the methodology for design and analysis of block ciphers is becoming more important. In this paper, we use the Mixed-Integer Linear Programming (MILP) based tools for automatic differential cryptanalysis in a clever way to find improved single-key and related-key differential characteristics for DESL (a lightweight variant of the well known Data Encryption Standard), and obtain tighter security bound for LBlock-s (a core component of an lightweight authenticated encryption algorithm submitted to the international competition for authenticated encryption – CAESAR) against related-key differential attack. To be more specific, in searching for improved characteristics, we restrict the differential patterns allowed in the first and last rounds of the characteristics in the feasible region of an MILP problem by imposing different constraints than other rounds, and we partition the differential patterns of the DESL S-box into different sets with 2-bit more information associated with each pattern according to their probabilities. In addition, we show how to use the Gurobi optimizer combined with a known good characteristic to speed up the characteristic searching and bound proving process. Using these techniques, we managed to find the currently known best 9-round related-key differential characteristic for DESL, and the first published nontrivial related-key and single-key differential characteristics covering 10 rounds of DESL. Also, we obtain the currently known tightest security bound for LBlock-s against related-key differential attack. These techniques should be useful in analysis and design of other lightweight block ciphers.

Keywords: Automatic cryptanalysis · Security evaluation · Related-key differential attack · Authenticated encryption · Mixed-Integer Linear Programming

© Springer International Publishing Switzerland 2015
K. Tanaka and Y. Suga (Eds.): IWSEC 2015, LNCS 9241, pp. 97–111, 2015.
DOI: 10.1007/978-3-319-22425-1_7

1   Introduction

Cryptography plays a central role in protecting today's information system, and block cipher is one of the most important types of cryptographic algorithms. Moreover, with the development of the ubiquitous computing and large-scale information processing systems, the demand for lightweight block ciphers which are suitable for resource constrained computing devices such as sensor nodes, and RFID tags is increasing. Therefore, design and analysis of lightweight block ciphers draw much attention from the researchers in applied cryptography.

Differential cryptanalysis [20], introduced by Eli Biham and Adi Shamir in the late 1980s, is one of the most effective attacks on modern block ciphers. Moreover, many cryptanalytic techniques, such as the related-key differential attack [2,6,19], truncated differential attack [24], statistical saturation attack [11,16], impossible differential attack [1,23], (probabilistic) higher order differential attack [24,43], boomerang attack [18], multiple differential attack [8,9,15,17], differential-linear cryptanalysis [42], multiple linear attack [5,14,28–30] and so on so forth, are essentially based on differential attack.

Typically, the first step in differential attack is to find a differential characteristic with high (or the highest possible) probability. Hence, a method used for searching for good (or the best) differential characteristic is of great importance. For a designer, such method can be used to obtain a proven security bound against differential attack, which is a necessary part of the design of a block cipher. In fact, a large part of the design document of a modern block cipher is devoted to the security evaluation of differential attack. For an attacker, such method can be used to find high probability differential characteristics of a cipher which lead to distinguishers or key recovery attacks.

Matusi's branch-and-bound depth-first search algorithm [31] is a classic method for finding the best differential characteristic of a cipher. Several works were devoted to improving the efficiency of Matsui's approach. The concept of *search pattern* was introduced in [34] to reduce the search complexity of Matusi's algorithm by detecting unnecessary search candidates. Further improvements were obtained by Aoki *et al.* [10] and Bao *et al.* [48]. Remarkably, significant efficiency improvement on Matsui's approach was observed for specific ciphers in [48].

Despite its guarantee for finding the best single-key differential characteristic for a cipher (given unlimited computational power), Matsui's algorithm and its variants has some important limitations making it not practically applicable in many situations. Firstly, for most ciphers, the original algorithm of Matsui is not practically applicable in the related-key model. Even though there exits Matsui's variant (see Biryukov *et al.*'s work [3]) for finding related-key differential characteristics, this method is not very useful for ciphers with nonlinear key schedule algorithms, whereas ciphers with nonlinear key schedule algorithms are plentiful. Moreover, with the developement of new techniques for cryptanalysis (*e.g.*, the differential fault attack [13,21,47] and biclique attack [7]), the related-key model is becoming more important and highly relevant to the design and analysis of symmetric-key cryptographic algorithms. Secondly, Matsui's approach is inefficient in finding the best characteristics for many ciphers, and some speeding-up techniques for Matusi's approach were intimately related to the special properties of the specific ciphers under consideration, making it difficult to implement and far from being a generic and convenient tool for cryptanalysis.

For ciphers that cannot be analyzed by Matsui's approach, the cryptanalysts turn to other methods which can be employed to find reasonably good characteristics. Although the characteristics found by these methods are not guaranteed to be the best, they do produce currently the best known results for many ciphers. In [4], Biryukov et al.extend Matsui's algorithm by using the *partial* (rather than the full) difference distribution table (pDDT) to prevent the number of explored candidates from exploding and at the same time keep the total probability of the resulting characteristic high. In [35], truncated differentials with the minimum number of active S-boxes are found by a breadth-first search based on the Dijkstra's algorithm, and then these truncated differentials are instantiated with actual differences.

Another line of research is to model the differential behavior of a cipher as an SAT or Mixed-Integer Linear Programming (MILP) problem which can be solved automatically by SAT or MILP solvers. Compared with other methods, these methods are easier to implement and more flexible. In [32,40,41], SMT/SAT solvers are employed to find differential characteristics of Salsa and other ciphers. Mouha et al. [33], Wu et al. [36], and Sun et al. [37] translated the problem of counting the minimum number of differentially active S-boxes into an MILP problem which can be solved automatically with open source or commercially available optimizers. This method has been applied in evaluating the security against (related-key) differential attacks of many symmetric-key schemes. However, this tool cannot be used to find the actual differential characteristics directly. In Asiacrypt 2014, two systematic methods for generating linear inequalities describing the differential properties of an arbitrary S-box were given in [39]. With these inequalities, the authors of [39] were able to construct an MILP model whose feasible region is a more accurate description of the differential behavior of a given cipher. Based on such MILP models, the authors of [39] proposed a heuristic algorithm for finding actual (related-key) differential characteristics, which is applicable to a wide range of block ciphers. In [38], Sun et al. get rid of the heuristic argument in [39] by constructing MILP models whose feasible regions are exactly the sets of all (related-key) differential characteristics.

These MILP based methods [37,39] mainly focus on finding characteristics with the minimum (or reasonably small) number of active S-boxes. However, it is well possible that a characteristic with more active S-boxes is better than a characteristic with a smaller number of active S-boxes. Even though a method for finding the best characteristic of a cipher by encoding the probability information into the differential patterns is proposed in [38], this method is only applicable to ciphers with 4×4 S-boxes and infeasible when the number of rounds is large. Therefore, by using these methods, we may miss some better characteristics. In this paper, we mainly focus on how to use the MILP based methods in a clever way such that better characteristics can be found.

Our Contribution. Based on Sun *et al.*'s MILP framework for automatic differential analysis presented in [37–39], we propose several techniques which are useful for finding improved characteristics. To be more specific, we restrict the differential patterns allowed in the first and last rounds to be those with relatively high probability in the differential distribution table, which makes sure that the active S-boxes in the first and last rounds assume relatively high probabilities. We also partition the differential patterns into different sets. For each of these sets, we associate 2-bit more information into its differential patterns, and try to find a characteristic maximizing a special objective function rather than maximizing the number of differentially active S-boxes. Moreover, after we find a good characteristic with N_A active S-boxes, we use the tool presented in [38] to enumerate all characteristics with N_A, $N_A + 1$ and $N_A + 2$ active S-boxes, from which we may find better characteristics than the original one. We also present some tricks in using the Gurobi [22] optimizer which may speed up the solving process.

With these techniques, we find a related-key characteristic covering 9 rounds of DESL whose probability is $2^{-41.89}$, while the best previously published 9-round related-key differential characteristic with probability $2^{-44.06}$ is given in [38]. Note that in [3], only the upper bound of the probability of the related-key differential characteristics covering 9 rounds of DESL is given. We also present a 10-round single-key and related-key differential characteristics of DESL with probabilities $2^{-52.25}$ and $2^{-51.85}$ respectively. Moreover, we give so far the tightest security bound of the full LBlock-s with respect to related-key differential attack.

Organization of the Paper. In Sect. 2, we introduce the MILP framework for automatic differential cryptanalysis. In Sect. 3, we present several techniques for finding improved (related-key) differential characteristics. Then we apply these techniques to DESL and LBlock-s in Sect. 4. Section 5 is the conclusion and discussion.

2 MILP Based Framework for Automatic Differential Cryptanalysis

A brief introduction of Sun *et al.*'s method is given below. Sun *et al.*'s method [37–39] is an extension of Mouha *et al.*'s technique [33] based on Mixed-Integer Linear Programming, which can be used to search for (related-key) differential characteristics and obtain security bounds of a cipher with respect to the (related-key) differential attack automatically.

Sun *et al.*'s method is applicable to ciphers involving the following three operations:

- bitwise XOR;
- bitwise permutation L which permutes the bit positions of an n dimensional vector in $\mathbb{F}_2^n$;
- S-box, $\mathcal{S} : \mathbb{F}_2^\omega \to \mathbb{F}_2^\nu$.

Note that a general linear transformation $T : \mathbb{F}_2^n \rightarrow \mathbb{F}_2^m$ can be treated as some XOR summations and bitwise permutations of the input bits. In Sun *et al.*'s methods, a new variable x_i is introduced for every input and output bit-level differences, where $x_i = 1$ means the XOR difference at this position is 1 and $x_i = 0$ if there is no difference.

Also, for every S-box involved in the cipher, introduce a new 0–1 variable A_j such that

$$A_j = \begin{cases} 1, & \text{if the input word of the Sbox is nonzero,} \\ 0, & \text{otherwise.} \end{cases}$$

Now, we are ready to describe Sun *et al.*'s method by clarifying the objective function and constraints in the MILP model. Note that we assume that all variables involved are 0–1 variables.

Objective Function. The objective function is to minimize the sum of all variables A_j indicating the activities of the S-boxes: $\sum_j A_j$.

Constraints. Firstly, for every XOR operation $a \oplus b = c \in \{0,1\}$, include the following constraints

$$\begin{cases} a + b + c \geq 2d_\oplus \\ a + b + c \leq 2 \\ d_\oplus \geq a, \ d_\oplus \geq b, \ d_\oplus \geq c \end{cases} \tag{1}$$

where $d_\oplus$ is a dummy variable.

Assuming $(x_{i_0}, \ldots, x_{i_{\omega-1}})$ and $(y_{i_0}, \ldots, y_{i_{\nu-1}})$ are the input and output differences of an $\omega \times \nu$ S-box marked by A_t, we have

$$\begin{cases} A_t - x_{i_k} \geq 0, \ k \in \{0, \ldots, \omega - 1\} \\ -A_t + \sum\limits_{j=0}^{\omega-1} x_{i_j} \geq 0 \end{cases} \tag{2}$$

and

$$\begin{cases} \sum\limits_{k=0}^{\omega-1} x_{i_k} + \sum\limits_{k=0}^{\nu-1} y_{j_k} \geq \mathcal{B}_S d_S \\ d_S \geq x_{i_k}, \ \ 0 \leq k \leq \omega - 1 \\ d_S \geq y_{j_k}, \ \ 0 \leq k \leq \nu - 1 \end{cases} \tag{3}$$

where d_S is a dummy variable, and the branch number $\mathcal{B}_S$ of an S-box $\mathcal{S}$, is defined as $\min_{a \neq b}\{\text{wt}((a \oplus b)\|(\mathcal{S}(a) \oplus \mathcal{S}(b))) : a, b \in \mathbb{F}_2^\omega\}$. For an bijective S-box we have

$$\begin{cases} \omega \sum\limits_{k=0}^{\nu-1} y_{j_k} - \sum\limits_{k=0}^{\omega-1} x_{i_k} \geq 0 \\ \nu \sum\limits_{k=0}^{\omega-1} x_{i_k} - \sum\limits_{k=0}^{\nu-1} y_{j_k} \geq 0 \end{cases} \tag{4}$$

Then, treat every possible input-output differential pattern $(x_0, \ldots, x_{\omega-1}) \rightarrow (y_0, \ldots, y_{\nu-1})$ of an $\omega \times \nu$ S-box as an $(\omega + \nu)$-dimensional vector $(x_0, \ldots, x_{\omega-1}, y_0, \ldots, y_{\nu-1}) \in \{0,1\}^{\omega+\nu} \subseteq \mathbb{R}^{\omega+\nu}$, and compute the H-representation of the convex hull of all possible input-output differential patterns of the S-box. From the H-representation select a small number of linear inequalities using the greedy algorithm presented in [38] which can be used to exactly describe the differential behavior of the S-box. Finally, relate the input and output variables of the S-box using these inequalities. Now, if we require that all the variables involved are 0–1 variables, then the feasible region of the resulting MILP model is exactly the set of all differential characteristics. We mention here that all the constraints in (3) and (4) can be omitted if we have already use the constraints from the critical set, since these constraints remove all impossible patterns.

3   Techniques for Obtaining Better Characteristics

In [37–39], MILP models are constructed and solved to search for characteristics with a small or the minimal number of active S-boxes. The main reason preventing the solution of such MILP models from leading to better characteristics is that the objective function in the MILP model is to minimize the number of differentially or linearly active S-boxes. Under this setting, an MILP optimizer, say Gurobi, is constantly trying to find a characteristic with a smaller number of active S-boxes. In this process, some characteristics with higher probability but larger numbers of active S-boxes are lost. Therefore, the method presented in [37–39] may fail to find some better characteristics. In this section, we show how to mitigate this situation such that improved characteristics can be obtained automatically.

Technique 1. Finding Characteristics with More Active S-Boxes. For an iterative r-round block cipher E, Sun *et al.*'s methods can be used to find a characteristic with the minimal or a reasonably small number of active S-boxes. Assuming that such an r-round characteristic with N_A active S-boxes has been found, then we add the constraint $N_A \leq \sum_j S_j \leq N_A + m$ to the MILP model, where S_j's are the variables marking the activities of the involved S-boxes. Then we try to enumerate all related-key differential characteristics satisfying all the constraints in the model, where m is a small positive integer typically chosen to be 1 or 2. From these characteristics we may find better ones. This is in fact the same heuristic employed by Biham *et al.* in [12], where they try to find differential characteristics with higher probabilities for the amplified boomerang attack. They try to accomplish this by adding an active S-box in the first round. This might seem a bad thing (as this increase the number of active S-boxes), but they find out that in exchange they get 3 more differentials of the active S-boxes with probability 2^{-2} instead of 2^{-3}.

Technique 2. Imposing Different Constraints for Different Rounds. Another technique for getting better characteristic is to allow only those differential patterns in the first and last rounds of a characteristic to take relatively high probabilities. This is because the input of the first round and the output of the last round is relatively free when compared to other rounds. In fact, for every characteristic we get, we can always manually modify its input and output differences (such that high probability differential patterns are used in the first and last rounds) to get a characteristic which is at least not worse than the original one. This manual process can be done automatically in the MILP framework by, in the first and last rounds, using the constraints generated from the critical set of the convex hull of all differential patterns with probability higher than a threshold value p_T we choose, rather than the convex hull of all possible differential patterns. Note that it is important to do this automatically since the enumeration process may return thousands of characteristics.

Technique 3. Encoding More Information into the Differential Patterns of an S-Box. In Sect. 5 of [38], an MILP based method for constructing an MILP model which can be used to search for the best (related-key) differential characteristic of a block cipher with 4×4 S-box is proposed by encoding the probability information of the differentials of a 4×4 S-box into the differential patterns.

Take the PRESENT S-box S for example. For every possible differential pattern $(x_0, x_1, x_2, x_3) \to (y_0, y_1, y_2, y_3)$, a corresponding *differential pattern with probability information* can be constructed as follows

$$(x_0, x_1, x_2, x_3, y_0, y_1, y_2, y_3; p_0, p_1) \in \{0, 1\}^{8+2}$$

where the two extra bits (p_0, p_1) are used to encode the differential probability $\Pr_S[(x_0, \ldots, x_{\omega-1}) \to (y_0, \ldots, y_{\nu-1})]$ as follows

$$\begin{cases} (p_0, p_1) = (0, 0), & \text{if } \Pr_S[(x_0, \ldots, x_{\omega-1}) \to (y_0, \ldots, y_{\nu-1})] = 1; \\ (p_0, p_1) = (0, 1), & \text{if } \Pr_S[(x_0, \ldots, x_{\omega-1}) \to (y_0, \ldots, y_{\nu-1})] = 2^{-2}; \quad (5) \\ (p_0, p_1) = (1, 1), & \text{if } \Pr_S[(x_0, \ldots, x_{\omega-1}) \to (y_0, \ldots, y_{\nu-1})] = 2^{-3}. \end{cases}$$

Hence, the probability of the differential pattern $(x_0, x_1, x_2, x_3) \to (y_0, y_1, y_2, y_3)$ is $2^{-(p_0+2p_1)}$. We refer the reader to [38] for more information of the technique.

This technique is only feasible for ciphers for 4×4 S-boxes because there are only 3 different probabilities for all differential patterns of a typical 4×4 S-boxes and hence we need only $\lceil \log_2 3 \rceil = 2$ extra bits to encode the probability information for each differential pattern. For an $\omega \times \mu$ S-box, if d extra bits are needed to encode the differential probability information, then we need to compute the H-representation of the convex hull of a subset in $\mathbb{R}^{\omega+\mu+d}$. For the PRESENT S-box, we need to compute the H-representation of a convex hull of a subset in $\mathbb{R}^{4+4+2} = \mathbb{R}^{10}$. For the S-box of DESL, this technique is infeasible since there are 9 different probabilities for the differentials of the DESL S-box and we need at least $\lceil \log_2 9 \rceil = 4$ extra bits to encode the probability information. This will force us to compute the H-representation of a convex hull of a set in $\mathbb{R}^{6+4+4} = \mathbb{R}^{14}$ which leads to MILP models with too many constraints to be solved in practical time. In the following, we propose a technique which partitions the differential patterns into several sets according to their probabilities and encodes their probability information with less extra bits.

Definition 1. *Define $\mathcal{D}_S^{[p]}$ to be the set of all differential patterns of an $\omega \times \mu$ S-box with probability p, that is*

$$\mathcal{D}_S^{[p]} = \{(x_0, \cdots, x_{\omega-1}, y_0, \cdots, y_{\mu-1}) : \Pr_S[(x_0, \ldots, x_{\omega-1}) \rightarrow (y_0, \ldots, y_{\mu-1})] = p\},$$

and we use $\mathcal{D}_S^{[p_1, \cdots, p_t]}$ to denote the set $\mathcal{D}_S^{[p_1]} \cup \cdots \cup \mathcal{D}_S^{[p_t]}$.

Take the DESL S-box for example. For every possible differential pattern $(x_0, \cdots, x_5) \rightarrow (y_0, \cdots, y_3)$, we can construct a corresponding pattern

$$(x_0, \cdots, x_5, y_0, \cdots, y_3; \theta_0, \theta_1) \in \mathbb{R}^{6+4+2}$$

such that

$$\begin{cases} (\theta_0, \theta_1) = (0,0), & \text{if } (x_0, \cdots, x_5, y_0, \cdots, y_3) \in \mathcal{D}_S^{[\frac{64}{64}]}; \\ (\theta_0, \theta_1) = (1,0), & \text{if } (x_0, \cdots, x_5, y_0, \cdots, y_3) \in \mathcal{D}_S^{[\frac{16}{64}, \frac{14}{64}, \frac{12}{64}]}; \\ (\theta_0, \theta_1) = (0,1), & \text{if } (x_0, \cdots, x_5, y_0, \cdots, y_3) \in \mathcal{D}_S^{[\frac{10}{64}, \frac{8}{64}, \frac{6}{64}]}; \\ (\theta_0, \theta_1) = (1,1), & \text{if } (x_0, \cdots, x_5, y_0, \cdots, y_3) \in \mathcal{D}_S^{[\frac{4}{64}, \frac{2}{64}]}. \end{cases} \qquad (6)$$

In this technique, the constraints for S-boxes are the critical sets of all patterns with the above encoding scheme, and the objective function is chosen to be minimizing $\sum(\theta_0 + \lambda\theta_1)$, where λ is a positive constant. Note that the differential patterns in $D^{[p_1, \cdots, p_t]}$ with larger p_i will lead to a smaller $\theta_0 + \lambda\theta_1$, and therefore tend to make the objective function $\sum(\theta_0 + \lambda\theta_1)$ smaller.

Note that this method is heuristic in nature. Firstly, unlike the case of PRESENT S-box, the encoding scheme does not represent the exact probability of a differential. Secondly, the solution which minimizes the objective function is not necessarily corresponding to the best characteristic. Finally, the partition of the differential patterns and the selection of λ are rather *ad-hoc* (we choose $\lambda = 3$ when applied this technique to DESL). All these problems deserve further investigation. In the next section, we will show that although this technique is heuristic and rather *ad-hoc*, it does produce the currently known best results for DESL.

Finally, we would also like to point out a feature provided by the MILP optimizer Gurobi [22] which may be useful in speeding up the searching process of better characteristics. In Gurobi, an MILP start (MST) file is used to specify an initial solution for a mixed integer programming model. The file lists values to assign to the variables in the model. If an MILP start file has been imported into an MILP model before optimization begins, the Gurobi optimizer will attempt to build a feasible solution from the specified start values. A good initial solution often speeds up the solution of the MILP model, since it provides an early bound on the optimal value, and also since the specified solution can be used to seed the local search heuristics employed by the MILP solver. An MILP start file consists of variable-value pairs, each on its own line. Any line that begins with the hash sign (#) is a comment line and is ignored. The following is a simple example:

```
# MIP start
x1   1
x2   0
x3   1
```

Therefore, by converting known good characteristics into an MST file, and importing it into our MILP model before optimization begins, we may speed up the searching process.

4   Application to DESL and LBlock-s

The techniques presented in Sect. 3 are implemented in a Python [44] framework, and we show its applications in the following.

4.1   Improved Single-Key and Related-Key Differential Characteristics for DESL

DESL [25] is a lightweight variant of the well known block cipher DES (the Data Encryption Standard), which is almost the same as DES except that it uses a single S-box instead of eight different S-boxes as in DES. This S-box has a special design criteria to discard high probability (single-key) differential characteristics. This simple modification makes DESL much stronger than DES with respect to differential attack. In [3], Alex Biryukov *et al.* observed that Matsui's tool is infeasible for finding the best differential characteristics for DESL. However, Matsui's tool can find the best characteristic of the full DES in no more than several hours on a PC. To the best of our knowledge, there is no published single-key or related-key differential characteristics covering 10 rounds of DESL, and in fact, even in the design document of DESL, there is no concrete security bounds provided for DESL.

We first generate an MILP model for 9-round DESL in the related-key model according to the technique 2 and technique 3 presented in Sect. 3. By solving this model using the Gurobi optimizer [22], we find a 9-round related-key differential characteristic for DESL with probability $2^{-41.89}$, which is the best published 9-round related-key differential characteristic so far. The concrete results are given in Tables 1 and 2.

Subsequently, we construct an MILP model for 10-round DESL in the related-key model. Before we start to solve this model, we import the 9-round related-key differential characteristic found previously as an MILP start file. Finally, we find a 10-round related-key differential characteristic of DESL with probability $2^{-51.85}$ and 14 active S-boxes (see Tables 3 and 4). Then by employing the technique 1 of Sect. 3, we search for characteristics with 14, 15 or 16 active S-boxes. Finally, we find a single-key (a special case of the related-key model where there is no key difference) differential characteristic with probability $2^{-52.25}$ and 15 active S-boxes, which is given in Table 5. Note that this is the first published nontrivial single-key differential characteristic covering 10 rounds of DESL.

Table 1. A 9-round related-key differential characteristic for DESL with probability $2^{-41.89}$ (characteristic in the encryption process)

Rounds	Left	Right
0	00000000000000010000000000000000	00000000010000000000000000000000
1	00000000010000000000000000000000	00000000000000000000000000000000
2	00000000000000000000000000000000	00000000010000000000000000000000
3	00000000010000000000000000000000	00000100000000000000000000000000
4	00000100000000000000000000000000	00000000010000000100000000000000
5	00000000010000000100000000000000	00100000000000000000000110000000
6	00100000000000000000000110000000	00000010110000000100000000000000
7	00000010110000000100000000000000	00000000000001000000000000000000
8	00000000000001000000000000000000	00000010100000000100000000000000
9	00000010100000000100000000000000	01100000000000000100000110010000

Table 2. A 9-round related-key differential characteristic for DESL with probability $2^{-41.89}$ (characteristic in the key schedule algorithm)

Rounds	The differences in the key register
1	000000000000000010000000000000000000000000000000
2	000000000000000000000000000000000000000000000000
3	000000000000100000000000000000000000000000000000
4	000000000010000000000000000000000000000000000000
5	000000000000010000000000000000000000000000000000
6	010000000000000000000000000000000000000000000000
7	000000000100000000000000000000000000000000000000
8	000000000000000000000010000000000000000000000000
9	000000000000000100000000000000000000000000000000

Table 3. A 10-round related-key differential characteristic for DESL with probability $2^{-51.85}$ (characteristic in the key schedule algorithm)

Rounds	The differences in the key register
1	111111111111111111111111111111111111111111111101
2	111111111111111111111111111111111111111011111111
3	111111111111111111111111111111111111111111111111
4	111111111111111111111111101111111111111111111111
5	111111111111111111111111111111111111111111110111
6	111111111111111111111111111111111110111111111111
7	111111111111111111111111111111111111111111011111
8	111111111111111111111111111111111111111011111111
9	111111111111111111111111111111111111111111111111
10	111111111111111111111101111111111111111111111111

Table 4. A 10-round related-key differential characteristic for DESL with probability $2^{-51.85}$ (characteristic in the encryption process)

Rounds	Left	Right
0	11111111111111111111111111011101	11011111111111111111111111111110
1	11011111111111111111111111111110	11111111111111111111111111011111
2	11111111111111111111111111011111	11011111111111111111111111111110
3	11011111111111111111111111111110	11111111111110101111111111111111
4	11111111111110101111111111111111	01011111111111111111111111111110
5	01011111111111111111111111111110	11111111111110101110101111111111
6	11111111111110101110101111111111	11111111111111111111111111111110
7	11111111111111111111111111111110	11111111111111111111111111011111
8	11111111111111111111111111011111	11111111111111111111111111111111
9	11111111111111111111111111111111	11111111111111111111111111011111
10	11111111111111111111111111011111	11011111111011111111101101111111

Table 5. A 10-round single-key differential characteristic for DESL with probability $2^{-52.25}$

Rounds	Left	Right
0	00000000000000010000000000000000	00000000010000000000000000000000
1	00000000010000000000000000000000	00000000000000000000000000000000
2	00000000000000000000000000000000	00000000010000000000000000000000
3	00000000010000000000000000000000	00000100000000000000000000000000
4	00000100000000000000000000000000	00000000010000000100000000000000
5	00000000010000000100000000000000	00100000000000000000000110000000
6	00100000000000000000000110000000	00000010110000000100000000000000
7	00000010110000000100000000000000	00000000000001000000000000000000
8	00000000000001000000000000000000	00000010100000000100000000000000
9	00000010100000000100000000000000	01100000000000000100000110010000

4.2   Tighter Security Bound for LBlock-s

LBlock is a lightweight block cipher proposed by Wu *et al.* in ACNS 2011 [46]. It is a Feistel Network with a 64-bit block size and a 80-bit key size. Since its publication, LBlock received extensive cryptanalysis, such as [27] and [45]. According to [45], the security of LBlock against biclique attack is not strong enough due to its relatively weak diffusion of the key schedule algorithm. So a new key schedule algorithm is proposed in [45]. The LBlock with this improved key schedule is called LBlock-s, which is a core component of the authenticated encryption LAC [26] submitted to the CAESAR competition (Competition for Authenticated Encryption: Security, Applicability, and Robustness). Also, instead of using 10 different S-boxes in LBlock, LBlock-s uses only one S-box to reduce the cost of hardware implementation. For a detailed description of the cipher LBlock-s we refer the reader to [45] and [26] for more information. In this section, we apply the technique presented in this paper and the method presented in [38,39] to LBlock-s, and we obtain so far the tightest security bound for the full LBlock-s.

To obtain the security bound for LBlock-s against related-key differential attack, we generate two MILP instances for 10-round and 11-round LBlock-s using the method presented in [38,39]. Then we use the Gurobi optimizer to solve these models. The results indicate that there are at least 10 active S-boxes for 10-round LBlock-s, and 11 active S-boxes for 11-round LBlock-s. However, when we solve the 11-round model, we use the 10-round related-key differential characteristic found by the Gurobi model by solving the 10-round model as an MILP start file (see Sect. 3), and import it into the 11-round model. Finally, we observe a roughly 7 % speed up compared with the case without using the MILP start file. Then, we employ the **technique 2** presented in Sect. 4.1 of [39] to generate an MILP model for 11-round LBlock-s such that only the differential patterns of the S-box with probability greater than or equal to 2^{-2} are allowed. By solving this model, we prove that there are at least 12 active S-boxes for 11-round LBlock-s in the related-key model if only those S-box differential patterns with probability greater than or equal to 2^{-2} are allowed. These results indicate that the probability of any related-key differential characteristic for the 11-round LBlock-s is at most $(2^{-2})^{10} \times 2^{-3} = 2^{-23}$. Consequently, the probability of the full round LBlock-s (32 rounds in total) is upper bounded by $2^{-23} \times 2^{-23} \times (2^{-2})^{10} = 2^{-66}$. Note that this is so far the tightest security bound for full LBlock-s against related-key differential attack.

5   Conclusion and Discussion

In this paper, we use the MILP based methods in a clever way to find better (related-key) differential characteristics of DESL and obtain tighter security bound for LBlock-s. The key idea is to force the active S-boxes in the first and last rounds of a characteristic to take the differentials with relatively high probabilities, encode more information into the differential patterns, and to find better characteristics by enumerating all characteristics with their objective value (number of active S-boxes) close to the minimum number of active S-boxes. Moreover, we show how to use Gurobi and a known good characteristic to speed up the searching process. Finally, we would like to propose some problems deserving further investigation. Firstly, how to find the related-key differential characteristic of DESL with the maximal probability automatically by using MILP technique? Secondly, how to exploit the special features and tune the available parameters of the Gurobi optimizer to speed up the solving process further?

Acknowledgements. The authors would like to thank the anonymous reviewers for their helpful comments and suggestions. The work of this paper was supported by the National Key Basic Research Program of China (2013CB834203), the National Natural Science Foundation of China (Grants 61472417, 61402469 and 61472415), the Strategic Priority Research Program of Chinese Academy of Sciences under Grant XDA06010702, and the State Key Laboratory of Information Security, Chinese Academy of Sciences.

References

1. Biryukov, A.: Impossible differential attack. In: van Tilborg, H.C.A., Jajodia, S. (eds.) Encyclopedia of Cryptography and Security, p. 597. Springer (2011)
2. Biryukov, A., Khovratovich, D.: Related-key cryptanalysis of the full AES-192 and AES-256. In: Matsui, M. (ed.) ASIACRYPT 2009. LNCS, vol. 5912, pp. 1–18. Springer, Heidelberg (2009)
3. Biryukov, A., Nikolić, I.: Search for Related-Key Differential Characteristics in DES-Like Ciphers. In: Joux, A. (ed.) FSE 2011. LNCS, vol. 6733, pp. 18–34. Springer, Heidelberg (2011)
4. Biryukov, A., Velichkov, V.: Automatic Search for Differential Trails in ARX Ciphers. In: Benaloh, J. (ed.) CT-RSA 2014. LNCS, vol. 8366, pp. 227–250. Springer, Heidelberg (2014)
5. Biryukov, A., De Cannière, C., Quisquater, M.: On multiple linear approximations. In: Franklin, M. (ed.) CRYPTO 2004. LNCS, vol. 3152, pp. 1–22. Springer, Heidelberg (2004)
6. Biryukov, A., Khovratovich, D., Nikolić, I.: Distinguisher and related-key attack on the full AES-256. In: Halevi, S. (ed.) CRYPTO 2009. LNCS, vol. 5677, pp. 231–249. Springer, Heidelberg (2009)
7. Bogdanov, A., Khovratovich, D., Rechberger, C.: Biclique Cryptanalysis of the Full AES. In: Lee, D.H., Wang, X. (eds.) ASIACRYPT 2011. LNCS, vol. 7073, pp. 344–371. Springer, Heidelberg (2011)
8. Canteaut, A., Fuhr, T., Gilbert, H., Naya-Plasencia, M., Reinhard, J.-R.: Multiple differential cryptanalysis of round-reduced PRINCE. In: Cid, C., Rechberger, C. (eds.) FSE 2014. LNCS, vol. 8540, pp. 591–610. Springer, Heidelberg (2015)
9. Canteaut, A., Fuhr, T., Gilbert, H., Naya-Plasencia, M., Reinhard, J.-R.: Multiple differential cryptanalysis of round-reduced PRINCE (full version). IACR Cryptology ePrint Archive, Report 2014/089 (2014). http://eprint.iacr.org/2014/089
10. Aoki, K., Kobayashi, K., Moriai, S.: Best differential characteristic search of FEAL. In: Biham, E. (ed.) FSE 1997. LNCS, vol. 1267, pp. 41–53. Springer, Heidelberg (1997)
11. Collard, B., Standaert, F.-X.: A statistical saturation attack against the block cipher PRESENT. In: Fischlin, M. (ed.) CT-RSA 2009. LNCS, vol. 5473, pp. 195–210. Springer, Heidelberg (2009)
12. Biham, E., Dunkelman, O., Keller, N.: The rectangle attack - rectangling the serpent. In: Pfitzmann, B. (ed.) EUROCRYPT 2001. LNCS, vol. 2045, pp. 340–357. Springer, Heidelberg (2001)
13. Boneh, D., DeMillo, R.A., Lipton, R.J.: On the importance of checking cryptographic protocols for faults. In: Fumy, W. (ed.) EUROCRYPT 1997. LNCS, vol. 1233, pp. 37–51. Springer, Heidelberg (1997)
14. Kaliski Jr, B.S., Robshaw, M.: Linear Cryptanalysis Using Multiple Approximations. In: Desmedt, Y.G. (ed.) CRYPTO 1994. LNCS, vol. 839, pp. 26–39. Springer, Heidelberg (1994)
15. Blondeau, C., Gérard, B.: Multiple differential cryptanalysis: theory and practice. In: Joux, A. (ed.) FSE 2011. LNCS, vol. 6733, pp. 35–54. Springer, Heidelberg (2011)
16. Blondeau, C., Nyberg, K.: Links between truncated differential and multidimensional linear properties of block ciphers and underlying attack complexities. In: Nguyen, P.Q., Oswald, E. (eds.) EUROCRYPT 2014. LNCS, vol. 8441, pp. 165–182. Springer, Heidelberg (2014)

17. Blondeau, C., Gérard, B., Nyberg, K.: Multiple differential cryptanalysis using LLR and χ^2 statistics. In: Visconti, I., De Prisco, R. (eds.) SCN 2012. LNCS, vol. 7485, pp. 343–360. Springer, Heidelberg (2012)

18. Wagner, D.: The boomerang attack. In: Knudsen, L.R. (ed.) FSE 1999. LNCS, vol. 1636, pp. 156–170. Springer, Heidelberg (1999)

19. Biham, E.: New types of cryptanalytic attacks using related keys. J. Cryptol. 7(4), 229–246 (1994)

20. Biham, E., Shamir, A.: Differential cryptanalysis of DES-like cryptosystems. J. Cryptol. 4(1), 3–72 (1991)

21. Biham, E., Shamir, A.: Differential fault analysis of secret key cryptosystems. In: Kaliski Jr, B.S. (ed.) CRYPTO 1997. LNCS, vol. 1294, pp. 513–525. Springer, Heidelberg (1997)

22. Optimization, G.: Gurobi optimizer reference manual (2013). http://www.gurobi.com

23. Knudsen, L.: DEAL-a 128-bit block cipher. Complexity 258(2), 216 (1998)

24. Knudsen, L.R.: Truncated and higher order differentials. In: Preneel, B. (ed.) FSE 1994. LNCS, vol. 1008, pp. 196–211. Springer, Heidelberg (1995)

25. Leander, G., Paar, C., Poschmann, A., Schramm, K.: New lightweight DES variants. In: Biryukov, A. (ed.) FSE 2007. LNCS, vol. 4593, pp. 196–210. Springer, Heidelberg (2007)

26. Zhang, L., Wu, W., Wang, Y., Wu, S., Zhang, J.: LAC: a lightweight authenticated encryption cipher. CAESAR submission (2014). http://competitions.cr.yp.to/round1/lacv1.pdf

27. Liu, Y., Gu, D., Liu, Z., Li, W.: Impossible differential attacks on reduced-round LBlock. In: Ryan, M.D., Smyth, B., Wang, G. (eds.) ISPEC 2012. LNCS, vol. 7232, pp. 97–108. Springer, Heidelberg (2012)

28. Hermelin, M., Nyberg, K.: Linear cryptanalysis using multiple linear approximations. IACR Cryptology ePrint Archive, Report 2011/93 (2011). https://eprint.iacr.org/2011/093

29. Hermelin, M., Cho, J.Y., Nyberg, K.: Multidimensional linear cryptanalysis of reduced round serpent. In: Mu, Y., Susilo, W., Seberry, J. (eds.) ACISP 2008. LNCS, vol. 5107, pp. 203–215. Springer, Heidelberg (2008)

30. Hermelin, M., Cho, J.Y., Nyberg, K.: Multidimensional extension of Matsui's algorithm 2. In: Dunkelman, O. (ed.) FSE 2009. LNCS, vol. 5665, pp. 209–227. Springer, Heidelberg (2009)

31. Matsui, M.: On correlation between the order of S-Boxes and the strength of DES. In: De Santis, A. (ed.) EUROCRYPT 1994. LNCS, vol. 950, pp. 366–375. Springer, Heidelberg (1995)

32. Mouha, N., Preneel, B.: Towards finding optimal differential characteristics for ARX: application to Salsa20. IACR Cryptology ePrint Archive, Report 2013/328 (2013). http://eprint.iacr.org/2013/328

33. Mouha, N., Wang, Q., Gu, D., Preneel, B.: Differential and linear cryptanalysis using mixed-integer linear programming. In: Wu, C.-K., Yung, M., Lin, D. (eds.) Inscrypt 2011. LNCS, vol. 7537, pp. 57–76. Springer, Heidelberg (2012)

34. Ohta, K., Moriai, S., Aoki, K.: Improving the search algorithm for the best linear expression. In: Coppersmith, D. (ed.) CRYPTO 1995. LNCS, vol. 963, pp. 157–170. Springer, Heidelberg (1995)

35. Fouque, P.-A., Jean, J., Peyrin, T.: Structural evaluation of AES and chosen-key distinguisher of 9-round AES-128. In: Canetti, R., Garay, J.A. (eds.) CRYPTO 2013, Part I. LNCS, vol. 8042, pp. 183–203. Springer, Heidelberg (2013)

36. Wu, S., Wang, M.: Security evaluation against differential cryptanalysis for block cipher structures. IACR Cryptology ePrint Archive, Report 2011/551 (2011). https://eprint.iacr.org/2011/551

37. Sun, S., Hu, L., Song, L., Xie, Y., Wang, P.: Automatic security evaluation of block ciphers with S-bP structures against related-key differential attacks. In: Lin, D., Xu, S., Yung, M. (eds.) Inscrypt 2013. LNCS, vol. 8567, pp. 39–51. Springer, Heidelberg (2014)

38. Sun, S., Hu, L., Wang, M., Wang, P., Qiao, K., Ma, X., Shi, ., Song, L., Fu, K.: Towards finding the best characteristics of some bit-oriented block ciphers and automatic enumeration of (related-key) differential and linear characteristics with predefined properties. Cryptology ePrint Archive, Report 2014/747 (2014). http://eprint.iacr.org/2014/747

39. Sun, S., Hu, L., Wang, P., Qiao, K., Ma, X., Song, L.: Automatic security evaluation and (related-key) differential characteristic search: application to SIMON, PRESENT, LBlock, DES(L) and other bit-oriented block ciphers. In: Sarkar, P., Iwata, T. (eds.) ASIACRYPT 2014. LNCS, vol. 8873, pp. 158–178. Springer, Heidelberg (2014)

40. Kölbl, S.: CryptoSMT - an easy to use tool for cryptanalysis of symmetric primitives likes block ciphers or hash functions. https://github.com/kste/cryptosmt

41. Kölbl, S., Leander, G., Tiessen, T.: Observations on the SIMON block cipher family. Cryptology ePrint Archive, Report 2015/145 (2015). http://eprint.iacr.org/2015/145

42. Langford, S.K., Hellman, M.E.: Differential-linear cryptanalysis. In: Desmedt, Y.G. (ed.) CRYPTO 1994. LNCS, vol. 839, pp. 17–25. Springer, Heidelberg (1994)

43. Iwata, T., Kurosawa, K.: Probabilistic higher order differential attack and higher order bent functions. In: Lam, K.-Y., Okamoto, E., Xing, C. (eds.) ASIACRYPT 1999. LNCS, vol. 1716, pp. 62–74. Springer, Heidelberg (1999)

44. Van Rossum, G., et al.: Python programming language. In: USENIX Annual Technical Conference (2007)

45. Wang, Y., Wu, W., Yu, X., Zhang, L.: Security on LBlock against biclique cryptanalysis. In: Lee, D.H., Yung, M. (eds.) WISA 2012. LNCS, vol. 7690, pp. 1–14. Springer, Heidelberg (2012)

46. Wu, W., Zhang, L.: LBlock: a lightweight block cipher. In: Lopez, J., Tsudik, G. (eds.) ACNS 2011. LNCS, vol. 6715, pp. 327–344. Springer, Heidelberg (2011)

47. Zhao, X.J., Wang, T., Guo, S.Z.: Fault-propagation pattern based dfa on spn structure block ciphers using bitwise permutation, with application to PRESENT and printcipher. Technical report, IACR Cryptology ePrint Archive, Report 2011/086 (2011)

48. Bao, Z., Zhang, W., Lin, D.: Speeding Up the search algorithm for the best differential and best linear trails. In: Lin, D., Yung, M., Zhou, J. (eds.) Inscrypt 2014. LNCS, vol. 8957, pp. 259–285. Springer, Heidelberg (2015)

Provable Security

NM-CPA Secure Encryption with Proofs of Plaintext Knowledge

Ben Smyth[1]([⊠]), Yoshikazu Hanatani[2], and Hirofumi Muratani[2]

[1] Mathematical and Algorithmic Sciences Lab, France Research Center,
Huawei Technologies Co. Ltd., Issy-les-Moulineaux, France
huawei@bensmyth.com
[2] Toshiba Corporation, Kawasaki, Japan

Abstract. NM-CPA secure asymmetric encryption schemes which prove plaintext knowledge are sufficient for secrecy and verifiability in some domains, for example, ballot secrecy and end-to-end verifiability in electronic voting. In these domains, some applications derive encryption schemes by coupling malleable IND-CPA secure ciphertexts with proofs of plaintext knowledge, without evidence that the sufficient condition is satisfied nor an independent security proof. Consequently, it is unknown whether these applications satisfy the desired secrecy and verifiability properties. In this paper, we propose a generic construction for such a coupling and prove that our construction produces NM-CPA secure encryption schemes which prove plaintext knowledge. Accordingly, we facilitate the development of applications satisfying their secrecy and verifiability objectives and, moreover, we make progress towards security proofs for existing applications.

1 Introduction

Asymmetric encryption schemes satisfying NM-CPA are known to be sufficient to achieve secrecy properties in some domains, for example, an NM-CPA secure encryption scheme is sufficient for ballot secrecy in electronic voting systems [8,9]. In these domains, some applications – including the electronic voting systems by Hirt [26], Damgård, Jurik & Nielsen [21] and Adida *et al.* [1] – derive encryption schemes on blocks of messages by coupling malleable IND-CPA secure ciphertexts with proofs of knowledge, without evidence that the sufficient condition is satisfied (i.e., without evidence that the derived encryption scheme satisfies NM-CPA) nor an independent security proof, hence, it is unknown whether these applications are secure. We stress that malleable encryption schemes and proofs of knowledge are used, rather than more efficient cryptographic primitives, to achieve properties other than secrecy, for example, electronic voting systems use homomorphic properties of encryption schemes for tallying [14,27,29] and proofs of knowledge to achieve end-to-end verifiability [28,35]. Moreover, the requirement for an encryption scheme on blocks of messages reduces complexity and ensures efficiency, for instance, compare the multi-candidate election scheme proposed by Cramer *et al.* [19] which encrypts a single message, with the schemes

© Springer International Publishing Switzerland 2015
K. Tanaka and Y. Suga (Eds.): IWSEC 2015, LNCS 9241, pp. 115–134, 2015.
DOI: 10.1007/978-3-319-22425-1_8

by Benaloh & Yung [5], Hirt [26] and Damgård, Jurik & Nielsen [20] which
encrypt blocks of messages.

Contribution. First, we propose a generic construction for NM-CPA secure
encryption schemes on blocks of messages from homomorphic IND-CPA schemes
using proofs of knowledge (as hinted, proofs of knowledge are used – at the cost
of efficiency – to enhance functionality, in particular, we can achieve additional
security properties). Our construction is inspired by Adida *et al.* [1]; it differs by
including ciphertexts and block numbers in challenges, to prevent attacks discovered by Cortier & Smyth [15–17]. Secondly, we prove the security of our result.
Finally, we discuss assumptions that are sufficient to achieve ballot secrecy in
electronic voting systems and make progress towards a computational security
proof of ballot secrecy for a variant of the Helios voting scheme by Adida *et al.* [1]
(the original Helios scheme is insecure [8,17]), in particular, we demonstrate that
it is sufficient to prove an argument by Smyth & Bernhard [33, Sect. 6] to show
that a variant of Helios using our construction satisfies ballot secrecy.

2   Preliminaries

2.1   Public Key Encryption

We begin by recalling the syntax for public key encryption schemes.

Definition 1 (Asymmetric encryption scheme). *An* asymmetric encryption scheme *is a tuple of algorithms* (Gen, Enc, Dec) *such that:*

- *The* key generation algorithm Gen *takes the security parameter κ as input and
  outputs a public key pk, a private key sk, and a message space* $\mathfrak{m}$.
- *The* encryption algorithm Enc *takes as input a public key pk, message space*
  $\mathfrak{m}$, *and a message $m \in \mathfrak{m}$. It outputs a ciphertext c.*
- *The* decryption algorithm Dec *takes as input a public key pk, a private key
  sk and a ciphertext c. It outputs a message m or a special symbol $\perp$ denoting
  failure.*

*Moreover, the scheme must be correct: for all $(pk, sk, \mathfrak{m}) \leftarrow$ Gen(κ), we have for
all messages $m \in \mathfrak{m}$ and ciphertexts $c \leftarrow$ Enc$_{pk}(m)$, that Dec$_{sk}(c) = m$ with
overwhelming probability.*[1]

An encryption scheme is homomorphic *if there exists binary operators $\otimes$, $\oplus$
and $\odot$ such that for all $(pk, sk, \mathfrak{m}) \leftarrow$ Gen(κ), messages $m_1, m_2 \in \mathfrak{m}$ and coins
r_1 and r_2, we have* Enc$_{pk}(m_1; r_1) \otimes$ Enc$_{pk}(m_2; r_2) =$ Enc$_{pk}(m_1 \oplus m_2; r_1 \odot r_2)$.

IND-CPA [2] is a standard computational security model used to evaluate
the suitability of encryption schemes. Intuitively, if an encryption scheme satisfies IND-CPA, then an arbitrary adversary is unable to distinguish between

---

[1] Let A$(x_1, \ldots, x_n; r)$ denote the output of probabilistic algorithm A on input
$x_1, \ldots, x_n$ and random coins r. We write A$(x_1, \ldots, x_n)$ for A$(x_1, \ldots, x_n; r)$, where r
is chosen uniformly at random. Assignment of α to x is written $x \leftarrow \alpha$.

ciphertexts. This notion can be captured by a cryptographic game between an adversary and a challenger. First, the challenger constructs a key pair (pk, sk) and message space $\mathfrak{m}$; the challenger publishes the public key pk and message space $\mathfrak{m}$. Secondly, the adversary $\mathcal{A}$ executes the algorithm A on the public key pk and outputs the triple (m_0, m_1, s), where m_0 and m_1 are valid plaintexts (that is, $m_0, m_1 \in \mathfrak{m}$) and s is some state information (possibly including pk and $\mathfrak{m}$). Thirdly, the challenger randomly selects a plaintext m_b and derives a ciphertext c using the public key pk. Finally, the adversary executes algorithm D which attempts to derive b given the ciphertext c, plaintexts m_0 and m_1, and the state information s. An encryption scheme satisfies IND-CPA if no adversary has non-negligible advantage over guessing the plaintext encrypted by the challenger. Formally, Definition 2 recalls the cryptographic game IND-CPA by Bellare *et al.* [2] and a variant IND-k-CPA by Bellare & Sahai [4] which allows the adversary k parallel decryption queries, we remark that IND-0-CPA = IND-CPA.

Definition 2 (IND-CPA and IND-k-CPA). *Let $\Pi = (\mathsf{Gen}, \mathsf{Enc}, \mathsf{Dec})$ be an asymmetric encryption scheme and let $\mathcal{A} = (A, D)$ be an adversary. For* atk $\in$ *{ CPA, k-CPA $\mid k \in \mathbb{N}_0$} let* IND-atk$_{\mathcal{A}, \Pi}(\kappa)$ *be the probability defined below, where κ is the security parameter.*[2]

$$2 \cdot Pr[(pk, sk, \mathfrak{m}) \leftarrow \mathsf{Gen}(\kappa);\ (m_0, m_1, s) \leftarrow A(pk, \mathfrak{m});$$
$$b \leftarrow_R \{0, 1\};\ c \leftarrow \mathsf{Enc}_{pk}(m_b) : D^{\mathcal{O}}(m_0, m_1, s, c) = b] - 1$$

where

$$\mathcal{O}(\cdot) = \begin{cases} \mathsf{Dec}_{sk}(\cdot) & \textit{if } \textsf{atk} = k\textit{-CPA} \\ \epsilon & \textit{otherwise} \end{cases}$$

In the above game we insist $m_0, m_1 \in \mathfrak{m}$, $|m_0| = |m_1|$ and if atk $=k$-CPA, *then the oracle will respond to at most k parallel queries and will return $\perp$ if the input vector contains c. We say Π satisfies* IND-atk *if for all probabilistic polynomial-time adversaries $\mathcal{A}$, there exists a negligible function* negl, *such that for all security parameters κ, we have* IND-atk$_{\mathcal{A}, \Pi}(\kappa) \leq$ negl(κ).

NM-CPA [2,4] is an alternative computational security model used to evaluate the suitability of encryption schemes. Intuitively, if an encryption scheme satisfies NM-CPA, then an arbitrary adversary is unable to construct a ciphertext *"meaningfully related"* to a challenge ciphertext, thereby capturing the notion that ciphertexts are tamper-proof. This notion can be captured by a pair of cryptographic games – namely, $\mathsf{Succ}_{\mathcal{A}, \Pi}$ and $\mathsf{Succ}_{\mathcal{A}, \Pi, \$}$ – between an adversary and a challenger. The first three steps of both games are identical. First, the challenger constructs a key pair (pk, sk) and message space $\mathfrak{m}$; the challenger publishes the public key pk and message space $\mathfrak{m}$. Secondly, the adversary $\mathcal{A}$ executes the algorithm A on the public key pk and outputs the pair $(\mathfrak{m}', s)$, where $\mathfrak{m}'$ is a message space and s is some state information. Thirdly, the challenger randomly selects a plaintext x from the message space; at this point, the challenger in $\mathsf{Succ}_{\mathcal{A}, \Pi, \$}$

---

[2] The assignment of a random element from set S to x is written $x \leftarrow_R S$.

performs an additional step, namely, the challenger samples a second plaintext x'. Fourthly, the challenger constructs a ciphertext $y \leftarrow \mathsf{Enc}_{pk}(x)$. In the fifth step, the adversary executes algorithm D which attempts to define a meaningful relation between x and $\mathbf{x}$, where $\mathbf{x} \leftarrow \mathsf{Dec}_{sk}(\mathbf{y})$ for some $\mathbf{y}$ such that $y \notin \mathbf{y}$. Finally, the challenger decrypts $\mathbf{y}$. The encryption scheme satisfies NM-CPA if no adversary can construct a meaningful relation with non-negligible probability. Formally, Definition 3 recalls the cryptographic game NM-CPA by Bellare & Sahai [4], which slightly strengthens the game by Bellare *et al.* [2].

Definition 3 (NM-CPA). *Let* $\Pi = (\mathsf{Gen}, \mathsf{Enc}, \mathsf{Dec})$ *be an asymmetric encryption scheme,* $\mathcal{A} = (A, D)$ *be an adversary, and* NM-CPA$_{\mathcal{A},\Pi}(\kappa) = |\mathsf{Succ}_{\mathcal{A},\Pi}(\kappa) - \mathsf{Succ}_{\mathcal{A},\Pi,\$}(\kappa)|$ *be a probability, where* $\mathsf{Succ}_{\mathcal{A},\Pi}(\kappa)$ *and* $\mathsf{Succ}_{\mathcal{A},\Pi,\$}(\kappa)$ *are defined below, and* κ *is the security parameter.*[3]

$$\mathsf{Succ}_{\mathcal{A},\Pi}(\kappa) = Pr[(pk, sk, \mathfrak{m}) \leftarrow \mathsf{Gen}(\kappa); (\mathfrak{m}', s) \leftarrow A(pk, \mathfrak{m}); \; x \leftarrow_R \mathfrak{m}';$$
$$y \leftarrow \mathsf{Enc}_{pk}(x); (R, \mathbf{y}) \leftarrow D(\mathfrak{m}', s, y); \; \mathbf{x} \leftarrow \mathsf{Dec}_{sk}(\mathbf{y}) : y \notin \mathbf{y} \wedge R(x, \mathbf{x})]$$

$$\mathsf{Succ}_{\mathcal{A},\Pi,\$}(\kappa) = Pr[(pk, sk, \mathfrak{m}) \leftarrow \mathsf{Gen}(\kappa); (\mathfrak{m}', s) \leftarrow A(pk, \mathfrak{m}); \; x, \tilde{x} \leftarrow_R \mathfrak{m}';$$
$$y \leftarrow \mathsf{Enc}_{pk}(x); (R, \mathbf{y}) \leftarrow D(\mathfrak{m}', s, y); \; \mathbf{x} \leftarrow \mathsf{Dec}_{sk}(\mathbf{y}) : y \notin \mathbf{y} \wedge R(\tilde{x}, \mathbf{x})]$$

In the above game we insist the message space $\mathfrak{m}'$ *is valid (that is,* $\mathfrak{m}' \subseteq \mathfrak{m}$ *and* $|x| = |x'|$ *for any* x *and* x' *given non-zero probability in the message space* $\mathfrak{m}'$*) and samplable in polynomial time, and the relation* R *is computable in polynomial time. We say* Π *satisfies NM-CPA if for all probabilistic polynomial-time adversaries* $\mathcal{A}$*, there exists a negligible function* negl*, such that for all security parameters* κ*, we have* NM-CPA$_{\mathcal{A},\Pi}(\kappa) \leq \mathsf{negl}(\kappa)$*.*

Proving that an encryption scheme satisfies NM-CPA can be problematic and the following result by Bellare & Sahai [4] can simplify proofs.

Proposition 1 (IND-1-CPA = NM-CPA). *Given an asymmetric encryption scheme* Π*, we have* Π *satisfies IND-1-CPA if and only if* Π *satisfies NM-CPA.*

2.2   Proofs of Knowledge

We begin by recalling the syntax for interactive proof systems.

Definition 4 (Sigma protocol). *A* sigma protocol *for a language* $\{s \mid \exists w : L(s, w) = 1\}$ *is a tuple of algorithms* $(\mathsf{Comm}, \mathsf{Chal}, \mathsf{Resp}, \mathsf{Verify})$ *such that:*

---

[3] Vectors are denoted using boldface, for example, $\mathbf{x}$. We write $|\mathbf{x}|$ to denote the length of a vector $\mathbf{x}$ and $\mathbf{x}[i]$ for the ith component of the vector, where $\mathbf{x} = (\mathbf{x}[1], \ldots, \mathbf{x}[|\mathbf{x}|])$. We extend set membership notation to vectors: we write $x \in \mathbf{x}$ (respectively, $x \notin \mathbf{x}$) if x is an element (respectively, x is not an element) of the set $\{\mathbf{x}[i] : 1 \leq i \leq |\mathbf{x}|\}$. We also extend notation for assignment: we write $\mathbf{x} \leftarrow A(\mathbf{y})$ for $\mathbf{x}[1] \leftarrow A(\mathbf{y}[1]); \ldots; \mathbf{x}[|\mathbf{y}|] \leftarrow A(\mathbf{y}[|\mathbf{y}|])$. We write $R(x, \mathbf{x})$ for $R(x, \mathbf{x}[1], \ldots, \mathbf{x}[t-1])$, where R is a relation of arity t, $|\mathbf{x}| = t - 1$, and t is polynomial in the security parameter.

- *The* commitment algorithm Comm *takes a statement s and a witness w as input, and outputs a commitment* comm *and state information t.*
- *The* challenge algorithm Chal *takes some input and outputs a challenge* chal.
- *The* response algorithm Resp *takes a challenge* chal *and state information t as input, and outputs a response* resp.
- *The* verification algorithm Verify *takes a statement s and transcript (*comm, chal, resp*) as input, and outputs* ⊤ *or* ⊥*.*

Moreover, the scheme must be sound: for all statements s, commitments comm*, challenges* chal *and responses* resp*, if* Verify$(s, ($comm, chal, resp$)) = \top$, *then there exists a witness w such that $L(s, w) = 1$ with overwhelming probability. Furthermore, the scheme must be complete: for all statements s and witnesses w, if $L(s, w) = 1$, then for all challenges* chal *we have* Verify$(s, ()$comm, chal, resp$) = \top$ *with overwhelming probability, where (*comm, t$) \leftarrow$ Comm(s, w) *and* resp $\leftarrow$ Resp$($chal, t$)$.*

Our construction will make use of sigma protocols satisfying *special soundness* and *special honest verifier zero knowledge* and, henceforth, we shall only consider *sigma protocols for plaintext knowledge*, that is, sigma protocols that demonstrate knowledge of a plaintext encrypted by a ciphertext.

Definition 5 (Special soundness). *A sigma protocol* (Comm, Chal, Resp, Verify) *for the language $\{s \mid \exists \, w : L(s, w) = 1\}$ satisfies special soundness if there exists a polynomial time algorithm, such that for all statements s and pairs of transcripts (*comm, chal$_1$, resp$_1$) *and (*comm, chal$_2$, resp$_2$)*, the polynomial time algorithm can compute w such that $L(s, w) = 1$, where* Verify$(s, ($comm, chal$_1$, resp$_1$)$) = \top$, Verify$(s, ($comm, chal$_2$, resp$_2$)$) = \top$ *and* chal$_1 \neq$ chal$_2$*.*

Definition 6 (Special honest verifier zero knowledge). *A sigma protocol* (Comm, Chal, Resp, Verify) *for the language $\{s \mid \exists \, w : L(s, w) = 1\}$ satisfies special honest verifier zero knowledge if there exists a polynomial time algorithm, such that for all statements s and challenges* chal$_1$*, the algorithm that can compute* comm$_1$ *and* resp$_1$ *such that* Verify$(s, ()$comm$_1$, chal$_1$, resp$_1$) $= \top$ *and for any w such that $L(s, w) = 1$ we have (*comm$_1$, chal$_1$, resp$_1$) *is indistinguishable from (*comm$_2$, chal$_2$, resp$_2$) *for all challenges* chal$_2$*, where (*comm$_2$, t$) \leftarrow$ Comm(s, w) *and* resp$_2 \leftarrow$ Resp$($chal$_2$, t$)$.*

Definition 7 (Sigma protocol for plaintext knowledge). *Given an asymmetric encryption scheme* (Gen, Enc, Dec)*, sigma protocol Σ for the language $\{s \mid \exists w : L(s, w) = 1\}$ and a message space $\mathfrak{M}$, we say Σ is a sigma protocol for plaintext knowledge in $\mathfrak{m}$ if for all $(pk, sk, \mathfrak{m}) \leftarrow$ Gen(κ), messages $m \in \mathfrak{m}$ and coins r, we have $L((pk, c), (m, r)) = 1$ iff $c =$ Enc$_{pk}(m; r) \wedge m \in \mathfrak{M}$.*

We can derive non-interactive proofs with zero-knowledge from sigma protocols using the *Fiat-Shamir heuristic* [8, 22], which replaces the verifier's challenge with a hash of the prover's commitment and statement, optionally concatenated with a message.

3   Our Scheme

We present our construction (Definition 8) for NM-CPA secure encryption schemes on blocks of messages from homomorphic IND-CPA schemes and sigma protocols for plaintext knowledge.

Definition 8. *Let $\Pi = (\mathsf{Gen}, \mathsf{Enc}, \mathsf{Dec})$ be a homomorphic asymmetric encryption scheme, $\Sigma = (\mathsf{Comm}, \mathsf{Chal}, \mathsf{Resp}, \mathsf{Verify})$ be a sigma protocol for plaintext knowledge in $\mathfrak{M}$ such that Chal is a random oracle, and k be a positive integer. We define $\Gamma(\Pi, \Sigma, k) = (\mathsf{Gen}', \mathsf{Enc}', \mathsf{Dec}')$ as follows:*

- *Gen' takes the security parameter κ as input. The algorithm computes $(pk, sk, \mathfrak{m}) \leftarrow \mathsf{Gen}(\kappa)$ where $\mathfrak{M} \subseteq \mathfrak{m}$. Let $\mathfrak{m}' = \{(m_1, \ldots, m_k) \mid m_1, \ldots, m_k \in \mathfrak{M} \wedge m_1 \oplus \cdots \oplus m_k \in \mathfrak{M}\}$. The algorithm outputs $(pk, sk, \mathfrak{m}')$.*
- *Enc' takes a public key pk, message space $\mathfrak{m}'$, and a message $m' \in \mathfrak{m}'$ as input. The algorithm parses m' as a block of messages $(m_1, \ldots, m_k)$. For each $1 \leq i \leq k$ compute $c_i \leftarrow \mathsf{Enc}_{pk}(m_i; r_i)$; $(comm_i, t_i) \leftarrow \mathsf{Comm}((pk, c_i), (m_i, r_i))$; $chal_i \leftarrow \mathsf{Chal}(i, pk, c_i, comm_i)$; $resp_i \leftarrow \mathsf{Resp}(chal_i, t_i)$, where r_i is randomly selected. Let the homomorphic combination of ciphertexts $c \leftarrow c_1 \otimes \cdots \otimes c_k$, messages $m \leftarrow m_1 \oplus \cdots \oplus m_k$, and coins $r \leftarrow r_1 \odot \cdots \odot r_k$. Compute $(comm, t) \leftarrow \mathsf{Comm}((pk, c), (m, r))$; $chal \leftarrow \mathsf{Chal}(k + 1, pk, c, comm)$; $resp \leftarrow \mathsf{Resp}(chal, t)$. The algorithm outputs $(c_1, comm_1, resp_1, \ldots, c_k, comm_k, resp_k, comm, resp)$.*
- *Dec' takes a public key pk, a private key sk and a ciphertext c' as input. The algorithm parses c' as $(c_1, comm_1, resp_1, \ldots, c_k, comm_k, resp_k, comm, resp)$. If the input cannot be parsed, then the algorithm outputs $\bot$, otherwise, the algorithm proceeds as follows. Compute the homomorphic combination of ciphertexts $c \leftarrow c_1 \otimes \cdots \otimes c_k$. Let $chal \leftarrow \mathsf{Chal}(k + 1, pk, c, comm)$ and, similarly, for each $1 \leq i \leq k$ let $chal_i \leftarrow \mathsf{Chal}(i, pk, c_i, comm_i)$. The algorithm outputs $\bot$, if $\mathsf{Verify}((pk, c), (comm, chal, resp)) = \bot \vee \bigvee_{1 \leq i \leq k} \mathsf{Verify}((pk, c_i), (comm_i, chal_i, resp_i)) = \bot$, and outputs $(\mathsf{Dec}_{sk}(c_1), \ldots, \mathsf{Dec}_{sk}(c_k))$, otherwise.*

In the above construction, witness that our message space is restricted such that for all ciphertexts $(c_1, comm_1, resp_1, \ldots, c_k, comm_k, resp_k, comm, resp)$ generated by Enc' using the public key pk, we have $(pk, c_1), \ldots, (pk, c_k), (pk, c_1 \otimes \ldots \otimes c_k) \in \mathcal{L}$, where $\mathcal{L}$ is the language associated with the sigma protocol used to construct Enc'. Moreover, since sigma protocols are complete, it follows that schemes generated using our construction satisfy the correctness property of asymmetric encryption schemes, that is, Γ constructs asymmetric encryption schemes.

Lemma 1 (Correctness). *Suppose Π is a homomorphic asymmetric encryption scheme, Σ is a sigma protocol for plaintext knowledge in $\mathfrak{M}$ such that Σ's challenge algorithm is a random oracle, and k be a positive integer, we have $\Gamma(\Pi, \Sigma, k)$ is an asymmetric encryption scheme.*

The proof of Lemma 1 follows by inspection of Definitions 1, 4, 7 and 8.

Our construction builds upon homomorphic encryption schemes and sigma protocols for plaintext knowledge in $\mathfrak{M}$ to enhance the functionality of the resulting encryption scheme. For example, homomorphic operations can be performed on the IND-CPA ciphertexts encapsulated in an NM-CPA ciphertext, that is, given an NM-CPA ciphertext $(c_1, \mathsf{comm}_1, \mathsf{resp}_1, \ldots, c_k, \mathsf{comm}_k, \mathsf{resp}_k, \mathsf{comm}, \mathsf{resp})$, homomorphic operations can be performed on the encapsulated IND-CPA ciphertext $c_1, \ldots, c_k$. Moreover, the IND-CPA ciphertexts contain plaintexts in $\mathfrak{M}$.

Lemma 2. *Suppose* $\Pi = (\mathsf{Gen}, \mathsf{Enc}, \mathsf{Dec})$ *is a homomorphic asymmetric encryption scheme,* $\Sigma = (\mathsf{Comm}, \mathsf{Chal}, \mathsf{Resp}, \mathsf{Verify})$ *is a sigma protocol for plaintext knowledge in* $\mathfrak{M}$ *such that* Chal *is a random oracle, and* k *is a positive integer. Let* $\Gamma(\Pi, \Sigma, k) = (\mathsf{Gen}', \mathsf{Enc}', \mathsf{Dec}')$. *We have for all keys* $(pk, sk, \mathfrak{m}') \leftarrow \mathsf{Gen}(\kappa)'$ *and all ciphertexts* $(c_1, \mathsf{comm}_1, \mathsf{resp}_1, \ldots, c_k, \mathsf{comm}_k, \mathsf{resp}_k, \mathsf{comm}, \mathsf{resp})$, *that* $\mathsf{Verify}((pk, c), (\mathsf{comm}, \mathsf{chal}, \mathsf{resp})) = \top \wedge \bigwedge_{1 \leq i \leq k} \mathsf{Verify}((pk, c_i), (\mathsf{comm}_i, \mathsf{chal}_i, \mathsf{resp}_i)) = \top$ *implies the existence of messages* $m_1, \ldots, m_k \in \mathfrak{M}$ *and coins* $r_1, \ldots, r_k$ *such that* $m_1 \oplus \cdots \oplus m_k \in \mathfrak{M}$ *and* $c_i = \mathsf{Enc}_{pk}(m_i; r_i)$ *for all* $1 \leq i \leq k$ *with overwhelming probability, where* $c \leftarrow c_1 \otimes \cdots \otimes c_k$, $\mathsf{chal} \leftarrow \mathsf{Chal}(k+1, pk, c, \mathsf{comm})$, *and* $\mathsf{chal}_i \leftarrow \mathsf{Chal}(i, pk, c_i, \mathsf{comm}_i)$ *for each* $1 \leq i \leq k$.

Proof sketch. The proof follows by inspection of Definitions 4, 8 and 7, in particular, observe that ciphertext $c_1 \otimes \cdots \otimes c_k$ is associated with the proof $\mathsf{comm}, \mathsf{chal}, \mathsf{resp}$ demonstrating plaintext knowledge in $\mathfrak{M}$ and for all integers $1 \leq i \leq k$ that the ciphertext c_i is associated with the proof $\mathsf{comm}_i, \mathsf{chal}_i, \mathsf{resp}_i$ demonstrating plaintext knowledge in $\mathfrak{M}$. □

We believe the enhanced functionality offered by encryption schemes derived from our construction justifies the efficiency cost incurred from the use of sigma protocols. Indeed, encryption schemes derived using our construction technique are useful for privacy preserving applications and, in Sect. 5, we will demonstrate the applicability of our results in the context of electronic voting systems. First, we prove the security of our construction.

4   Security Proof

Our construction Γ (Definition 8) can be used to derive an asymmetric encryption scheme from an IND-CPA encryption scheme and a sigma protocol for plaintext knowledge satisfying special soundness and special honest verifier zero knowledge. The derived scheme satisfies NM-CPA in the random oracle model [3].

Theorem 1. *Let* $\Pi = (\mathsf{Gen}, \mathsf{Enc}, \mathsf{Dec})$ *be a homomorphic asymmetric encryption scheme,* $\Sigma = (\mathsf{Comm}, \mathsf{Chal}, \mathsf{Resp}, \mathsf{Verify})$ *be a sigma protocol for plaintext knowledge in* $\mathfrak{M}$ *such that* Chal *is a random oracle, and* k *be a positive integer. If* Π *satisfies IND-CPA and* Σ *satisfies special soundness and special honest verifier zero knowledge, then the asymmetric encryption scheme* $\Gamma(\Pi, \Sigma, k)$ *satisfies NM-CPA.*

Proof sketch. Suppose $\mathcal{A} = (A_1, A_2)$ is a successful NM-CPA adversary against $\Gamma(\Pi, \Sigma, k)$. Since NM-CPA = IND-1-CPA (Proposition 1), we can suppose $\mathcal{A}$ is an IND-1-CPA adversary. By contradiction, we construct an IND-CPA adversary $\mathcal{B} = (B_1, B_2)$ against Π as follows:

Algorithm B_1. The algorithm takes pk and $\mathfrak{m}$ as input and proceeds as follows. First, the algorithm derives $\Gamma(\Pi, \Sigma, k)$'s message space: suppose $\{s \mid \exists w : L(s, w) = 1\}$ is the language used by the sigma protocol Σ and let $\mathfrak{m}' = \{(m_1, \ldots, m_k) \mid m_1, \ldots, m_k \in \mathfrak{M} \wedge m_1 \oplus \ldots \oplus m_k \in \mathfrak{M} \wedge \exists c_1,$ $\ldots, c_k, r_1, \ldots, r_k : L((pk, c_1 \otimes \ldots \otimes c_k), (m_1 \oplus \ldots \oplus m_k, r_1 \odot \ldots \odot r_k)) = 1 \wedge \bigwedge_{1 \leq i \leq k} L((pk, c_i), (m_i, r_i)) = 1\}$. Secondly, the algorithm computes $(M_0, M_1, s) \leftarrow A_1(pk, \mathfrak{m}')$, where $M_0, M_1 \in \mathfrak{m}'$. It follows that $M_0 = (m_{0,1}, \ldots, m_{0,k})$ and $M_1 = (m_{1,1}, \ldots, m_{1,k})$ for some $m_{0,1}, m_{1,1}, \ldots, m_{0,k},$ $m_{1,k} \in \mathfrak{M} \subseteq \mathfrak{m}$. Thirdly, the algorithm randomly selects an integer i^* such that $1 \leq i^* \leq k$, and outputs $(m_{0,i^*}, m_{1,i^*}, (M_0, M_1, s, i^*, k))$.

Algorithm B_2. The algorithm takes $(m_{0,i^*}, m_{1,i^*}, (M_0, M_1, s, i^*, k), c_{i^*})$ as input, where $M_0 = (m_{0,1}, \ldots, m_{0,k})$, $M_1 = (m_{1,1}, \ldots, m_{1,k})$, and c_{i^*} is the challenge ciphertext. For each $j \in \{1, \ldots, i^* - 1\}$ the algorithm computes

$$c_j \leftarrow \mathsf{Enc}_{pk}(m_{0,j}; r_j); (\mathsf{comm}_j, t_j) \leftarrow \mathsf{Comm}((pk, c_j), (m_{0,j}, r_j));$$
$$\mathsf{chal}_j \leftarrow \mathsf{Chal}(j, pk, c_j, \mathsf{comm}_j); \mathsf{resp}_j \leftarrow \mathsf{Resp}(\mathsf{chal}_j, t_j)$$

where r_j is randomly selected. For each $j \in \{i^* + 1, \ldots, k\}$ the algorithm computes

$$c_j \leftarrow \mathsf{Enc}_{pk}(m_{1,j}; r_j); (\mathsf{comm}_j, t_j) \leftarrow \mathsf{Comm}((pk, c_j), (m_{1,j}, r_j));$$
$$\mathsf{chal}_j \leftarrow \mathsf{Chal}(j, pk, c_j, \mathsf{comm}_j); \mathsf{resp}_j \leftarrow \mathsf{Resp}(\mathsf{chal}_j, t_j)$$

where r_j is randomly selected. The algorithm also computes $c_{k+1} \leftarrow c_1 \otimes \cdots \otimes c_k$. By the honest verifier zero-knowledge property of Σ, the algorithm computes $(\mathsf{comm}_{i^*}, \mathsf{resp}_{i^*})$ and $(\mathsf{comm}_{k+1}, \mathsf{resp}_{k+1})$ – without knowing the plaintext encrypted by the ciphertext c_{i^*} – such that $\mathsf{Verify}((pk, c_{i^*}), (\mathsf{comm}_{i^*}, \mathsf{chal}_{i^*}, \mathsf{resp}_{i^*})) = \top \wedge \mathsf{Verify}((pk, c_{k+1}), (\mathsf{comm}_{k+1}, \mathsf{chal}_{k+1}, \mathsf{resp}_{k+1})) = \top$, where $\mathsf{chal}_{i^*} \leftarrow \mathsf{Chal}(i^*, pk, c_{i^*}, \mathsf{comm}_{i^*})$ and $\mathsf{chal}_{k+1} \leftarrow \mathsf{Chal}(k + 1, pk, c_{k+1}, \mathsf{comm}_{k+1})$. The algorithm simulates the challenge ciphertext C^*:

$$C^* \leftarrow (c_1, \mathsf{comm}_1, \mathsf{resp}_1, \ldots, c_k, \mathsf{comm}_k, \mathsf{resp}_k, \mathsf{comm}_{k+1}, \mathsf{resp}_{k+1})$$

The algorithm computes $b \leftarrow A_2(M_0, M_1, s, C^*)$, simulating the decryption oracle using an extractor derived from the special soundness property in the random oracle model (see Appendix A.4 for details). Finally, the algorithm outputs the bit b.

We have c_{i^*} is embedded in C^* and, by a standard hybrid argument, we can show that $\mathcal{B}$ has an advantage in the IND-CPA game if A_2 notices the embedding (see Appendix A.4 for details).   □

The full proof of Theorem 1 appears in Appendix A.

5   Applications: Electronic Voting

Paper-based elections derive ballot secrecy from physical characteristics of the real-world, for example, ballots are constructed in isolation inside polling booths and complete ballots are deposited into locked ballot boxes. (See Schneier [30] for a informal security analysis of Papal elections.) By comparison, in electronic voting systems, ballots are sent using publicly readable communication channels and, in end-to-end verifiable elections, stored on a publicly readable bulletin board. These attributes make the provision of ballot secrecy difficult in a digital setting, but, nevertheless, ballot secrecy is a *de facto* standard property of electronic voting systems. In this section, we shall discuss assumptions that are sufficient to achieve ballot secrecy in electronic voting systems, and we show how our results make progress towards a computational security proof of ballot secrecy for a variant of the Helios electronic voting system.

5.1   Election Schemes

We adopt syntax for election schemes from Bernhard & Smyth [10].

Definition 9 (Election scheme). *An* election scheme *is a tuple of efficient algorithms* (Setup, Vote, BB, Tally) *such that:*

- Setup *takes a security parameter κ as input and outputs a vote space $\mathfrak{m}$, public key pk, and private key sk, where $\mathfrak{m}$ is a set.*
- Vote *takes a public key pk and vote $v \in \mathfrak{m}$ as input, and outputs a ballot b.*
- BB *takes a bulletin board $\mathfrak{bb}$ and ballot b as input. It outputs $\mathfrak{bb} \cup \{b\}$ if successful (i.e., b is added to $\mathfrak{bb}$) or $\mathfrak{bb}$ to denote failure (i.e., b is not added). This algorithm must be deterministic.*
- Tally *takes a private key sk and bulletin board $\mathfrak{bb}$ as input. It outputs a multiset $\mathfrak{v}$ representing the election outcome if successful or the empty multiset to denote failure. It also outputs auxiliary data aux.*

Moreover, the scheme must satisfy correctness: for any $(\mathfrak{m}, pk, sk)$ output by Setup(κ) *and any bulletin board $\mathfrak{bb}$, the following conditions are satisfied.*[4]

1. *If computing* Tally$_{sk}(\mathfrak{bb})$ *twice produces $(\mathfrak{v}, aux)$ and $(\mathfrak{v}', aux')$, then $\mathfrak{v} = \mathfrak{v}'$.*

Let algorithm τ be defined as follows: $\tau_{sk}(\mathfrak{bb})$ computes $(\mathfrak{v}, aux) \leftarrow$ Tally$_{sk}(\mathfrak{bb})$ and outputs $\mathfrak{v}$. By Condition 1, τ is deterministic.

2. *If b is output by* Vote$_{pk}(v)$ *and $b \notin \mathfrak{bb}$, then* BB$(\mathfrak{bb}, b) = \mathfrak{bb} \cup \{b\}$.
3. *If $\mathfrak{bb} \neq \emptyset$ and $\tau_{sk}(\mathfrak{bb}) = \emptyset_M$ (i.e., $\mathfrak{bb}$ is invalid), then for all ballots b we have $\tau_{sk}(\mathfrak{bb} \cup \{b\}) = \emptyset_M$ too.*
4. *If $\mathfrak{bb} = \emptyset$ or $\tau_{sk}(\mathfrak{bb}) \neq \emptyset_M$ (i.e., $\mathfrak{bb}$ is valid), then for any vote $v \in \mathfrak{m}$ and any ballot b output by* Vote$_{pk}(v)$ *such that $b \notin \mathfrak{bb}$, we have $\tau_{sk}(\mathfrak{bb} \cup \{b\}) = \tau_{sk}(\mathfrak{bb}) \cup_M \{| v |\}$.*
5. *If $\tau_{sk}(\mathfrak{bb}) \neq \emptyset_M$, then $|\tau_{sk}(\mathfrak{bb})| = |\mathfrak{bb}|$.*

---

[4] We write "for any x output by $A(x_1, \ldots, x_n)$" for the universal quantification over x such that x is a result of running probabilistic algorithm A on input $x_1, \ldots, x_n$, i.e., $x = A(x_1, \ldots, x_n; r)$ for some coins r.

We denote multisets as $\{| x_1, \ldots, x_n |\}$ and write $\emptyset_M$ for the empty multiset. The multiset union operator is denoted $\cup_M$ and the multiset intersection operator is denoted $\cap_M$. We write $|S|$ for the cardinality of multiset S.

We recall a construction (Enc2Vote) for election schemes from encryption scheme [7,9,33,34] to demonstrate the applicability of Definition 9. Real voting systems, such as Helios, can also be modelled as election schemes [7,9].

Definition 10 (Enc2Vote). *Given an asymmetric encryption scheme $\Pi = (\mathsf{Gen}, \mathsf{Enc}, \mathsf{Dec})$, election scheme $\mathsf{Enc2Vote}(\Pi)$ is defined as follows.*

- Setup *takes a security parameter κ as input and outputs $(\mathfrak{m}, pk, sk)$, where $(pk, sk) \leftarrow \mathsf{Gen}(\kappa)$ and $\mathfrak{m}$ is the encryption scheme's message space.*
- Vote *takes a public key pk and vote $v \in \mathfrak{m}$ as input, and outputs $\mathsf{Enc}_{pk}(v)$.*
- BB *takes a bulletin board $\mathfrak{bb}$ and ballot b as input, where $\mathfrak{bb}$ is a set. It outputs $\mathfrak{bb} \cup \{b\}$.*
- Tally *takes as input a private key sk and a bulletin board $\mathfrak{bb}$, where $\mathfrak{bb}$ is a set. It outputs multiset $\{\!|\ \mathsf{Dec}_{sk}(b) \mid b \in \mathfrak{bb}\ |\!\}$ and auxiliary data $\perp$.*

We recall the security definition for ballot secrecy from Smyth & Bernhard [10,34].

Definition 11 (Ballot secrecy). *Given an election scheme $\Gamma = (\mathsf{Setup}, \mathsf{Vote}, \mathsf{BB}, \mathsf{Tally})$, a security parameter κ and an adversary $\mathcal{A} = (A_1, A_2)$, let quantity $\mathsf{IND\text{-}SEC}_{\mathcal{A}, \Gamma}(\kappa)$ be defined as follow:*

$$2 \cdot \Pr \left[\begin{array}{l} M_0 \leftarrow \emptyset_M; M_1 \leftarrow \emptyset_M; (\mathfrak{m}, pk, sk) \leftarrow \mathsf{Setup}(\kappa); \\ \mathfrak{bb}_0 \leftarrow \emptyset; \mathfrak{bb}_1 \leftarrow \emptyset; \beta \leftarrow_R \{0,1\}; s \leftarrow A_1^{\mathcal{O}}(\mathfrak{m}, pk); \\ \text{if } M_0 = M_1 \text{ then } \{(\mathfrak{v}, aux) \leftarrow \mathsf{Tally}_{sk}(\mathfrak{bb}_\beta)\} \\ \text{else } \{aux \leftarrow \perp; (\mathfrak{v}, aux') \leftarrow \mathsf{Tally}_{sk}(\mathfrak{bb}_0)\} \\ \quad : A_2(\mathfrak{v}, aux, s) = \beta \end{array} \right] - 1$$

Oracle $\mathcal{O}$ is defined as follows:

$\mathcal{O}()$: *output $\mathfrak{bb}_\beta$.*
$\mathcal{O}(b)$: *$\mathfrak{bb}'_\beta \leftarrow \mathfrak{bb}_\beta$; $\mathfrak{bb}_\beta \leftarrow \mathsf{BB}(\mathfrak{bb}_\beta, b)$; if $\mathfrak{bb}_\beta \neq \mathfrak{bb}'_\beta$ then $\mathfrak{bb}_{1-\beta} \leftarrow \mathsf{BB}(\mathfrak{bb}_{1-\beta}, b)$.*
$\mathcal{O}(v_0, v_1)$: *$M_0 \leftarrow M_0 \cup_M \{\!| v_0 |\!\}$; $M_1 \leftarrow M_1 \cup_M \{\!| v_1 |\!\}$; $b_0 \leftarrow \mathsf{Vote}_{pk}(v_0)$; $b_1 \leftarrow \mathsf{Vote}_{pk}(v_1)$; $\mathfrak{bb}_0 \leftarrow \mathsf{BB}(\mathfrak{bb}_0, b_0)$; $\mathfrak{bb}_1 \leftarrow \mathsf{BB}(\mathfrak{bb}_1, b_1)$. We assume $v_0, v_1 \in \mathfrak{m}$.*

We say Γ satisfies ballot secrecy *(IND-SEC) if for all probabilistic polynomial time adversaries $\mathcal{A}$ we have* $\mathsf{IND\text{-}SEC}_{\mathcal{A}, \Gamma}(\kappa)$ *is negligible in κ.*

5.2   Sufficient Conditions for Ballot Secrecy

Wikström [37] has shown that an encryption scheme satisfying IND-CCA2 is sufficient for ballot secrecy in the Universal Composability framework and Bernhard *et al.* [7] have achieved similar results using cryptographic games. Moreover, Bernhard, Pereira & Warinschi [8,9] have shown that an encryption scheme satisfying NM-CPA is sufficient for ballot secrecy, however, this result is restricted to referendums in a class of electronic voting systems known as *minivoting schemes*. Smyth & Bernhard [34] generalise the result to multi-candidate elections, using a notion of *ballot independence*.

Proposition 2. *Given an encryption scheme Π satisfying NM-CPA, the election scheme* Enc2Vote(Π) *satisfies ballot independence.*

Theorem 2. *Let $\Gamma = (\mathsf{Setup}, \mathsf{Vote}, \mathsf{BB}, \mathsf{Tally})$ be an election scheme. Suppose there exists a constant symbol $\perp$ such that for all $(\mathfrak{m}, pk, sk)$ output by $\mathsf{Setup}(\kappa)$, all multisets $\mathfrak{bb}$, and all $(\mathfrak{v}, aux)$ output by $\mathsf{Tally}_{sk}(\mathfrak{bb})$, we have $aux = \perp$. It follows that Γ satisfies ballot secrecy if and only if Γ satisfies ballot independence.*

It follows immediately that ballot secrecy is preserved.

Corollary 1. *Given an encryption scheme Π satisfying NM-CPA, the election scheme* Enc2Vote(Π) *satisfies ballot secrecy.*

Moreover, our results aid the construction of such schemes.

Corollary 2. *Suppose Π is ElGamal and Σ is the proof of knowledge scheme for equality between discrete logs by Cramer, Damgård & Schoenmakers, we have* Enc2Vote($\Gamma(\Pi, \Sigma)$) *satisfies ballot secrecy.*

The proof of Corollary 2 follows from Smyth & Bernhard (Corollary 1) and Theorem 1.

5.3   Towards a Secure Variant of Helios

Helios [1] is an open-source web-based electronic voting system which uses homomorphic encryption and is claimed to satisfy ballot secrecy and end-to-end verifiability. Helios is particularly significant due to its real-world deployment: the International Association of Cryptologic Research (IACR) have used Helios annually since 2010 to elect its board members, following a successful trial in a non-binding poll; the Catholic University of Louvain adopted the system to elect the university president; and Princeton University have used Helios to elect several student governments. Unfortunately, Cortier & Smyth [15–17], Smyth [32] and Bernhard, Pereira & Warinschi [8] have shown that the scheme does not satisfy ballot secrecy, because ballots are malleable. (See [28,31,35,36] for analysis of verifiability in Helios.) Accordingly, we propose a variant of Helios which defines ballots using the NM-CPA secure encryption scheme derived from our construction parametrised with ElGamal [12] and the proof of knowledge scheme for equality between discrete logs by Cramer, Damgård & Schoenmakers [18]. Our variant of Helios differs from original scheme by including ciphertexts and block numbers in challenges. It follows that our variant is as efficient as the original.

Smyth & Bernhard [33, Sect. 6] argue that Theorem 2 can be generalised to schemes in which tallying outputs the election outcome along with some data that can be simulated. Our variant of Helios satisfies this criterion, since tallying outputs partial ElGamal decryptions and proofs demonstrating knowledge of discrete logarithms, which can be simulated.

Remark 1. Our variant of Helios satisfies ballot secrecy.

The proof of Remark 1 follows from the argument by Smyth & Bernhard [33, Sect. 6] and Theorem 1, hence, it is sufficient to formalise and prove Smyth & Bernhard's argument for a proof that our variant of Helios satisfies ballot secrecy. Formally proving this result is a direction for future work.

6 Related Work

Hirt [26] proposes a construction for NM-CPA secure encryption schemes, from schemes satisfying IND-CPA, for a block of messages $m_1, \ldots, m_k \in \{0, 1\}$, such that the homomorphic combination of messages in the block is between 1 and max: a ciphertext c_i is generated on message m_i for $1 \leq i \leq k$, and ciphertexts $c_{k+1}, \ldots, c_{k+max}$ are generated on *dummy messages* $m_{k+1}, \ldots, m_{k+max} \in \{0, 1\}$ such that $max = \sum_{j=1}^{k+max} m_j$, in addition, proofs of knowledge are used to demonstrate that each ciphertext contains plaintext 0 or 1 and the homomorphic combination of ciphertexts $c_1, \ldots, c_{k+max}$ contains plaintext max, where non-interactive proofs use a common challenge derived from the ciphertexts and commitments. Concurrently, Damgård, Jurik & Nielsen [21] propose a similar construction using Paillier encryption but their work is reliant on unique identifiers to achieve non-malleability and it is unclear what security guarantees can be achieved when state is not maintained. Damgård, Jurik & Nielsen also propose an optimisation to the scheme by Hirt which reduces the number of dummy ciphertexts to one: a ciphertext c_i is generated on each message m_i for $1 \leq i \leq k$, as before, and a ciphertext c_{k+1} is generated on plaintext $m_{k+1} = max - \sum_{i=1}^{k} m_i$ for the dummy candidate, in addition, proofs of knowledge are used to demonstrate that the ciphertexts $c_1, \ldots, c_k$ contain plaintext 0 or 1, the dummy ciphertext c_{k+1} contains a plaintext between 0 and max, and the homomorphic combination of ciphertexts $c_1, \ldots, c_{k+1}$ contains plaintext max. Adida *et al.* [1] generalise Hirt's scheme to consider cases where the homomorphic combination of messages $m_1, \ldots, m_k$ is between min and max, this is achieved by removing the dummy ciphertexts and proving that the homomorphic combination of ciphertexts $c_1, \ldots, c_k$ contains a plaintext between min and max (the non-interactive proofs proposed by Adida *et al.* do not include ciphertexts in challenges).

The aforementioned constructions were developed for electronic voting systems and have not been shown to satisfy either IND-CPA or NM-CPA[5]. Moreover, Cortier & Smyth [15–17] have shown that the construction proposed by Adida *et al.* does not satisfy NM-CPA because the adversary can submit two blocks of messages during the challenge, receive ciphertexts on each message in one of the blocks (along with proofs of knowledge for each individual ciphertext and a proof of knowledge for the homomorphic combination of ciphertexts), and then submit a permutation of the ciphertexts (coupled with the proofs of knowledge for each individual ciphertext, also permuted, and the proof of knowledge

---

[5] Groth [24,25] proves that a special case of the construction proposed by Damgård, Jurik & Nielsen satisfies a notion of secrecy used in electronic voting systems.

for the homomorphic combination of ciphertexts) to the challenger for decryption. Moreover, Smyth [32] has shown that rejecting permutations is insufficient, since the adversary may also submit the homomorphic combination of challenge ciphertexts (along with the proof of knowledge for the homomorphic combination of ciphertexts) to the challenger for decryption. Furthermore, a variant of these attacks is possible, namely, an adversary submits a single ciphertext (along with proofs of knowledge for that ciphertext), selected from the challenge ciphertexts (and associated proofs), that the challenger will decrypt. In addition, Bernhard, Pereira & Warinschi [8] show that the non-interactive proofs proposed by Adida *et al.* belong to the class of *weak Fiat-Shamir transformations* and demonstrate a further variant of these attacks. Nonetheless, the aforementioned constructions provide the inspiration for our work.

As an immediate consequence of these vulnerabilities, Helios does not satisfy ballot secrecy. In the context of electronic voting, Cortier & Smyth advocate rejecting ciphertexts where the associated proofs of knowledge have been previously observed (Bernhard [6] and, independently, Clark [13] suggest it is sufficient to check if challenges have been previously observed, moreover, Bernhard argues that it is necessary) to fix the vulnerability in the construction by Adida *et al.*, however, this solution is infeasible in a more general setting because maintaining state is impractical, in particular, the solution does not satisfy NM-CPA.

Bernhard, Pereira & Warinschi [8, Theorem 2] prove that an IND-CPA encryption scheme and sigma protocol satisfying special soundness, special honest verifier zero knowledge and unique responses, can be combined to derive NM-CPA security. This enables the following two constructions.

1. A (secure) variant of the construction by Adida *et al.* for the special case $k = 1$ (i.e., for blocks of messages containing a single message) satisfying NM-CPA.

The aforementioned vulnerabilities only apply in cases where the block size is greater than one ($k > 1$), hence, this construction only provides vacuous protection against the attacks discussed. By comparison, our scheme develops defences against the aforementioned attacks. Furthermore, the restriction on block size limits application, in particular, it cannot be applied to multi-candidate electronic voting systems. By comparison, our scheme can.

2. Assuming the IND-CPA encryption scheme is for blocks of messages, a construction for an NM-CPA secure encryption scheme on blocks of messages.

In comparison to Bernhard, Pereira & Warinschi, we focus on the construction of NM-CPA secure encryption schemes for arbitrary block sizes from homomorphic IND-CPA encryption schemes, without assuming the IND-CPA encryption scheme is a scheme for blocks of messages. It follows that our construction can be instantiated with ElGamal and the proof of knowledge scheme for equality between discrete logs by Cramer, Damgård & Schoenmakers [18], hence, we can construct our variant of Helios, whereas Bernhard, Pereira & Warinschi cannot.

In complimentary work, constructions for IND-CCA2 secure encryption schemes from schemes satisfying IND-CPA have been presented. For instance, IND-CCA2 secure encryption schemes can be derived from trapdoor one-way functions [23] and identity-based encryption schemes [11]. These results are orthogonal to our work, since we explicitly focus on schemes using proofs of knowledge to achieve additional security properties. The aforementioned schemes do not.

7   Conclusion

We deliver a construction for NM-CPA asymmetric encryption schemes from homomorphic IND-CPA schemes and sigma protocols satisfying special soundness and special honest verifier zero knowledge in the random oracle model. Our use of proofs of knowledge enhances the functionality of encryption schemes derived using our construction, for example, an arbitrary observer can check that ciphertexts contain plaintexts from a particular message space, without the private key. Moreover, the use of homomorphic IND-CPA encryption schemes permits homomorphic operations to be performed on the IND-CPA ciphertexts encapsulated in an NM-CPA ciphertext. We believe that the enhanced functionality justifies the efficiency cost. In the context of electronic voting, we demonstrate the application of our construction by presenting a variant of Helios which satisfies ballot secrecy, assuming Smyth & Bernhard's argument can be formalised and proved.

Acknowledgements. We are grateful to Ben Adida, David Bernhard, Véronique Cortier, Olivier Pereira, Elizabeth Quaglia and Bogdan Warinschi for extensive discussion leading to this result. We are also grateful to the anonymous reviewers for their helpful suggestions. Smyth's work was largely conducted as part of the Toshiba Fellowship Programme atToshiba Corporation, Kawasaki, Japan.

A   Proof of Theorem 1

Suppose $\Gamma(\Pi, \Sigma, k)$ does not satisfy IND-1-CPA, hence $\mathsf{IND\text{-}1\text{-}CPA}_{\mathcal{A}, \Gamma(\Pi, \Sigma, k)}(\kappa)$ $\geq \mathsf{negl}(\kappa)$ for some adversary $\mathcal{A} = (A_1, A_2)$, negligible function f, and security parameter κ. We construct an adversary $\mathcal{B} = (B_1, B_2)$ against IND-CPA using $\mathcal{A}$. Let S_i be an event such that $b^* = b$ in the game i.

A.1   Game 0: IND-1-CPA

Game 0 is derived from IND-1-CPA by replacing the challenger with oracles:

1. A_1 takes $(pk, \mathfrak{m}')$ from the key generation oracle KG.
2. A_1 chooses $M_0, M_1 \in \mathfrak{m}'$ such that $|M_0| = |M_1|$ and sends M_0 and M_1 to the challenge oracle E. A_1 outputs (M_0, M_1, s), where s is some state.

3. A_2 takes (M_0, M_1, s, C^*) from E, where C^* is a challenge ciphertext.
4. A_2 sends ciphertexts $(C'_1, \ldots C'_m)$ to the decryption oracle D and D responds with the corresponding plaintexts $(M'_1, \ldots, M'_m)$ or the error symbol $\bot$.
5. Finally, A_2 outputs $b' \in [0, 1]$.

The challenge and decryption oracles are defined in Tables 1 and 2, and the key generation and random oracle are defined as follows.

Key generation oracle KG. The oracle takes a security parameter κ as input, computes $(pk, sk, \mathsf{m}') \leftarrow \mathsf{Gen}'(\kappa)$, and outputs (pk, m').

Random oracle $Chal$. The oracle takes $(i, pk, c, \mathsf{comm})$ as input and if $((i, pk, c, \mathsf{comm}), \mathsf{chal}) \in L_{Chal}$, then $Chal$ outputs chal, otherwise, $Chal$ chooses $\mathsf{chal} \in \mathcal{CH}$ uniformly at random, where $\mathcal{CH}$ is the range of Chal, outputs chal and adds $((i, pk, c, \mathsf{comm}), \mathsf{chal})$ to L_{Chal}.

By definition of IND-1-CPA, we have: $|\Pr[\mathsf{S}_0] - \frac{1}{2}| = \mathsf{IND}\text{-}1\text{-}\mathsf{CPA}_{\mathcal{A}, \Gamma}(\kappa)$.

A.2   Game 1: Simulate Decryption Oracle

Game 1 uses simulation sound extractability to simulate the decryption oracle without knowledge of sk. Formally, the simulation of the decryption oracle is presented in Table 2, where $i^* \in [1, k]$ is chosen by $\mathcal{B}$ uniformly at random. In the case $c'_{i,j} = c_{b,i} \wedge i \neq i^*$, the simulator knows the plaintext $m_{b,i}$ corresponding to $c_{b,i}$. In the case $c'_{i,j} \neq c_{b,i} \wedge i \neq i^*$, due to [8, Theorem 1], a plaintext $m'_{i,j}$ and a coin $r'_{i,j}$ can be extracted from the ciphertext $c'_{i,j}$ with non-negligible probability using an extractor $\mathcal{K}$. In the remaining case, $m'_{k+1,j} = m'_{1,j} \oplus \cdots \oplus m'_{i^*-1,j} \oplus m'_{i^*,j} \oplus m'_{i^*+1,j} \oplus \cdots \oplus m'_{k,j}$ and $m'_{i^*,j}$ can be computed as $m'_{k+1,j} \oplus^{-1} (m'_{1,j} \oplus \cdots \oplus m'_{i^*-1,j} \oplus m'_{i^*+1,j} \oplus \cdots \oplus m'_{k,j})$.

Let F_1 be the event that occurs if $\mathcal{K}$ cannot extract $m'_{i,j}$ and $r'_{i,j}$, i.e., $c'_{i,j} \neq \mathsf{Enc}_{pk}(m'_{i,j}; r'_{i,j})$ or $\mathcal{K}$ halts with no output. Game 0 and Game 1 are the same when F_1 does not occur. Since F_1 never occurs in Game 0, we have $\Pr[\mathsf{S}_0] = \Pr[\mathsf{S}_1 | \neg \mathsf{F}_1]$. Moreover, we have an extractor such that $\Pr[\neg \mathsf{F}_1] = \delta_1$, where δ_1 is a non-negligible (i.e., δ_1 is a success probability of $\mathcal{K}$). $\mathcal{B}$ can detect whether F_1 occurs, by checking $c'_{i,j} = \mathsf{Enc}_{pk}(m'_{i,j}, r'_{i,j})$. Let $\mathcal{B}$ decide $b' \in \{0, 1\}$ uniformly at random if F_1 occurs, hence, $\Pr[\mathsf{S}_1 | \mathsf{F}_1] = 1/2$. We have:

$$\Pr[\mathsf{S}_1] = \Pr[\mathsf{S}_1 \wedge \neg \mathsf{F}_1] + \Pr[\mathsf{S}_1 \wedge \mathsf{F}_1] = \Pr[\neg \mathsf{F}_1] \cdot \Pr[\mathsf{S}_1 | \neg \mathsf{F}_1] + \Pr[\mathsf{F}_1] \cdot \Pr[\mathsf{S}_1 | \mathsf{F}_1]$$
$$= \Pr[\neg \mathsf{F}_1] \cdot \Pr[\mathsf{S}_0] + 1/2 \cdot \Pr[\mathsf{F}_1]$$
$$= \delta_1 \cdot (1/2 + \mathsf{IND}\text{-}1\text{-}\mathsf{CPA}_{\mathcal{A}, \Gamma}(\kappa)) + 1/2 \cdot (1 - \delta_1)$$
$$= 1/2 + \delta_1 \cdot \mathsf{IND}\text{-}1\text{-}\mathsf{CPA}_{\mathcal{A}, \Gamma}(\kappa).$$

Table 1. Challenge oracle and simulator E (Assume $M_0 = (m_{0,1}, \ldots, m_{0,k})$ and $M_1 = (m_{1,1}, \ldots, m_{1,k})$ are vectors of plaintexts, and parsing always succeeds. Let $[1, k]$ be $\{1, 2, \ldots, k\}$, where k is a positive integer. Let $\mathcal{R}$ be the random number space of Π and Σ, and let $\mathcal{CH}$ be the range of Chal.)

Game0, Game1	Game2	Game3
Challenge $E(M_0, M_1)$	Challenge $E(M_0, M_1)$	Challenge $E(M_0, M_1)$
Choose $b \in [0,1]$ at random	Choose $b \in [0,1]$ at random	
Parse M_0 as $(m_{0,1}, \ldots, m_{0,k})$	Parse M_0 as $(m_{0,1}, \ldots, m_{0,k})$	Parse M_0 as $(m_{0,1}, \ldots, m_{0,k})$
Parse M_1 as $(m_{1,1}, \ldots, m_{1,k})$	Parse M_1 as $(m_{1,1}, \ldots, m_{1,k})$	Parse M_1 as $(m_{1,1}, \ldots, m_{1,k})$
For each $i \in [1, k]$	For each $j \in [1, k]$	For each $j \in [1, i^* - 1]$
Choose $r_i \in \mathcal{R}$ at random	Choose $r_j \in \mathcal{R}$ at random	Choose $r_j \in \mathcal{R}$ at random
$c_{b,i} \leftarrow \mathsf{Enc}_{pk}(m_{b,i}; r_i)$	$c_{b,j} \leftarrow \mathsf{Enc}_{pk}(m_{b,j}; r_j)$	$c_{0,j} \leftarrow \mathsf{Enc}_{pk}(m_{0,j}; r_j)$
$(\mathsf{comm}_{b,i}, t_{b,i}) \leftarrow \mathsf{Comm}((pk, c_{b,i}), (m_{b,i}; r_i))$	For each $j \in [1, k] \setminus \{i^*\}$	$(\mathsf{comm}_{0,j}, t_{0,j}) \leftarrow \mathsf{Comm}((pk, c_{0,j}), (m_{0,j}; r_j))$
$\mathsf{chal}_{b,i} \leftarrow Chal(i, pk, c_{b,i}, \mathsf{comm}_{b,i})$	$(\mathsf{comm}_{b,j}, t_{b,j}) \leftarrow \mathsf{Comm}((pk, c_{b,j}), (m_{b,j}; r_j))$	$\mathsf{chal}_{0,j} \leftarrow Chal(j, pk, c_{0,j}, \mathsf{comm}_{0,j})$
$\mathsf{resp}_{b,i} \leftarrow \mathsf{Resp}(\mathsf{chal}_{b,i}; t_{b,i})$	$\mathsf{chal}_{b,j} \leftarrow Chal(j, pk, c_{b,j}, \mathsf{comm}_{b,j})$	$\mathsf{resp}_{0,j} \leftarrow \mathsf{Resp}(\mathsf{chal}_{0,j}; t_{0,j})$
$C_{b,i} \leftarrow (c_{b,i}, \mathsf{comm}_{b,i}, \mathsf{resp}_{b,i})$	$\mathsf{resp}_{b,j} \leftarrow \mathsf{Resp}(\mathsf{chal}_{b,j}; t_{b,j})$	$C_{0,j} \leftarrow (c_{0,j}, \mathsf{comm}_{0,j}, \mathsf{resp}_{0,j})$
	$C_{b,j} \leftarrow (c_{b,j}, \mathsf{comm}_{b,j}, \mathsf{resp}_{b,j})$	For each $j \in [i^* + 1, k]$
		Choose $r_j \in \mathcal{R}$ at random
		$c_{1,j} \leftarrow \mathsf{Enc}_{pk}(m_{1,j}; r_j)$
		$(\mathsf{comm}_{1,j}, t_{1,j}) \leftarrow \mathsf{Comm}((pk, c_{1,j}), (m_{1,j}; r_j))$
		$\mathsf{chal}_{1,j} \leftarrow Chal(j, pk, c_{1,j}, \mathsf{comm}_{1,j})$
		$\mathsf{resp}_{1,j} \leftarrow \mathsf{Resp}(\mathsf{chal}_{1,j}; t_{1,j})$
		$C_{1,j} \leftarrow (c_{1,j}, \mathsf{comm}_{1,j}, \mathsf{resp}_{1,j})$
		Call challenge oracle E' of Π with m_{0,i^*} and m_{1,i^*}, receive response c_{b,i^*}
$c_b \leftarrow c_{b,1} \otimes \cdots \otimes c_{b,k}$	$c_b \leftarrow c_{b,1} \otimes \cdots \otimes c_{b,k}$	$c_b \leftarrow c_{0,1} \otimes \cdots \otimes c_{1,k}$
$m_b \leftarrow m_{b,1} \oplus \cdots \oplus m_{b,k}$	Choose $\mathsf{chal}_{b,i^*}, \mathsf{chal}_b \in \mathcal{CH}$ at random	Choose $\mathsf{chal}_{b,i^*}, \mathsf{chal}_b \in \mathcal{CH}$ at random
$r \leftarrow r_1 \odot \cdots \odot r_k$	Compute $(\mathsf{comm}_{b,i^*}, \mathsf{resp}_{b,i^*})$ for chal_{b,i^*} and	Compute $(\mathsf{comm}_{b,i^*}, \mathsf{resp}_{b,i^*})$ for chal_{b,i^*} and
$(\mathsf{comm}_b, t_b) \leftarrow \mathsf{Comm}((pk, c_b), (m_b, r))$	$(\mathsf{comm}_b, \mathsf{resp}_b)$ for chal_b by HVZK	$(\mathsf{comm}_b, \mathsf{resp}_b)$ for chal_b by HVZK
$\mathsf{chal}_b \leftarrow Chal(k+1, pk, c, \mathsf{comm}_b)$	Set the random oracle Chal as	Set the random oracle Chal as
$\mathsf{resp}_b \leftarrow \mathsf{Resp}(\mathsf{chal}_b, t_b)$	$\mathsf{chal}_{b,i^*} \leftarrow Chal(i^*, pk, c_{b,i^*}, \mathsf{comm}_{b,i^*})$	$\mathsf{chal}_{b,i^*} \leftarrow Chal(i^*, pk, c_{b,i^*}, \mathsf{comm}_{b,i^*})$
	$\mathsf{chal}_b \leftarrow Chal(k+1, pk, c_b, \mathsf{comm}_b)$	$\mathsf{chal}_b \leftarrow Chal(k+1, pk, c_b, \mathsf{comm}_b)$
$C^* \leftarrow (C_{b,1}, \ldots, C_{b,k}, \mathsf{comm}_b, \mathsf{resp}_b)$	$C^* \leftarrow (C_{b,1}, \ldots, C_{b,k}, \mathsf{comm}_b, \mathsf{resp}_b)$	$C^* \leftarrow (C_{b,1}, \ldots, C_{b,k}, \mathsf{comm}_b, \mathsf{resp}_b)$
Return C^*	Return C^*	Return C^*

Table 2. Decryption oracle and simulator D (Assume $\mathbf{C}'$ is a vector $(C'_1, \ldots, C'_m)$ for some positive integer m represented by polynomial of the security parameter κ. Let $[1, k]$ be $\{1, \ldots, k\}$, where k is a positive integer. Assume each C'_j is a vector $(c'_{1,j}, \mathsf{comm}'_{1,j}, \mathsf{resp}'_{1,j}, \ldots, c'_{k,j}, \mathsf{comm}'_{k,j}, \mathsf{resp}'_{k,j}, \mathsf{comm}'_{k+1,j}, \mathsf{resp}'_{k+1,j})$. We say $c'_{i,j}$ is valid if $\mathsf{Verify}((pk, c'_{i,j}), (\mathsf{comm}'_{i,j}, \mathsf{chal}'_{i,j}, \mathsf{resp}'_{i,j})) = \top$. Let $m_{b,i}$ be the plaintext encapsulated inside challenge ciphertext $c_{b,i}$. The decryption oracle can obtain $m_{b,i}$, because $m_{b,i}$ is chosen by the challenge oracle, which is part of the simulator. Let $\oplus^{-1}$ be an inverse operator of $\oplus$.)

Game0	Game1, Game2, Game3
Decryption $D(\mathbf{C}')$	Decryption $D(\mathbf{C}')$
Parse $\mathbf{C}'$ as $C'_1, C'_2, \ldots, C'_m$.	Parse $\mathbf{C}'$ as $C'_1, C'_2, \ldots, C'_m$.
For each $j \in [1, m]$	For each $j \in [1, m]$
$\quad$ Parse C'_j as $(c'_{1,j}, \ldots, \mathsf{resp}'_{k+1,j})$	$\quad$ Parse C'_j as $(c'_{1,j}, \ldots, \mathsf{resp}'_{k+1,j})$
$\quad c'_{k+1,j} \leftarrow c'_{1,j} \otimes \cdots \otimes c'_{k,j}$	$\quad c'_{k+1,j} \leftarrow c'_{1,j} \otimes \cdots \otimes c'_{k,j}$
$\quad$ For each $i \in [1, k+1]$	$\quad$ For each $i \in [1, k+1]$
$\quad\quad \mathsf{chal}'_{i,j} \leftarrow Chal(i, pk, c'_{i,j}, \mathsf{comm}'_{i,j})$	$\quad\quad \mathsf{chal}'_{i,j} \leftarrow Chal(i, pk, c'_{i,j}, \mathsf{comm}'_{i,j})$
$\quad\quad m'_{i,j} \leftarrow \mathsf{Dec}_{sk}(c'_{i,j})$	$\quad\quad$ If $c'_{i,j} = c_{b,i} \wedge i = i^*$
	$\quad\quad\quad m'_{i,j} \leftarrow m_{b,i}$
$\quad$ If $C'_j \neq C^* \wedge \forall i \in [1, k+1] : c'_{i,j}$ is valid	$\quad\quad$ else if $c'_{i,j} \neq c_{b,i} \wedge i \neq i^*$
$\quad\quad M'_j \leftarrow (m'_{1,j}, \ldots, m'_{k+1,j})$	$\quad\quad\quad$ extract $m'_{i,j}$ from $c'_{i,j}$ using an extractor
$\quad$ else	$\quad\quad$ else if $i = i^*$
$\quad\quad M'_j \leftarrow \bot$	$\quad\quad\quad m'_j \leftarrow m'_{1,j} \oplus \cdots \oplus m'_{i^*-1,j} \oplus m'_{i^*+1,j} \oplus \cdots \oplus m'_{k,j}$
return $(M'_1, \ldots, M'_m)$	$\quad\quad\quad m'_{i^*,j} \leftarrow m'_{k+1,j} \oplus^{-1} m'_j$
	$\quad$ If $C'_j \neq C^* \wedge \forall i \in [1, k+1] : c'_{i,j}$ is valid
	$\quad\quad M'_j \leftarrow (m'_{1,j}, \ldots, m'_{k+1,j})$
	$\quad$ else
	$\quad\quad M'_j \leftarrow \bot$
	return $(M'_1, \ldots, M'_m)$

A.3 Game 2: Simulate the Challenge Oracle

Game 2 uses the special honest verifier zero knowledge (special HVZK) property to simulate the challenge oracle. Table 1 formalises the simulator. By Definition 6, $\mathcal{B}$ can compute $(\mathsf{comm}, \mathsf{resp})$ from a correct ciphertext c and challenge chal such that $\mathsf{Verify}((pk, c), (\mathsf{comm}, \mathsf{chall}, \mathsf{resp})) = \top$. But, if the random oracle has already been queried with $(i^*, pk, c_{b,i^*}, \mathsf{comm}_{b,i^*})$ or $(k+1, pk, c_b, \mathsf{comm}_b)$, then $\mathcal{B}$ fails to compute the challenge ciphertext. Let F_2 be the event that $\mathcal{B}$ fails. Game 1 and Game 2 are the same, when F_2 does not occur. Since F_2 never occurs in Game 1, we have $\Pr[\mathsf{S}_1] = \Pr[\mathsf{S}_2|\neg\mathsf{F}_2]$. Let $\Pr[\neg\mathsf{F}_2] = \delta_2$. Since coins $\{r_j\}_{j\in[1,k]}$ are chosen from a large space, δ_2 is non-negligible. Let $\mathcal{B}$ decide $b' \in \{0, 1\}$ uniformly at random if F_2 occurs, hence, $\Pr[\mathsf{S}_2|\mathsf{F}_2] = 1/2$. We have

$$
\begin{aligned}
\Pr[\mathsf{S}_2] &= \Pr[\mathsf{S}_2 \wedge \mathsf{F}_2] + \Pr[\mathsf{S}_2 \wedge \neg\mathsf{F}_2] = \Pr[\mathsf{F}_2] \cdot \Pr[\mathsf{S}_2|\mathsf{F}_2] + \Pr[\neg\mathsf{F}_2] \cdot \Pr[\mathsf{S}_2|\neg\mathsf{F}_2] \\
&= 1/2 \cdot \Pr[\mathsf{F}_2] + \Pr[\neg\mathsf{F}_2] \cdot \Pr[\mathsf{S}_1] \\
&= 1/2 \cdot (1 - \delta_2) + \delta_2 \cdot (1/2 + \delta_1 textsf{IND}\text{-}1\text{-}\mathsf{CPA}_{\mathcal{A},\Gamma}(\kappa)) \\
&= 1/2 + \delta_1 \cdot \delta_2 \cdot \mathsf{IND}\text{-}1\text{-}\mathsf{CPA}_{\mathcal{A},\Gamma}(\kappa)
\end{aligned}
\tag{1}
$$

A.4   Game 3: Embed a Challenge Ciphertext

Game 3 embeds $\mathcal{B}$'s challenge ciphertext as the i^*th ciphertext in the vector of challenge ciphertexts sent to $\mathcal{A}$. Formally, the embedding is handled by the decryption oracle (Table 2), where $i^* \in [1, k]$ is chosen by $\mathcal{B}$.

Let H_0 be Game 2 modified such that b is always 1 and let H_n be Game 2 when b is always 0. For $0 < i < k$, let H_i be H_{i-1} modified such that the first $3 \cdot i$ elements of the challenge ciphertext are generated from M_0 and the remaining elements of the challenge ciphertext are generated from M_1. If $b^* = 1$, then the challenge ciphertext that $\mathcal{B}$ inputs to $\mathcal{A}$ is the same as the hybrid game H_{i^*-1}, since $c^* = c_{1,i^*} = \mathsf{Enc}_{pk}(m_{1,i^*}; r_{i^*})$. Otherwise ($b^* = 0$), $\mathcal{B}$'s input to $\mathcal{A}$ is the same as the game H_{i^*}, since $c^* = c_{0,i^*} = \mathsf{Enc}_{pk}(m_{0,i^*}; r_{i^*})$. Let E_i be an event that occurs if $\mathcal{A}$ outputs 1 in H_i, then $|\Pr[E_{i-1}] - \Pr[E_i]| \leq \mathsf{IND\text{-}CPA}_{\mathcal{B},\Pi}(\kappa)$ holds. By a hybrid argument, we have

$$|\Pr[E_0] - \Pr[E_k]| \leq \sum_{i=1}^{k} |\Pr[E_{i-1}] - \Pr[E_i]| \leq k \cdot \mathsf{IND\text{-}CPA}_{\mathcal{B},\Pi}(\kappa) \qquad (2)$$

Moreover, since

$$\begin{aligned}
|2 \cdot \Pr[\mathsf{S}_2] - 1| &= |2 \cdot (\Pr[\mathcal{A} \to 1 \text{ in Game } 2 \wedge b = 1] \\
&\quad + \Pr[\mathcal{A} \to 0 \text{ in Game } 2 \wedge b = 0]) - 1| \\
&= |2 \cdot (\Pr[b = 1] \cdot \Pr[\mathcal{A} \to 1 \text{ in Game } 2|b = 1] \\
&\quad + \Pr[b = 0] \cdot \Pr[\mathcal{A} \to 0 \text{ in Game } 2|b = 0]) - 1| \\
&= |\Pr[\mathcal{A} \to 1 \text{ in Game } 2|b = 1] \\
&\quad + (1 - \Pr[\mathcal{A} \to 1 \text{ in Game } 2|b = 0]) - 1| \\
&= |\Pr[E_0] + (1 - \Pr[E_k]) - 1| \\
&= |\Pr[E_0] - \Pr[E_k]|
\end{aligned}$$

We have, by (1), that

$$|\Pr[E_0] - \Pr[E_k]| = 2 \cdot \delta_1 \cdot \delta_2 \cdot \mathsf{IND\text{-}1\text{-}CPA}_{\mathcal{A},\Gamma}(\kappa) \qquad (3)$$

We have $2 \cdot \delta_1 \cdot \delta_2 \cdot \mathsf{IND\text{-}1\text{-}CPA}_{\mathcal{A},\Gamma}(\kappa)$ is non-negligible. By (2) and (3), we have $|\Pr[E_{i-1}] - \Pr[E_i]|$ is non-negligible too, i.e., $\mathsf{IND\text{-}1\text{-}CPA}_{\mathcal{B},\Pi}(\kappa)$ is non-negligible. It follows by Proposition 1 that Γ satisfies NM-CPA.   □

References

1. Adida, B., Marneffe, O., Pereira, O., Quisquater, J.: Electing a University President using open-audit voting: analysis of real-world use of Helios. In: EVT/WOTE 2009: Electronic Voting Technology Workshop/Workshop on Trustworthy Elections. USENIX Association (2009)
2. Bellare, M., Desai, A., Pointcheval, D., Rogaway, P.: Relations among notions of security for public-key encryption schemes. In: Krawczyk, H. (ed.) CRYPTO 1998. LNCS, vol. 1462, p. 26. Springer, Heidelberg (1998)

3. Bellare, M., Rogaway, P.: Random oracles are practical: a paradigm for designing efficient protocols. In: CCS 1993: 1st ACM Conference on Computer and Communications Security, pp. 62–73. ACM Press (1993)
4. Bellare, M., Sahai, A.: Non-malleable encryption: equivalence between two notions, and an indistinguishability-based characterization. Cryptology ePrint Archive, Report 2006/228 (2006)
5. Benaloh, J., Yung, M.: Distributing the power of a government to enhance the privacy of voters. In: PODC 1986: 5th Principles of Distributed Computing Symposium, pp. 52–62. ACM Press (1986)
6. Bernhard, D.: Private email communication, 15th March 2012
7. Bernhard, D., Cortier, V., Pereira, O., Smyth, B., Warinschi, B.: Adapting Helios for provable ballot privacy. In: Atluri, V., Diaz, C. (eds.) ESORICS 2011. LNCS, vol. 6879, pp. 335–354. Springer, Heidelberg (2011)
8. Bernhard, D., Pereira, O., Warinschi, B.: How not to prove yourself: pitfalls of the Fiat-Shamir Heuristic and applications to Helios. In: Wang, X., Sako, K. (eds.) ASIACRYPT 2012. LNCS, vol. 7658, pp. 626–643. Springer, Heidelberg (2012)
9. Bernhard, D., Pereira, O., Warinschi, B.: On Necessary and sufficient conditions for private ballot submission. Cryptology ePrint Archive, Report 2012/236 (2012)
10. Bernhard, D., Smyth, B.: Ballot secrecy with malicious bulletin boards. Cryptology ePrint Archive, Report 2014/822 (2014)
11. Canetti, R., Halevi, S., Katz, J.: Chosen-ciphertext security from identity-based encryption. In: Cachin, C., Camenisch, J.L. (eds.) EUROCRYPT 2004. LNCS, vol. 3027, pp. 207–222. Springer, Heidelberg (2004)
12. Chaum, D., Pedersen, T.P.: Wallet databases with observers. In: Brickell, E.F. (ed.) CRYPTO 1992. LNCS, vol. 740, pp. 89–105. Springer, Heidelberg (1993)
13. Clark, J.: Private email communication, 4th April 2012
14. Cohen, J.D., Fischer, M.J.: A robust and verifiable cryptographically secure election scheme. In: FOCS 1985: 26th Symposium on Foundations of Computer Science, pp. 372–382. IEEE Computer Society (1985)
15. Cortier, V., Smyth, B.: Attacking and fixing Helios: an analysis of ballot secrecy. In: CSF'11: 24th Computer Security Foundations Symposium, pp. 297–311. IEEE Computer Society (2011)
16. Cortier, V., Smyth, B.: Attacking and fixing Helios: an analysis of ballot secrecy. Cryptology ePrint Archive, Report 2010/625 (version 20111110:012334) (2011)
17. Cortier, V., Smyth, B.: Attacking and fixing Helios: an analysis of ballot secrecy. J. Comput. Secur. **21**(1), 89–148 (2013)
18. Cramer, R., Damgård, I.B., Schoenmakers, B.: Proof of partial knowledge and simplified design of witness hiding protocols. In: Desmedt, Y.G. (ed.) CRYPTO 1994. LNCS, vol. 839, pp. 174–187. Springer, Heidelberg (1994)
19. Cramer, R., Franklin, M.K., Schoenmakers, B., Yung, M.: Multi-authority secret-ballot elections with linear work. In: Maurer, U.M. (ed.) EUROCRYPT 1996. LNCS, vol. 1070, pp. 72–83. Springer, Heidelberg (1996)
20. Damgård, I., Jurik, M.: A generalisation, a simplification and some applications of Paillier's probabilistic public-key system. In: Kim, K. (ed.) PKC 2001. LNCS, vol. 1992. Springer, Heidelberg (2001)
21. Damgård, I., Jurik, M., Nielsen, J.B.: A generalization of Paillier's public-key system with applications to electronic voting. Int. J. Inf. Secur. **9**(6), 371–385 (2010)
22. Fiat, A., Shamir, A.: How to prove yourself: practical solutions to identification and signature problems. In: Odlyzko, A.M. (ed.) CRYPTO 1986. LNCS, vol. 263, pp. 186–194. Springer, Heidelberg (1987)

23. Fujisaki, E., Okamoto, T.: How to enhance the security of public-key encryption at minimum cost. In: Imai, H., Zheng, Y. (eds.) PKC 1999. LNCS, vol. 1560, p. 53. Springer, Heidelberg (1999)

24. Groth, J.: Extracting witnesses from proofs of knowledge in the random oracle model. Technical report RS-01-52, Basic Research in Computer Science (BRICS) (2001)

25. Groth, J.: Evaluating security of voting schemes in the universal composability framework. In: Jakobsson, M., Yung, M., Zhou, J. (eds.) ACNS 2004. LNCS, vol. 3089, pp. 46–60. Springer, Heidelberg (2004)

26. Hirt, M.: Receipt-free K-out-of-L voting based on ElGamal encryption. In: Chaum, D., Jakobsson, M., Rivest, R.L., Ryan, P.Y.A., Benaloh, J., Kutylowski, M., Adida, B. (eds.) Towards Trustworthy Elections. LNCS, vol. 6000, pp. 64–82. Springer, Heidelberg (2010)

27. Hirt, M., Sako, K.: Efficient receipt-free voting based on homomorphic encryption. In: Preneel, B. (ed.) EUROCRYPT 2000. LNCS, vol. 1807, pp. 539–556. Springer, Heidelberg (2000)

28. Kremer, S., Ryan, M., Smyth, B.: Election verifiability in electronic voting protocols. In: Gritzalis, D., Preneel, B., Theoharidou, M. (eds.) ESORICS 2010. LNCS, vol. 6345, pp. 389–404. Springer, Heidelberg (2010)

29. Sako, K., Kilian, J.: Secure voting using partially compatible homomorphisms. In: Desmedt, Y.G. (ed.) CRYPTO 1994. LNCS, vol. 839, pp. 411–424. Springer, Heidelberg (1994)

30. Schneier, B.: Hacking the Papal Election (2013)

31. Smyth, B.: Formal verification of cryptographic protocols with automated reasoning. Ph.D. thesis, School of Computer Science, University of Birmingham (2011)

32. Smyth, B.: Replay attacks that violate ballot secrecy in Helios. Cryptology ePrint Archive, Report 2012/185 (2012)

33. Smyth, B., Bernhard, D.: Ballot secrecy and ballot independence coincide. In: Crampton, J., Jajodia, S., Mayes, K. (eds.) ESORICS 2013. LNCS, vol. 8134, pp. 463–480. Springer, Heidelberg (2013)

34. Smyth, B., Bernhard, D.: Ballot secrecy and ballot independence: definitions and relations. Cryptology ePrint Archive, Report 2013/235 (version: 20141010:082554) (2014)

35. Smyth, B., Frink, S., Clarkson, M.R.: Computational election verifiability: definitions and an analysis of Helios and JCJ. Cryptology ePrint Archive, Report 2015/233 (2015)

36. Smyth, B., Ryan, M., Kremer, S., Kourjieh, M.: Towards automatic analysis of election verifiability properties. In: Armando, A., Lowe, G. (eds.) ARSPA-WITS 2010. LNCS, vol. 6186, pp. 146–163. Springer, Heidelberg (2010)

37. Wikström, D.: Simplified submission of inputs to protocols. In: Ostrovsky, R., De Prisco, R., Visconti, I. (eds.) SCN 2008. LNCS, vol. 5229, pp. 293–308. Springer, Heidelberg (2008)

Improvement of UC Secure Searchable Symmetric Encryption Scheme

Shunsuke Taketani and Wakaha Ogata[✉]

Tokyo Institute of Technology, Meguro, Japan
{taketani.s.aa,ogata.w.aa}@m.titech.ac.jp

Abstract. Searchable symmetric encryption refers to a system whereby clients store encrypted documents in a server that can be searched by keywords without revealing private information. In this paper, we demonstrate that the UC-secure SSE scheme proposed by Kurosawa and Ohtaki is inefficient under certain scenarios, and we propose a modified scheme. Our scheme has reliability and privacy, where privacy is slightly weaker than the original Kurosawa-Ohtaki scheme. Therefore, our scheme offers UC-security with slightly weaker privacy. More precisely, the additional information our scheme leaks is only the size of a set of keywords. On the other hand, the index size for our scheme is much smaller than the original scheme when the set of keywords is a very sparse subset of l-bit strings for some l. The UC-secure Kurosawa-Ohtaki scheme is improved with the proposed scheme by introducing a new tag for proving "non-existence." The proposal is an example of how an SSE scheme can be effectively converted into a verifiable SSE scheme.

1 Introduction

In recent years, many storage services have become available with which clients can store documents or files on the service provider's server. By using such services, clients can access their information at any time and from anywhere and any device. If the number of stored files increases, a keyword search is desirable to find particular files. On the other hand, the client can encrypt files to avoid leaking confidential information to the service provider. Searchable symmetric encryption (SSE) enables the client to search a large number of encrypted files with encrypted keywords.

The concept for SSE was introduced by Song et al. [21], and many SSE schemes have since been proposed [2,9,11,12,14,18]. Most SSE schemes offer privacy; e.g., the server cannot learn anything about the stored files or keywords. However, such schemes do not have any mechanism to verify search results; that is, it is assumed that the server is honest and that it always follows protocol. Thus, Kurosawa and Ohtaki [18] introduced enhanced security notions, *reliability* and universally composable security (UC-security) for SSE schemes. Reliability ensures the validity of search results even if the server is malicious. Kurosawa

© Springer International Publishing Switzerland 2015
K. Tanaka and Y. Suga (Eds.): IWSEC 2015, LNCS 9241, pp. 135–152, 2015.
DOI: 10.1007/978-3-319-22425-1_9

and Ohtaki showed that an SSE scheme that has both privacy and reliability has UC-security[1], and proposed a concrete SSE scheme which has UC-security.

In this paper, we evaluate the scheme proposed by Kurosawa and Ohtaki [18], and show that it is inefficient under some scenarios. We then propose a modified scheme for which privacy is somewhat weaker than the original Kurosawa-Ohtaki scheme. Therefore, our scheme provides UC-security with slightly weaker privacy. More precisely, the additional information our scheme leaks is merely the size of a set of keywords. Yet, the index size for our scheme is much smaller than the original scheme when the set of keywords is a very sparse subset of l-bit strings for some l. The index size is reduced by eliminating dummy elements from the original index and introducing a new tag that proves that the eliminated elements do not exist in the reduced index.

Related works. Various SSE schemes have been proposed. Some support additional search functions, such as multi-keyword searching [1,13], ranked searching [7], and fuzzy searching [3,22]. Others support adding and removing documents [15–17,19].

In [10], a verifiable SSE scheme is proposed. The scheme has two modes: "privacy preferred" and "efficiency preferred." However, the former mode, which is relatively more secure in terms of privacy, requires a very large index. Moreover, no formal security proof is provided for this scheme.

[19] proposed another verifiable SSE scheme. However, their scheme does not assume that a client will query a keyword that is not contained in the set of keywords used to build the index. Assume that a client generates an (encrypted) index based on a certain keyword set $\mathcal{W}$, forgets $\mathcal{W}$, and then searches for $w \notin \mathcal{W}$. In this case, the server has no choice but to return "no document hits" without any proof. This means that the server can forge the search results by answering "no document hits" at any time.

2   Verifiable Searchable Symmetric Encryption

In this section, we define a (verifiable) SSE scheme and its security. Basically, we follow the notation used in [8,18,20].

- Let $\mathcal{D} = \{D_0, \ldots, D_{N-1}\}$ be a set of documents.
- Let $\mathcal{W}$ be a set of keywords.
- For a keyword $w \in \mathcal{W}$, let $\mathcal{D}(w)$ denote the set of documents that contain w.

We consider a system model that has two components: a client and a server. Roughly speaking, in SSE schemes, clients encrypt all documents in $\mathcal{D}$ before storing them on a server. Clients can then search through these documents using a keyword $w \in \mathcal{W}$ from the set $\mathcal{D}$, the output for which is derived as follows:

$$\mathcal{C}(w) = \{C_i \mid C_i \text{ is a ciphertext of } D_i \in \mathcal{D}(w)\}. \tag{1}$$

---

[1] In [20], it was shown that *strong* reliability rather than ordinary reliability is required to be US-security.

In response to a search query, the server returns $\mathcal{C}(w)$. If there is a mechanism to verify the validity of the response, the scheme is a verifiable SSE (vSSE).

Hereafter, $|X|$ denotes the bit length of X for a bit string X, and $|X|$ denotes the cardinality of X for a set X. Furthermore, "PPT" refers to the probabilistic polynomial time.

2.1   System Model

Formally, a vSSE scheme has two phases: the store phase (executed only once) and the search phase (executed a polynomial number of times). Such a scheme consists of the following six polynomial-time algorithms:[2]

$$\mathsf{vSSE} = (\mathsf{Gen}, \mathsf{Enc}, \mathsf{Trpdr}, \mathsf{Search}, \mathsf{Dec}, \mathsf{Verify})$$

such that

- $K \leftarrow \mathsf{Gen}(1^\lambda)$: a probabilistic algorithm that generates a key K, where λ is a security parameter. This algorithm is run by the client during the store phase, and K is kept secret.
- $(\mathcal{I}, \mathcal{C}) \leftarrow \mathsf{Enc}(K, \mathcal{D}, \mathcal{W})$: a probabilistic encryption algorithm that outputs an encrypted index $\mathcal{I}$ and $\mathcal{C} = \{C_0, \ldots, C_{N-1}\}$, where C_i is the ciphertext for D_i. This algorithm is run by the client during the store phase, and $(\mathcal{I}, \mathcal{C})$ are sent to the server.
- $t(w) \leftarrow \mathsf{Trpdr}(K, w)$: an algorithm that outputs a trapdoor $t(w)$ for a keyword w. This is run by the client during the search phase, and $t(w)$ is sent to the server.
- $(\tilde{\mathcal{C}}(w), Proof) \leftarrow \mathsf{Search}(\mathcal{I}, \mathcal{C}, t(w))$: a deterministic search algorithm, where $\tilde{\mathcal{C}}(w)$ is the search result and $Proof$ is its proof. This algorithm is run by the server during the search phase, and $(\tilde{\mathcal{C}}(w), Proof)$ is sent to the client.
- $\mathsf{accept/reject} \leftarrow \mathsf{Verify}(K, t(w), \tilde{\mathcal{C}}(w), Proof)$: a deterministic verification algorithm that determines the validity of $\tilde{\mathcal{C}}(w)$ based on $Proof$. This algorithm is run by the client.
- $D \leftarrow \mathsf{Dec}(K, C)$: a deterministic decryption algorithm. The client uses this algorithm for all $C \in \tilde{\mathcal{C}}(w)$, when $\mathsf{Verify}(K, t(w), \tilde{\mathcal{C}}(w), Proof) = \mathsf{accept}$.

Correctness entails the following from the scheme for the set of documents $\mathcal{D}$ and a keyword $w \in \mathcal{W}$:

- $D_i = \mathsf{Dec}(K, C_i)$ if $\mathcal{C} = \{C_0, \ldots, C_{N-1}\}$ is the output of $\mathsf{Enc}(K, \mathcal{D}, \mathcal{W})$.
- $\mathsf{Verify}(K, t(w), \tilde{\mathcal{C}}(w), Proof) = \mathsf{accept}$, if $(\mathcal{I}, \mathcal{C})$ is outputted by $\mathsf{Enc}(K, \mathcal{D}, \mathcal{W})$, $t(w)$ is outputted by $\mathsf{Trpdr}(K, w)$, and $(\tilde{\mathcal{C}}(w), Proof)$ is outputted by $\mathsf{Search}(\mathcal{I}, \mathcal{C}, t(w))$.

2.2   Security Definition

We next define some security conditions that should be satisfied by a vSSE scheme.

---

[2] If the search result does not need to be verified, $Proof$ and Verify can be omitted.

- Adversary **A** chooses $(\mathcal{D}, \mathcal{W})$ and sends them to challenger **C**.
- **C** generates $K \leftarrow \text{Gen}(1^k)$ and sends $(\mathcal{I}, \mathcal{C}) \leftarrow \text{Enc}(K, \mathcal{D}, \mathcal{W})$ to **A**.
- For $i = 0, \ldots, q-1$, do:
  1. **A** chooses a keyword $w_i \in \mathcal{W}$ and sends it to **C**.
  2. **C** sends the trapdoor $t(w_i) \leftarrow \text{Trpdr}(K, w_i)$ back to **A**.
- **A** outputs bit b.

Fig. 1. Real game Game_{real}

- Adversary **A** chooses $(\mathcal{D}, \mathcal{W})$ and sends them to challenger **C**.
- **C** sends $L_1(\mathcal{D}, \mathcal{W})$ to simulator **S**.
- **S** computes $(\mathcal{I}', \mathcal{C}')$ from $L_1(\mathcal{D}, \mathcal{W})$, and sends them to **C**.
- **C** relays $(\mathcal{I}', \mathcal{C}')$ to **A**
- For $i = 0, \ldots, q-1$, do:
  1. **A** chooses $w_i \in \mathcal{W}$ and sends it to **C**.
  2. **C** sends $L_2(\mathcal{D}, \mathcal{W}, \mathbf{w}, w_i)$ to **S**, where $\mathbf{w} = (w_1, \ldots, w_{i-1})$.
  3. **S** computes $t'(w_i)$ from $L_2(\mathcal{D}, \mathcal{W}, \mathbf{w}, w_i)$ and sends it to **C**.
  4. **C** relays $t'(w_i)$ to **A**.
- **A** outputs bit b.

Fig. 2. Simulation game Game_{sim}^L

Privacy. In a vSSE, the server should learn as little information as possible regarding $\mathcal{D}, \mathcal{W}$, and the queried keyword w. Let $L = (L_1, L_2)$ be a pair of leakage functions, such that $L_1(\mathcal{D}, \mathcal{W})$ (and respectively, $L_2(\mathcal{D}, \mathcal{W}, \mathbf{w}, w)$) denote the information the user permits the server to learn during the store phase (and respectively, the search phase). Here, $\mathbf{w} = (w_1, w_2, \ldots)$ is the list of keywords queried in past searches, and w is the keyword queried now. The client's privacy is defined by using two games: a real game Game_{real}, and a simulation game Game_{sim}^L, as shown in Figs. 1 and 2, respectively. Game_{real} is played by a challenger **C** and an adversary **A**, and Game_{sim}^L is played by **C**, **A** and a simulator **S**.

Definition 1 (L-privacy). *We say that a vSSE scheme has L-privacy, if there exists a PPT simulator **S** such that*

$$|\Pr(\mathbf{A} \text{ outputs } b = 1 \text{ in } \text{Game}_{real}) - \Pr(\mathbf{A} \text{ outputs } b = 1 \text{ in } \text{Game}_{sim}^L)| \quad (2)$$

*is negligible for any PPT adversary **A**.*

In most existing SSE schemes, $L_1(\mathcal{D}, \mathcal{W})$ includes $(|D_0|, \ldots, |D_{N-1}|)$ and some information about $\mathcal{W}$, such as $|\mathcal{W}|$ or the length of the keywords. On the other hand, $L_2(\mathcal{D}, \mathcal{W}, \mathbf{w}, w)$ consists of

$$\text{List}(w) = \{j \mid D_j \in \mathcal{D} \text{ contains } w\}$$

(Store phase)

- $\mathbf{A}_1$ chooses $(\mathcal{D}, \mathcal{W})$ and sends them to $\mathbf{C}$.
- $\mathbf{C}$ generates $K \leftarrow \mathtt{Gen}(1^\lambda)$, and sends $(\mathcal{I}, \mathcal{C}) \leftarrow \mathtt{Enc}(K, \mathcal{D}, \mathcal{W})$ to $\mathbf{A}_2$.

(Search phase)

- For $i = 0, \ldots, q-1$, do
    1. $\mathbf{A}_1$ chooses a keyword w_i and sends it to $\mathbf{C}$.
    2. $\mathbf{C}$ sends the trapdoor $t(w_i) \leftarrow \mathtt{Trpdr}(K, w_i)$ to $\mathbf{A}_2$.
    3. $\mathbf{A}_2$ returns $(\tilde{C}(w_i)^*, Proof_i^*)$ to $\mathbf{C}$.
    4. $\mathbf{C}$ computes

$$\mathtt{accept/reject} \leftarrow \mathtt{Verify}(K, t(w_i), \tilde{C}(w_i)^*, Proof_i^*)$$

    and returns a set $\tilde{D}(w_i)^*$ of plaintexts of documents in $\tilde{C}(w_i)^*$ to $\mathbf{A}_1$ if the result is $\mathtt{accept}$, otherwise sends $\mathtt{reject}$ to $\mathbf{A}_1$.
- If there exists i, such that both $\mathtt{Verify}(K, t(w_i), \tilde{C}(w_i)^*, Proof_i^*) = \mathtt{accept}$ and $(\tilde{C}(w_i)^*, Proof_i^*) \neq (C(w_i), Proof_i)$ hold, then $\mathbf{A}$ (strongly) wins; otherwise $\mathbf{A}$ loses.

Fig. 3. $\mathtt{Game}_{reli}$

and the search pattern

$$\mathtt{SPattern}((w_1, \ldots, w_{q-1}), w) = (sp_1, \ldots, sp_{q-1}), \quad sp_j = \begin{cases} 1 & \text{if } w_j = w \\ 0 & \text{if } w_j \neq w \end{cases}$$

that reveals the past queries that are the same as w.

Reliability. In an SSE scheme, a malicious server should not cheat a client by returning a false result $\tilde{C}(w)^*(\neq C(w))$ during the search phase. We generally call this notion (weak) reliability. In [20], *strong reliability* was also defined, and a relation between strong reliability and universal composability was discussed. Strong reliability is formulated by considering the game $\mathtt{Game}_{reli}$ shown in Fig. 3. This game is played by an adversary $\mathbf{A} = (\mathbf{A}_1, \mathbf{A}_2)$ (malicious server) and a challenger $\mathbf{C}$. We assume that $\mathbf{A}_1$ and $\mathbf{A}_2$ can communicate freely.

Definition 2 ((Strong)Reliability). *We say that a vSSE scheme satisfies (strong) reliability if for any PPT adversary $\mathbf{A}$, $\Pr(\mathbf{A}$ wins$)$ is negligible for any $(\mathcal{D}, \mathcal{W})$ and any search queries $w_0, \ldots, w_{q-1}$.*

From now on, we will say just *Reliability* for what we mean *Strong Reliability*.

Universally Composable Security. It is known that if protocol Σ is secure in the universally composable (UC) security framework, then the security of Σ is maintained even if it is combined with other protocols. The security in the

Store: Upon receiving the input $(\mathbf{store}, sid, D_0, \ldots, D_{N-1}, \mathcal{W})$ from the (dummy) client, verify that this is the first input from the client with $(\mathbf{store}, sid)$.
If it is, then store $\mathcal{D} = \{D_0, \ldots, D_{N-1}\}$, and send $L_1(\mathcal{D}, \mathcal{W})$ to $\mathbf{S}$. Otherwise, ignore this input.
Search: Upon receiving $(\mathbf{search}, sid, w)$ from the client, send $L_2(\mathcal{D}, \mathcal{W}, \mathbf{w}, w)$ to $\mathbf{S}$.
  1. If $\mathbf{S}$ returns "OK," then send $\mathcal{D}(w)$ to the client.
  2. If $\mathbf{S}$ returns $\bot$, then send $\bot$ to the client.

Fig. 4. Ideal functionality $\mathcal{F}_{vSSE}^{L}$

UC framework is defined by associating it with a given *ideal functionality* $\mathcal{F}$. Refer to [4–6] for the formal definition of the UC framework. Kurosawa and Ohtaki introduced an ideal functionality of vSSE [18,20]. Here, we generalize the definition in order to handle the general leakage functions $L = (L_1, L_2)$ as shown in Fig. 4. Note that the server does not interact with $\mathcal{F}_{vSSE}^{L}$, because it does not have its own input and output.

Definition 3 (UC-Security with Leakage L). *We say that vSSE scheme has universally composable (UC) security with leakage L, if it realizes[3] the ideal functionality $\mathcal{F}_{vSSE}^{L}$.*

The following theorem can be proved in the same way as the theorem in [20].

Theorem 1. vSSE *has UC security with leakage L against non-adaptive adversaries if the* vSSE *scheme satisfies L-privacy and reliability.*

2.3   Kurosawa-Ohtaki Scheme (KO-Scheme)

We next review the UC-secure SSE scheme proposed by Kurosawa and Ohtaki, KO-scheme [18]. In this scheme, the set of searchable keywords is $\mathcal{W} \subseteq \{0,1\}^l$ for some l.

Let $\mathtt{SKE} = (G, E, E^{-1})$ be a symmetric encryption scheme, where G is a key-generation algorithm, E is an encryption algorithm, and E^{-1} is the corresponding decryption algorithm. For a security parameter λ, let $\pi : \{0,1\}^\lambda \times \{0,1\}^{l+1+\log N} \to \{0,1\}^{l+1+\log N}$ be a pseudorandom permutation, where N denotes the number of documents in $\mathcal{D}$, and let $\mathtt{MAC} : \{0,1\}^\lambda \times \{0,1\}^* \to \{0,1\}^n$ be a tag-generation function. For simplicity, we write $y = \pi(x)$ rather than $y = \pi(K, x)$, and $\mathtt{MAC}(m)$ rather than $\mathtt{MAC}(K, m)$, where K is a key.
The KO-scheme proceeds as follows:

---

[3] That is, if there does not exist any environment $\mathcal{Z}$ that can distinguish the real world and the ideal world by interacting with the real-world adversary or the ideal-world adversary.

$\text{Gen}(1^\lambda)$: Run G to generate a key K_0 for SKE. Randomly choose a key $K_1 \in \{0,1\}^\lambda$ for π and a key $K_2 \in \{0,1\}^\lambda$ for MAC. Output $K = (K_0, K_1, K_2)$.

$\text{Enc}(K, \mathcal{D}, \mathcal{W})$: First, compute $C_i = E(K_0, D_i)$ for each $D_i \in \mathcal{D}$, and let $\mathcal{C} = \{C_0, \ldots, C_{N-1}\}$. Let $\mathcal{I}$ be an array of size $2 \times 2^l N$. We write $\mathcal{I}[i]$ for the i-th element of $\mathcal{I}$.

  1. Let
$$\mathcal{I}[i] \leftarrow (\text{dummy}, \text{MAC}(i\|\text{dummy}))$$
    for all $i = 0, \ldots, 2 \times 2^l N - 1$.

  2. For each $w \in \{0,1\}^l$, suppose that $\mathcal{D}(w) = (D_{s_1}, \ldots, D_{s_m})$. Then for $j = 1, \ldots, m$, let
$$addr = \pi(0, w, j)$$
$$tag_{w,j} = \text{MAC}(addr\|C_{s_j})$$
$$\mathcal{I}[addr] \leftarrow (s_j, tag_{w,j}).$$

  3. For each $D_k \in \mathcal{D}$, suppose that document number k appears N_k times in $\mathcal{I}$. Then for $j = 1, \ldots, 2^l - N_k$, let
$$addr = \pi(1, j, k)$$
$$tag_{j,k} = \text{MAC}(addr\|C_k)$$
$$\mathcal{I}[addr] \leftarrow (k, tag_{j,k}).$$

Finally, output $(\mathcal{I}, \mathcal{C})$.

$\text{Trpdr}(K, w)$: Output
$$t(w) = (\pi(0, w, 0), \ldots, \pi(0, w, N-1))$$

$\text{Search}(\mathcal{I}, \mathcal{C}, t(w))$: Parse $t(w)$ as $t(w) = (addr_0, \ldots, addr_{N-1})$. Suppose that
$$\mathcal{I}[addr_i] = (s_i, tag_i)$$

for $i = 0, \ldots, N-1$. First, set $\tilde{\mathcal{C}}(w) \leftarrow \text{empty}$. Then, for $i = 0, \ldots, N-1$, add C_{s_i} to $\tilde{\mathcal{C}}(w)$, if $s_i \neq \text{dummy}$. Set $Proof = (tag_0, \ldots, tag_{N-1})$. Finally, output $(\tilde{\mathcal{C}}(w), Proof)$.

$\text{Verify}(K, t(w), \tilde{\mathcal{C}}(w), Proof)$: Parse $t(w), \tilde{\mathcal{C}}(w)$, and $Proof$ as
$$t(w) = (addr_0, \ldots, addr_{N-1})$$
$$\tilde{\mathcal{C}}(w) = (\tilde{C}_0, \ldots, \tilde{C}_{m-1})$$
$$Proof = (tag_0, \ldots, tag_{N-1}).$$

Then, verify the validity of the result with the following steps.

  1. Let $X_i \leftarrow \tilde{C}_i$ for $i = 0, \ldots, m-1$.

  2. Let $X_i \leftarrow \text{dummy}$ for all $i = m, \ldots, N-1$.

  3. If $tag_i = \text{MAC}(addr_i\|X_i)$ for all $i = 0, \ldots, N-1$, then output accept. Otherwise output reject.

$\text{Dec}(K, C)$: Output a document $D = E_{K_0}^{-1}(C)$ for a ciphertext C.

Proposition 1. *For* $\mathcal{W} \subseteq \{0,1\}^l$, *let*

$$L^{KO} = (L_1^{KO}, L_2^{KO})$$
$$L_1^{KO}(\mathcal{D}, \mathcal{W}) = (|D_0|, \ldots, |D_{N-1}|, l)$$
$$L_2^{KO}(\mathcal{D}, \mathcal{W}, \mathbf{w}, w) = (\mathtt{List}(\mathcal{D}, w), \mathtt{SPattern}(\mathbf{w}, w))$$

The above scheme has L^{KO}-privacy and reliability. Therefore, it also has UC security with leakage L^{KO}.

Remark 1. In [18], Kurosawa and Ohtaki claimed that only $\mathtt{List}(\mathcal{D}, w)$ is leaked during the search phase. However, it is obvious that the trapdoor leaks a search pattern, since $\mathtt{Trpdr}$ is deterministic.

2.4   Inefficiency of KO-Scheme

With the KO-scheme, the index size is $\mathcal{O}(2^l N)$, where l denotes the bit-length of the keywords. Here we consider a case where $\mathcal{W}$ is a set of English words that includes a long word, namely "indistinguishability." The bit-length of this set of keywords is $l = 8 \times 20$. With such a keyword set, then, the index will become very large.

Let l be the maximum length of the keywords expressed as bit strings, and $\mathcal{W}_0 = \{0,1\}^l$. Then, $\mathcal{W} \subseteq \mathcal{W}_0$ holds. In general, the KO-scheme is inefficient whenever $|\mathcal{W}| \ll |\mathcal{W}_0|$, and it is not uncommon.

An easy solution to this problem is to transform each word into a short bit string as follows: Let $l' = \lceil \log_2 |\mathcal{W}| \rceil$. First, the client numbers the keywords in $\mathcal{W}$ from 0 to $|\mathcal{W}| - 1$. That is, $\mathcal{W} = \{w_0, \ldots, w_{|\mathcal{W}|-1}\}$. For the $\mathtt{Enc}$ algorithm, the client does not use the keyword $w_i \in \mathcal{W}$, but rather its index $i \in [0, 2^{l'} - 1]$. Then, the index size is $\mathcal{O}(2^{l'} N) \approx \mathcal{O}(|\mathcal{W}|N)$, even if $|\mathcal{W}| \ll 2^l$.

For this solution, however, the client must keep $\mathcal{W}$ on hand in order to translate the keyword w into its index i when searching w.

In the next section, we provide another solution to reduce the size of the index.

3   Improvement of KO-Scheme

In this section, we propose a new vSSE scheme.

The idea of reducing the index size is elimination of **dummy** elements from the index, and introduction of a new tag that proves that the eliminated elements do not exist in the constructed index.

Let $M = |\mathcal{W}|$ be the number of keywords. The index of our scheme is much smaller than the index of the KO-scheme, if $M \ll 2^l$. This means that the computation cost to generate an index and to search it is also reduced.

3.1 Concrete Description of Our Scheme

Here, $\mathtt{SKE} = (G, E, E^{-1}), \mathcal{D}, \pi$, and $\mathtt{MAC}$ follow the denotations from Sect. 2.3. Our vSSE scheme is as follows:

$\mathtt{Gen}(1^\lambda)$: Run G to generate a key K_0 for $\mathtt{SKE}$. Randomly choose a key $K_1 \in \{0,1\}^\lambda$ for π and a key $K_2 \in \{0,1\}^\lambda$ for $\mathtt{MAC}$. Output $K = (K_0, K_1, K_2)$.

$\mathtt{Enc}(K, \mathcal{D}, \mathcal{W})$: First, compute $C_i = E(K_0, D_i)$ for each $D_i \in \mathcal{D}$ and let $\mathcal{C} = \{C_0, \ldots, C_{N-1}\}$.

Our index $\mathcal{I}$ is an array of size $MN + 1$, and each element of $\mathcal{I}$ has four fields:

$$(\mathtt{addr}, \mathtt{ID}, \mathtt{tag}, \mathtt{Ntag}).$$

Hereafter, $\mathcal{I}[i]$ denotes the i-th element in $\mathcal{I}$, and $\mathcal{I}[i].\mathtt{addr}$ denotes the $\mathtt{addr}$ field of the i-th element in $\mathcal{I}$. Furthermore, we will use the same notation for the other three fields. $\mathcal{I}$ is constructed as follows.

1. Set

$$\mathcal{I}[MN].\mathtt{addr} \leftarrow 2^{l+1+\log N}$$
$$\mathcal{I}[MN].\mathtt{ID} \leftarrow \mathtt{dummy}$$
$$\mathcal{I}[MN].\mathtt{tag} \leftarrow \mathtt{dummy}.$$

2. Let

$$p_{i,j} = \begin{cases} \pi(0, w_i, j) & \text{if } D_j \in \mathcal{D}(w_i) \\ \pi(1, w_i, j) & \text{if } D_j \notin \mathcal{D}(w_i) \end{cases}$$

for $w_i \in \mathcal{W}$ and $D_j \in \mathcal{D}$, and set

$$\mathcal{I}[Ni + j].\mathtt{addr} \leftarrow p_{i,j}$$
$$\mathcal{I}[Ni + j].\mathtt{ID} \leftarrow j$$
$$\mathcal{I}[Ni + j].\mathtt{tag} \leftarrow \mathtt{MAC}(0\|p_{i,j}\|C_j)$$

for $i = 0, \ldots, M - 1$ and $j = 0, \ldots, N - 1$. At this time, each element of $\mathcal{I}$ is

$\mathcal{I}[0]$	$= (p_{0,0},$	$0,$ $\mathtt{MAC}(0\|p_{0,0}\|C_0),$	$\mathtt{undefined})$
$\mathcal{I}[1]$	$= (p_{0,1},$	$1,$ $\mathtt{MAC}(0\|p_{0,1}\|C_1),$	$\mathtt{undefined})$

$$\vdots$$

$\mathcal{I}[N-1]$	$= (p_{0,N-1},$	$N-1, \mathtt{MAC}(0\|p_{0,N-1}\|C_{N-1}),$	$\mathtt{undefined})$
$\mathcal{I}[N]$	$= (p_{1,0},$	$0,$ $\mathtt{MAC}(0\|p_{1,0}\|C_0),$	$\mathtt{undefined})$

$$\vdots$$

$\mathcal{I}[2N-1]$	$= (p_{1,N-1},$	$N-1, \mathtt{MAC}(0\|p_{1,N-1}\|C_{N-1}),$	$\mathtt{undefined})$

$$\vdots$$

$$\mathcal{I}[MN-1] = (p_{M-1,N-1}, N-1, \mathtt{MAC}(0\|p_{M-1,N-1}\|C_{N-1}), \mathtt{undefined}).$$

Note that all values in the $\mathtt{addr}$ field are distinct, because π is a permutation.

3. Sort $\mathcal{I}[0], \ldots, \mathcal{I}[MN]$ based on the addr field, such that

$$\mathcal{I}[0.\text{addr} < \mathcal{I}[1].\text{addr} < \cdots < \mathcal{I}[MN].\text{addr}.$$

4. For $r = 0, \ldots, MN$, compute

$$Ntag_r = \begin{cases} \text{MAC}(1\|0\|\mathcal{I}[r].\text{addr}) & \text{if } r = 0 \\ \text{MAC}(1\|\mathcal{I}[r-1].\text{addr}\|\mathcal{I}[r].\text{addr}) & \text{if } r \neq 0 \end{cases}$$

and set
$$\mathcal{I}[r].\text{Ntag} \leftarrow Ntag_r.$$

Finally, output $(\mathcal{I}, \mathcal{C})$.

$\text{Trpdr}(K, w)$: Output

$$t(w) = (\pi(0, w, 0), \ldots, \pi(0, w, N-1)).$$

$\text{Search}(\mathcal{I}, \mathcal{C}, t(w))$: First set $\tilde{\mathcal{C}}(w) \leftarrow$ empty. Parse $l(w)$ as $l(w) = (addr_0, \ldots, addr_{N-1})$. For $i = 0, \ldots, N-1$, search $addr_i$ from the addr field in $\mathcal{I}$, and follow the steps below:

- If $addr_i = \mathcal{I}[r_i].\text{addr}$ for some $r_i \in [0, MN-1]$, then set

$$\tilde{\mathcal{C}}(w) \leftarrow \tilde{\mathcal{C}}(w) \cup C_{\mathcal{I}[r_i].\text{ID}}$$
$$pr_i \leftarrow \mathcal{I}[r_i].\text{tag}.$$

- If $addr_i < \mathcal{I}[0].\text{addr}$, set

$$pr_i \leftarrow (0\|\mathcal{I}[0].\text{addr}, \ \mathcal{I}[0].\text{Ntag}).$$

- If $\mathcal{I}[r_i-1].\text{addr} < addr_i < \mathcal{I}[r_i].\text{addr}$ for some $r_i \in [1, MN]$, then set

$$pr_i \leftarrow (\mathcal{I}[r_i-1].\text{addr}\|\mathcal{I}[r_i].\text{addr}, \ \mathcal{I}[r_i].\text{Ntag}).$$

Set $Proof = (pr_0, \ldots, pr_{N-1})$. Finally, output $(\tilde{\mathcal{C}}(w), Proof)$.

$\text{Verify}(K, t(w), \tilde{\mathcal{C}}(w), Proof)$: Parse $\tilde{\mathcal{C}}(w)$, $Proof$, and $t(w)$ as

$$\tilde{\mathcal{C}}(w) = (\tilde{C}_0, \ldots, \tilde{C}_{m-1})$$
$$Proof = (pr_0, \ldots, pr_{N-1})$$
$$t(w) = (addr_0, \ldots, addr_{N-1})$$

and follow the steps below to verify the validity of the search result.

1. If m is not equal to the number of pr_js that are *tags*—meaning that they do not consist of a pair of *addr* and *Ntag*—then output reject.
2. For each pr_j that is a *tag*, if there does not exist a distinct $i \in \{0, \ldots, m-1\}$, such that

$$\text{MAC}(0\|addr_j\|\tilde{C}_i) = pr_j, \tag{3}$$

then output reject.

3. For a pr_j that is not a tag but rather a pair of $addr$ and $Ntag$, assume that $pr_j = (addr'_{j,1} \| addr'_{j,2}, Ntag_j)$. If the following two statements are true for all such j, then output accept.

$$addr'_{j,1} < addr_j < addr'_{j,2} \tag{4}$$
$$\texttt{MAC}(1\|addr'_{j,1}\|addr'_{j,2}) = Ntag_j \tag{5}$$

Otherwise, output reject.

$\texttt{Dec}(K, C)$: Output document $D = E_{K_0}^{-1}(C)$ for a ciphertext C.

3.2 Security

Let

$$L^{new} = (L_1^{new}, L_2^{new})$$
$$L_1^{new}(\mathcal{D}, \mathcal{W}) = L_1^{KO}(\mathcal{D}, \mathcal{W}) \cup |\mathcal{W}| = (|D_0|, \dots, |D_{N-1}|, l, |\mathcal{W}|)$$
$$L_2^{new}(\mathcal{D}, \mathcal{W}, \mathbf{w}, w) = L_2^{KO}(\mathcal{D}, \mathcal{W}, \mathbf{w}, w) = (\texttt{List}(\mathcal{D}, w), \texttt{SPattern}(\mathbf{w}, w))$$

We can prove that the above scheme satisfies L^{new}-privacy and reliability, and therefore UC-security with leakage L^{new}.

Theorem 2. *If the symmetric encryption scheme* $\texttt{SKE} = (G, E, E^{-1})$ *is secure in terms of indistinguishability against chosen-plaintext attacks (IND-CPA), and if π is a pseudorandom permutation, then the proposed scheme satisfies L^{new}-privacy.*

Proof. We construct the simulator $\mathbf{S}$ as follows. First, $\mathbf{S}$ receives $L_1(\mathcal{D}, \mathcal{W}) = (|D_0|, \dots, |D_{N-1}|, l, |\mathcal{W}|)$.

1. $\mathbf{S}$ runs $\texttt{Gen}(1^\lambda)$ to generate key $K = (K_0, K_1, K_2)$.
2. Let $C'_j = E(K_0, 0^{|D_j|})$ for $j = 0, \dots, N-1$, and let $\mathcal{C}' = \{C'_0, \dots, C'_{N-1}\}$.
3. Choose $w'_0, \dots, w'_{M-1}$ from $\{0,1\}^l$ randomly, and set $\mathcal{W}' = \{w'_0, \dots, w'_{M-1}\}$. Compute the index $\mathcal{I}'$ as if all of the documents $D_0, \dots, D_{N-1} \in \mathcal{D}$ include all of the keywords in $\mathcal{W}'$. That is, for $i = 0, \dots, M-1$ and $j = 0, \dots, N-1$,

$$\mathcal{I}'[r'_{i,j}] = (\pi(0, w'_i, j), j, tag'_{i,j}, \texttt{undefined}) \tag{6}$$
$$\mathcal{I}'[MN] = (2^{l+1+\log N}, \texttt{dummy}, \texttt{MAC}(0\|\texttt{dummy}), \texttt{undefined})$$

where

$$r'_{i,j} = Ni + j$$
$$tag'_{i,j} = \texttt{MAC}(0\|\pi(0, w'_i, j)\|C'_j).$$

Next, sort the elements based on the addr field, and set

$$\mathcal{I}'[r].\texttt{Ntag} \leftarrow \texttt{MAC}(1\|\mathcal{I}'[r-1].\texttt{addr}\|\mathcal{I}'[r].\texttt{addr}).$$

4. Return $(\mathcal{I}', \mathcal{C}')$.

During the i-th search iteration, $\mathbf{S}$ is given

$$\texttt{List}(w_i) = \{s_0, \ldots, s_{m-1}\}$$

and

$$\texttt{SPattern}(\mathbf{w}, w_i) = (sp_1, \ldots, sp_{i-1})$$

(but neither w_i nor $\mathbf{w}$). $\mathbf{S}$ simulates the trapdoor as follows.

1. If $sp_j = 1$ for some $j < i$, then $\mathbf{S}$ sets $t_i' = t_j'$ and returns t_i'.
2. If $\texttt{List}(w_i) = \emptyset$, then $\mathbf{S}$ randomly chooses $w' \in \{0,1\}^l \backslash (\mathcal{W}' \cup \mathcal{W}_{\text{used}})^4$, otherwise, $\mathbf{S}$ randomly chooses $w' \in \mathcal{W}' \backslash \mathcal{W}_{\text{used}}$, where $\mathcal{W}_{\text{used}}$ is initially empty. Then, $\mathbf{S}$ sets

$$\mathcal{W}_{\text{used}} = \mathcal{W}_{\text{used}} \cup \{w'\}$$

$$addr_j' = \begin{cases} \pi(0, w', j) & \text{if } j \in \texttt{List}(w_i) \\ \pi(1, w', j) & \text{if } j \notin \texttt{List}(w_i) \end{cases}$$

for $j = 0, \ldots, N-1$, and returns

$$t_i' = (addr_0', \ldots, addr_{N-1}').$$

We will prove that there is no adversary $\mathbf{A}$ who can distinguish the games Game_{real} and Game_{sim} by using six games $\mathsf{Game}_0, \ldots, \mathsf{Game}_5$. Let $\mathsf{Game}_0 = \mathsf{Game}_{real}$. Hereafter, we write

$$P_i = \Pr(\mathbf{A} \text{ outputs } b = 1 \text{ in } \mathsf{Game}_i)$$

for simplicity.

- Game_1 is equivalent to Game_0, except that each $C_j = E(K_0, D_j)$ is replaced with $C_j' = E(K_0, 0^{|D_j|})$ for $j = 0, \ldots, N-1$. From the assumption for SKE, $|P_0 - P_1|$ is negligible.
- Game_2 uses a real random permutation π_2 for computing $addr$s rather than pseudorandom permutation π as with Game_1. Then, $|P_1 - P_2|$ is negligible, owing to the pseudorandomness of π.
- Game_3 is equivalent to Game_2, except that the set of keywords is changed from $\mathcal{W}$ to $\mathcal{W}'(|\mathcal{W}'| = M)$, and the random permutation is changed to π_3, whose output for a keyword $w_i \in \mathcal{W}$ is the same as the output of π_2 for input $w_i' \in \mathcal{W}'$, and the output for $w_i' \in \mathcal{W}$ is the same as the output of π_2 for $w_i \in \mathcal{W}$ for all i. Then, π_3 is a random permutation, as is π_2, and the constructed indexes for Game_2 and Game_3 are identical. Hence, $|P_2 - P_3|$ is negligible.

---

[4] If $\{0,1\}^l \backslash (\mathcal{W}' \cup \mathcal{W}_{\text{used}}) = \emptyset$, w' is chosen from $\mathcal{W}' \backslash \mathcal{W}_{\text{used}}$.

– Game_4 is equivalent to Game_3, except that the $\text{List}(w')$ in Game_3 is replaced by $\text{List}'(w') = \{0, \dots, M-1\}$ for all $w' \in \mathcal{W}'$, and π_3 is replaced by π_4, which satisfies

$$\pi_4(0\|w'\|j) = \begin{cases} \pi_3(0\|w'\|j) & \text{if } j \in \text{List}(w') \\ \pi_3(1\|w'\|j) & \text{if } j \notin \text{List}(w') \end{cases}$$

$$\pi_4(1\|w'\|j) = \begin{cases} \pi_3(1\|w'\|j) & \text{if } j \in \text{List}(w') \\ \pi_3(0\|w'\|j) & \text{if } j \notin \text{List}(w') \end{cases}$$

for all $j = 0, \dots, N-1$ and all w'. Then, π_4 is also a random permutation, and the constructed indexes for Game_3 and Game_4 are identical. Hence, $|P_3 - P_4|$ is negligible.
– In Game_5, we use pseudorandom permutation π, rather than π_4, and this is the only difference between Game_5 and Game_4. Because π_4 is a random permutation, $|P_4 - P_5|$ is negligible, owing to the pseudorandomness of π.

From the above, $|P_0 - P_5|$ is negligible. Since it is obvious that $\text{Game}_5 = \text{Game}_{sim}$, Game_{real} and Game_{sim} are indistinguishable for any adversary $\mathbf{A}$. □

Theorem 3. *If* MAC *is existentially unforgeable against chosen-message attacks, our scheme satisfies reliability.*

Proof. Suppose that for $(\mathcal{D}, \mathcal{W})$ and search queries $w_0, \dots, w_{q-1}$, there exists an adversary $\mathbf{A} = (\mathbf{A}_1, \mathbf{A}_2)$ who can break the reliability. We show that a forger $\mathbf{B}$ against MAC can be constructed using $\mathbf{A}$.

$\mathbf{B}$ behaves like a client. When $\mathbf{B}$ receives $(\mathcal{D}, \mathcal{W})$ from $\mathbf{A}_1$ during the store phase, it creates $\mathcal{I}$ and $\mathcal{C}$ ordinarily, except that $\mathbf{B}$ does not choose the key for MAC, but rather uses its own MAC oracle to compute $\mathcal{I}$. Here, $\mathbf{B}$ will send queries to its MAC oracle only when constructing $\mathcal{I}$.then $\mathbf{B}$ sends $(\mathcal{I}, \mathcal{C})$ to $\mathbf{A}_2$. We note that $\mathbf{B}$ will send queries to its MAC oracle only when constructing $\mathcal{I}$.

In the search phase, $\mathbf{A}_1$ sends w_i to $\mathbf{B}$ for q times. $\mathbf{B}$ calculates a trapdoor $t(w_i)$ for w_i normally and sends it to $\mathbf{A}_2$. $\mathbf{A}_2$ outputs $(\tilde{\mathcal{C}}(w_i)^*, Proof_i^*)$ and sends it back to $\mathbf{B}$. While this step, $\mathbf{B}$ also runs the Search algorithm and gets $(\mathcal{C}(w_i), Proof_i)$ for its own.

For each i, $\mathbf{A}_2$'s output $(\tilde{\mathcal{C}}(w_i)^*, Proof_i^*)$ is either of the following three types:

Type 1 $(\tilde{\mathcal{C}}(w_i)^*, Proof_i^*) = (\mathcal{C}(w_i), Proof_i)$.
Type 2 $(\tilde{\mathcal{C}}(w_i)^*, Proof_i^*) \neq (\mathcal{C}(w_i), Proof_i)$ and:
  Type 2-1 the Verify algorithm outputs reject.
  Type 2-2 the Verify algorithm outputs accept.

For each output of **Type 1**, $\mathbf{B}$ returns $\mathcal{D}(w_i)$ to $\mathbf{A}_1$.

For each output of **Type 2**, $\mathbf{B}$ has to return reject if it is **Type 2-1**, and a plaintext $\tilde{\mathcal{D}}(w_i)^*$ of $\tilde{\mathcal{C}}(w_i)^*$ if it is **Type 2-2**. However, $\mathbf{B}$ cannot distinguish **Type 2-1** and **Type 2-2**, since $\mathbf{B}$ does not have the key for the MAC itself.

For this problem, $\mathbf{B}$ randomly chooses J from $[1, q]$ at the beginning of the search phase. This J is the prediction by $\mathbf{B}$ of i such that i-th output is the first **Type 2-2** output. Based on this J, $\mathbf{B}$ performs as follows.

For outputs of $\mathbf{A}_2$ before the J-th output in the search phase, $\mathbf{B}$ considers all **Type 2** outputs to be **Type 2-1**, and returns reject to $\mathbf{A}_1$.

If the J-th output of $\mathbf{A}_2$ is not **Type 2**, $\mathbf{B}$ fails to forge a MAC and aborts. Otherwise, $\mathbf{B}$ considers it as **Type 2-2**, and determines its output as below.

Case a: $\mathcal{C}(w_J) \neq \tilde{\mathcal{C}}(w_J)^*$

Suppose that

$$t(w_J) = (addr_0, \ldots, addr_{N-1})$$
$$\mathcal{C}(w_J) = (C_{q_0}, \ldots, C_{q_{m-1}})$$
$$\tilde{\mathcal{C}}(w_J)^* = (C_0^*, \ldots, C_{m^*-1}^*)$$
$$Proof_J^* = (pr_0^*, \ldots, pr_{N-1}^*).$$

Note that $addr_j = \pi(0, w_J, j)$, and that $\mathbf{B}$ knows all of the above values. Because $\tilde{\mathcal{C}}(w_J)^* \neq \mathcal{C}(w_J)$, we need to consider only the following three cases.

Case a-1: $m^* = m$, and there exists C_i^* in $\tilde{\mathcal{C}}(w_J)^*$ but not in $\mathcal{C}(w_J)$

Case a-2: $m^* > m$

Case a-3: $m^* < m$.

In both **Cases a-1** and **a-2**, there exists C_i^* in $\tilde{\mathcal{C}}(w_J)^*$ but not in $\mathcal{C}(w_J)$. Then, $\mathbf{B}$ randomly chooses $pr_j^* \in Proof_J^*$ that is a *tag*, and outputs a forged message-tag pair $((0\|addr_j\|C_i^*), pr_j^*)$. Here we assume that $\mathbf{A}$ wins in Game_{reli} and that $\mathbf{B}$ successfully predicts J, that is,

$$\mathsf{Verify}(K, t(w_J), \tilde{\mathcal{C}}(w_J)^*, Proof_J^*) = \mathsf{accept} \tag{7}$$

holds with a non-negligible probability. We show that $\mathbf{B}$'s output shown above is a valid forgery against MAC with a non-negligible probability. This means that $pr_j^* = \mathsf{MAC}(0\|addr_j\|C_i^*)$, and that $\mathbf{B}$ did not send the query $(0\|addr_j\|C_i^*)$ to its own MAC oracle.

First, from Eqs. (3) and (7), there exists $pr_{j'}^*$ such that $\mathsf{MAC}(0\|addr_{j'}\|C_i^*) = pr_{j'}^*$, in $Proof_J^*$. $j = j'$ holds with at least probability $1/m$. Next, we can see that $\mathbf{B}$ has never queried $(0\|addr_{j''}\|C_i^*)$ for any j'' to its own MAC oracle when computing $\mathcal{I}$, because $C_i^* \notin \mathcal{C}(w_J)$. Therefore, $\mathbf{B}$ has succeeded in forging a valid tag with non-negligible probability.

In **Case a-3**, there exists C_{q_i} in $\mathcal{C}(w_J)$, but not in $\tilde{\mathcal{C}}(w_J)^*$. Then, $pr_{q_i}^*$ consists of a pair of *addrs* and *Ntag*. Let $pr_{q_i}^* = (addr_1^*\|addr_2^*, Ntag^*)$. $\mathbf{B}$ outputs $((1\|addr_1^*\|addr_2^*), Ntag^*)$ as a forgery. Assume that $\mathbf{A}$ wins in Game_{reli}. From Eqs.(4) and (5),

$$addr_1^* < addr_{q_i} < addr_2^*$$
$$\mathsf{MAC}(1\|addr_1^*\|addr_2^*) = Ntag^*$$

hold. Further, $C_{q_i} \in \mathcal{C}(w_J)$ implies that $addr_{q_i}$ appears in the addr field of $\mathcal{I}$. Therefore, $\mathbf{B}$ did not query $(1\|addr_1'\|addr_2')$ to the MAC oracle for any $(addr_1', addr_2')$ such that $addr_1' < addr_{q_i} < addr_2'$. This means that if $\mathbf{A}$ wins in Game_{reli} with non-negligible probability, then $\mathbf{B}$ succeeds with a non-negligible probability. This contradicts the assumption about the security of MAC.

Case b: $\mathcal{C}(w_J) = \tilde{\mathcal{C}}(w_J)^*$

From $(\mathcal{C}(w_J), Proof_J) \neq (\tilde{\mathcal{C}}(w_J)^*, Proof_J^*)$, it is obvious that $Proof_J \neq Proof_J^*$.

Suppose that

$$Proof_J = (pr_1, \ldots, pr_N),$$
$$Proof_J^* = (pr_1^*, \ldots, pr_N^*)$$

Then, there exists an i s.t. $pr_i \neq pr_i^*$. Since pr_i and pr_i^* are either a tag or a pair $(addr_1 \| addr_2, Ntag)$ of $addr$s and $Ntag$, every case will be either of the following four cases.

Case b-1: both pr_i and pr_i^* are $tags$

Case b-2: both pr_i and pr_i^* are pairs $(addr_1 \| addr_2, Ntag)$

Case b-3: pr_i is a tag, and pr_i^* is a pair $(addr_1 \| addr_2, Ntag)$

Case b-4: pr_i is a pair $(addr_1 \| addr_2, Ntag)$, and pr_i^* is a tag.

In **Case b-1**, **B** knows the C_j which satisfies $pr_i = \text{MAC}(0\|addr_i\|C_j)$. So it chooses another $C_{j'}$ from $\mathcal{C}(w_J)$ randomly, and outputs $(0\|addr_i\|C_{j'})$ and pr_i^*.

In **Cases b-2** and **b-3**, **B** outputs $((1\|addr_1^*\|addr_2^*), Ntag^*)$, where $pr_i^* = (addr_1^*\|addr_2^*, Ntag^*)$.

In **Case b-4**, if there exists an $i'(\neq i)$ where **Case b-3** is occurring, then **B** applies exactly the same method as in **Case b-3** to $pr_{i'}^*$ instead of pr_i^*[5].

If **A** succeeds in breaking the reliability, **B** successfully predicts J in probability $1/q$. When **A** wins and **B** successfully predicts J, then **B** successfully forges a MAC with probability at least $1/N$. Therefore, we obtain

$$\Pr(\mathbf{B}\,\text{succeeds}) \geq \Pr(\mathbf{A}\,\text{wins in}\,\texttt{Game}_{reli}) \times \frac{1}{qN}.$$

Note that q and N are polynomials of security parameter λ.

As a result, our scheme satisfies reliability if MAC is unforgeable against chosen message attack.   □

From Theorems 1, 2, and 3, our scheme is UC-secure.

Corollary 1. *If the symmetric encryption scheme* SKE $= (G, E, E^{-1})$ *is IND-CPA secure, and if π is a pseudorandom permutation, and if* MAC *is existentially unforgeable against chosen-message attacks, then the above scheme is UC-secure with leakage L^{new}.*

---

[5] When $Proof_J^*$ is accepted by `Verify`, such i' will always exist because the `Verify` algorithm starts with a step to check whether the number of the $tags$ in $Proof$ is equal to the numbers of encrypted documents in the search result $\mathcal{C}(w_J) = \mathcal{C}(w_J)$, and output `reject` if not.

Table 1. Comparison of the efficiency of proposed scheme with the KO-scheme

		KO-scheme [18]	Proposed scheme
Index Size(bits)		$\mathcal{O}(2^l N \log N)$	$\mathcal{O}(MN \log N)$
Complexity	Enc	$\mathcal{O}(2^l N)$	$\mathcal{O}(MN)$
	Search	$\mathcal{O}(N)$	$\mathcal{O}(N \log MN)$
	Verify	$\mathcal{O}(N)$	$\mathcal{O}(N + m^2)$

N : number of documents
l : (maximum) length of keywords
M : number of keywords $|\mathcal{W}|$ $(M \leq 2^l)$
m : number of documents in the search result

The leaked information under our scheme is slightly more than the leakage from the KO-scheme. In particular, our scheme leaks the number of keywords to the server.

Should the client prefer to avoid leaking the exact number of keywords, dummy keywords can be added to $\mathcal{W}$. Of course, the more dummy keywords that are added, the larger the index grows. This is constitutes a trade-off between security and computational costs.

Nevertheless, the proposed scheme can modify the maximum length of a keyword l by replacing the permutation π with a collision resistant hash function.

3.3 Comparison

Table 1 shows the index size and the computational cost for each algorithm, comparing the KO-scheme with the proposed scheme. In the estimation on the cost of Enc algorithm, we ignored the cost of sorting, which we see it as negligible compared to MAC. We eliminated the rows for algorithms Gen, Trpdr, Dec, inasmuch as they are exactly the same in both schemes.

We can see that our Enc algorithm is much more efficient and that the index size is much smaller than the KO-scheme, when $|\mathcal{W}| \ll 2^l$. However, our Search and Verify algorithms are less efficient than those in the KO-scheme, because our scheme requires an extra step to search the *addr*s from the addr field in Search, and to search the C_is corresponding to *tag*s in Verify.

4   Conclusion

In this paper, we provided generalized definitions for the privacy and UC-security of SSE schemes, and we proposed a vSSE scheme as a modified version of the KO-scheme. Whereas the privacy of the proposed scheme is slightly weaker than the original, the index size is much smaller when the set of keywords is a very sparse subset of l-bit strings for some l.

Importantly, the idea of *Ntag* to prove "non-existence" can be applied widely. For example, the weak point of the (dynamic) vSSE scheme [19] can be overcome by adding an $Ntag$ after sorting the elements in $\mathcal{I}$ based on $\texttt{label}$. Similarly, most SSE schemes might be converted to vSSE schemes by including a tag and an $Ntag$.

References

1. Ballard, L., Kamara, S., Monrose, F.: Achieving efficient conjunctive keyword searches over encrypted data. In: Qing, S., Mao, W., López, J., Wang, G. (eds.) ICICS 2005. LNCS, vol. 3783, pp. 414–426. Springer, Heidelberg (2005)
2. Bellare, M., Boldyreva, A., O'Neill, A.: Deterministic and efficiently searchable encryption. In: Menezes, A. (ed.) CRYPTO 2007. LNCS, vol. 4622, pp. 535–552. Springer, Heidelberg (2007)
3. Boldyreva, A., Chenette, N.: Efficient Fuzzy search on encrypted data. IACR Cryptology ePrint Archive 2014/235
4. Canetti, R.: Universally composable security: "A New Paradigm for Cryptographic," protocols. Revision 1 of ECCC Report TR01-016 (2001)
5. Canetti, R.: Universally composable signatures, certification and authentication. Cryptology ePrint Archive, Report 2003/239 (2003). http://eprint.iacr.org/
6. Canetti, R.: Universally composable security: a new paradigm for cryptographic protocols. Cryptology ePrint Archive, Report 2000/067 (2005). http://eprint.iacr.org/
7. Cao, N., Wang, C., Li, M., Ren, K., Lou, W.: Privacy-preserving multi-keyword ranked search over encrypted cloud data. IEEE Trans. Parallel Distrib. Syst. **25**, 222–233 (2014)
8. Cash, D., Jarecki, S., Jutla, C., Krawczyk, H., Roşu, M.-C., Steiner, M.: Highly-scalable searchable symmetric encryption with support for Boolean queries. In: Canetti, R., Garay, J.A. (eds.) CRYPTO 2013, Part I. LNCS, vol. 8042, pp. 353–373. Springer, Heidelberg (2013)
9. Chang, Y.-C., Mitzenmacher, M.: Privacy preserving keyword searches on remote encrypted data. In: Ioannidis, J., Keromytis, A.D., Yung, M. (eds.) ACNS 2005. LNCS, vol. 3531, pp. 442–455. Springer, Heidelberg (2005)
10. Chai, Q., Gong, G.: Verifiable symmetric searchable encryption for semi-honest-but-curious cloud servers. In: 2012 IEEE International Conference on Communications (ICC), pp. 917–922 (2012)
11. Curtmola, R., Garay, J.A., Kamara, S., Ostrovsky, R.: Searchable symmetric encryption: improved definitions and efficient constructions. In: ACM Conference on Computer and Communications Security, pp. 79–88 (2006). Full version: Cryptology ePrint Archive, Report 2006/210. http://eprint.iacr.org/
12. Goh, E.-J.: Secure indexes. Technical Report 2003/216, IACR ePrint Cryptography Archive (2003)
13. Golle, P., Staddon, J., Waters, B.: Secure conjunctive keyword search over encrypted data. In: Jakobsson, M., Yung, M., Zhou, J. (eds.) ACNS 2004. LNCS, vol. 3089, pp. 31–45. Springer, Heidelberg (2004)
14. Kamara, S., Lauter, K.: Cryptographic cloud storage. In: Sion, R., Curtmola, R., Dietrich, S., Kiayias, A., Miret, J.M., Sako, K., Sebé, F. (eds.) RLCPS, WECSR, and WLC 2010. LNCS, vol. 6054, pp. 136–149. Springer, Heidelberg (2010)
15. Kamara, S., Papamanthou, C., Roeder, T.: CS2: a searchable cryptographic cloud storage system. MSR Technical Report no. MSR-TR-2011-58. Microsoft (2011)

16. Kamara, S., Roeder, T.: Dynamic searchable symmetric encryption. In: Proceedings of the 2012 ACM Conference on Computer and Communications Security, pp. 965–976 (2012)

17. Kamara, S., Papamanthou, C.: Parallel and dynamic searchable symmetric encryption. In: Sadeghi, A.-R. (ed.) FC 2013. LNCS, vol. 7859, pp. 258–274. Springer, Heidelberg (2013)

18. Kurosawa, K., Ohtaki, Y.: UC-secure searchable symmetric encryption. In: Keromytis, A.D. (ed.) FC 2012. LNCS, vol. 7397, pp. 285–298. Springer, Heidelberg (2012)

19. Kurosawa, K., Ohtaki, Y.: How to update documents *Verifiably* in searchable symmetric encryption. In: Abdalla, M., Nita-Rotaru, C., Dahab, R. (eds.) CANS 2013. LNCS, vol. 8257, pp. 309–328. Springer, Heidelberg (2013)

20. Kurosawa, K., Ohtaki, Y.: How to construct UC-secure searchable symmetric encryption scheme. Cryptology ePrint Archive, Report 2015/251 (2015). http://eprint.iacr.org/2015/251

21. Song, D., Wagner, D., Perrig, A.: Practical techniques for searches on encrypted data. In: IEEE Symposium on Security and Privacy, pp. 44–55 (2000)

22. Wang, C., Ren, K., Yu, S., Urs, K.M.R.: Achieving usable and privacy-assured similarity search over outsourced cloud data. In: Proceedings of INFOCOM 2012, pp. 451–459 (2012)

Fully Leakage-Resilient Non-malleable Identification Schemes in the Bounded-Retrieval Model

Tingting Zhang[1,2,3](✉) and Hongda Li[1,2,3]

[1] State Key Laboratory of Information Security, Institute of Information Engineering of Chinese Academy of Sciences, Beijing, China
[2] Data Assurance and Communication Security Research Center of Chinese Academy of Sciences, Beijing, China
[3] University of Chinese Academy of Sciences, Beijing, China
{zhangtingting,lihongda}@iie.ac.cn

Abstract. Alwen, Dodis and Wichs first formulated the security notions of identification (ID) schemes resilient to key-leakage attacks, which is called leakage-resilient ID schemes. In fact, the notions they considered are the so-called active security where the adversary is only allowed to interact with the prover before the impersonation attempt. However, recently, there has been a huge emphasis on stronger attacks, such as man-in-the-middle (MIM) attacks. So can we extend the results about leakage-resilient ID schemes to man-in-the-middle security? Besides, we consider the setting where the adversary is allowed to perform leakage attacks on the *entire state* of the honest prover during the lifetime of the system, which is called *full leakage attacks*. Clearly, this type of leakage attacks is stronger and more meaningful than key-leakage attacks.

In conclusion, we study the design of ID schemes resilient to MIM attacks and *fully leakage attacks* at the same time, which means that while attempting to impersonate a prover, the adversary can interact with an honest prover and obtain arbitrary bounded leakage on the entire state of the honest prover during the lifetime of the system. Informal speaking, an ID scheme secure against this type of attacks is said to be *fully leakage-resilient non-malleable*.

To obtain fully leakage-resilient non-malleable ID schemes, we propose two variants of the so-called Knowledge-of-Exponent Assumption (KEA) over bilinear groups, called tag based Knowledge-of-Exponent Assumption (TagKEA) and Selective-tag based Knowledge-of-Exponent Assumption (Selective-TagKEA). To argue for believing in these two assumptions, we demonstrate that KEA implies TagKEA and is equivalent to Selective-TagKEA.

Keywords: Identification schemes · Man-in-the-middle attacks · Fully leakage-resilient · Leakage attacks · Knowledge-of-exponent assumption · Tag

This research is supported by the National Natural Science Foundation of China (Grant No. 61003276) and the Strategy Pilot Project of Chinese Academy of Sciences (Grant No. Y2W0012203).

K. Tanaka and Y. Suga (Eds.): IWSEC 2015, LNCS 9241, pp. 153–172, 2015.
DOI: 10.1007/978-3-319-22425-1_10

1   Introduction

An identification (ID) scheme enables a prover to prove its identity to a verifier. Generally, the prover holds a secret key and the verifier holds the corresponding public key. They interact for some rounds doing some necessary computations until the verifier can decide to accept or reject.

As regards security notions for ID schemes, one wants to prevent adversaries from impersonating a prover. Adversaries may perform various attacks such as active attacks [8], concurrent attacks [8], man-in-the-middle (MIM) attacks [19] and reset attacks [7]. Among these attacks, we consider the relatively stronger attack, namely MIM attack, in which an adversary is allowed to interact with an honest prover while attempting to impersonate a prover.

We note that the traditional security notions for ID schemes assume that an adversary is given only black-box access to the prover's algorithms, which means that the prover is able to maintain its internal state, consisting of the secret key and the random coins, perfectly hidden from the adversary. Unfortunately, when considering attacks in the real world, this assumption may be unrealistic. Because the adversary is able to learn some partial information about the internal state of the prover by performing *side channel attacks*. Motivated by such a scenario, Alwen et al. [4] initiated a research on ID schemes resilient to key-leakage attacks. Note that the security notions suggested by [4] are reminiscent of so-called active security in which the adversary is only allowed to interact with the prover before the impersonation attempt. However, there has been a huge emphasis on stronger attacks, such as MIM attacks. Besides, [4] only considered key-leakage attacks where the adversary only can perform leakage attacks on the secret key. However, in reality, the adversary may perform leakage attacks on the *entire state* of the prover (consisting of its secret key and random coins used during the protocol execution), which are called *fully leakage attacks*. In conclusion, compared with previous works, we study the design of ID schemes resilient to MIM attacks and *fully leakage attacks* at the same time.

1.1   Our Contribution

In this paper, we study the design of ID schemes secure against *fully leakage attacks* and MIM attacks at the same time, which means that while attempting to impersonate a prover, the adversary can interact with an honest prover and obtain arbitrary bounded leakage on the entire state of the honest prover during the lifetime of the system. Informal speaking, an ID scheme secure against this type of attacks is said to be *fully leakage-resilient non-malleable* (FLRNM).

We first give the security notion. The attack game consists of two stages, called key stage and impersonation stage. In the key stage, the public parameters pub, a pubic key pk and a secret key sk are generated. In the impersonation stage, the MIM adversary $\mathcal{M}$, on input (pub, pk), interacts with an honest prover $\mathcal{P}(\text{pk}, \text{sk})$ by playing the role of cheating verifier, and simultaneously impersonates a prover to an honest verifier $\mathcal{V}(\text{pk})$. Besides, $\mathcal{M}$ is also allowed to obtain arbitrary leakage on the *entire state* of $\mathcal{P}(\text{pk}, \text{sk})$, as long as the total number of bits leaked during the lifetime of the system is bounded by some leakage para-meter ℓ. Let π denote a transcript of the interaction between $\mathcal{P}(\mathsf{pk}, \mathsf{sk})$ and $\mathcal{M}$, and $\tilde{\pi}$ denote a transcript of the interaction between $\mathcal{M}$ and $\mathcal{V}(\mathsf{pk})$. Informally speaking, an ID scheme is said to be FLRNM if the probability that $\pi \neq \tilde{\pi}$ and $\mathcal{M}$ makes $\mathcal{V}(\mathsf{pk})$ accept in the above attack game is negligible.

Alwen et al. in [4] presented a leakage-resilient ID scheme, which also satisfies our requirement for fully leakage-resilient property. However, their construction is not secure against MIM attacks. Recall the scheme in [4]: the public parameters $\mathsf{pub} = (p, \mathbb{G}, g, g_1, \ldots, g_m)$, the secret key $\mathsf{sk} = (s_1, \ldots, s_m)$ and the pubic key $\mathsf{pk} = \prod_{j=1}^{m} \{g_j\}^{s_j}$, where $\mathbb{G}$ is a cyclic group of order p and g is its generator. Let $(R, c, \mathbf{z})$ denote the three message exchanged between $\mathcal{P}(\mathsf{pk}, \mathsf{sk})$ and $\mathcal{M}$. To impersonate a prover, $\mathcal{M}$ acts as following:

- After receiving R from $\mathcal{P}(\mathsf{pk}, \mathsf{sk})$, $\mathcal{M}$ generates $\tilde{R} = R \cdot \prod_{j=1}^{m} \{g_j\}^{\tilde{r}_j}$ where $\tilde{r}_j \leftarrow \mathbb{Z}_p$ for $j = 1, \ldots, m$. Send $\tilde{R}$ to $\mathcal{V}(\mathsf{pk})$.
- After receiving $\tilde{c}$ from $\mathcal{V}(\mathsf{pk})$, $\mathcal{M}$ sets $c = \tilde{c}$ and sends it to $\mathcal{P}(\mathsf{pk}, \mathsf{sk})$.
- After receiving $\mathbf{z} = (z_1, \ldots, z_m)$ from $\mathcal{P}(\mathsf{pk}, \mathsf{sk})$, $\mathcal{M}$ sets $\tilde{z}_j = z_j + \tilde{r}_j$ for $j = 1, \ldots, m$. Send $\tilde{\mathbf{z}} = (\tilde{z}_1, \ldots, \tilde{z}_m)$ to $\mathcal{V}(\mathsf{pk})$.

Clearly, if $\mathcal{P}(\mathsf{pk}, \mathsf{sk})$ succeeds, $\mathcal{M}$ can also make $\mathcal{V}(\mathsf{pk})$ accept with $\pi \neq \tilde{\pi}$.

Now, we study how to extend this leakage-resilient ID scheme to MIM secu-rity. Note that to generate a valid answer for a challenge $\tilde{c}$, $\mathcal{M}$ must generate c depending on $\tilde{c}$. So if we can break this dependence, we may achieve non-malleability. Let $(n, \bar{\mathbb{G}}, \mathbb{H}, e, \bar{g})$ be a bilinear group (defined in Sect. 2.1). From [1], we find that given $X, Y \in \bar{\mathbb{G}}$ and two different tags $t, \tilde{t} \in \mathbb{Z}_n$, it is difficult for any adversary to compute a pair (A, D) so $D = (X^t Y)^a$ without knowing a so $A = \bar{g}^a$, even though the adversary has got a pair $(\tilde{A} = \bar{g}^{\tilde{a}}, (X^{\tilde{t}} Y)^{\tilde{a}})$. Fol-lowing the idea of Knowledge-of-Exponent Assumption (KEA) [11,18], we can propose a similar assumption, called tag based Knowledge-of-Exponent Assump-tion (TagKEA), where we assume that there exists an extractor that can extract the value a so $A = \bar{g}^a$ and $D = (X^t Y)^a$. With this assumption, the Discrete-Log Assumption and a strong one-time signature scheme, we can achieve the non-malleability by generating the challenge c as following:

- Firstly, we also assume that there exists an injective and efficiently computable map $f : \bar{\mathbb{G}} \rightarrow \mathbb{Z}_p$, which is used to transform an element in $\bar{\mathbb{G}}$ to a challenge in $\mathbb{Z}_p$.
- Every interaction is labeled with a tag t. In the formal construction, t is set to be the verification key of the one-time signature scheme.
- The challenge c is jointly generated by the prover and the verifier: first, the verifier generates a pair $(A = \bar{g}^a, D = (X^t Y)^a)$ where $a \in \mathbb{Z}_n$ and $X, Y \in \bar{\mathbb{G}}$, then the prover randomly generates $b \in \mathbb{Z}_n$, and finally the challenge c is equal to $f(A^b)$.

Details are presented in Sect. 5. Note that if $t \neq \tilde{t}$, a MIM adversary cannot control the challenge $\tilde{c}$ of the right interaction. Since to make the challenge be equal to a value $\tilde{c}^* = f(\tilde{\alpha}^*)$, $\mathcal{M}$ must compute $\tilde{b}^*$ so that $\tilde{\alpha}^* = \tilde{A}^{\tilde{b}^*}$, which is difficult under the Discrete-Log Assumption. Therefore, we break the dependence of c and $\tilde{c}$.

The use of one-time signature scheme can be removed by using a tag framework, which we call tag based fully leakage-resilient non-malleable ID schemes. To achieve this, we have to modify the TagKEA assumption to the selective-tag based Knowledge-of-Exponent Assumption (Selective-TagKEA).

Besides, to argue for believing in these two assumptions, we demonstrate that KEA implies TagKEA and is equivalent to Selective-TagKEA.

1.2 Related Work

Leakage-resilient identification schemes. In this work, we focus on bounded-retrieval model (BRM) [4,5] where we assume that there is an external natural bound ℓ on the length of the maximum tolerated leakage. [4] studied ID schemes resilient to key leakage attacks in the BRM. They considered two notions of security: a weaker notion called pre-impersonation security and a stronger notion called anytime security. Following these two notions of security, a series of leakage-resilient ID schemes were constructed [5,12,13,24]. However, we stress that all these works only considered active attacks and key-leakage attacks.

Non-malleable identification schemes. Bellare, et al. [7] gave a definition of security secure against MIM attacks. But their definition of security is much stronger because they allowed the adversary to perform reset attacks. Motivated by the work of [7], Katz [19,20] considered a weaker definition of security where they allowed the adversary $\mathcal{M}$ to interact with only a single honest prover while attempting to impersonate a prover, and they did not consider reset attacks. Gennaro [14] considered concurrent man-in-the-middle (CMIM) attacks where the adversary is allowed to interact with prover clones in arbitrarily interleaved order of messages while attempting to impersonate a prover, and they constructed an ID scheme of (fully) concurrently non-malleable proof of knowledge by employing a multi-trapdoor commitment. Anada and Arita [1–3] studied the construction of efficient ID schemes secure against CMIM attacks.

Knowledge-of-Exponent Assumption. KEA was firstly introduced and used by Damgård in [11]. And later, Hada and Tanaka [18] extended the KEA to KEA2. But after that, Bellare and Palacio [9] showed that KEA2 was false, and presented a new extended version called KEA3 for saving Hada and Tanaka's results. In past few years, there have been a number of interesting research papers to apply these knowledge of exponent assumptions to interactive proofs [6,16,17].

2   Preliminaries

We first give some standard notations used in this paper. The set of natural numbers is represented by $\mathbb{N}$. In the following section, we will use $\lambda \in \mathbb{N}$ to denote the security parameter, and we always implicitly require the size of the input of all the algorithms in this paper to be bounded by some polynomial in λ. For non-uniform probabilistic poly-time (PPT) algorithms, we mean that the algorithms, besides taking as input 1^λ, also get some poly-size auxiliary input aux_λ. For brevity, in some cases, we may leave the input 1^λ (and aux_λ) implicit. We write $y \leftarrow A(x)$ to denote that algorithm A is executed on input x (and 1^λ, in the non-uniform case, aux_λ) and the output is y. Similarly, for any finite set $\mathcal{S}$, the notation $y \leftarrow \mathcal{S}$ means that y is sampled uniformly from $\mathcal{S}$. A probability of an event E is denoted by $\Pr[E]$.

Besides, we assume familiarity with some notions such as ID schemes with completeness [4], leakage oracle [4], Discrete-Log Assumption (DLA) [6], Knowledge-of-Exponent Assumption (KEA) [6,9], strong one-time signature [14].

2.1 Bilinear Groups and Hardness Assumptions

Let $\mathcal{G}$ be a group sampling algorithm which, on input 1^λ, outputs a tuple $G = (p, \mathbb{G}, g)$ where $\mathbb{G}$ is a cyclic group of prime order p with g is its generator. Besides, we also need a bilinear group with a map f.

Definition 1 (Bilinear Groups with Special property). *Let $\mathcal{BGG}$ be a bilinear group generator that takes as input the security parameter 1^λ and a prime p, and then outputs a tuple $BG = (n, \bar{\mathbb{G}}, \mathbb{H}, \bar{g}, e, f)$, where $\bar{\mathbb{G}}$ is a cyclic group of prime order n, $\bar{g}$ is a generator of $\bar{\mathbb{G}}$ and $e : \bar{\mathbb{G}} \times \bar{\mathbb{G}} \to \mathbb{H}$ is a non-degenerate bilinear map, i.e., $\forall X, Y \in \bar{\mathbb{G}}, \forall a, b \in \mathbb{Z}_n : e(X^a, Y^b) = e(X, Y)^{ab}$, and $e(\bar{g}, \bar{g})$ generates $\mathbb{H}$. Besides, we require that f is an injective and efficiently computable map from $\bar{\mathbb{G}}$ to $\mathbb{Z}_p$.*

We emphasize that the map f is easy to be constructed in some cases. For example, $\bar{\mathbb{G}}$ is a subgroup of the group of points on an elliptic curve E over $\mathbb{Z}_q$ (q is a prime). Then for every $(x, y) \in \bar{\mathbb{G}}$, f can just return the concatenation of the binary representations of x and y. If $p \geq 2^{2 \log q}$, clearly, f is an injective and efficiently computable map from $\bar{\mathbb{G}}$ to $\mathbb{Z}_p$. Therefore, the primes p should be large enough so that the prime n can be also large enough to make our assumptions hold. In the following section, we require that all these parameters are set appropriately, and we omit this requirement.

Now, we present our new assumptions, namely tag based Knowledge-of-Exponent Assumption (TagKEA) and Selective-tag based Knowledge-of-Exponent Assumption (Selective-TagKEA). Similar to KEA, these two assumptions informally state that for any two different tags $t, \tilde{t}$, given $(BG, X, Y, \bar{g}^{\tilde{a}}, (X^{\tilde{t}}Y)^{\tilde{a}})$, it is infeasible to create (A, D) so $e(\bar{g}, D) = e(X^t Y, A)$ without knowing a so $A = \bar{g}^a$.

Assumption 1 (TagKEA). *For every non-uniform PPT algorithm $\mathcal{A}$, consider the following game:*

- *Choose $BG \leftarrow \mathcal{BGG}$, $X, Y \leftarrow \bar{\mathbb{G}}$, set $\tilde{A} = \perp$, $\tilde{D} = \perp$ and send (BG, X, Y) to $\mathcal{A}$;*

- In case that $\mathcal{A}$ requires a DDH tuple, the challenger selects a tag $\tilde{t}$, $\tilde{a} \leftarrow \mathbb{Z}_n$, sets $\tilde{A} = \bar{g}^{\tilde{a}}$, $\tilde{D} = (X^{\tilde{t}}Y)^{\tilde{a}}$, and gives $(\tilde{t}, \tilde{A}, \tilde{D})$ to $\mathcal{A}$. In this paper, we restrict that $\tilde{A}$ only can obtain one such DDH tuple.
- Finally, the adversary $\mathcal{A}$, on input $(\mathsf{BG}, X, Y, \tilde{t}, \tilde{A}, \tilde{D})$, outputs a tuple (t, A, D).

We say that the TagKEA assumption holds if for every such non-uniform PPT algorithm $\mathcal{A}$, there exists a non-uniform PPT algorithm $\mathcal{X}_{\mathcal{A}}$, the extractor, which, on input $(\mathsf{BG}, X, Y, \tilde{t}, \tilde{A}, \tilde{D})$, uses the same random tape with $\mathcal{A}$ and outputs $a \in \mathbb{Z}_n$, such that

$$\Pr[t \neq \tilde{t} \wedge e(\bar{g}, D) = e(X^t Y, A) \wedge A \neq \bar{g}^a] \leq negl(\lambda).$$

In some case, we consider the selective-tag attack [21] where the tags $t, \tilde{t}$ are priori given to the challenger. Thus, we give the selective-tag version of the TagKEA assumption, called Selective-TagKEA.

Assumption 2 (Selective-TagKEA). For every non-uniform PPT algorithm $\mathcal{A}$, consider the following game:

- Choose $\mathsf{BG} \leftarrow \mathcal{BGG}$, and give BG to $\mathcal{A}$.
- $t, \tilde{t} \leftarrow \mathcal{A}(\mathsf{BG})$;
- Select $X, Y \leftarrow \bar{\mathbb{G}}$, set $\tilde{A} = \perp$, $\tilde{D} = \perp$ and send (X, Y) to $\mathcal{A}$;
- In case that $\mathcal{A}$ requires a DDH tuple, the challenger selects $\tilde{a} \leftarrow \mathbb{Z}_n$, sets $\tilde{A} = \bar{g}^{\tilde{a}}$, $\tilde{D} = (X^{\tilde{t}}Y)^{\tilde{a}}$, and gives $(\tilde{A}, \tilde{D})$ to $\mathcal{A}$. Besides, we restrict that $\tilde{A}$ only can obtain one such DDH tuple.
- Finally, the adversary $\mathcal{A}$, on input $(\mathsf{BG}, X, Y, t, \tilde{t}, \tilde{A}, \tilde{D})$, outputs a pair (A, D).

We say that the Selective-TagKEA assumption holds if for every such non-uniform PPT algorithm $\mathcal{A}$, there exists a non-uniform PPT algorithm $\mathcal{X}_{\mathcal{A}}$, the extractor, which, on input $(\mathsf{BG}, X, Y, t, \tilde{t}, \tilde{A}, \tilde{D})$, uses the same random tape with $\mathcal{A}$ and outputs $a \in \mathbb{Z}_n$, such that

$$\Pr[t \neq \tilde{t} \wedge e(\bar{g}, D) = e(X^t Y, A) \wedge A \neq \bar{g}^a] \leq negl(\lambda).$$

To argue for believing in TagKEA and Selective−TagKEA, we show the following two propositions.

Proposition 1. KEA *implies* TagKEA.

Proof. Let $\mathcal{A}$ be an adversary (non-uniform PPT algorithm) for TagKEA. We now show that if KEA holds, there exists a non-uniform PPT algorithm $\mathcal{X}_{\mathcal{A}}$ such that $\mathcal{X}_{\mathcal{A}}$ is a TagKEA extractor for $\mathcal{A}$.

Firstly, We construct an adversary $\mathcal{A}'$ for KEA from $\mathcal{A}$. On input (BG, h), $\mathcal{A}'$ acts as following:

- Select $\tilde{t}, r \leftarrow \mathbb{Z}_n$, and set $X = h$, $Y = X^{-\tilde{t}}\bar{g}^r$, $\tilde{A} = \perp$, $\tilde{D} = \perp$. Give (BG, X, Y) to $\mathcal{A}$.

– In case that $\mathcal{A}$ requires a DDH tuple, select $\tilde{A} \leftarrow \bar{\mathbb{G}}$, compute $\tilde{D} = \tilde{A}^r$, and give $(\tilde{t}, \tilde{A}, \tilde{D})$ to $\mathcal{A}$.
– If the adversary $\mathcal{A}$ outputs (t, A, D) such that $e(A, X^t Y) = e(\bar{g}, D)$, $\mathcal{A}'$ computes $C = (D/A^r)^{1/(t-\tilde{t})}$. Else, set $(A, C) = (\bot, \bot)$. Return (A, C).

It is easy to show that $\mathcal{A}'$ simulates the TagKEA challenger perfectly. Since $Y = X^{-\tilde{t}}\bar{g}^r$ is still a random group element in group $\bar{\mathbb{G}}$, and $\tilde{D}$ is equal to $(X^{\tilde{t}}Y)^{\tilde{a}}$ since $(X^{\tilde{t}}Y)^{\tilde{a}} = (X^{\tilde{t}}X^{-\tilde{t}}\bar{g}^r)^{\tilde{a}} = \tilde{A}^r = \tilde{D}$. Besides, if $D = (X^t Y)^a$, then $C = h^a$. This is because $D/A^r = (X^t Y/\bar{g}^r)^a = X^{(t-\tilde{t})a} = h^{(t-\tilde{t})a}$.

From the KEA assumption, we know that there exists an extractor $\mathcal{X}_{\mathcal{A}'}$ for the KEA adversary $\mathcal{A}'$. We now define an extractor $\mathcal{X}_{\mathcal{A}}$ for the TagKEA adversary $\mathcal{A}$ from the KEA extractor $\mathcal{X}_{\mathcal{A}'}$. On input $(\mathsf{BG}, X, Y, \tilde{t}, \tilde{A}, \tilde{D})$, $\mathcal{X}_{\mathcal{A}}$ does as following:

– Set $h = X$, and run $\mathcal{X}_{\mathcal{A}'}$ on the input (BG, h) and get the output a. Then, $\mathcal{X}_{\mathcal{A}}$ returns a.

Clearly, if $\mathcal{X}_{\mathcal{A}'}$ is a KEA extractor for $\mathcal{A}'$, there must have $A = \bar{g}^a$, $C = h^a$ where a is the output of $\mathcal{X}_{\mathcal{A}'}$. This implies that $D = C^{(t-\tilde{t})} A^r = h^{a(t-\tilde{t})} A^r = (X^{(t-\tilde{t})}\bar{g}^r)^a = (X^t Y)^a$. Therefore, $\mathcal{X}_{\mathcal{A}}$ is also successful. This completes the proof of this Proposition. $\qquad \square$

Proposition 2. KEA *is equivalent to* Selective-TagKEA.

Proof. Following the above proof, it is easy to prove that KEA implies Selective-TagKEA. Thus, we only prove that if Selective-TagKEA holds, KEA also holds.

Let $\mathcal{A}$ be an adversary for KEA. Similarly, we first use $\mathcal{A}$ to construct an adversary $\mathcal{A}'$ for Selective-TagKEA. $\mathcal{A}'$ acts as following:

– After receiving BG from an external challenger, $\mathcal{A}'$ generates $t, \tilde{t} \leftarrow \mathbb{Z}_n$.
– After receiving (X, Y) from the challenger, $\mathcal{A}'$ sets $h = X^t Y$ and runs the KEA adversary $\mathcal{A}$ on input (BG, h). If $\mathcal{A}$ outputs a pair (A, C), $\mathcal{A}'$ sets $D = C$ and returns (A, D).

Clearly, $\mathcal{A}'$ simulates $\mathcal{A}$ perfectly. From the Selective-TagKEA assumption, we know that there exists an extractor $\mathcal{X}_{\mathcal{A}'}$ for $\mathcal{A}'$. We now show how to construct a KEA extractor $\mathcal{X}_{\mathcal{A}}$ for the adversary $\mathcal{A}$ from $\mathcal{X}_{\mathcal{A}'}$. The KEA extractor $\mathcal{X}_{\mathcal{A}}$, on input (BG, h), do as following:

– Select $t, \tilde{t} \leftarrow \mathbb{Z}_n$, $X \leftarrow \bar{\mathbb{G}}$, and set $Y = X^{-t}h$, $(\tilde{A}, \tilde{D}) = (\bot, \bot)$. Run $\mathcal{X}_{\mathcal{A}'}$ on input $(\mathsf{BG}, X, Y, t, \tilde{t}, \tilde{A}, \tilde{D})$. If $\mathcal{X}_{\mathcal{A}'}$ outputs a, $\mathcal{X}_{\mathcal{A}}$ returns a. Else, return $\bot$.

Note that it is reasonable to set $(\tilde{A}, \tilde{D}) = (\bot, \bot)$ since in our construction of $\mathcal{A}'$, it does not require a DDH tuple from the external challenger.

If the Selective-TagKEA assumption holds, $\mathcal{X}_{\mathcal{A}'}$ can successfully output a such that $A = \bar{g}^a$ and $D = (X^t Y)^a$. This implies that $C = D = (X^t Y)^a = (X^t X^{-t}h)^a = h^a$. Therefore, the KEA extractor $\mathcal{X}_{\mathcal{A}}$ is successful as well. $\qquad \square$

3   Definition of Fully Leakage Resilient Non-malleable Identification Schemes

In this section we present the definition of fully leakage-resilient non-malleable identification schemes.

Let $\Pi = (\mathsf{ParamGen}, \mathsf{KeyGen}, \mathcal{P}, \mathcal{V})$ be an ID scheme with completeness [4]. As discussed above, we consider man-in-the-middle attacks where an adversary $\mathcal{M}$, on input a public key pk, is simultaneously participating in two interactions. These two interactions are respectively called the *left* and *right* interaction. In the left interaction, $\mathcal{M}$ interacts with the honest prover $\mathcal{P}(\mathsf{pk}, \mathsf{sk})$ by playing the role of cheating verifier. In the right interaction, $\mathcal{M}$ try to impersonate the prover to the honest verifier $\mathcal{V}(\mathsf{pk})$. And $\mathcal{M}$ has the ability to control over the scheduling of the messages sent by $\mathcal{P}(\mathsf{pk}, \mathsf{sk})$ and $\mathcal{V}(\mathsf{pk})$. We say that an ID scheme is non-malleable if it is secure against this type of MIM attacks.

Besides, we also consider the *fully leakage attacks* where the adversary is allowed to learn leakage on the *entire state* of the prover. We now recall some descriptions about leakage attacks from [4,15]. Let ℓ be a leakage parameter used to bound the total number of bits leaked during the lifetime of the system. We assume that the random coins used by a party in any particular round are determined only at the beginning of that round. Let a variable state denote the "current state" of the prover at any point during the protocol execution. state is initialized to the secret key sk of the prover. At the completion of each round i, $\mathcal{P}$ updates $\mathsf{state} := \mathsf{state} \parallel r_i$, where r_i denotes the random coins used for that round. A leakage query on prover's state consists of an efficiently computable function $F(\cdot)$, to which the prover may respond with $F(\mathsf{state})$. That is, after receiving a leakage function $F_i : \{0, 1\}^* \to \{0, 1\}^{\alpha_i}$, $\mathcal{P}$ checks if the sum of α_i, over all queries received so far, exceeds the leakage parameter ℓ. If this is the case, $\mathcal{P}$ ignores the query. Else, $\mathcal{P}$ computes the function $F_i(\mathsf{state})$ for at most $poly(\lambda)$ steps, and if the computation completes, responds with the output. Otherwise, $\mathcal{P}$ responds with the dummy value 1^{α_i}. Besides, the verifier is allowed to make any arbitrary polynomial number of leakage queries on $\mathcal{P}$'s state.

We now present the notion of fully leakage-resilient non-malleable (FLRNM) ID schemes, which is modeled by the following attack game $\mathsf{LRNMGame}_\ell^\lambda(\mathcal{M})$.

$\mathsf{LRNMGame}_\ell^\lambda(\mathcal{M})$

1. **Key Stage:** Select $\mathsf{pub} \leftarrow \mathsf{ParamGen}(1^\lambda)$, $(\mathsf{pk}, \mathsf{sk}) \leftarrow \mathsf{KeyGen}()$ and give $(\mathsf{pub}, \mathsf{pk})$ to the adversary $\mathcal{M}$.
2. **Impersonation Stage:** The adversary $\mathcal{M}$ is simultaneously interacting with both an honest prover $\mathcal{P}(\mathsf{pk}, \mathsf{sk})$ and an honest verifier $\mathcal{V}(\mathsf{pk})$, and is also allowed to make leakage queries on $\mathcal{P}$'s state state with ℓ as the leakage parameter.

Let π and $\tilde{\pi}$ respectively denote the transcripts of the left and right interactions. The *advantage* of an adversary $\mathcal{M}$ in the game $\mathsf{LRNMGame}_\ell^\lambda(\mathcal{M})$ is the probability that the verifier $\mathcal{V}$ accepts in the impersonation stage and $\pi \neq \tilde{\pi}$.

Definition 2 (FLRNM ID Schemes). *An ID scheme* $\Pi =$ (ParamGen, KeyGen, $\mathcal{P}, \mathcal{V}$) *with completeness, parameterized by security parameter* λ, *is said to be* ℓ-fully-leakage-resilient non-malleable, *if the advantage of any PPT adversary* $\mathcal{M}$ *in the attack game* LRNMGame$_\ell^\lambda(\mathcal{M})$ *is negligible in* λ.

3.1 Tag-Based Fully Leakage-Resilient Non-malleable Identification Schemes

We now consider a "tag-based" variant of non-malleability [23] where we require that every interaction is labeled by a *tag t*. Denote by Γ the set of all valid tags, which is associated to the scheme. In this paper, we consider the selective-tag attack [21], which means that the tag t is selected before (pk, sk) is generated. Let t denote the tag of the left interaction, and $\tilde{t}$ denote the tag of the right interaction. We define the attack game as following:

TagLRNMGame$_\ell^\lambda(\mathcal{M})$

1. **Public Parameters Stage:** Let pub $\leftarrow$ ParamGen(1^λ). Give pub to the adversary $\mathcal{M}$. (This stage may be omitted if the public parameters has been given to all users.)
2. **Tag Stage:** $t, \tilde{t} \leftarrow \mathcal{M}(1^\lambda)$ where $t, \tilde{t} \in \Gamma$.
3. **Key Stage:** Select (pk, sk) $\leftarrow$ KeyGen() and give pk to the adversary $\mathcal{M}$.
4. **Impersonation Stage:** The adversary $\mathcal{M}$ is simultaneously interacting with both an honest prover $\mathcal{P}(t, \text{pk}, \text{sk})$ and an honest verifier $\mathcal{V}(\tilde{t}, \text{pk})$, and is also allowed to make leakage queries on $\mathcal{P}$'s state state with ℓ as the leakage parameter.

The *advantage* of an adversary $\mathcal{M}$ in the game TagLRNMGame$_\ell^\lambda(\mathcal{M})$ is the probability that the verifier $\mathcal{V}$ accepts in the impersonation stage and $t \neq \tilde{t}$.

Definition 3 (Tag-Based FLRNM ID Schemes). *An ID scheme* $\Pi =$ (ParamGen, KeyGen, $\mathcal{P}, \mathcal{V}$) *with completeness, parameterized by security parameter* λ, *is said to be* ℓ-fully-leakage-resilient non-malleable with respect to tag set Γ, *if the advantage of any PPT adversary* $\mathcal{M}$ *in the attack game* TagLRNMGame$_\ell^\lambda(\mathcal{M})$ *is negligible in* λ.

4 A Tag-Based Fully Leakage-Resilient Non-malleable Identification Scheme

We now proceed to give our construction of a tag-based fully leakage-resilient non-malleable ID scheme as per Definition 3. Roughly speaking, we will reuse the basic components of the scheme in [4] as building-blocks, which can make us achieve the leakage-resilient property. Besides, to achieve the non-malleability, we challenge the method of generating the challenge c where we let c be jointly generated by the prover and the verifier, i.e., the verifier first generates a pair $(A = \bar{g}^a, D = (X^t Y)^a)$, then the prover randomly generates $b \in \mathbb{Z}_n$, and the final challenge c is set to $f(A^b)$. The construction is illustrated in Fig. 1.

Public Parameters Genration:

- $\mathsf{ParamGen}(1^\lambda)$: $\mathsf{G} = (p, \mathbb{G}, g) \leftarrow \mathcal{G}(1^\lambda)$, $\mathsf{BG} = (n, \bar{\mathbb{G}}, \bar{g}, e, f) \leftarrow \mathcal{BGG}(1^\lambda, p)$, $g_1, ..., g_m \leftarrow \mathbb{G}$. Set $\mathsf{pub} = (\mathsf{G}, \mathsf{BG}, g_1, ..., g_m)$.

Tag-Receiving:

- $\mathcal{P}$ and $\mathcal{V}$ receive a tag $t \in \mathbb{Z}_n$ by the first round.

Key Genration:

- $\mathsf{KeyGen}()$: $X, Y \leftarrow \bar{\mathbb{G}}$, $\mathsf{sk} = (s_1, ..., s_m) \leftarrow (\mathbb{Z}_p)^m$, and set $\mathsf{h} = \prod_{j=1}^m \{g_j\}^{s_j}$, $\mathsf{pk} = (X, Y, \mathsf{h})$. Output $(\mathsf{pk}, \mathsf{sk})$.

Interaction: The machines $\mathcal{P}$ and $\mathcal{V}$ interact as following:

- $\mathcal{P}$: Choose $\boldsymbol{r} = (r_1, ..., r_m) \leftarrow (\mathbb{Z}_p)^m$, compute $R = \prod_{j=1}^m \{g_j\}^{r_j}$, and send R to $\mathcal{V}$.
- $\mathcal{V}$: Choose $a \leftarrow \mathbb{Z}_n$, and compute $A = \bar{g}^a$, $D = (X^t Y)^a$. Send (A, D) to $\mathcal{P}$.
- $\mathcal{P}$: If $e(A, X^t Y) \neq e(\bar{g}, D)$ then abort. Else,
    - Choose $b \leftarrow \mathbb{Z}_n$, set $c = f(A^b)$,
    - Compute $z_j = r_j + cs_j$ for $j = 1, ..., m$, and set $\mathbf{z} = (z_1, ..., z_m)$.
    Send $(b, \mathbf{z})$ to $\mathcal{V}$.
- $\mathcal{V}$: Compute $c = f(A^b)$, and output 1 iff $\prod_{j=1}^m \{g_j\}^{z_j} = R \cdot \mathsf{h}^c$.

Fig. 1. A Tag-Based fully Leakage-Resilient Non-Malleable ID Scheme Π_{tag}

Theorem 1. *If the* Selective–TagKEA *and* DLA *assumptions hold, then* Π_{tag} *is a ℓ-fully-leakage-resilient non-malleable ID scheme with respect to tag set* $\mathbb{Z}_n$, *where* $\ell = \frac{1}{2}((m-1)\log p - \omega(\log \lambda))$.

4.1   Proof of Theorem 1

We now argue that our scheme Π_{tag} is complete and fully leakage-resilient non-malleable.

The completeness property is obvious. To analyze the security, we define the relation $\mathcal{R} = \{(\mathsf{h}, \mathsf{sk}) \mid \mathsf{sk} = (s_1, ..., s_m) \leftarrow (\mathbb{Z}_p)^m, \mathsf{h} = \prod_{j=1}^m \{g_j\}^{s_j}\}$. Before giving the formal proof of security, we first present a property about the relation $\mathcal{R}$, which can be directly generalized from [4,13].

Lemma 1. *Let* $\ell' = 2\ell = (m-1)\log(p) - \omega(\log \lambda)$. *If the* DLA *assumption holds, for all PPT adversaries* $\mathcal{A}^{\mathcal{O}_{\mathsf{sk}}^{\lambda, \ell'}}$ *with access to the leakage oracle* $\mathcal{O}_{\mathsf{sk}}^{\lambda, \ell'}$, *we have that:*

$$\Pr\left[(\mathsf{h}, \mathsf{sk}^*) \in \mathcal{R} \mid (\mathsf{h}, \mathsf{sk}) \leftarrow \mathcal{R}, \mathsf{sk}^* \leftarrow \mathcal{A}^{\mathcal{O}_{\mathsf{sk}}^{\lambda, \ell'}}(\mathsf{h})\right] \leq negl(\lambda),$$

where we assume that pub *has been given to all users and omit them from the description.*

Now, we proceed to give our formal proof of security. Assume that there exists a PPT adversary $\mathcal{M}$ which has non-negligible advantage ρ in the game

$\mathsf{TagLRNMGame}_\ell^\lambda(\mathcal{M})$. Then, we construct an adversary $\mathcal{A}$ which runs in poly-time and

$$\Pr\left[(\mathsf{h},\mathsf{sk}^*) \in \mathcal{R} \mid (\mathsf{h},\mathsf{sk}) \leftarrow \mathcal{R}, \mathsf{sk}^* \leftarrow \mathcal{A}^{\mathcal{O}_{\mathsf{sk}}^{\lambda,\ell'}}(\mathsf{h})\right] \geq \rho^2 - negl(\lambda).$$

The adversary $\mathcal{A}$ uses $\mathcal{M}$ as a subroutine. After getting the input pub, h and the access to a leakage oracle $\mathcal{O}_{\mathsf{sk}}^{\lambda,\ell'}$, $\mathcal{A}$ acts as following:

1. Give pub to $\mathcal{M}$.
2. Run $\mathcal{M}$ on input 1^λ and pub, and get the tags $t, \tilde{t} \in \mathbb{Z}_n$.
3. Select $X, Y \leftarrow \bar{\mathbb{G}}$, and set $\mathsf{pk} = (X, Y, \mathsf{h})$. Give pk to $\mathcal{M}$.
4. When $\mathcal{M}$ reaches the impersonation stage, $\mathcal{A}$ simulates the prover $\mathcal{P}(t, \mathsf{pk}, \mathsf{sk})$ and the verifier $\mathcal{V}(\tilde{t}, \mathsf{pk})$ for $\mathcal{M}$, and computes sk^* as following:
    (a) *For the left interaction*, $\mathcal{A}$ needs to execute the protocol with $\mathcal{M}$ as a prover and answer its leakage queries. To deal with the leakage queries, we define a special deterministic function $\mathbf{R}(\mathsf{sk})$ (described in [15]) which, on input sk, outputs randomness r' such that an honest prover when used along with r' and sk will result in the exact same messages generated by $\mathcal{A}$. $\mathbf{R}(\cdot)$ is initialized with the null string. Besides, we also define a variable sum to count the number of bits leaked during the protocol execution. sum is initialized to 0.
    Answer $\mathcal{M}$'s queries on the prover's messages:
    - In case that $\mathcal{M}$ queries the prover $\mathcal{P}$ for the first messages, $\mathcal{A}$ chooses $\mathbf{z} = (z_1, \ldots, z_m) \leftarrow (\mathbb{Z}_p)^m$, $d \leftarrow \mathbb{Z}_n$, and sets $c = f(\bar{g}^d)$, $R = \prod_{j=1}^m \{g_j\}^{z_j}/\mathsf{h}^c$. Send R to $\mathcal{M}$. Besides, the output of the function $\mathbf{R}(\mathsf{sk})$ is defined as a vector $\mathbf{r} = (r_1, \ldots, r_m)$ where for $j = 1, \ldots, m$, $r_j = z_j - cs_j$ ($(s_1, \ldots, s_m)$ is the secret key sk corresponding to the leakage oracle $\mathcal{O}_{\mathsf{sk}}^{\lambda,\ell'}$).
    - In case that $\mathcal{M}$ sends a pair (A, D) to the prover, $\mathcal{A}$ first checks whether $e(\bar{g}, D) = e(A, X^t Y)$ holds. If it does not hold, then $\mathcal{A}$ sets $(b, \mathbf{z}) = (\perp, \perp)$. Otherwise, $\mathcal{A}$ runs the Selective-TagKEA extractor $\mathcal{X}_\mathcal{M}$ for $\mathcal{M}$. If $\mathcal{X}_\mathcal{M}$ fails, $\mathcal{A}$ outputs abort1. Else, $\mathcal{A}$ gets a so that $A = \bar{g}^a$, and then computes $b = da^{-1}$. Send $(b, \mathbf{z})$ to $\mathcal{M}$. Besides, update the function $\mathbf{R}(\cdot)$ as $\mathbf{R}(\cdot) \parallel b$.
    Answer $\mathcal{M}$'s leakage queries on the prover's state:
    - In case that $\mathcal{M}$ sends a leakage function $F_i : \{0,1\}^* \rightarrow \{0,1\}^{\alpha_i}$, $\mathcal{A}$ ignores the query if $\mathsf{sum} + \alpha_i > \ell$. Otherwise, $\mathcal{A}$ first creates a new query F_i' such that $F_i'(\mathsf{sk}) = F_i(\mathsf{sk}, \mathbf{R}(\mathsf{sk}))$. It then queries $\mathcal{O}_{\mathsf{sk}}^{\lambda,\ell'}$ with F_i', and returns the answer to $\mathcal{M}$. Set $\mathsf{sum} = \mathsf{sum} + \alpha_i$.
    (b) *For the right interaction*, $\mathcal{A}$ simulates the verifier $\mathcal{V}(\tilde{t}, \mathsf{pk})$ as following:
    - In case that $\mathcal{M}$ sends the first message $\tilde{R}$, $\mathcal{A}$ chooses $\tilde{a} \leftarrow \mathbb{Z}_n$, and sets $\tilde{A} = \bar{g}^{\tilde{a}}$ and $\tilde{D} = (X^{\tilde{t}}Y)^{\tilde{a}}$. Send $(\tilde{A}, \tilde{D})$ to $\mathcal{M}$.
    (c) Now, we have completed the description of $\mathcal{A}$'s simulation for $\mathcal{M}$. Then we start to describe how to compute sk^*. If $\mathcal{M}$ does not give an accepted message $(\tilde{b}, \tilde{\mathbf{z}})$ (corresponding to $(\tilde{A}, \tilde{D})$), output $\perp$. Otherwise, $\mathcal{A}$ rewinds $\mathcal{M}$ by sending a new pair $(\tilde{A}', \tilde{D}')$. Let sum' be equal to the total number of bits learned by $\mathcal{M}$ excatly at the point that $\mathcal{A}$ rewinds $\mathcal{M}$. Then, the left interaction is simulated just as Step 4a except setting $\mathsf{sum} = \mathsf{sum}'$.

(d) If $\mathcal{M}$ does not give another accepted message $(\tilde{b}', \tilde{\mathbf{z}}')$ (corresponding to $(\tilde{A}', \tilde{D}')$), output $\bot$. Otherwise, let $\tilde{c} = f(\tilde{A}^{\tilde{b}})$, $\tilde{c}' = f(\tilde{A}'^{\tilde{b}'})$. Do as following.
  - If $\tilde{c} = \tilde{c}'$, $\mathcal{A}$ outputs Fail1.
  - If $\tilde{c} \neq \tilde{c}'$, $\mathcal{A}$ computes sk^* from $(\tilde{R}, \tilde{c}, \tilde{\mathbf{z}})$ and $(\tilde{R}, \tilde{c}', \tilde{\mathbf{z}}')$, i.e., for $j = 1, \ldots, m$, computes $s_j^* = (\tilde{z}_j - \tilde{z}_j')/(\tilde{c} - \tilde{c}')$. Output $\mathsf{sk}^* = (s_1^*, \ldots, s_m^*)$.

Now, we have completed the description of $\mathcal{A}$. Clearly, $\mathcal{A}$ runs in poly-time because both $\mathcal{M}$ and $\mathcal{O}_{\mathsf{sk}}^{\lambda, \ell'}$ run in poly-time. And the total number of bits needed by $\mathcal{A}$ to answer $\mathcal{M}$'s leakage queries does not exceed ℓ' since it has limited that in every execution, the total number of bits learned by $\mathcal{M}$ is bounded by $\ell = \frac{1}{2}\ell'$.

Then, we start to analyze the probability that $\mathcal{A}$ succeeds. Let E_1 be the event that $\mathcal{A}$ outputs Fail1, and E_2 be the event that $\mathcal{A}$ successfully gets two valid messages $(\tilde{b}, \tilde{\mathbf{z}})$ and $(\tilde{b}', \tilde{\mathbf{z}}')$. Thus, the success probability of the adversary $\mathcal{A}$ is $\Pr[\mathsf{E}_2] - \Pr[\mathsf{E}_1]$. To evaluate this probability, we consider the following two claims.

Claim 1. *The probability of the event E_1 is negligible.*

Proof. Note that E_1 means that $\mathcal{M}$ can make $\tilde{c}$ be equal to $\tilde{c}'$. Assume that the event E_1 happens with non-negligible probability at least ρ'. Then we can construct an adversary $\mathcal{B}$ that breaks the DLA assumption with probability at least ρ'. The adversary $\mathcal{B}$, on input (BG, h), acts as following:

- $\mathcal{B}$ first honestly generate $(\mathsf{G}, g_1, \ldots, g_m)$, set $\mathsf{pub} = (\mathsf{G}, \mathsf{BG}, g_1, \ldots, g_m)$ and give pub to $\mathcal{M}$.
- After receiving the tags $t, \tilde{t}$, select $\mathsf{sk} = (s_1, \ldots, s_m) \leftarrow (\mathbb{Z}_p)^m$, $X \leftarrow \tilde{\mathbb{G}}$, $r \leftarrow \mathbb{Z}_n$, and set $\mathsf{h} = \prod_{j=1}^{m}\{g_j\}^{s_j}$, $Y = X^{-\tilde{t}}\bar{g}^r$. Give pk to $\mathcal{M}$.
- $\mathcal{B}$ uses sk to construct the leakage oracle $\mathcal{O}_{\mathsf{sk}}^{\lambda, \ell'}$. When $\mathcal{M}$ reaches the impersonation stage, $\mathcal{B}$ acts similarly to the adversary $\mathcal{A}$:
  - For the left interaction, simulate the prover $\mathcal{P}(t, \mathsf{pk}, \mathsf{sk})$ just the same as the adversary $\mathcal{A}$.
  - For the right interaction, simulate the verifier $\mathcal{V}(\tilde{t}, \mathsf{pk})$ by sending $(\tilde{A} = \bar{g}^{\tilde{a}}, \tilde{D} = (X^{\tilde{t}}Y)^{\tilde{a}})$ with $\tilde{A} \neq h$.
  - Then, if $\mathcal{B}$ does not get an accepted message $(\tilde{b}, \tilde{\mathbf{z}})$, $\mathcal{B}$ outputs $\bot$. Otherwise, $\mathcal{B}$ rewinds $\mathcal{M}$ by sending $(\tilde{A}' = h, \tilde{D}' = h^r)$. The left interaction are handled just as $\mathcal{A}$.
  - If $\mathcal{M}$ successfully gives another accepted message $(\tilde{b}', \tilde{\mathbf{z}}')$ such that $\tilde{A}^{\tilde{b}} = \tilde{A}'^{\tilde{b}'}$, $\mathcal{B}$ outputs $w = \tilde{a}\tilde{b}/\tilde{b}'$. Otherwise, $\mathcal{B}$ outputs $\bot$.

Clearly, the running time of $\mathcal{B}$ is polynomial. And the simulation of $\mathcal{B}$ is identical to the simulation of $\mathcal{A}$ since the distribution of $(\tilde{A}' = h, \tilde{D}' = h^r)$ is equal to that of $(\tilde{A}', \tilde{D}')$ generated by $\mathcal{A}$. To see it, let $\tilde{A}' = \bar{g}^{\tilde{a}'}$. Then, we have $\tilde{D}' = \tilde{A}'^r = (X^{\tilde{t}}X^{-\tilde{t}}\bar{g}^r)^{\tilde{a}'} = (X^{\tilde{t}}Y)^{\tilde{a}'}$.

Besides, since f is injective, it is easy to see that $\tilde{c} = \tilde{c}'$ implies $\tilde{A}^{\tilde{b}} = \tilde{A}'^{\tilde{b}'}$. Thus, the success probability of $\mathcal{B}$ is equal to $\Pr[\mathsf{E}_1]$. This completes the proof of this claim. □

Claim 2. *The probability of the event E_2 is $\Pr[E_2] \geq \rho^2 - negl(\lambda)$.*

Proof. Note that the event E_2 means that the adversary $\mathcal{M}$ successfully completes the right interaction both in the first execution and in the rewinding. Recall that we have assumed that in the game $\mathsf{TagLRNMGame}_\ell^\lambda(\mathcal{M})$, the probability that $\mathcal{M}$ successfully convinces the honest verifier $\mathcal{V}(\tilde{t}, \mathsf{pk})$ is at least ρ. Let $\mathsf{TagSimExe}$ denote the simulation of $\mathcal{A}$ for $\mathcal{M}$. Now, we consider the differences between $\mathsf{TagSimExe}$ and $\mathsf{TagLRNMGame}_\ell^\lambda(\mathcal{M})$:

1. In $\mathsf{TagSimExe}$, the prover's first message R is generated by randomly selecting $c, \mathbf{z}$ and then computing $R = \prod_{j=1}^m \{g_j\}^{z_j} / \mathsf{h}^c$. However, the distribution of R generated here is equal to that of R generated in $\mathsf{TagLRNMGame}_\ell^\lambda(\mathcal{M})$. This is because R is still a random group element in $\mathbb{G}$.
2. In $\mathsf{TagSimExe}$, after receiving a valid message (A, D) from $\mathcal{M}$, $\mathcal{A}$ outputs abort1 if the Selective- TagKEA extractor fails. This happens with negligible probability since the Selective- TagKEA assumption holds.
3. In $\mathsf{TagSimExe}$, the leakage queries are answered by querying the leakage oracle $\mathcal{O}_{\mathsf{sk}}^{\lambda, \ell'}$, where the adversary $\mathcal{A}$ does not hold the secret key sk and can not efficiently compute the vector $\mathbf{r}$ such that $R = \prod_{j=1}^m \{g_j\}^{r_j}$. But the vector $\mathbf{r}$ comes from the same distribution as in $\mathsf{TagLRNMGame}_\ell^\lambda(\mathcal{M})$, and with access to the leakage oracle $\mathcal{O}_{\mathsf{sk}}^{\lambda, \ell'}$, $\mathcal{A}$ can evaluate the leakage queries correctly. Because the leakage queries responded in $\mathsf{TagSimExe}$ is in exactly the same manner as $\mathsf{TagLRNMGame}_\ell^\lambda(\mathcal{M})$. Finally, the leakage queries responded in $\mathsf{TagSimExe}$ and $\mathsf{TagLRNMGame}_\ell^\lambda(\mathcal{M})$ are identically distributed.

Thus, we can conclude that $\Pr[E_2] \geq \rho^2 - negl(\lambda)$.   □

From the above three claims, we can conclude that,

$$\Pr\left[(\mathsf{h}, \mathsf{sk}^*) \in \mathcal{R} \mid (\mathsf{h}, \mathsf{sk}) \leftarrow \mathcal{R}, \mathsf{sk}^* \leftarrow \mathcal{A}^{\mathcal{O}_{\mathsf{sk}}^{\lambda, \ell'}}(\mathsf{h})\right] \geq \rho^2 - negl(\lambda),$$

which contradicts Lemma 1. This completes the proof of Theorem 1.

5   A Fully Leakage-Resilient Non-malleable Identification Scheme

In this section, we present our construction of a fully leakage-resilient non-malleable identification scheme Π_{lrnm}. To remove the use of tags, we add an one-time signature to the scheme Π_{tag}. Let $\mathsf{OTS} = (\mathsf{SigGen}, \mathsf{Sig}, \mathsf{Ver})$ be a strong one-time signature [14] such that the verification key vk is in $\mathbb{Z}_n$, and the message space of the signature scheme contains the sets $\mathbb{G} \times \bar{\mathbb{G}} \times \bar{\mathbb{G}}$. Roughly speaking, Π_{lrnm} is just like Π_{tag} except that we use vk to replace the tag t. The details are presented in Fig. 2.

Theorem 2. *If the TagKEA and DLA assumptions hold and if $\mathsf{OTS} = (\mathsf{SigGen}, \mathsf{Sig}, \mathsf{Ver})$ is a strong one-time signature scheme, then the scheme Π_{lrnm} is a ℓ-fully-leakage-resilient non-malleable ID scheme, where $\ell = \frac{1}{2}((m-1)\log p - \omega(\log \lambda))$.*

Public Parameters Genration:
- $\mathsf{ParamGen}(1^\lambda)$: $\mathsf{G} = (p, \mathbb{G}, g) \leftarrow \mathcal{G}(1^\lambda)$, $\mathsf{BG} = (n, \bar{\mathbb{G}}, \mathbb{H}, \bar{g}, e, f) \leftarrow \mathcal{BGG}(1^\lambda, p)$, $g_1, ..., g_m \leftarrow \mathbb{G}$. Set $\mathsf{pub} = (\mathsf{G}, \mathsf{BG}, g_1, ..., g_m)$.

Key Genration:
- $\mathsf{KeyGen}()$: $X, Y \leftarrow \bar{\mathbb{G}}$, $\mathsf{sk} = (s_1, ..., s_m) \leftarrow (\mathbb{Z}_p)^m$, and set $\mathsf{h} = \prod_{j=1}^m \{g_j\}^{s_j}$, $\mathsf{pk} = (X, Y, \mathsf{h})$. Output $(\mathsf{pk}, \mathsf{sk})$.

Interaction: The machines $\mathcal{P}$ and $\mathcal{V}$ interact as following:
- $\mathcal{P}$: Choose $\boldsymbol{r} = (r_1, ..., r_m) \leftarrow (\mathbb{Z}_p)^m$, and compute $R = \prod_{j=1}^m \{g_j\}^{r_j}$. Send R to $\mathcal{V}$.
- $\mathcal{V}$: Choose $(\mathsf{sgk}, \mathsf{vk}) \leftarrow \mathsf{SigGen}(1^\lambda)$, $a \leftarrow \mathbb{Z}_n$, and set $A = \bar{g}^a$, $D = (X^{\mathsf{vk}}Y)^a$, $\sigma \leftarrow \mathsf{Sig}_{\mathsf{sgk}}(R, A, D)$. Send $(\mathsf{vk}, (A, D), \sigma)$ to $\mathcal{P}$.
- $\mathcal{P}$: If $\mathsf{Ver}_{\mathsf{vk}}((R, A, D), \sigma) \neq 1$ or $e(A, X^{\mathsf{vk}}Y) \neq e(\bar{g}, D)$ then abort. Else,
  - Choose $b \leftarrow \mathbb{Z}_n$, and compute $c = f(A^b)$,
  - Compute $z_i = r_i + cs_i$ for $i = 1, ..., m$, and set $\mathbf{z} = (z_1, ..., z_m)$,
  - Send $(b, \mathbf{z})$ to $\mathcal{V}$.
- $\mathcal{V}$: Compute $c = f(A^b)$, and output 1 iff $\prod_{j=1}^m \{g_j\}^{z_j} = R \cdot \mathsf{h}^c$.

Fig. 2. Fully Leakage-Resilient Non-Malleable ID Scheme Π_{lrnm}

5.1 Proof of Security

It is easy to verify the completeness property. For security, we follow the technique used in Sect. 4.1, except that we need to handle the case when the tag of the left interaction vk is equal to the tag of the right interaction $\widetilde{\mathsf{vk}}$. Similarly, we first define the relation (with respect to pub):

$$\mathcal{R} = \{(\mathsf{h}, \mathsf{sk}) \mid \mathsf{sk} = (s_1, ..., s_m) \leftarrow (\mathbb{Z}_p)^m, \mathsf{h} = \prod_{j=1}^m \{g_j\}^{s_j}\}.$$

Assume that there exists a PPT adversary $\mathcal{M}$ which can win in the game $\mathsf{LRNMGame}_\ell^\lambda(\mathcal{M})$ with non-negligible advantage ρ. Then, we construct an adversary $\mathcal{A}$ which runs in poly-time and

$$\Pr\big[(\mathsf{h}, \mathsf{sk}^*) \in \mathcal{R} \mid (\mathsf{h}, \mathsf{sk}) \leftarrow \mathcal{R}, \mathsf{sk}^* \leftarrow \mathcal{A}^{\mathcal{O}_{\mathsf{sk}}^{\lambda, \ell'}}(\mathsf{h})\big] \geq \mu - negl(\lambda),$$

where μ depends on the advantage of $\mathcal{M}$ in the game $\mathsf{LRNMGame}_\ell^\lambda(\mathcal{M})$. The adversary $\mathcal{A}$ uses $\mathcal{M}$ as a subroutine. After getting input $(\mathsf{pub}, \mathsf{h})$ and access to a leakage oracle $\mathcal{O}_{\mathsf{sk}}^{\lambda, \ell'}$, $\mathcal{A}$ acts as following:

1. Select $X, Y \leftarrow \bar{\mathbb{G}}$, and set $\mathsf{pk} = (X, Y, \mathsf{h})$. Give $\mathsf{pub}, \mathsf{pk}$ to $\mathcal{M}$.
2. When $\mathcal{M}$ reaches the impersonation stage, $\mathcal{A}$ simulates the prover $\mathcal{P}(\mathsf{pk}, \mathsf{sk})$ of the left interaction and the verifier $\mathcal{V}(\mathsf{pk})$ of the right interaction for $\mathcal{M}$, and uses it to compute sk^* as following:
   (a) *For the left interaction*, $\mathcal{A}$ needs to execute the protocol with $\mathcal{M}$ as a prover and answer its leakage queries. As described in Sect. 4.1, we define

a function $\mathbf{R}(\mathsf{sk})$ and a variable sum, where $\mathbf{R}$ is initialized with the null string, and sum is initialized to 0.

Answer $\mathcal{M}$'s queries on the prover's messages:

- In case that $\mathcal{M}$ queries the prover $\mathcal{P}$ for the first messages, choose $\mathbf{z} = (z_1, \ldots, z_m) \leftarrow (\mathbb{Z}_p)^m, d \leftarrow \mathbb{Z}_n$, and set $c = f(\bar{g}^d)$, $R = \prod_{j=1}^{m}\{g_j\}^{z_j}/\mathsf{h}^c$. Send R to $\mathcal{M}$. Besides, the output of the function $\mathbf{R}(\mathsf{sk})$ is defined as a vector $\mathbf{r} = (r_1, \ldots, r_m)$ where for $j = 1, \ldots, m$, $r_j = z_j - cs_j$.

- In case that $\mathcal{M}$ sends $(\mathsf{vk}, (A, D), \sigma)$ to $\mathcal{P}$, $\mathcal{A}$ first checks whether $\mathsf{Ver}_{\mathsf{vk}}((R, A, D), \sigma) = 1$ and $e(A, X^{\mathsf{vk}}Y) = e(\bar{g}, D)$ hold. If one of them does not hold then $\mathcal{A}$ sets $(b, \mathbf{z}) = (\bot, \bot)$. Otherwise,

  - In case that in the right interaction, $\mathcal{A}$ (acting as a verifier) has not sent its message $(\widetilde{\mathsf{vk}}, (\tilde{A}, \tilde{D}), \tilde{\sigma})$, $\mathcal{A}$ runs the TagKEA extractor $\mathcal{X}_{\mathcal{M}}$ for $\mathcal{M}$. If $\mathcal{X}_{\mathcal{M}}$ fails, $\mathcal{A}$ outputs abort1. Else, $\mathcal{A}$ gets a so $A = \bar{g}^a$, and then computes $b = da^{-1}$. Send $(b, \mathbf{z})$ to $\mathcal{M}$.

  - In case that in the right interaction, $\mathcal{A}$ (acting as a verifier) has sent its message $(\widetilde{\mathsf{vk}}, (\tilde{A}, \tilde{D}), \tilde{\sigma})$:

    * if $\mathsf{vk} \neq \widetilde{\mathsf{vk}}$, $\mathcal{A}$ runs the TagKEA extractor $\mathcal{X}_{\mathcal{M}}$ for $\mathcal{M}$. If $\mathcal{X}_{\mathcal{M}}$ fails, $\mathcal{A}$ outputs abort1. Else, $\mathcal{A}$ gets a so $A = \bar{g}^a$, and then computes $b = da^{-1}$. Send $(b, \mathbf{z})$ to $\mathcal{M}$.

    * if $\mathsf{vk} = \widetilde{\mathsf{vk}}$, $(R, A, D) = (\tilde{R}, \tilde{A}, \tilde{D})$ and $\sigma = \tilde{\sigma}$, $\mathcal{A}$ now has $\tilde{a}$ so $A = \tilde{A} = \bar{g}^{\tilde{a}}$ (from the right interaction). Thus, $\mathcal{A}$ can compute $b = d\tilde{a}^{-1}$, and then sends $(b, \mathbf{z})$ to $\mathcal{M}$.

    * if $\mathsf{vk} = \widetilde{\mathsf{vk}}$ but $(R, A, D) \neq (\tilde{R}, \tilde{A}, \tilde{D})$ or $\sigma \neq \tilde{\sigma}$, $\mathcal{A}$ outputs abort2. Besides, update the function $\mathbf{R}(\cdot)$ as $\mathbf{R}(\cdot) \| b$.

Answer $\mathcal{M}'$ s leakage queries on the prover's state :

- In case that $\mathcal{M}$ sends a leakage function $F_i : \{0,1\}^* \rightarrow \{0,1\}^{\alpha_i}$, $\mathcal{A}$ ignores the query if $\mathsf{sum} + \alpha_i > \ell$. Otherwise, $\mathcal{A}$ first creates a new query F_i' such that $F_i'(\mathsf{sk}) = F_i(\mathsf{sk}, \mathbf{R}(\mathsf{sk}))$. It then queries $\mathcal{O}_{\mathsf{sk}}^{\lambda, \ell'}$ with F_i', and returns the answer to $\mathcal{M}$. Set $\mathsf{sum} = \mathsf{sum} + \alpha_i$.

(b) *For the right interaction,* $\mathcal{A}$ interacts with $\mathcal{M}$ by acting as a verifier $\mathcal{V}(\mathsf{pk})$ as following:

- In case that $\mathcal{M}$ sends the first message $\tilde{R}$, choose $(\widetilde{\mathsf{sgk}}, \widetilde{\mathsf{vk}}) \leftarrow \mathsf{SigGen}(1^\lambda)$, $\tilde{a} \leftarrow \mathbb{Z}_n$, set $\tilde{A} = \bar{g}^{\tilde{a}}$, $\tilde{D} = (X^{\widetilde{\mathsf{vk}}}Y)^{\tilde{a}}$, $\sigma \leftarrow \mathsf{Sig}_{\widetilde{\mathsf{sgk}}}(\tilde{R}, \tilde{A}, \tilde{D})$, and sends $(\widetilde{\mathsf{vk}}, (\tilde{A}, \tilde{D}), \tilde{\sigma})$ to $\mathcal{M}$.

(c) Now, we have completed the description of $\mathcal{A}$'s simulation for $\mathcal{M}$. Then, we start to describe how to compute sk^*. Let $\pi = (R, (\mathsf{vk}, (A, D), \sigma), (b, \mathbf{z}))$ denote the transcript of the left interaction, where a message in π is set to $\bot$ if it has not been exchanged when the right interaction is completed. Similarly, let $\tilde{\pi}$ denote the transcript of the right interaction. If $\mathcal{M}$ does not give an accepted message $(\tilde{b}, \tilde{\mathbf{z}})$ or gives an accepted message $(\tilde{b}, \tilde{\mathbf{z}})$ but $\pi = \tilde{\pi}$, output $\bot$. Otherwise,

  i. If $(b, \mathbf{z}) = (\bot, \bot)$, $\mathcal{A}$ outputs Fail1.

  ii. Else, consider the following two cases:

- if $\mathsf{vk} = \widetilde{\mathsf{vk}}$, let $c = f(A^b)$ and $\tilde{c} = f(\tilde{A}^{\tilde{b}})$. Consider the following two cases:
  - If $c = \tilde{c}$, $\mathcal{A}$ outputs Fail2.
  - If $c \neq \tilde{c}$, from Step 2a, we know that $(R, A, D) = (\tilde{R}, \tilde{A}, \tilde{D})$. Thus, we have two accepted messages $(R, c, \mathbf{z})$ and $(\tilde{R}, \tilde{c}, \tilde{\mathbf{z}})$ with $R = \tilde{R}$ and $c \neq \tilde{c}$. Thus, for $j = 1, \ldots, m$, we can compute $s_j^* = (z_j - \tilde{z}_j)/(c - \tilde{c})$. $\mathcal{A}$ outputs $\mathsf{sk}^* = (s_1^*, \ldots, s_m^*)$.
- if $\mathsf{vk} \neq \widetilde{\mathsf{vk}}$, $\mathcal{A}$ acts as following:
  - $\mathcal{A}$ rewinds $\mathcal{M}$ by sending a new message $(\widetilde{\mathsf{vk}}', (\tilde{A}', \tilde{D}'), \tilde{\sigma}')$. Let sum' be equal to the total number of bits learned by $\mathcal{M}$ exactly at the point that $\mathcal{A}$ rewinds $\mathcal{M}$. Then, the left interaction is simulated just as Step 2a except setting $\mathsf{sum} = \mathsf{sum}'$.
  - If $\mathcal{M}$ does not give another accepted message $(\tilde{b}', \tilde{\mathbf{z}}')$ or gives an accepted message $(\tilde{b}', \tilde{\mathbf{z}}')$ but $\pi' = \tilde{\pi}'$, output $\perp$. Otherwise, let $\tilde{c} = f(\tilde{A}^{\tilde{b}})$ and $\tilde{c}' = f(\tilde{A}'^{\tilde{b}'})$. $\mathcal{A}$ acts as following:
    * if $\tilde{c} = \tilde{c}'$, $\mathcal{A}$ outputs Fail3.
    * if $\tilde{c} \neq \tilde{c}'$, for $j = 1, \ldots, m$, $\mathcal{A}$ computes $s_j^* = (\tilde{z}_j - \tilde{z}_j')/(\tilde{c} - \tilde{c}')$. Output $\mathsf{sk}^* = (s_1^*, \ldots, s_m^*)$.

Now, we have completed the description of $\mathcal{A}$. Clearly, $\mathcal{A}$ runs in poly-time because both $\mathcal{M}$ and $\mathcal{O}_{\mathsf{sk}}^{\lambda, \ell'}$ run in poly-time. And the total number of bits needed by $\mathcal{A}$ to answer $\mathcal{M}$'s leakage queries does not exceed ℓ' since it has limited that in every execution, the total number of bits learned by $\mathcal{M}$ is bounded by $\ell = \frac{1}{2}\ell'$.

Then, we start to analyze the probability that $\mathcal{A}$ succeeds. Let E_1, E_2 and E_3 respectively denote the event that $\mathcal{A}$ outputs Fail1, Fail2 and Fail3. Let E_4 denote the event that $\mathcal{M}$ successfully outputs a valid message $(\tilde{b}, \tilde{\mathbf{z}})$ with $\pi \neq \tilde{\pi}$. Let E_5 denote the event that $\mathcal{M}$ successfully outputs another valid message $(\tilde{b}', \tilde{\mathbf{z}}')$ with $\pi' \neq \tilde{\pi}'$. Thus, the success probability of the adversary $\mathcal{A}$ is:

$$\Pr[\mathsf{vk} = \widetilde{\mathsf{vk}} \wedge (b, \mathbf{z}) \neq (\perp, \perp) \wedge \mathsf{E}_4] + \Pr[\mathsf{vk} \neq \widetilde{\mathsf{vk}} \wedge (b, \mathbf{z}) \neq (\perp, \perp) \wedge \mathsf{E}_4 \wedge \mathsf{E}_5] - \Pr[\mathsf{E}_2] - \Pr[\mathsf{E}_3]$$

To evaluate this probability, we consider the following four claims.

Claim 3. *The probability of the event E_1 is negligible.*

Proof. This event E_1 means that $\mathcal{M}$ does not require $(b, \mathbf{z})$ or always sends incorrect $(\mathsf{vk}, (A, D), \sigma)$, which implies that $\mathcal{M}$ can win without the help of the left interaction. Assume that $\Pr[\mathsf{E}_1] \geq \varepsilon$. Then we construct an adversary $\mathcal{B}$ which runs in poly-time and

$$\Pr\big[(\mathsf{h}, \mathsf{sk}^*) \in \mathcal{R} \mid (\mathsf{h}, \mathsf{sk}) \leftarrow \mathcal{R}, \mathsf{sk}^* \leftarrow \mathcal{B}^{\mathcal{O}_{\mathsf{sk}}^{\lambda, \ell'}}(\mathsf{h})\big] \geq \varepsilon^2 - negl(\lambda).$$

Similar to $\mathcal{A}$, $\mathcal{B}$ uses $\mathcal{M}$ as a subroutine, receives the input $(\mathsf{pub}, \mathsf{h})$ and the access to a leakage oracle $\mathcal{O}_{\mathsf{sk}}^{\lambda, \ell'}$, and acts as following:

- Select $X, Y \leftarrow \bar{\mathbb{G}}$, set $\mathsf{pk} = (X, Y, \mathsf{h})$, and give $\mathsf{pub}, \mathsf{pk}$ to $\mathcal{M}$.

- When $\mathcal{M}$ reaches the impersonation stage, $\mathcal{B}$ needs to simulates the prover of left interaction and the verifier of the right interaction for $\mathcal{M}$, and uses it to compute sk^* as following:
    - *For the left interaction*, $\mathcal{B}$ acts in the same way as the adversary $\mathcal{A}$, except that when he needs to generate a valid message $(b, \mathbf{z})$ before $(\tilde{b}, \tilde{\mathbf{z}})$ is sent, $\mathcal{B}$ simply aborts.
    - *For the right interaction*, $\mathcal{B}$ simply acts in the same way as the adversary $\mathcal{A}$.
    - If $\mathcal{M}$ successfully sends a valid message $(\tilde{b}, \tilde{\mathbf{z}})$, $\mathcal{B}$ rewinds $\mathcal{M}$ by sending a new message $(\tilde{\mathsf{vk}}', (\tilde{A}', \tilde{D}'), \tilde{\sigma}')$. The left interaction is simulated just as above.
    - If $\mathcal{M}$ successfully sends another valid message $(\tilde{b}', \tilde{\mathbf{z}}')$, let $\tilde{c} = f(\tilde{A}^{\tilde{b}})$ and $\tilde{c}' = f(\tilde{A}'^{\tilde{b}'})$.
        * if $\tilde{c} = \tilde{c}'$, $\mathcal{B}$ outputs Fail.
        * if $\tilde{c} \neq \tilde{c}'$, $\mathcal{B}$ computes $s_j^* = (\tilde{z}_j - \tilde{z}_j')/(\tilde{c} - \tilde{c}')$ for $j = 1, \ldots, m$. Output $\mathsf{sk}^* = (s_1^*, \ldots, s_m^*)$.

This completes the description of $\mathcal{B}$. Clearly, the success probability of $\mathcal{B}$ depends on $\Pr[\mathsf{E}_1]$, and the probability that $\tilde{c} \neq \tilde{c}'$. From Claim 1, it is easy to get that the probability that $\tilde{c} = \tilde{c}'$ is negligible. Thus, the success probability of $\mathcal{B}$ is at least $\Pr[\mathsf{E}_1]^2 - negl(\lambda) \geq \varepsilon^2 - negl(\lambda)$. Combined with Lemma 1, we can conclude that the probability of the event E_1 is negligible. □

Claim 4. *The probability of the event E_2 is negligible.*

Proof. This event means that $\mathcal{M}$ can successfully complete the right interaction with $\mathsf{vk} = \tilde{\mathsf{vk}}$, $b = \tilde{b}$ and $\mathbf{z} \neq \tilde{\mathbf{z}}$. From the simulation of the left interaction, we know that $\mathsf{vk} = \tilde{\mathsf{vk}}$ implies that $R = \tilde{R}$, $A = \tilde{A}$ and $D = \tilde{D}$. Thus, we have that $c = \tilde{c}$ and,

$$\prod_{j=1}^{m} \{g_j\}^{z_j} = R(\mathsf{h})^c = \tilde{R}(\mathsf{h})^{\tilde{c}} = \prod_{j=1}^{m} \{g_j\}^{\tilde{z}_j},$$

which implies that for $Z = \prod_{j=1}^{m} \{g_j\}^{z_j}$, we can get two different vectors $\mathbf{z} \neq \tilde{\mathbf{z}}$ for Z. From the Lemma 3 of [4], we know that this can be used to break the DLA assumption. We refer the reader to [4] for a formal proof. □

Claim 5. *The probability of the event E_3 is negligible.*

Proof. This event E_3 means that $\mathcal{M}$ can make $\tilde{c}$ be equal to $\tilde{c}'$. From the proof of Claim 1, we can easily prove that this contradicts the DLA assumption (Please refer to Claim 1 for more details). □

Claim 6. *The probability that $\mathcal{M}$ successfully completes the right interaction in the simulation is at least $\rho - negl(\lambda)$.*

Proof. Recall that we have assumed that the advantage of $\mathcal{M}$ in the game $\mathsf{LRNMGame}_\ell^\lambda(\mathcal{M})$ is at least ρ. Therefore, we only need to prove that the simulation of $\mathcal{A}$ for $\mathcal{M}$ is indistinguishable from the real attack game $\mathsf{LRNMGame}_\ell^\lambda(\mathcal{M})$. Let SimExe denote the simulation of $\mathcal{A}$ for $\mathcal{M}$.

Note that SimExe is different from $\mathsf{LRNMGame}_\ell^\lambda(\mathcal{M})$ only in the way of generating the prover's messages and answers the leakage queries.

1. In SimExe, the prover's first message R is generated by randomly generating $c, \mathbf{z}$ and then computing $R = \prod_{j=1}^m \{g_j\}^{z_j}/(\mathsf{h})^c$. The distribution of R generated here is equal to that of R generated in $\mathsf{LRNMGame}_\ell^\lambda(\mathcal{M})$. This is because R is still a random group element in $\mathbb{G}$.
2. In SimExe, after receiving a valid message $(\mathsf{vk}, (A, D), \sigma)$ from $\mathcal{M}$, $\mathcal{A}$ may fail to generate the prover's last message. This is because of the following two reasons:
   - $\mathcal{A}$ outputs abort1 when the TagKEA extractor fails. This event happens with negligible probability because the TagKEA assumption holds.
   - $\mathcal{A}$ outputs abort2 when $\mathsf{vk} = \tilde{\mathsf{vk}}$ but $(R, A, D) \neq (\tilde{R}, \tilde{A}, \tilde{D})$ or $\sigma \neq \tilde{\sigma}$. This event also happens with negligible probability because (SigGen, Sig, Ver) is a strong one-time signature scheme, which guarantees that after receiving a valid signature $\tilde{\sigma}$ for $\tilde{M}$, no PPT algorithm can output a pair (M, σ) such that $\mathsf{Ver}_{\mathsf{vk}}(M, \sigma) = 1$ but $\tilde{M} \neq M$ or $\tilde{\sigma} \neq \sigma$.
3. In SimExe, the leakage queries are answered by querying the leakage oracle $\mathcal{O}_{\mathsf{sk}}^{\lambda, \ell'}$ where $\mathcal{A}$ does not know the secret key sk and the random string $\mathbf{r}$. As analyzed in Claim 2 of Sect. 4.1, we know that the leakage queries responded in SimExe and $\mathsf{LRNMGame}_\ell^\lambda(\mathcal{M})$ are identically distributed.

Therefore, SimExe and $\mathsf{LRNMGame}_\ell^\lambda(\mathcal{M})$ are computationally indistinguishable. Thus, the probability that $\mathcal{M}$ successfully completes the right interaction in SimExe is at least $\rho - negl(\lambda)$. That is, $\Pr[\mathsf{E}_4] \geq \rho - negl(\lambda)$.   □

Now, we go back to evaluate the success probability of $\mathcal{A}$, which is equal to

$$\Pr[\mathsf{vk} = \tilde{\mathsf{vk}} \wedge (b, \mathbf{z}) \neq (\perp, \perp) \wedge \mathsf{E}_4] + \Pr[\mathsf{vk} \neq \tilde{\mathsf{vk}} \wedge (b, \mathbf{z}) \neq (\perp, \perp) \wedge \mathsf{E}_4 \wedge \mathsf{E}_5] - \Pr[\mathsf{E}_2] - \Pr[\mathsf{E}_3]$$

From the above analysis, we know that

$$\begin{aligned}
\Pr[\mathsf{E}_4] &= \Pr[\mathsf{vk} = \tilde{\mathsf{vk}} \wedge (b, \mathbf{z}) \neq (\perp, \perp) \wedge \mathsf{E}_4] + \Pr[\mathsf{vk} \neq \tilde{\mathsf{vk}} \wedge (b, \mathbf{z}) \neq (\perp, \perp) \wedge \mathsf{E}_4] \\
&\quad + \Pr[(b, \mathbf{z}) = (\perp, \perp) \wedge \mathsf{E}_4] \\
&\geq \rho - negl(\lambda).
\end{aligned}$$

Let $\rho_1 = \Pr[\mathsf{vk} = \tilde{\mathsf{vk}} \wedge (b, \mathbf{z}) \neq (\perp, \perp) \wedge \mathsf{E}_4]$, and $\rho_2 = \Pr[\mathsf{vk} \neq \tilde{\mathsf{vk}} \wedge (b, \mathbf{z}) \neq (\perp, \perp) \wedge \mathsf{E}_4]$. Then, we have that $\rho_1 + \rho_2 \geq \rho - negl(\lambda)$ because $\Pr[(b, \mathbf{z}) = (\perp, \perp) \wedge \mathsf{E}_4] = \Pr[\mathsf{E}_1] = negl(\lambda)$.

Besides, we know that the events E_4 and E_5 are independent of each other. Thus, the success probability of $\mathcal{A}$ is $\rho_1 + \rho\rho_2 - negl(\lambda)$. Because ρ is non-negligible, one of ρ_1 and ρ_2 must be non-negligible. Thus, $\rho_1 + \rho\rho_2 - negl(\lambda)$ is non-negligible. That is, $\mathcal{A}$ can succeed with at least non-negligible probability, which contradicts Lemma 1. This completes the proof of Theorem 2.

References

1. Anada, H., Arita, S.: Identification schemes of proofs of ability secure against concurrent man-in-the-middle attacks. In: Heng, S.-H., Kurosawa, K. (eds.) ProvSec 2010. LNCS, vol. 6402, pp. 18–34. Springer, Heidelberg (2010)

2. Anada, H., Arita, S.: Identification schemes from key encapsulation mechanisms. In: Nitaj, A., Pointcheval, D. (eds.) AFRICACRYPT 2011. LNCS, vol. 6737, pp. 59–76. Springer, Heidelberg (2011)

3. Anada, H.: A study on efficient identification schemes secure against concurrent man-in-the-middle attacks. Doctoral dissertation, Institute of Information Security (2012)

4. Alwen, J., Dodis, Y., Wichs, D.: Leakage-resilient public-key cryptography in the bounded-retrieval model. In: Halevi, S. (ed.) CRYPTO 2009. LNCS, vol. 5677, pp. 36–54. Springer, Heidelberg (2009)

5. Alwen, J., Dodis, Y., Wichs, D.: Survey: leakage resilience and the bounded retrieval model. In: Kurosawa, K. (ed.) Information Theoretic Security. LNCS, vol. 5973, pp. 1–18. Springer, Heidelberg (2010)

6. Abe, M., Fehr, S.: Perfect NIZK with adaptive soundness. In: Vadhan, S.P. (ed.) TCC 2007. LNCS, vol. 4392, pp. 118–136. Springer, Heidelberg (2007)

7. Bellare, M., Fischlin, M., Goldwasser, S., Micali, S.: Identification protocols secure against reset attacks. In: Pfitzmann, B. (ed.) EUROCRYPT 2001. LNCS, vol. 2045, pp. 495–511. Springer, Heidelberg (2001)

8. Bellare, M., Palacio, A.: GQ and Schnorr identification schemes: proofs of security against impersonation under active and concurrent attacks. In: Yung, M. (ed.) CRYPTO 2002. LNCS, vol. 2442, pp. 162–177. Springer, Heidelberg (2002)

9. Bellare, M., Palacio, A.: The knowledge-of-exponent assumptions and 3-round zero-knowledge protocols. In: Franklin, M. (ed.) CRYPTO 2004. LNCS, vol. 3152, pp. 273–289. Springer, Heidelberg (2004)

10. Boyle, E., Segev, G., Wichs, D.: Fully leakage-resilient signatures. In: Paterson, K.G. (ed.) EUROCRYPT 2011. LNCS, vol. 6632, pp. 89–108. Springer, Heidelberg (2011)

11. Damgård, I.B.: Towards practical public key systems secure against chosen ciphertext attacks. In: Feigenbaum, J. (ed.) CRYPTO 1991. LNCS, vol. 576, pp. 445–456. Springer, Heidelberg (1992)

12. Dodis, Y., Haralambiev, K., Lpez-Alt, A., Wichs, D.: Cryptography against continuous memory attacks. In: FOCS, pp. 511–520 (2010)

13. Dodis, Y., Haralambiev, K., López-Alt, A., Wichs, D.: Efficient public-key cryptography in the presence of key leakage. In: Abe, M. (ed.) ASIACRYPT 2010. LNCS, vol. 6477, pp. 613–631. Springer, Heidelberg (2010)

14. Gennaro, R.: Multi-trapdoor commitments and their applications to proofs of knowledge secure under concurrent man-in-the-middle attacks. In: Franklin, M. (ed.) CRYPTO 2004. LNCS, vol. 3152, pp. 220–236. Springer, Heidelberg (2004)

15. Garg, S., Jain, A., Sahai, A.: Leakage-resilient zero knowledge. In: Rogaway, P. (ed.) CRYPTO 2011. LNCS, vol. 6841, pp. 297–315. Springer, Heidelberg (2011)

16. Groth, J.: Short pairing-based non-interactive zero-knowledge arguments. In: Abe, M. (ed.) ASIACRYPT 2010. LNCS, vol. 6477, pp. 321–340. Springer, Heidelberg (2010)

17. Gupta, D., Sahai, A.: On constant-round concurrent zero-knowledge from a knowledge assumption. Cryptology ePrint Archive, Report 2012/572 (2012)

18. Hada, S., Tanaka, T.: On the existence of 3-round zero-knowledge protocols. In: Krawczyk, H. (ed.) CRYPTO 1998. LNCS, vol. 1462, pp. 408–423. Springer, Heidelberg (1998)
19. Katz, J.: Efficient cryptographic protocols preventing "Man-in-the-Middle" attacks. Doctor of Philosophy Dissertation, Columbia University, USA (2002)
20. Katz, J.: Efficient and non-malleable proofs of plaintext knowledge and applications. In: Biham, E. (ed.) EUROCRYPT 2003. LNCS, vol. 2656, pp. 211–228. Springer, Heidelberg (2003)
21. Kiltz, E.: Chosen-ciphertext security from tag-based encryption. In: Halevi, S., Rabin, T. (eds.) TCC 2006. LNCS, vol. 3876, pp. 581–600. Springer, Heidelberg (2006)
22. Kawachi, A., Tanaka, K., Xagawa, K.: Concurrently secure identification schemes based on the worst-case hardness of lattice problems. In: Pieprzyk, J. (ed.) ASIACRYPT 2008. LNCS, vol. 5350, pp. 372–389. Springer, Heidelberg (2008)
23. MacKenzie, P.D., Reiter, M.K., Yang, K.: Alternatives to non-malleability: definitions, constructions, and applications. In: Naor, M. (ed.) TCC 2004. LNCS, vol. 2951, pp. 171–190. Springer, Heidelberg (2004)
24. Nielsen, J.B., Venturi, D., Zottarel, A.: Leakage-resilient signatures with graceful degradation. In: Krawczyk, H. (ed.) PKC 2014. LNCS, vol. 8383, pp. 362–379. Springer, Heidelberg (2014)

LWE-Based Encryption

LWE-Based FHE with Better Parameters

Fuqun Wang[1,2,3(✉)], Kunpeng Wang[1,2], and Bao Li[1,2]

[1] State Key Laboratory of Information Security, Institute of Information Engineering, Chinese Academy of Sciences, Beijing, China
{fqwang,kpwang,lb}@is.ac.cn
[2] Data Assurance and Communication Security Research Center, Chinese Academy of Sciences, Beijing, China
[3] University of Chinese Academy of Sciences, Beijing, China

Abstract. Fully homomorphic encryption allows a remote server to perform computable functions over the encrypted data without decrypting them first. We propose a public-key fully homomorphic encryption with better parameters from the learning with errors problem. It is a public key variant of Alperin-Sheriff and Peikert's symmetric-key fully homomorphic encryption (CRYPTO 2014), a simplified symmetric version of Gentry, Sahai and Waters' (CRYPTO 2013), and is based on a minor variant of Regev's public-key encryption scheme (STOC 2005) which may be of independent interest. Meanwhile, it can not only homomorphically handle "NAND" circuits, but also conform to the traditional wisdom that circuit evaluation procedure should be a naive combination of homomorphic additions and multiplications.

Keywords: Fully homomorphic encryption · Learning with errors · Subgaussian random variable

1 Introduction

Fully homomorphic encryption (FHE), allowing anyone to evaluate any efficiently computable function upon the encrypted data without using the knowledge of secret key, was first constructed by Gentry [14,15] in 2009. And ever since then, it has been being a hot area of research in cryptography for its valuable applications in cloud computing setting. Plenty of FHE schemes [3–5,8–10,12,13,16–21,23, 26,27] have been proposed.

Originally, Gentry's blueprint [14,15] for building an FHE scheme consists of three steps: constructing a somewhat homomorphic encryption, squashing the decryption circuit and bootstrapping. In this wise, we can obtain an unbounded FHE scheme combined with an additional circular security assumption.

All FHE schemes followed the Gentry's blueprint, until a leveled one without bootstrapping was proposed by Brakerski, Gentry and Vaikuntanathan [4]. But it

This work is supported in part by the National Nature Science Foundation of China (Grant No. 61272040 and No. 61379137), and in part by the National Basic Research Program of China (973 project) (Grant No. 2013CB338001).

K. Tanaka and Y. Suga (Eds.): IWSEC 2015, LNCS 9241, pp. 175–192, 2015.
DOI: 10.1007/978-3-319-22425-1_11

yet needs evaluation key, i.e., the encrypted secret-key, to perform the successive "*dimension-modulus switching*" which appears in [9] at the earliest.

In CRYPTO 2013, Gentry, Sahai and Waters [17] proposed the first leveled FHE without evaluation key (denoted as GSW) from learning with errors (LWE) problem using "*approximation eigenvector*" method. Specifically, as the ciphertexts are matrices and the homomorphic multiplications (additions) are roughly the multiplications (additions) of ciphertext matrices, GSW does not use the evaluation key to aid homomorphic evaluation, which naturally leads the first identity-based and attribute-based FHE schemes to appear.

Shortly after observing that the errors in GSW grow with quasi-additivity under asymmetric homomorphic multiplications, Brakerski and Vaikuntanathan [10] derived the first FHE scheme from GSW based on approximation shortest vector problem (GapSVP) with low-degree polynomial approximation factors. As the depth of the decryption circuit of arbitrary LWE-based FHE scheme is $c \log \lambda$, where $c \geq 1$ is a constant and λ is the security parameter, the number of homomorphic operations in one-time bootstrapping is roughly $\Omega(\lambda^{2c})$ by Barrington's theorem.

Recently, Alperin-Sheriff and Peikert [3] proposed an original and essential bootstrapping algorithm that can homomorphically and directly evaluate the decryption function instead of the decryption circuit transformed from the native decryption function. Their bootstrapping algorithm uses a simplified GSW's symmetric-key fully homomorphic encryption (SK-FHE) scheme with simpler and tighter analysis of errors, and results in a quasi-linear $\tilde{O}(\lambda)$ number of homomorphic operations in one-time bootstrapping. The most important of all is that the crafted bootstrapping algorithm can be essentially applied in any LWE-based FHE scheme without loss of efficiency caused by the transformation from the native decryption function to the decryption circuit, which is undoubtedly a new milestone in the study of (unbounded) fully homomorphic cryptography.

1.1   Motivation and Techniques

Alperin-Sheriff and Peikert [3] simplified Gentry, Sahai and Waters' SK-FHE [17] to construct a bootstrapping algorithm that can bootstrap essentially any LWE-based FHE scheme. In order to bootstrap, an SK-FHE scheme is sufficient as demonstrated by Alperin-Sheriff and Peikert [3]. However, their techniques, using the standard subgaussian random variable and perfect re-randomization to analyze the behavior of errors under homomorphic operations, are simpler and tighter than that in [4,8–10,17]. Therefore, it is an interesting work to construct a PK-FHE roughly following their novel techniques, especially with better parameters.

Our first observation is that it is not well-content to transform directly the Alperin-Sheriff and Peikert's SK-FHE to a PK-FHE following *approximation eigenvector* method and then analyze the behavior of errors under homomorphic computations following Alperin-Sheriff and Peikert's techniques. The reason is that if we do so, we can not obtain an optimal scheme with better error analysis and hence smaller parameters remaining secure with 2^λ security under standard LWE assumption.

More precisely, for a fresh ciphertext $\mathbf{C}$, we can get $\mathbf{C} \cdot \mathbf{s} = \mu\mathbf{Gs} + \mathbf{Re}$ (mod q) following directly Alperin-Sheriff and Peikert's construction, where $\mathbf{R} \xleftarrow{\$} \{0,1\}^{N \times m}$ is a random matrix used in encryption. Although $\mathbf{Re}$ is zero mean and its entries are mutually independent, which conform with the novel techniques of noise analysis used by Alperin-Sheriff and Peikert, the drawback is that the spectral norm $s_1(\mathbf{R})$ of $\mathbf{R} \in \{0,1\}^{N \times m}$ may be as large as m ($N \leq m$). On the contrary, we choose $\mathbf{R}$ from a distribution $\chi_2^{N \times m}$ defined in Sect. 3. For $\mathbf{R} \leftarrow \chi_2^{N \times m}$, it is easy to see that $\mathbf{Re}$ remains zero mean and its entries are mutually independent, and that $s_1(\mathbf{R}) \leq O(\sqrt{m})$ with overwhelming probability. However, we also bring in a new problem: the scheme using such new randomness in encryption is secure or not. We give a positive answer in Sects. 3 and 4. We propose a minor variant of Regev's public-key encryption scheme and show that it is semantically secure in Sect. 3.

The second observation is that all the schemes in [3, 10, 17] only can perfectly compute NAND-circuits, deviating from the conventional wisdom that circuit evaluation procedure should be a naive combination of homomorphic additions and multiplications [10]. In fact, they enable themselves to conform with the conventional wisdom at a cost of using an "*extended*" decryption procedure, i.e., a sequence of original decryption, to decrypt bit by bit. However, complicating the decryption procedure is a bad news to FHE with bootstrapping. Worse still, all the schemes in [3, 10, 17] with some parameters even can not decrypt correctly after one-time homomorphic addition while maintaining the original decryption procedure. The behavior is very different from them and the scheme constructed by Brakerski [8], on which are built. The reason bringing in the difference is roughly that their decryption procedures are different, although they look a lot like.

More precisely, for some secret key $\mathbf{s} \in \mathbb{Z}_q^d$ and a binary ciphertext $\mathbf{c} \in \{0,1\}^d$, the decryption function of GSW can be written as

$$\mathsf{Dec}_{\mathsf{GSW}}(\mathbf{c}, \mathbf{s}) = \left[\frac{\lfloor \langle \mathbf{c}, \mathbf{s} \rangle \rceil_q}{p} \right]_2,$$

where certain $p \in [q/4, q/2)$, while the decryption function of Brakerski's scheme [8], i.e., original Regev's decryption, can be written as

$$\mathsf{Dec}_{\mathsf{Regev}}(\mathbf{c}, \mathbf{s}) = \left[\frac{\lfloor \langle \mathbf{c}, \mathbf{s} \rangle \rceil_q}{q/2} \right]_2.$$

It is very easy to check that, given two ciphertexts $\mathbf{C}_i = \mathsf{Enc}_{\mathsf{Regev}}(pk, 1)$ with small errors, $i = 1, 2$, we can decrypt $\mathbf{C}_1 + \mathbf{C}_2$ and gain a plaintext 0. But if $\mathbf{C}_i = \mathsf{Enc}_{\mathsf{GSW}}(pk, 1)$ and $p \approx \frac{3}{8}q$, it is likely to happen that we decrypt $\mathbf{C}_1 + \mathbf{C}_2$ to a wrong plaintext "1", since

$$[\langle \mathbf{c}_1 + \mathbf{c}_2, \mathbf{s} \rangle]_q \approx [(\mathbf{e}_1 + \frac{3}{8}q) + (\mathbf{e}_2 + \frac{3}{8}q)]_q = (\mathbf{e}_1 + \mathbf{e}_2) - \frac{1}{4}q,$$

where c_i is the second row of C_i. This is the reason of that the FHE schemes in [3,10,17] only can homomorphically compute an addition-gate with at most one ciphertext encrypting "1" when bootstrapping, and that they can perfectly be bootstrapped with circuits of NAND gates as it always maintains the message belonging to $\{0,1\}$.

Our FHE scheme can partly resolve this drawback, because it can handle depth-$o(\log \lambda)$ circuits with additions and multiplications. In lots of practical application scenarios, a somewhat homomorphic encryption is sufficient. Moreover, as the schemes in [3,10,17], our scheme also has the merit of quasi-additivity of errors growth under sequentially homomorphic multiplications, which is convenient to compute homomorphically multivariate polynomials with degree-poly(λ) and poly(λ) monomials.

1.2   Contribution

Our main contribution is to build a public-key fully homomorphic encryption (PK-FHE) with better parameters. The proposed PK-FHE scheme is a non-trivial public-key variation of the simplified symmetric-key GSW proposed by Alperin-Sheriff and Peikert [3], and follows their analytical methods – using standard subgaussian random variable in conjunction with perfect re-randomization – which result in simpler and tighter analysis of errors. It also relies on a slight variant of Regev's public-key encryption designed in Sect. 3 which may be of independent interest. More specifically, to reduce parameters further (smaller parameters, higher efficiency), we choose randomness used in the encryption algorithm from a distribution χ_2 defined in Sect. 3 instead of the uniform distribution over $\{0,1\}$, which will roughly decrease modulus q by a factor of $O(\sqrt{m})$ where m is the dimension of the underlying lattice.

Conventionally (before GSW), circuit evaluation process is a naive combination of homomorphic additions and multiplications [10]. However, we observe that, if we require that the receiver must use the decryption circuit to decrypt a ciphertext,[1] the FHE schemes built in [3,10,17] can not correctly compute some circuits of additions and multiplications, unlike the former scheme [8] on which they are built. The bootstrapping algorithms in [3,10] rely heavily on this fact and they have to assure that the message always belong to $\{0,1\}$ throughout the bootstrapping procedure. One merit of our scheme is that it is able to handle homomorphic additions and multiplications without changing the decryption circuit. Although this power is very limited, it is convenient and efficient to homomorphically compute multivariate polynomials with degree-poly(λ) and poly(λ) monomials, which are sufficient in lots of practical applications.

Moreover, our scheme is able to handle NAND-circuits as [3,10,17] as well. We can then gain an unbounded FHE scheme with polynomial modulus q using Alperin-Sheriff and Peikert's Bootstrapping algorithm [3].

---

[1] Although in fact we do not do so, we stress on the difference of the decryption procedures between GSW [17] and Brakerski's FHE [8].

1.3   Other Related Works

Not long ago, Ducas and Micciancio [13] proposed an efficient ring-variant of GSW from the scheme in [3] to bootstrap LWE-based FHEs with most significant bit message encoding in less than a second. But, since our scheme is an LWE-based FHE with least significant bit message encoding, we can not use their crafted bootstrapping algorithm. It is an interesting problem to design a bootstrapping algorithm following [13] in the least significant bit message encoding setting. The main difficulty is how to homomorphically *"extract"* a ciphertext of the least significant bit after the homomorphic inner product computation of the targeted ciphertext and the encrypted secret-key.

Recently, Hiromasa, Abe and Okamoto [18] proposed a packed variant of GSW additionally assuming circular security assumption and an optimized bootstrapping algorithm from Alperin-Sheriff and Peikert's one [3] to bootstrap any LWE-based FHEs including the proposed FHE scheme in this work.

1.4   Roadmap

We begin with backgrounds and preliminaries in Sect. 2. We then formally describe a variation of RPKE in Sect. 3, from which we present the proposed leveled FHE in Sect. 4. Finally, we conclude in Sect. 5.

2   Preliminaries

We denote matrices by bold-face upper-case letters (e.g., $\mathbf{A}, \mathbf{B}$) and vectors by bold-face lower-case letters (e.g., $\mathbf{r}, \mathbf{e}$). We use $[\mathbf{A}||\mathbf{B}]$ to denote the concatenation of two matrices. For an integer q, we define $\mathbb{Z}_q \triangleq (-q/2, q/2] \cap \mathbb{Z}$. For any $a \in \mathbb{Z}$, we let $b = [a]_q$ denote the unique value $b \in (-q/2, q/2]$ such that $b = a \pmod{q}$.

We use $x \xleftarrow{\$} \mathcal{S}$ to denote that x was sampled uniformly at random from a set $\mathcal{S}$ and $x \leftarrow \mathcal{D}$ to denote that x was sampled from a distribution $\mathcal{D}$.

Throughout, we let λ denote the *security parameter*. When we say a negligible function $\mathrm{negl}(\lambda)$, we mean a function that grows slower than λ^{-c} for any constant $c > 0$ and sufficiently large values of λ. When we say an event happens with overwhelming probability, we mean that it happens with probability at least $1 - \mathrm{negl}(\lambda)$ for some negligible function $\mathrm{negl}(\lambda)$. Moreover, we use $\mathcal{X} \approx_s \mathcal{Y}$ to denote *statistical indistinguishability* and $\mathcal{X} \approx_c \mathcal{Y}$ to denote *computational indistinguishability*.

2.1   Homomorphism

In this subsection, we will give some definitions related to homomorphic encryption. However, we only give the definitions of homomorphic encryption and semantic security for limited space. The other important notions and properties (e.g., leveled FHE, unbounded (or pure) FHE and bootstrapping theorem) used in this work can be found in [9,14,26] and hence are omitted.

Definition 1. *A public-key homomorphic encryption scheme* HE = (HE.KeyGen, HE.Enc, HE.Dec, HE.Eval) *consists of a quadruple of probabilistic polynomial time* (PPT) *algorithms with syntax as follows:*

- **KeyGeneration** $(pk, evk, sk) \leftarrow$ HE.KeyGen(1^λ): *Output a public encryption key pk, a public evaluation key evk and a secret decryption key sk.*
- **Encryption** $c \leftarrow$ HE.Enc(μ, pk): *For a bit message* $\mu \in \{0,1\}$, *output a ciphertext c using the public key pk.*
- **Decryption** $\mu^* \leftarrow$ HE.Dec(c, sk): *Using the secret key sk, decrypt a ciphertext c to a message* $\mu^* \in \{0,1\}$.
- **Homomorphic Evaluation** $c_f \leftarrow$ HE.Eval$(evk, f, c_1, c_2, \ldots, c_t)$: *Using the evaluation key evk, apply a circuit* $f : \{0,1\}^t \rightarrow \{0,1\}$ *to* $c_1, c_2, \ldots, c_t$, *and output a ciphertext* c_f.

The notion of security we consider in this work is CPA security, defined as follows.

Definition 2 (IND-CPA Security). *An* HE *scheme is* IND-CPA *secure if any* PPT *adversary who gets a valid public key pk at first, then chooses two messages* m_0, m_1 *with identical length, and then obtains* HE.Enc$_{pk}(m_b)$ *for uniformly random b, finally can only guess b with probability at most* $1/2 + \text{negl}(\lambda)$.

We also consider a stronger INDr-CPA security notion where a ciphertext is computationally indistinguishable from a random membership in ciphertext space.

2.2 Useful Tools

Here we recall some results about the spectral norm of a matrix, subgaussian random variable/vector/matrix and gadget matrix.

Spectral Norm. The spectral norm (the biggest singular value) $s_1(\mathbf{A})$ of a real matrix $\mathbf{A}$ is defined as $s_1(\mathbf{A}) \triangleq \max_{\mathbf{y}} ||\mathbf{A}\mathbf{y}||_2 / ||\mathbf{y}||_2$ for all $\mathbf{y} \neq \mathbf{0}$ (Note that $||\mathbf{y}||_2 = \sqrt{\sum y_i^2} = s_1(\mathbf{y})$ if $\mathbf{y} \in \mathbb{R}^n$). It also represents that spectral norm of a real matrix is induced by the Euclidean norm, so it is apparently sub-multiplicative.

Lemma 1 ([7], Sub-multiplication). *For two real matrices* $\mathbf{A} \in \mathbb{R}^{n \times k}, \mathbf{B} \in \mathbb{R}^{k \times m}$, *it holds that* $s_1(\mathbf{AB}) \leq s_1(\mathbf{A}) \cdot s_1(\mathbf{B})$.

Subgaussian Random Variable. As in the work [3], it is beneficial to analyze the behavior of noise using the standard notion of subgaussian random variable in this paper. A detailed material is given in [28]. A real random variable X is subgaussian with parameter $s \geq 0$ if for all $t \geq 0$, it holds that $\Pr[|X| > t] \leq 2\exp(-\pi t^2/s^2)$. It is obvious to see that the Gaussian tails also imply subgaussianity if $\mathbb{E}[X] = 0$. Any B-bounded zero-mean random variable X is subgaussian with parameter $B\sqrt{2\pi}$.

In our analysis of behavior of noise term we will use the following lemma.

Lemma 2 ([7]). *Let* $X_1, X_2, \ldots, X_k$ *be independent, 0-mean, real subgaussian random variables with parameter* s *and* $\mathbf{a} = (a_1, a_2, \ldots, a_k) \in \mathbb{R}^k$. *Then* $\sum_i (a_i X_i)$ *is a subgaussian random variable with parameter* $s \cdot \|\mathbf{a}\|_2$.

As a matter of course, it can be extended the notion of subgaussianity to vectors: a random real vector $\mathbf{x}$ is subgaussian with parameter s if for all unit vector $\mathbf{u}$, the marginal $\langle \mathbf{x}, \mathbf{u} \rangle \in \mathbb{R}$ is subgaussian with parameter s. It follows directly from the definition that the concatenation of independent subgaussian random variables with identical s is also subgaussian with parameter s. Furthermore, the notion can be extended to subgaussian random matrix in direct way.

The following lemma is a standard result of subgaussianity [28].

Lemma 3 ([28]). *Let* $\mathbf{X} \in \mathbb{R}^{n \times m}$ *be a subgaussian random matrix with parameter* s, *then there exists a universal constant* $C > 0$ *such that for any* $t \geq 0$, $s_1(\mathbf{X}) \leq C \cdot s \cdot (\sqrt{m} + \sqrt{n} + t)$ *except with probability at most* $2 \exp(-\pi t^2)$. *In particular, for* $n \leq m$ *and* $t = \omega(\sqrt{\log n})$, *it holds that* $s_1(\mathbf{X}) \leq cs\sqrt{m}$ *for some quite small constant* c *with overwhelming probability.*

Gadget Matrix. For an odd integer q, let the *gadget matrix* $\mathbf{G} = \mathbf{I}_n \otimes \mathbf{g} \in \mathbb{Z}_q^{(n \cdot \lceil \log q \rceil) \times n}$, where $\mathbf{g}^T = (1, 2, 2^2, \ldots, 2^{\lfloor \log q \rfloor})$ and $\mathbf{I}_n$ denotes the n-dimension identity matrix. Our work will rely heavily on the following *randomized subgaussian decomposition* with parameter $O(1)$ ($\leq \sqrt{2\pi}$) of a matrix $\mathbf{A}$, in contrast to the deterministic bit-decomposition used in most LWE-based FHE schemes.

Lemma 4 ([3,22]). *For any matrix* $\mathbf{A} \in \mathbb{Z}_q^{m \times n}$, *given the gadget matrix* $\mathbf{G}$ *defined above, there exists an efficient and randomized algorithm* (*denoted by* $\mathbf{G}^{-1}$) *to sample a subgaussian matrix* $\mathbf{X} \in \mathbb{Z}_q^{m \times (n \cdot \lceil \log q \rceil)}$ *with parameter* $O(1)$ ($\leq \sqrt{2\pi}$) *such that* $\mathbf{X} = \mathbf{G}^{-1}(\mathbf{A})$ (*i.e.,* $\mathbf{X} \cdot \mathbf{G} = \mathbf{A}$).

2.3   Learning with Errors

The LWE problem was proposed by Regev [25] as an extension of "learning parity with noise". In this section, we will define a variant of LWE problem which is already used in fully homomorphic cryptography [4,9] and give the relation to intractable worst-case lattice problems.

For a positive integer $n \geq 2$, an odd integer $q \geq 2$, an arbitrary vector $\mathbf{s} \in \mathbb{Z}_q^n$ and a probability distribution χ over $\mathbb{Z}$, let $\mathcal{A}_{\mathbf{s},\chi}$ denote a distribution obtained by choosing a vector $\mathbf{a} \xleftarrow{\$} \mathbb{Z}_q^n$ and an error term $e \leftarrow \chi$, and outputting $(\mathbf{a}, [\langle \mathbf{a}, \mathbf{s} \rangle + 2e]_q) \in \mathbb{Z}_q^n \times \mathbb{Z}_q$. The search learning with errors ($\mathsf{LWE}_{n,m,q,\chi}$) problem is, given $m = \mathrm{poly}(n)$ independent samples from $\mathcal{A}_{\mathbf{s},\chi}$, to find $\mathbf{s}$ for some random $\mathbf{s} \in \mathbb{Z}_q^n$. The decisional learning with errors ($\mathsf{DLWE}_{n,m,q,\chi}$) problem is, given m independent samples, to decide that, with non-negligible advantage, they are sampled from $\mathcal{A}_{\mathbf{s},\chi}$ for a uniformly random $\mathbf{s} \in \mathbb{Z}_q^n$, or from the uniform distribution over $\mathbb{Z}_q^n \times \mathbb{Z}_q$.

We often write $\mathsf{DLWE}_{n,m,q,\alpha}$ to denote $\mathsf{DLWE}_{n,m,q,\chi}$ for $\chi = \mathcal{D}_{\mathbb{Z},\alpha q}$. If there is not a priori bounded number of samples, we use $\mathsf{DLWE}_{n,q,\chi}$ to denote the variant where the attacker can get an oracle access to $\mathcal{A}_{\mathbf{s},\chi}$.

The following two propositions state quantum and classical reductions from GapSVP or shortest independent vector problem (SIVP) in the worst-case to $\mathsf{DLWE}_{n,q,\chi}$, where χ is a discrete Gaussian distribution $\mathcal{D}_{\mathbb{Z},\alpha q}$.

Proposition 1 ([10,22,24,25]). *Let $q = q(n) \in \mathbb{N}$ be either a prime power or a product of co-prime primes, and let $\alpha \geq 2\sqrt{n}/q$. If there exists an efficient algorithm that solves the average-case $\mathsf{DLWE}_{n,q,\alpha}$ problem, then:*

- *There is an efficient quantum algorithm for $\mathsf{GapSVP}_{\tilde{O}(n/\alpha)}$ and $\mathsf{SIVP}_{\tilde{O}(n/\alpha)}$ on arbitrary n-dimensional lattice.*
- *If $q \geq \tilde{O}(2^{n/2})$, then there is an efficient classical algorithm that solves $\mathsf{GapSVP}_{\tilde{O}(n/\alpha)}$ on arbitrary n-dimensional lattice.*

Proposition 2 ([6]). *Solving n-dimension DLWE with poly (n) modulus implies an equally efficient algorithm to GapSVP and SIVP in dimension $\sqrt{n}$.*

A useful claim [1] is that LWE is still hard if the secret key $\mathbf{s}$ is chosen from the error distribution χ, which applies far and wide in lattices-based cryptography including the construction in this work.

3   Building Block

In this section, we propose a slight variant of the Regev's public-key encryption scheme [25] (hereafter also called RPKE) as a cornerstone of our main construction. It follows the basic scheme (without homomorphism) in [4,9] in the same flavor except the randomness used in the encryption algorithm.

- RPKE.Setup(1^λ): Let $q = q(\lambda)$ (odd integer) be the LWE modulus and $n - 1 = n(\lambda) - 1$ be the LWE dimension, where λ is the security parameter. Set $\ell = \lceil \log q \rceil$ and $m = O(n \log q)$. Let χ_1 be the error distribution over $\mathbb{Z}$ for LWE which we assume to be a discrete (sub)gaussian distribution with parameter s. Let χ_2 be a discrete distribution over $\{-1, 0, 1\}$ from which we choose 0 with probability $\frac{1}{2}$, and 1 and -1 each with probability $\frac{1}{4}$. Let $params = (n, m, q, \chi_1, \chi_2)$.
- RPKE.SKGen($params$): Sample a vector $\mathbf{t} \leftarrow \chi_1^{n-1}$. Let secret key $sk = \mathbf{s} = (1, \mathbf{t}) \in \mathbb{Z}_q^n$. Note that here $\mathbf{t} \xleftarrow{\$} \mathbb{Z}_q^{n-1}$ is permissible. However, in next section, a smaller $\mathbf{t}$ is necessary for the FHE with Alperin-Sheriff and Peikert's bootstrapping algorithm.
- RPKE.PKGen($params, sk$): Choose a matrix $\mathbf{A} \xleftarrow{\$} \mathbb{Z}_q^{m \times (n-1)}$ and a vector $\mathbf{e} \leftarrow \chi_1^m$. Compute $\mathbf{b} = -\mathbf{A}\mathbf{t} + 2\mathbf{e}$. Set the public key $pk = \mathbf{P} = [\mathbf{b}\|\mathbf{A}] \in \mathbb{Z}_q^{m \times n}$. (Remark that $\mathbf{P} \cdot \mathbf{s} = 2\mathbf{e}$.)
- RPKE.Enc($\mu \in \{0,1\}, pk$): To encrypt a bit $\mu \in \{0,1\}$, choose a vector $\mathbf{r} \leftarrow \chi_2^m$ and output the ciphertext (column) vector

$$\mathbf{c} = \mu \cdot \hat{\mathbf{e}}_1 + \mathbf{P}^T \cdot \mathbf{r} \in \mathbb{Z}_q^n,$$

  where $\hat{\mathbf{e}}_1$ is the first standard n-dimension unit vector.
- RPKE.Dec($\mathbf{c}, sk$): To decrypt $\mathbf{c} \in \mathbb{Z}_q^n$ using secret key $sk = \mathbf{s}$, compute and output

$$\mu' = \langle \mathbf{c}, \mathbf{s} \rangle \bmod q \bmod 2.$$

Correctness. For arbitrary ciphertext $\mathbf{c}$, we will keep the invariant such that

$$\langle \mathbf{c}, \mathbf{s} \rangle = \mu + 2 \cdot e \pmod{q} \tag{1}$$

where μ is the plaintext, $\mathbf{s}$ is the secret key and $2e$ is the error. This leads us to define the error term (1-dimension vector) in arbitrary ciphertext $\mathbf{c}$ as follows.

Definition 3. Let $\mathbf{c} \in \mathbb{Z}_q^n$, $e \in \mathbb{Z}$ and $\mu \in \{0, 1\}$ such that the invariant form (1) holds. We define $2e$ is the error term, i.e., $\mathsf{Error}_\mu(\mathbf{c}) = 2e$.

We now proceed to analyze the error magnitude in a fresh ciphertext and show the correctness of decryption for low-error ciphertexts in following lemma.

Lemma 5. Let $params = \{n, m, q, \chi_1, \chi_2\}$ be parameters for RPKE. Then, for each $\mu \in \{0, 1\}$, if $\mathbf{s} \leftarrow \mathsf{RPKE.SKGen}(params)$, $\mathbf{P} \leftarrow \mathsf{RPKE.PKGen}(params, sk)$ and $\mathbf{c} \leftarrow \mathsf{RPKE.Enc}(\mu, pk)$, we have that error term $\mathsf{Error}_\mu(\mathbf{c})$ is subgaussian with parameter $2cs \cdot \sqrt{2\pi m}$ where c comes from Lemma 3.

In particular, we can decrypt correctly with overwhelming probability if the modulus q of the system is at least $2cs \cdot \sqrt{2\pi m} \cdot \omega(\sqrt{\log \lambda})$.

Proof. Given above parameters, we have

$$\langle \mathbf{c}, \mathbf{s} \rangle = \mu + 2 \cdot \langle \mathbf{P}^T \mathbf{r}, \mathbf{s} \rangle = \mu + 2 \cdot \langle \mathbf{r}, \mathbf{e} \rangle.$$

By Lemma 2, 3, we then have $2 \cdot \langle \mathbf{r}, \mathbf{e} \rangle$ is subgaussian with parameter

$$2 \cdot \|\mathbf{r}\|_2 \cdot s \le 2 \cdot c \cdot \sqrt{2\pi m} \cdot s = 2cs \cdot \sqrt{2\pi m}.$$

The correctness of decryption follows, which finishes the proof.   □

Security. In order to show the security of RPKE, we need the next lemma adapted from [2].

Lemma 6 ([2]). Let $\mathcal{Y} = \mathbb{Z}_q$ be a finite additive group and $m \ge 1$ be an integer. For $\mathbf{a} \in \mathcal{Y}^m$, define $h_\mathbf{a} : \chi_2^m \to \mathcal{Y}$ as $h_\mathbf{a} = \sum_{i=1}^m x_i a_i$. Then the family $\mathcal{H} = \{h_\mathbf{a}\}_{\mathbf{a} \in \mathcal{Y}^m}$ is 2-universal. It follows that $(h_\mathbf{a}, h_\mathbf{a}(x))$ is $\frac{1}{2}\sqrt{q/2^m}$-uniform.

By Lemma 6, we can show the below theorem claiming the security of RPKE.

Theorem 1. Let $params = \{n, m, q, \chi_1, \chi_2\}$ be parameters for RPKE. For each $\mu \in \{0, 1\}$, if $\mathbf{s} \leftarrow \mathsf{RPKE.SKGen}(params)$, $\mathbf{P} \leftarrow \mathsf{RPKE.PKGen}(params, sk)$ and $\mathbf{c} \leftarrow \mathsf{RPKE.Enc}(\mu, pk)$, it holds that the joint distribution $(\mathbf{P}, \mathbf{c})$ is computationally indistinguishable from uniform assuming the $\mathsf{DLWE}_{n,m,q,\chi_1}$ assumption. In other words, RPKE constructed above is INDr-CPA secure under $\mathsf{DLWE}_{n,m,q,\chi_1}$ assumption.

Proof. We now show that the scheme RPKE constructed above is INDr-CPA secure, i.e., a fresh ciphertext is computationally indistinguishable from a random membership in ciphertext space. To this end, we only need to show that a fresh ciphertext of 0 is indistinguishable from a random ciphertext.

For uniformly random values $\mathbf{u}, \mathbf{v}, v$, it holds that

$$(\mathbf{P}, \mathbf{c}) = (\mathbf{A}, -\mathbf{At} + 2\mathbf{e}, \mathbf{A}^T\mathbf{r}, (-\mathbf{At} + 2\mathbf{e})^T\mathbf{r})$$
$$\approx_c (\mathbf{A}, \mathbf{u}, \mathbf{A}^T\mathbf{r}, \mathbf{u}^T\mathbf{r})$$
$$\approx_s (\mathbf{A}, \mathbf{u}, \mathbf{v}, v)$$

where the equality holds by the definitions of algorithms in RPKE, the computational indistinguishability holds under the $\mathsf{DLWE}_{n,m,q,\chi_1}$ assumption and the statistical indistinguishability holds by Lemma 6.

Therefore, in the view of an INDr-CPA attacker, $\mathbf{c} \approx_c (\mathbf{v}, v)$ holds. Theorem 1 follows. $\qquad\square$

4   A New LWE-Based PK-FHE

In this section, we will describe formally a leveled PK-FHE (called YAP since it is yet AP-type FHE) as a simplification and improvement of GSW. We begin with the YAP scheme in Sect. 4.1 and show its homomorphic property in Sect. 4.2. In Sect. 4.3 we will overview bootstrapping and obtain an unbounded FHE with polynomial errors. The error analysis and bootstrapping closely follow [3].

4.1   FHE Scheme: YAP

Now, we present the scheme YAP based on RPKE constructed in Sect. 3, closely following the newly wonderful Gentry, Sahai and Waters' symmetric-key variant demonstrated by Alperin-Sheriff and Peikert [3].

Looking ahead, although the scheme YAP described in this section is very similar to Alperin-Sheriff and Peikert's, it is able to conform with the traditional wisdom. However, it is not good at operations with additions and multiplications. We thus consider NAND-circuits as well.

- YAP.Setup($1^\lambda, 1^L$): This algorithm is identical to RPKE's except another parameters L, i.e., the maximum depth of circuits which the scheme is able to handle, and $N = n \cdot \ell$ fully determined by n and q. Let yet $params = (n, m, q, \chi_1, \chi_2)$.
- YAP.SKGen($params$): This algorithm is identical to RPKE's, hence outputs $sk \leftarrow$ RPKE.SKGen. Recall that $sk = \mathbf{s} = (1, \mathbf{t}) \in \mathbb{Z}_q^n$.
- YAP.PKGen($params, sk$): This algorithm is identical to RPKE's, so outputs $pk \leftarrow$ RPKE.PKGen. Note that $pk = \mathbf{P} = [\mathbf{b}||\mathbf{A}] \in \mathbb{Z}_q^{m \times n}$ such that $\mathbf{Ps} = 2\mathbf{e}$.
- YAP.Enc(μ, pk): To encrypt a bit $\mu \in \{0, 1\}$, choose a random matrix $\mathbf{R} \leftarrow \chi_2^{N \times m}$ and output the ciphertext matrix

$$\mathbf{C} = \mu\mathbf{G} + \mathbf{RP} \in \mathbb{Z}_q^{N \times n},$$

where $\mathbf{G}$ is the gadget matrix defined in Subsect. 2.2.

– YAP.Dec($\mathbf{C}, sk$): Let $\mathbf{c}$ be the first row of $\mathbf{C}$. Using RPKE's decryption algorithm on $\mathbf{c}$, output

$$\mu' = \mathsf{RPKE.Dec}(\mathbf{c}, sk).$$

– YAP.Add($\mathbf{C}_1, \mathbf{C}_2$): Given two ciphertext matrices $\mathbf{C}_1, \mathbf{C}_2$ for two plaintexts μ_1, μ_2 respectively, output

$$\mathbf{C}_{\mathrm{Add}} = \mathbf{C}_1 \oplus \mathbf{C}_2 \triangleq \mathbf{C}_1 + \mathbf{C}_2.$$

– YAP.Multi($\mathbf{C}_1, \mathbf{C}_2$): Given two ciphertext matrices $\mathbf{C}_1, \mathbf{C}_2$ for two plaintexts μ_1, μ_2 respectively, we define a *left-associative* multiplication algorithm YAP.Multi which outputs

$$\mathbf{C}_{\mathrm{Multi}} \leftarrow \mathbf{G} \odot \mathbf{C}_1 \odot \mathbf{C}_2 \triangleq \mathbf{G}^{-1}(\mathbf{G}^{-1}(\mathbf{G}) \cdot \mathbf{C}_1) \cdot \mathbf{C}_2.$$

Note that $\mathbf{G}$ is a valid ciphertext of 1 with error 0 and that the algorithm YAP.Multi is random, since $\mathbf{G}^{-1}$ is a random algorithm.

– YAP.NAND($\mathbf{C}_1, \mathbf{C}_2$): Given two ciphertext matrices $\mathbf{C}_1, \mathbf{C}_2$ for two plaintexts μ_1, μ_2, output

$$\mathbf{C}_{\mathrm{NAND}} \leftarrow \mathbf{G} - \mathbf{G} \odot \mathbf{C}_1 \odot \mathbf{C}_2 \triangleq \mathbf{G} - \mathbf{G}^{-1}(\mathbf{G}^{-1}(\mathbf{G}) \cdot \mathbf{C}_1) \cdot \mathbf{C}_2.$$

Note that this algorithm is also random.

– YAP.Eval($f, \mathbf{C}_1, \mathbf{C}_2, \ldots, \mathbf{C}_t$): apply a circuit $f : \{0,1\}^t \to \{0,1\}$ combined by NAND gates (or by Add and Multi gates) to t ciphertexts $\mathbf{C}_1, \mathbf{C}_2, \ldots, \mathbf{C}_t$, and output a ciphertext $\mathbf{C}_f$.

Note that one should homomorphically compute a circuit gate-by-gate.

Correctness. Below, we will analyze briefly the magnitude of error vector in a fresh ciphertexts, which will lead to the correctness of YAP.

For arbitrary ciphertext $\mathbf{C}$, as the last section, we will keep the invariant such that

$$\mathbf{C} \cdot \mathbf{s} = \mu \mathbf{G} \mathbf{s} + 2\mathbf{e}^* \pmod{q} \tag{2}$$

where μ is the plaintext, $\mathbf{s}$ is the secret key and $2\mathbf{e}^*$ is error term. This leads us to define the error vector in arbitrary ciphertext $\mathbf{C}$ as follows.

Definition 4. *Let arbitrary ciphertext* $\mathbf{C} \in \{0,1\}^{N \times N}$, $\mathbf{e}^* = (e_1, e_2, \ldots, e_m) \in \mathbb{Z}^m$ *and* $\mu \in \{0,1\}$ *such that the invariant form* (2) *holds. Define error vector*

$$\mathsf{Error}_\mu(\mathbf{C}) = 2\mathbf{e}^*.$$

We now measure the error level in any fresh ciphertext and analyze the independence of entries of noise vector. The former leads to the correctness of decryption and the latter is very important to measure the behavior of noise under homomorphic evaluation.

Lemma 7. *Let params $= \{n, m, q, \chi_1, \chi_2\}$ be parameters for YAP. For each secret key $\mathbf{s} \in \mathbb{Z}_q^n$ and $\mu \in \{0, 1\}$, if $\mathbf{C} \leftarrow$ YAP.Enc(μ, pk), we then have that the entries of $\mathsf{Error}_\mu(\mathbf{C})$ are mutually independent and subgaussian with parameter $2cs \cdot \sqrt{2\pi m}$ (c comes from Lemma 3) with overwhelming probability over the coin tosses of all randomized algorithms involved.*

In particular, we can decrypt correctly for any fresh ciphertext if $q/4 \geq 2cs \cdot \sqrt{2\pi m} \cdot \omega(\sqrt{\log \lambda})$.

Proof. The independence of entries of error comes from the independence of the rows of $\mathbf{R}$ and the proof of other result follows the proof of Lemma 5. $\square$

Security. We can easily show that the semantic security of YAP without reference to the homomorphic operations is based on the security of RPKE in previous section using a standard hybrid argument. In other words, it is IND-CPA secure under the average-case $\mathsf{DLWE}_{n,m,q,\chi_1}$ assumption, and hence under the worse-case GapSVP assumption. Please see similar arguments in details in [9,17,25].

4.2 Homomorphic Property of YAP

Now, we show that we can decrypt correctly after kinds of homomorphic operations including YAP.Eval for proper parameters. We begin with the homomorphic addition and multiplication.

Lemma 8. *Let params $= \{n, m, q, \chi_1, \chi_2\}$ be parameters for YAP, where χ_1 is the discrete Gaussian distribution with parameter s. For some secret key $\mathbf{s} \in \mathbb{Z}_q^n$, let $\mu_i \in \{0, 1\}$, $\mathbf{C}_i \in \mathbb{Z}_q^{N \times n}$, $\mathbf{e}_i \leftarrow \chi_1^m$ and $\mathbf{R}_i \leftarrow \chi_2^{N \times m}$ such that $\mathbf{C}_i \cdot \mathbf{s} = \mu_i \mathbf{Gs} + 2\mathbf{R}_i \mathbf{e}_i \pmod{q}$, $i = 1, 2$. Then, for the constant c in Lemma 3, we have with overwhelming probability over the randomness of algorithms involved*

- $\mathsf{Error}_{\mu_1 \oplus \mu_2}(\text{YAP.Add}(\mathbf{C}_1, \mathbf{C}_2)) = 2(\mathbf{R}_1 \mathbf{e}_1 + \mathbf{R}_2 \mathbf{e}_2) \triangleq 2\mathbf{e}^*$, *where the entries of $\mathbf{e}^*$ are mutually independent and subgaussian with parameter $2cs\sqrt{\pi m}$.*
- $\mathsf{Error}_{\mu_1 \otimes \mu_2}(\text{YAP.Multi}(\mathbf{C}_1, \mathbf{C}_2)) = 2(\mu_2 \mathbf{X}_1 \mathbf{R}_1 \mathbf{e}_1 + \mathbf{X}_2 \mathbf{R}_2 \mathbf{e}_2) \triangleq 2\mathbf{e}$, *where $\mathbf{X}_1, \mathbf{X}_2$ are some subgaussian matrices and the entries of $\mathbf{e}$ are mutually independent and subgaussian with parameter $2\sqrt{2}\pi c^2 sm$.*

In particular, we can decrypt correctly after one-time homomorphic addition or multiplication if $q/4 \geq 2\sqrt{2}\pi c^2 sm \cdot \omega(\sqrt{\log \lambda})$.

Proof. The first claim is obvious by Lemma 2, 3 and the second is proved below.

For any two fresh ciphertexts $\mathbf{C}_1, \mathbf{C}_2$, let $\mathbf{C}_{\text{Multi}} = \text{YAP.Multi}(\mathbf{C}_1, \mathbf{C}_2)$. Then, we have

$$
\begin{aligned}
\mathbf{C}_{\text{Multi}} \cdot \mathbf{s} &= (\mathbf{G} \odot \mathbf{C}_1 \odot \mathbf{C}_2) \cdot \mathbf{s} \\
&= (\mathbf{G}^{-1}(\mathbf{G} \odot \mathbf{C}_1) \cdot \mathbf{C}_2) \cdot \mathbf{s} \\
&= \mathbf{X}_2 \cdot (\mathbf{C}_2 \cdot \mathbf{s}) \\
&= \mathbf{X}_2 \cdot (\mu_2 \mathbf{G} \mathbf{s} + 2\mathbf{R}_2 \mathbf{e}_2) \\
&= \mu_2 \cdot (\mathbf{G} \odot \mathbf{C}_1) \mathbf{s} + 2\mathbf{X}_2 \mathbf{R}_2 \mathbf{e}_2 \\
&= \mu_2 \cdot (\mathbf{G}^{-1}(\mathbf{G}) \cdot \mathbf{C}_1) \mathbf{s} + 2\mathbf{X}_2 \mathbf{R}_2 \mathbf{e}_2 \\
&= \mu_2 \cdot \mathbf{X}_1 \cdot (\mu_1 \mathbf{G} \mathbf{s} + 2\mathbf{R}_1 \mathbf{e}_1) + 2\mathbf{X}_2 \mathbf{R}_2 \mathbf{e}_2 \\
&= \mu_1 \mu_2 \mathbf{G} \mathbf{s} + 2(\mu_2 \mathbf{X}_1 \mathbf{R}_1 \mathbf{e}_1 + \mathbf{X}_2 \mathbf{R}_2 \mathbf{e}_2)
\end{aligned}
$$

Thus, it holds that

$$
s_1(\mathbf{X}_1 \mathbf{R}_1) \leq s_1(\mathbf{X}_1) s_1(\mathbf{R}_1) \leq (c \cdot \sqrt{2\pi m}) \cdot (c \cdot \sqrt{2\pi m}) = 2\pi c^2 m,
$$

where the first inequality follows by Lemma 1 and the second inequality follows by Lemma 3 and the fact that $\mathbf{X}_1$ and $\mathbf{R}_1$ are subgaussian random matrices with parameter $\sqrt{2\pi}$. Similarly, $s_1(\mathbf{X}_2 \mathbf{R}_2) \leq 2\pi c^2 m$ also holds. By Lemma 2, we then have that $\mathbf{e}$ is subgaussian with parameter $2\sqrt{2\pi} c^2 sm$. The independence of entries of $\mathbf{e}$ comes from the fact that plaintext μ_2 is 0 or 1 and that the rows of $\mathbf{X}_i$ and $\mathbf{R}_i$ are mutually independent, $i = 1, 2$. The correctness of decryption follows by above results and the definition of subgaussianity. We thus finish the proof. $\square$

As the observation in the introduction, to some extent, our scheme puts up with operating low-depth circuits composed of addition and multiplication gates, which conforms with the conventional wisdom. In order to show this, we will define a special arithmetic circuits class $\text{Arith}[L]$ as [9]. Looking ahead, we must set $L \ll \log \lambda$, because the noise grows double-exponentially with L.

Definition 5. *Let* $L = L(\lambda)$. *The class* $\text{Arith}[L]$ *is the class of arithmetic circuits over* $\text{GF}(2)$, *with* $\{+, \times\}$ *gates and the following layered structure. Each circuit contains exactly $2L$ layers of gates, and gates of layer $i+1$ are fed only by gates of layer i. The odd layers contain only "$+$" gates with fan-in 2, while the even layers contain only "$\times$" with fan-in 2.*

Theorem 2. *Let* $L = o(\log \lambda)$ *and* $params = \{n, m, q, \chi_1, \chi_2\}$ *be parameters for* YAP, *where* χ_1 *is the discrete Gaussian distribution with parameter s. Let* $\mu_i \in \{0, 1\}$, $i \in [t] = \{1, 2, \ldots, t\}$. *For* $\mathbf{C}_i \leftarrow \text{YAP.Enc}(\mu_i, sk)$ *and every arithmetic circuit* $f \in \text{Arith}[L]$, *set* $\mathbf{C}_f \leftarrow \text{YAP.Eval}(f, \mathbf{C}_1, \mathbf{C}_2, \ldots, \mathbf{C}_t)$. *Then, if* $q/4 > 2^{2^{L+1}} \cdot (2c\sqrt{\pi m})^L \cdot cs\sqrt{2\pi m} \cdot \omega(\sqrt{\log \lambda})$, *we have*

$$
\text{YAP.Dec}(\mathbf{C}_f, sk) = f(\mu_1, \mu_2, \ldots, \mu_t)
$$

with overwhelming probability over the randomness of all algorithms related.

Proof. For $k = 0, 1, \ldots, 2L$, let $\mu^{(k)}$ be a *quasi-plaintext* (may be large, not bounded) in level-k and S_k be the subgaussian parameter of entries of $\mathbf{e}^{(k)}$ in $\mathsf{Error}(\mathbf{C}^{(k)})$, where $\mathbf{C}^{(k)}$ is a ciphertext in level-k such that $\mathbf{C}^{(k)} \cdot \mathbf{s} = \mu^{(k)} \mathbf{Gs} + 2\mathbf{e}^{(k)} \pmod{q}$. Note that $S_0 = cs\sqrt{2\pi m}$ for some small constant c. It is very easy to see that $|\mu^{(2k-1)}| \leq 2^{2^{k-1}+2^{k-2}+\cdots+1}$ and $|\mu^{(2k)}| \leq 2^{2^k+2^{k-1}+\cdots+2}$ for $k = 1, 2, \ldots, L$.

Recall that ciphertexts as inputs of arbitrary gate have almost equal subgaussian parameters, since they come from the same layer.

Then, by Lemma 8, for $k \leq L$, a bound S_{2k} that is the subgaussian parameter of entries of $\mathbf{e}^{(2k)}$ in the outputs of layer $2k$ is gained by

$$S_{2k} \leq \left(\sqrt{1 + |\mu^{(2k-1)}|^2} \cdot c \cdot \sqrt{2\pi m}\right) \cdot \left(\sqrt{2} \cdot S_{2(k-1)}\right),$$

and therefore, inductively,

$$S_{2L} \leq \prod_{k=1}^{L} \sqrt{1 + |\mu^{(2k-1)}|^2} \cdot (2c\sqrt{\pi m})^L \cdot S_0 \leq 2^{2^{L+1}} \cdot (2c\sqrt{\pi m})^L \cdot cs\sqrt{2\pi m}.$$

Thus, by proper setting of the parameters, it holds that

$$\|\mathsf{Error}(\mathbf{C}_f)\|_\infty \leq 2^{2^{L+1}} \cdot (2c\sqrt{\pi m})^L \cdot cs\sqrt{2\pi m} \cdot \omega(\sqrt{\log \lambda}) < q/4$$

which ensures the correctness of decryption. $\square$

Through above theorem, we observe that in order to execute homomorphically a low-depth circuit combined by addition and multiplication gates, we need a very large modulus q, which drastically lowers the security of the scheme because of larger modulus lower security for roughly equal level of fresh error. This drawback is essential that the plaintexts involve in the growth of error and that the plaintexts grow double-exponentially with circuit depth L when executing homomorphic operations. It then is very important to retain the plaintexts small. To this end, we give two solutions: one is, in practical side, that we can compute homomorphically and efficiently multivariate polynomials with small coefficients, degree-poly(λ) and poly(λ) monomials, using the quasi-additivity of noise growth under a sequence of asymmetric multiplications; the other is, in theoretical side, that we employ the circuits of NANDs, as [10, 17].

In practical side, by Lemma 2, it is sufficient to show the way of evaluating a monomial as follows. The proof follows Lemma 8 trivially and thus is omitted. (Note that, it plays an important role in bootstrapping in [3]).

Lemma 9. *Let $params = \{n, m, q, \chi_1, \chi_2\}$ be parameters for* YAP, *where χ_1 is the discrete Gaussian distribution with parameter s. For some secret key $\mathbf{s} \in \mathbb{Z}_q^n$, let $\mu_i \in \{0, 1\}$, $\mathbf{C}_i \in \mathbb{Z}_q^{N \times n}$, $\mathbf{e}_i \leftarrow \chi_1^m$ and $\mathbf{R}_i \leftarrow \chi_2^{N \times m}$ such that $\mathbf{C}_i \cdot \mathbf{s} = \mu_i \mathbf{Gs} + 2\mathbf{R}_i \mathbf{e}_i \pmod{q}$, $i \in [k]$, define*

$$\mathsf{YAP.MULTI}(\mathbf{C}_1, \ldots, \mathbf{C}_k) \leftarrow \mathbf{G} \odot \mathbf{C}_1 \odot \cdots \odot \mathbf{C}_k$$
$$\triangleq \mathbf{G}^{-1}(\cdots (\mathbf{G}^{-1}(\mathbf{G}^{-1}(\mathbf{G}) \cdot \mathbf{C}_1) \cdot \mathbf{C}_2) \cdots) \cdot \mathbf{C}_k.$$

Then, for the constant c in Lemma 3, we have, with overwhelming probability over the coin tosses of all algorithms involved,

$$\mathsf{Error}(\mathsf{YAP.MULTI}(\mathbf{C}_1, \ldots, \mathbf{C}_k))$$
$$= 2(\mathbf{X}_k\mathbf{R}_k\mathbf{e}_k + \mu_k\mathbf{X}_{k-1}\mathbf{R}_{k-1}\mathbf{e}_{k-1} + \mu_2 \cdots \mu_k\mathbf{X}_1\mathbf{R}_1\mathbf{e}_1)$$
$$\triangleq 2\mathbf{e}$$

where the entries of $\mathbf{e}$ are mutually independent and subgaussian with parameter $2\sqrt{k}\pi c^2 sm$.

In theoretical side, we have following results.

Lemma 10. *For some secret key $\mathbf{s}$, let $\mu_i \in \{0,1\}$, $\mathbf{C}_i \in \mathbb{Z}_q^{N \times n}$, $\mathbf{e}_i \leftarrow \chi_1^m$ and $\mathbf{R}_i \leftarrow \chi_2^{N \times m}$ such that $\mathbf{C}_i \cdot \mathbf{s} = \mu_i\mathbf{Gs} + 2\mathbf{R}_i\mathbf{e}_i \pmod{q}$, $i = 1, 2$. We then have with overwhelming probability that*

$$\mathsf{Error}_{\mu_1\mathrm{NAND}\mu_2}(\mathsf{YAP.NAND}(\mathbf{C}_1, \mathbf{C}_2)) = -2(\mu_2\mathbf{X}_1\mathbf{R}_1\mathbf{e}_1 + \mathbf{X}_2\mathbf{R}_2\mathbf{e}_2) \triangleq 2\mathbf{e}, \quad (3)$$

where the entries of $\mathbf{e}$ are mutually independent and subgaussian with parameter $2\sqrt{2}\pi c^2 sm$ (c comes from Lemma 3).

Proof. For any two fresh ciphertexts $\mathbf{C}_1, \mathbf{C}_2$, let $\mathbf{C}_{\mathrm{NAND}} = \mathsf{YAP.NAND}(\mathbf{C}_1, \mathbf{C}_2)$. We have

$$\begin{aligned}\mathbf{C}_{\mathrm{NAND}} \cdot \mathbf{s} &= (\mathbf{G} - \mathbf{G} \odot \mathbf{C}_1 \odot \mathbf{C}_2) \cdot \mathbf{s}\\ &= \mathbf{Gs} - (\mathbf{G} \odot \mathbf{C}_1 \odot \mathbf{C}_2) \cdot \mathbf{s}\\ &= \mathbf{Gs} - (\mu_1\mu_2\mathbf{Gs} + 2(\mu_2\mathbf{X}_1\mathbf{R}_1\mathbf{e}_1 + \mathbf{X}_2\mathbf{R}_2\mathbf{e}_2))\\ &= (1 - \mu_1\mu_2)\mathbf{Gs} + (-2(\mu_2\mathbf{X}_1\mathbf{R}_1\mathbf{e}_1 + \mathbf{X}_2\mathbf{R}_2\mathbf{e}_2))\end{aligned}$$

where the third equality follows the proof of Lemma 8. The Eq. (3) then holds.

The claim that the entries of $\mathbf{e}$ are mutually independent and subgaussian with parameter $2\sqrt{2}\pi c^2 sm$ follows the proof of Lemma 8 too. This finishes the proof. $\qquad\square$

The following theorem claims that, for $L = \mathrm{poly}(\lambda)$, YAP can handle every depth $\leq L$ circuit of NANDs using successively above Lemma:

Theorem 3. *Let $L = \mathrm{poly}(\lambda)$ and params $= \{n, m, q, \chi_1, \chi_2\}$ be parameters for YAP, where χ_1 is the discrete Gaussian distribution with parameter s. Given a depth-L circuit f combined by NAND gates, if the inputs are fresh ciphertexts output by YAP.Enc and the output is $\mathbf{C}_f$, we then can decrypt it correctly for*
$$q/4 > (2c\sqrt{\pi m})^L \cdot (cs\sqrt{2\pi m}) \cdot \omega(\sqrt{\log \lambda}) = 2^{L+\frac{1}{2}}c^{L+1}(\pi m)^{\frac{L+1}{2}} \cdot \omega(\sqrt{\log \lambda}).$$

4.3   Bootstrapping and Unbounded PK-FHE

Recall that there are two efficient bootstrapping algorithms for any LWE-based FHEs. One is proposed by Brakerski and Vaikuntanathan [10] using the asymmetric multiplications and the Barrington's theorem. The other is designed by Alperin-Sheriff and Peikert [3] using an improved Gentry, Sahai and Waters' SK-FHE to evaluate homomorphically and directly the original decryption function. It is obvious that Alperin-Sheriff and Peikert's bootstrapping is more efficient than Brakerski and Vaikuntanathan's, because there hides a mild larger factor which impacts on the efficiency when transforming decryption function to a circuit, and then to a branching program in Brakerski and Vaikuntanathan's method. So, we can gain the following theorem by Alperin-Sheriff and Peikert's bootstrapping trick.

Theorem 4 (PK-FHE). *For $n = n(\lambda), q = \text{poly}(n)$, and $m = O(n \log q)$, there exists an unbounded PK-FHE scheme which is 2^λ secure under the $GapSVP_{\tilde{O}(n^2 m^{0.5}\lambda)}$ assumption and weakly circular secure assumption.*

Roughly speaking, the modulus q in our scheme is bigger than Alperin-Sheriff and Peikert's by a factor of $O(\sqrt{m})$, since noise in a fresh ciphertext is $2\mathbf{R}\mathbf{e}$ instead of $\mathbf{e}$ for $\mathbf{R} \leftarrow \chi_2^{N \times m}$. However, if we transform directly the Alperin-Sheriff and Peikert's SK-FHE to a PK-FHE, the fresh noise increases roughly by a factor of $O(m)$, which will result in a larger modulus q than ours by a factor of $O(\sqrt{m})$.

Remark. In Alperin-Sheriff and Peikert's bootstrapping procedure, it has to decide in advance if $x \in \mathbb{Z}_q$ such that $f(x) = 1$ for the rounding function $f(x)$ (please go to [3] for definition). However, YAP does not need this step, because all odd integers in the set $\{x \in \mathbb{Z}_q | x \text{ is odd}\}$ are exactly the integers such that $f(x) = 1$. In fact, we have to use a symmetric-key variant of our leveled FHE to bootstrap, because we encode message with least-significant-bit while Alperin-Sheriff and Peikert encode it with most-significant-bit.

5   Conclusions

We improved GSW to YAP with better parameters. Our scheme YAP enables to conform with and depart from the traditional wisdom that the circuits are naive combinations of additions and multiplications when without increasing the complexity of decryption function. Its security is based on a slight variation of Regev's PKE proposed in this work. Furthermore, we gained an unbounded PK-FHE scheme assuming $GapSVP_{\tilde{O}(n^2 m^{0.5}\lambda)}$ assumption, using the novel bootstrapping algorithm designed by Alperin-Sheriff and Peikert.

Acknowledgement. The authors would like to thank anonymous reviewers for their helpful comments and suggestions.

References

1. Applebaum, B., Cash, D., Peikert, C., Sahai, A.: Fast cryptographic primitives and circular-secure encryption based on hard learning problems. In: Halevi, S. (ed.) CRYPTO 2009. LNCS, vol. 5677, pp. 595–618. Springer, Heidelberg (2009)
2. Alwen, J., Peikert, C.: Generating shorter bases for hard random lattices. Theor. Comput. Syst. **48**(3), 535–553 (2011). Preliminary version in STACS 2009

3. Alperin-Sheriff, J., Peikert, C.: Faster bootstrapping with polynomial error. In: Garay, J.A., Gennaro, R. (eds.) CRYPTO 2014, Part I. LNCS, vol. 8616, pp. 297–314. Springer, Heidelberg (2014)
4. Brakerski, Z., Gentry, C., Vaikuntanathan, V.: Fully homomorphic encryption without bootstrapping. In: ITCS, pp. 309–325 (2012)
5. Berkoff, A., Liu, F.-H.: Leakage resilient fully homomorphic encryption. In: Lindell, Y. (ed.) TCC 2014. LNCS, vol. 8349, pp. 515–539. Springer, Heidelberg (2014)
6. Brakerski, Z., Langlois, A., Peikert, C., Regev, O., Stehlé, D.: Classal hardness of learning with errors. In: STOC, pp. 575–584 (2013)
7. Banerjee, A., Peikert, C.: New and improved key-homomorphic pseudorandom functions. In: Garay, J.A., Gennaro, R. (eds.) CRYPTO 2014, Part I. LNCS, vol. 8616, pp. 353–370. Springer, Heidelberg (2014)
8. Brakerski, Z.: Fully homomorphic encryption without modulus switching from classical GapSVP. In: Safavi-Naini, R., Canetti, R. (eds.) CRYPTO 2012. LNCS, vol. 7417, pp. 868–886. Springer, Heidelberg (2012)
9. Brakerski, Z., Vaikuntanathan, V.: Efficient fully homomorphic encryption from (standard) LWE. In: FOCS, pp. 97–106 (2011)
10. Brakerski, Z., Vaikuntanathan, V.: Lattice-based FHE as secure as PKE. In: ITCS, pp. 1–12 (2014)
11. Coron, J.-S., Mandal, A., Naccache, D., Tibouchi, M.: Fully homomorphic encryption over the integers with shorter public keys. In: Rogaway, P. (ed.) CRYPTO 2011. LNCS, vol. 6841, pp. 487–504. Springer, Heidelberg (2011)
12. Cheon, J.H., Stehlé, D.: Fully homomophic encryption over the integers revisited. In: Oswald, E., Fischlin, M. (eds.) EUROCRYPT 2015. LNCS, vol. 9056, pp. 513–536. Springer, Heidelberg (2015)
13. Ducas, L., Micciancio, D.: FHEW: bootstrapping homomorphic encryption in less than a second. In: Oswald, E., Fischlin, M. (eds.) EUROCRYPT 2015. LNCS, vol. 9056, pp. 617–640. Springer, Heidelberg (2015)
14. Gentry, C.: A fully homomorphic encryption scheme. Ph.D. thesis, Stanford University (2009). http://crypto.stanford.edu/craig
15. Gentry, C.: Fully homomorphic encryption using ideal lattices. In: STOC, pp. 169–178 (2009)
16. Goldwasser, S., Kalai, Y.T., Popa, R.A., Vaikuntanathan, V., Zeldovich, N.: How to run turing machines on encrypted data. In: Canetti, R., Garay, J.A. (eds.) CRYPTO 2013, Part II. LNCS, vol. 8043, pp. 536–553. Springer, Heidelberg (2013)
17. Gentry, C., Sahai, A., Waters, B.: Homomorphic encryption from learning with errors: conceptually-simpler, asymptotically-faster, attribute-based. In: Canetti, R., Garay, J.A. (eds.) CRYPTO 2013, Part I. LNCS, vol. 8042, pp. 75–92. Springer, Heidelberg (2013)
18. Hiromasa, R., Abe, M., Okamoto, T.: Packing messages and optimizing bootstrapping in GSW-FHE. In: Katz, J. (ed.) PKC 2015. LNCS, vol. 9020, pp. 699–715. Springer, Heidelberg (2015)
19. Halevi, S., Shoup, V.: Algorithms in HElib. In: Garay, J.A., Gennaro, R. (eds.) CRYPTO 2014, Part I. LNCS, vol. 8616, pp. 554–571. Springer, Heidelberg (2014)
20. Halevi, S., Shoup, V.: Bootstrapping for HElib. In: Oswald, E., Fischlin, M. (eds.) EUROCRYPT 2015. LNCS, vol. 9056, pp. 641–670. Springer, Heidelberg (2015)
21. López-Alt, A., Tromer, E., Vaikuntanathan, V.: On-the-fly multiparty computation on the cloud via multikey fully homomorphic encryption. In: STOC, pp. 1219–1234 (2012)

22. Micciancio, D., Peikert, C.: Trapdoors for lattices: simpler, tighter, faster, smaller. In: Pointcheval, D., Johansson, T. (eds.) EUROCRYPT 2012. LNCS, vol. 7237, pp. 700–718. Springer, Heidelberg (2012)

23. Nuida, K., Kurosawa, K.: (Batch) Fully homomorphic encryption over integers for non-binary message spaces. In: Oswald, E., Fischlin, M. (eds.) EUROCRYPT 2015. LNCS, vol. 9056, pp. 537–555. Springer, Heidelberg (2015)

24. Peikert, C.: Public key cryptosystems from the worst-case shortest vector problem. In: STOC, pp. 333–32 (2009)

25. Regev, O.: On lattices, learning with errors, random linear codes, and cryptography. In: STOC, pp. 84–93 (2005)

26. Wang, F., Wang, K.: Fully homomorphic encryption with auxiliary inputs. In: Lin, D., Yung, M., Zhou, J. (eds.) Inscrypt 2014. LNCS, vol. 8957, pp. 220–238. Springer, Heidelberg (2015)

27. van Dijk, M., Gentry, C., Halevi, S., Vaikuntanathan, V.: Fully homomorphic encryption over the integers. In: Gilbert, H. (ed.) EUROCRYPT 2010. LNCS, vol. 6110, pp. 24–43. Springer, Heidelberg (2010)

28. Vershynin, R.: Compressed Sensing, Theory and Applications, Chapter 5, pp. 210–268. Cambridge University Press, Cambridge (2012). http://www-personal.umich.edu/romanv/papers/non-asymptotic-rmt-plain.pdf

Improved Efficiency of MP12

Fuyang Fang[1,2,3(✉)], Bao Li[1,2], Xianhui Lu[1,2], and Xiaochao Sun[1,2,3]

[1] State Key Laboratory of Information Security, Institute of Information
Engineering, Chinese Academy of Sciences, Beijing 100093, China
{fyfang13,lb,xhlu,xchsun}@is.ac.cn
[2] Data Assurance and Communication Security Research Center,
Chinese Academy of Sciences, Beijing 100093, China
[3] University of Chinese Academy of Sciences, Beijing, China

Abstract. MP12 (proposed by Micciancio and Peikert in Eurocrypt 2012) presented a chosen ciphertext secure public key encryption (PKE) scheme based on the learning with errors (LWE) problem. In this paper, we improve the efficiency of the scheme and give two sets of parameters. In both sets of parameters, the size of secret key is $O(\log q)$ times less than the scheme in MP12. Furthermore, in the second case of $p = O(n)$, we can decrease the message expansion rate from $O(\log q)$ to $O(1)$. The main ideas in our scheme are we can hide the message by the knapsack LWE assumption and embed the message into the error item.

Keywords: Public key encryption · Tag-based encryption · Chosen ciphertext security · LWE

1 Introduction

Recently, MP12 [12] gives an LWE-based chosen-ciphertext (IND-CCA) secure PKE scheme with a new approach in designing the trapdoors for lattices. With the trapdoor, the decryption algorithm can efficiently and quickly reconstruct the witness from the ciphertext. In this scheme, the public key $[\mathbf{A}|\mathbf{AR}] \in \mathbb{Z}_q^{n \times (m+nk)}$ is statistical indistinguishable from uniform distribution which results in the sizes of public key and secret key still large, because leftover hash lemma [8] requires that $m = O(n \log q)$. The message $\boldsymbol{m}$ is mapped bijectively to the cosets of $\Lambda/2\Lambda$ for $\Lambda = \Lambda(\mathbf{G})$, that is, map $\boldsymbol{m}$ to encode$(\boldsymbol{m}) = \mathbf{S}\boldsymbol{m} \in \Lambda/2\Lambda$, where $\mathbf{S}$ is any basis of $\Lambda(\mathbf{G})$. We can observe that the encrypted message is restricted to binary strings and the message expansion rates are more than $O(\log q)$. To decrease the size of secret key, our modification is to replace the origin public key with $(\mathbf{A}|\mathbf{AT})$, where $(\mathbf{A}|\mathbf{AT})$ is computational indistinguishable from uniform distribution. With this method, the parameter m can be decreased, as well as

This research is supported by the National Nature Science Foundation of China (No. 61379137 and No. 61272040), the National Basic Research Program of China (973 project) (No. 2013CB338002), and IIE's Cryptography Research Project (No. Y4Z0061D03).

© Springer International Publishing Switzerland 2015
K. Tanaka and Y. Suga (Eds.): IWSEC 2015, LNCS 9241, pp. 193–210, 2015.
DOI: 10.1007/978-3-319-22425-1_12

the size of the trapdoor $\mathbf{T}$. However, the problem is that the trapdoor $\mathbf{T}$ cannot be used to answer decryption queries in the simulation of decryption. Fortunately, the problem can be solved completely by the double trapdoor technique provided in [10], in which one trapdoor is used to hide the challenge tag and the second trapdoor is used to answer the decryption queries. However, [10] had also mentioned that the scheme based on the LWE problem would be worse in ciphertext size than the scheme in [12], because the scheme obtained from [10] had to add an additional ciphertext which made the size of ciphertext large. Therefore, the goal of our work is to obtain a scheme with small secret key and small message expansion rate.

1.1 Our Contributions

In this work, we propose an efficient CCA-secure PKE scheme based on the LWE problem. The comparison between our scheme and [12] is as follows in Table 1:

Table 1. Comparison with [12], where k=$\lceil \log q \rceil$

Scheme	Secket key	Message	Ciphertext	Error rate $(1/\alpha)$	Message exp
Our scheme $(p = 2)$	$2n^2 \log q \omega(\sqrt{\log q})$	nk	$(2n + 2nk) \log q$	$O(n^{1.5})\omega(\log n)$	$O(\log q)$
Our scheme $(p = O(n))$	$2n^2 \log q \omega(\sqrt{\log q})$	$nk \log n$	$(2n + 2nk) \log q$	$O(n^{2.5})\omega(\log n)$	$O(1)$
The scheme in [12]	$O(n^2 \log^2 q)\omega(\sqrt{\log q})$	nk	$O(nk) \log q$	$O(nk)\omega(\sqrt{\log n})$	$O(\log q)$

The Improvement: For our scheme in the case of $p = 2$, the size of secret key is decreased by a factor of $O(\log q)$ without increasing the size of ciphertext. Furthermore, when $p = O(n)$, not only the size of secret key is $O(\log q)$ less than [12], but also the message expansion rate is decreased to $O(1)$, the cost is that the related error rate $(1/\alpha)$ is slightly increased by a factor $O(n^{1.5})/\omega(\sqrt{\log q})$.

Overview of Techniques: In the process of generating the public key, we use a computational assumption called the knapsack LWE problem instead of the leftover hash lemma. By this way, we can decrease the parameter m to $O(n)$ (with small constant factor). In our scheme, we do not encrypt message by adding an extra c_2 in [10], but adding the message directly to the items of c_0 and c_1. With Lemma 2 in this paper we can prove c_0 and c_1 in our scheme can be replaced by a random value over $\mathbb{Z}_q^{nk}$ respectively and the message can be hided by c_0 and c_1 completely. In this process, we also need the method of double trapdoor that one trapdoor is used to answer the decryption queries and the other trapdoor is used to hide the message. In order to decrypt the ciphertext correctly, we will use a new variant of the LWE problem, the form is $\boldsymbol{b} = \mathbf{A}\boldsymbol{s} + p\boldsymbol{e} \bmod q$, where p is relatively prime to q. The method of recovering message from ciphertext has been mentioned in [5], and the message is encrypted as $(\boldsymbol{a}, \langle \boldsymbol{a}, \boldsymbol{s} \rangle + 2e + m)$ $\bmod q$. In fact, we can regard the message and original error term as the whole "error"and can simply extract the message from the error in decryption. In our work, we can let 2 be p that is relatively prime to q. With the double trapdoor technique, the message is encrypted as:

$$(\mathbf{A}s + pe, \mathbf{T}_0\mathbf{A}s + \mathbf{T}_0'pe + m, \mathbf{T}_1\mathbf{A}s + \mathbf{T}_1'pe + m) \bmod q$$

the encrypted message can be regarded as a part of error, the $\log p$ least significant bits of every item of the "new" error. After reconstructing the random value s with the notion of $\mathbf{G}$-*trapdoor* for matrix $\mathbf{A}$ in [12], we can recover the message from the "new" error. In our scheme, we need to recover the message from the third part of ciphertext to testify the consistency of ciphertext with the witness reconstructed by the decryption algorithm, and this method is also different from the schemes in [10,12]. With the appropriate parameter for our scheme, we can let p large enough so as to increase the encrypted message space without enlarging the size of ciphertext, that is, we can encrypt up to $n \cdot k \cdot \log p$ bits message per ciphertext and the message expansion rate can be decreased.

1.2   Related Works

LWE-based CCA-Secure Scheme: Based on the LWE-based lossy trapdoor function, [15] gave the first chosen ciphertext secure scheme from the LWE problem. This scheme used a collection of *lossy trapdoor functions* and a collection of *all-but-one (ABO) trapdoor functions* and the length of encrypted message was depended on the total residual leakage of the lossy and ABO collection. Then through the approach of witness-recovering decryption of [15] and the notion of "correlated inputs" in [17], [14] proposed a LWE-based scheme directly from the LWE problem. In this scheme the message was concealed by generic *hard-core* bits, the sizes of public key and secret key were $\tilde{O}(n^3)$ and $\tilde{O}(n^2)$ respectively, and the security was based on the LWE error rate of $\alpha = \tilde{O}(1/n)$. With the generic technique proposed in [3] that conversing a selectively secure ID-based encryption to CCA-secure encryption, we can obtain a scheme from [1], in which the size of public key is $\tilde{O}(n^2)$ and the error rate $\alpha = \tilde{O}(1/n^2)$. For the above schemes, we can easily observe that all the encrypted message were binary strings and the message expansion rates are more than $O(\log q)$.

2   Preliminaries

Let n be the security parameter and we use $negl(n)$ to denote an arbitrary negligible function $f(n)$ such that $f(n) = o(n^{-c})$ for every fixed constant c. We say that a probability is *overwhelming* if it is $1 - negl(n)$. Let $poly(n)$ denotes an unspecified function $f(n) = O(n^c)$ for some constant c. We use $\tilde{O}(n)$ be a function $f(n)$ if $f(n) = O(n \cdot log^c n)$ for some fixed constant c. For an integer p, we denote $\{0, \ldots, p-1\}$ by $[p]$. Let $\mathbb{Z}_q$ be a q-ary finite field for a prime $q \geq 2$. We denote by $a \xleftarrow{\$} \mathbb{Z}_q$ that a is randomly chosen from $\mathbb{Z}_q$. We use PPT denotes *probability polynomial-time*.

2.1   Tag-Based Encryption

A tag-based encryption (TBE) scheme with message space $\mathcal{M}$ and tag space $\mathcal{T}$ can be defined by a tuple of PPT algorithms (**Gen, Enc, Dec**) as below:

- **Gen(1^n)** → (**pk, sk**) : The probabilistic algorithm **Gen(1^n)** generates a pair (**pk, sk**), which denotes public key and secret key respectively.
- **Enc(pk, τ, m)** → **C** : Given a message $\mathbf{m} \in \mathcal{M}$ and a tag $\tau \in \mathcal{T}$, the probabilistic algorithm **Enc** uses the public key **pk** to encrypt the message with respect to the tag and output a ciphertext **C**.
- **Dec(sk, τ, C)** → **m** or $\perp$: Given a ciphertext **C** with respect to tag $\tau \in \mathcal{T}$, the deterministic algorithm **Dec** uses the secret key **sk** to recover the message **m**. When the ciphertext **C** is invalid, the algorithm outputs $\perp$.

For a TBE system described above, the correctness is that: for any message $\mathbf{m} \in \mathcal{M}$, $\tau \in \mathcal{T}$ and (**pk, sk**) generated by **Gen(1^n)**, **C** is the ciphertext output by the **Enc(pk, τ, m)** algorithm, then the **Dec(sk, τ, C)** will output **m** with overwhelming probability.

2.2   Selective-Tag Weak CCA Security

For a TBE system, besides the requirement of correctness, it also needs to achieve other security requirements. In the following, we will simply define selective-tag weak CCA security [9] related to a TBE system. Let $\mathcal{A}$ be any non-uniform PPT adversary, then the advantage of $\mathcal{A}$ in the selective-tag weak CCA experiment [9] is defined as

$$\mathbf{Adv}_{TBE,\mathcal{A}}^{stw-cca} \stackrel{def}{=} \left| \mathbf{Pr} \left[b = b' : \begin{array}{c} (\tau^*, st_0) \leftarrow \mathcal{A}(1^n), (pk, sk) \leftarrow Gen(1^n); \\ (\mathbf{m}_0, \mathbf{m}_1, st) \leftarrow \mathcal{A}^{\mathcal{O}}(pk, st_0); b \xleftarrow{\$} \{0, 1\}; \\ c^* \leftarrow Enc(pk, \mathbf{m}_b, \tau^*); b' = \mathcal{A}^{\mathcal{O}}(c^*, st) \end{array} \right] - \frac{1}{2} \right|$$

where $\mathcal{O} = Dec(sk, \cdot)$ means that the adversary $\mathcal{A}$ can query the decryption oracle polynomial times with valid ciphertexts, then the oracle can return the correct plaintexts corresponding to the ciphertexts. The limitation is that the adversary $\mathcal{A}$ can not query decryption oracle $\mathcal{O}$ for tag τ^*. The TBE scheme is said to be selective-tag weak CCA secure if $\mathbf{Adv}_{TBE,\mathcal{A}}^{stw-cca} \leq negl(n)$.

Chosen Ciphertext Security from TBE: To convert any selective-tag weak CCA-secure TBE scheme with tag space $\mathcal{T}$ exponential in n into a CCA-secure PKE scheme, we can use an one-time signature [7,9] or a message authentication code plus a commitment [4,7].

2.3   Lattices

An m-dimensional lattice Λ is defined as a set generated by n linearly independent vectors $\mathbf{b}_1, \ldots, \mathbf{b}_n \in \mathbb{R}^m$, i.e.

$$\Lambda = \{\Sigma_{i=1}^n k_i \mathbf{b}_i | k_i \in \mathbb{Z}, i = 1, \ldots, n\}$$

where the $\mathbf{B} = (\boldsymbol{b}_1, \ldots, \boldsymbol{b}_n)$ is a basis of the lattice Λ.

In particular, we often use a class of lattices, called q-ary lattices, which contain a sub-lattice $q\mathbb{Z}^m$ for integer q. For some integers q, m, n and a matrix $\mathbf{A} \in \mathbb{Z}_q^{n \times m}$, we can define the following two m-dimensional q-ary lattices,

$$\Lambda^{\perp}(A) = \{\boldsymbol{y} \in \mathbb{Z}^m : \mathbf{A}\boldsymbol{y} = 0 \bmod q\}$$
$$\Lambda(A^t) = \{\boldsymbol{y} \in \mathbb{Z}^m : \boldsymbol{y} = \mathbf{A}^t\boldsymbol{s} \bmod q \ for \ some \ \boldsymbol{s} \in \mathbb{Z}^n\}$$

For linearly vectors $\mathbf{B} = (\boldsymbol{b}_1, \ldots, \boldsymbol{b}_n)$, its fundamental parallelepiped is

$$\mathcal{P}_{1/2}(\mathbf{B}) := \mathbf{B} \cdot [-\frac{1}{2}, \frac{1}{2})^n = \{\Sigma_{i \in [n]} c_i \cdot \boldsymbol{b}_i : c_i \in [-\frac{1}{2}, \frac{1}{2})\}$$

3 The Learning with Errors (LWE) Problem

Let m, n, q be positive integers, for a vector $\boldsymbol{s} \in \mathbb{Z}_q^n$ and an error distribution χ over $\mathbb{Z}_q$, we can define a distribution

$$\mathbf{A}_{s,\chi} = \{(\boldsymbol{a}, \langle \boldsymbol{a}, \boldsymbol{s} \rangle + e) | \boldsymbol{a} \xleftarrow{\$} \mathcal{U}(\mathbb{Z}_q^n), e \leftarrow \chi\}$$

Based on this definition, we can define two classes of the LWE problem: The search version of the LWE problem ($\text{CLWE}_{m,n,q,\chi}$) is said that for some random $\boldsymbol{s} \in \mathbb{Z}_q^n$, given polynomial samples from the distribution $\boldsymbol{A}_{s,\chi}$, it is hard to find $\boldsymbol{s}$. The decisional version of the LWE problem ($\text{DLWE}_{m,n,q,\chi}$) is said that one cannot distinguish $\boldsymbol{A}_{s,\chi}$ from the uniform distribution $\mathcal{U}(\mathbb{Z}_q^n \times \mathbb{Z})$ for a uniformly random $\boldsymbol{s} \in \mathbb{Z}_q^n$, given arbitrary m independent samples.

In the work of proving the hardness of the LWE problem, [16] firstly gave a quantum reduction from the GapSVP problem in the worst case to the search version of LWE, and then [6,14] gave the classical reductions respectively.

In this paper, we will use a variant of the standard LWE, and the distribution is defined as

$$D_{s,\chi} \stackrel{def}{=} \{(\boldsymbol{a}, \langle \boldsymbol{a}, \boldsymbol{s} \rangle + p \cdot e) | \boldsymbol{a} \xleftarrow{\$} \mathcal{U}(\mathbb{Z}_q^n), e \leftarrow \chi\}$$

where p is relatively prime to q. The hardness is equivalent to the standard LWE problem which was proved in [5].

3.1 Knapsack LWE

The LWE problem in knapsack form was firstly proposed in [11] and had been used in many cryptographic applications. Let $n, m \geq n + \omega(\log n), q, m_1$ be integers, $\chi = \bar{\Psi}_\alpha$ be an error distribution over $\mathbb{Z}_q$, the distribution of the m_1-fold knapsack LWE is defined as

$$D_{\text{KLWE}_{m,n,q,\chi}^{m_1}} \stackrel{def}{=} ((\mathbf{A}, \mathbf{TA}) | \mathbf{A} \xleftarrow{\$} \mathbb{Z}_q^{m \times (m-n)}, \mathbf{T} \leftarrow \bar{\Psi}_\alpha^{m_1 \times m})$$

The knapsack LWE problem is to distinguish the distribution $D_{\mathrm{KLWE}_{n,m,q,\chi}^{m_1}}$ from the distribution $(\mathbf{A}, \mathcal{U}(\mathbb{Z}_q^{m_1 \times (m-n)}))$. Let $\mathcal{A}_1$ be any PPT adversary to solve the knapsack LWE problem, we can define the advantage of $\mathcal{A}_1$ as

$$\mathbf{Adv}_{\mathrm{KLWE}_{m,n,q}^{m_1}}^{\mathcal{A}_1} \overset{def}{=} |\mathbf{Pr}[\mathcal{A}_1(\mathbf{A}, \mathbf{TA}) = 1] - \mathbf{Pr}[\mathcal{A}_1(\mathbf{A}, \mathbf{B}) = 1]|$$

where $(\mathbf{A}, \mathbf{TA}) \leftarrow D_{\mathrm{KLWE}_{m,n,q,\chi}^{m_1}}$ and $\mathbf{B} \overset{\$}{\leftarrow} \mathcal{U}(\mathbb{Z}_q^{m_1 \times (m-n)})$.

In [11] the knapsack LWE problem was proved as hard as the standard LWE problem when $m_1 = 1$. Due to the standard hybrid argument, we can let m_1 be any polynomial number and directly obtain the m_1-fold knapsack LWE distribution as above. Therefore, the knapsack LWE problem is also as hard as the LWE problem and only lose a factor of m_1 by the following lemma.

Lemma 1. *Let $\mathcal{A}_1$ be the adversary attacking the $KLWE_{m,n,q}^{m_1}$, then there exists an adversary $\mathcal{S}_1$ attacking the $LWE_{m,n,q}$, and the advantage of $\mathcal{S}_1$ satisfies*

$$\mathbf{Adv}_{KLWE_{m,n,q}^{m_1}}^{\mathcal{A}_1} \leq m_1 \cdot \mathbf{Adv}_{LWE_{m,n,q}}^{\mathcal{S}_1}$$

From Lemma 1, we observe that the $KLWE_{m,n-1,q}^{m_1}$ problem is as hard as the $LWE_{m,n-1,q}$ problem. Then we can obtain a lemma as follows:

Lemma 2. *If the $KLWE_{m,n-1,q}^{m_1}$ problem and the $KLWE_{m,n,q}^{m_1}$ problem are both hard, then for any PPT adversary $\mathcal{D}$ we have*

$$|\mathbf{Pr}[\mathcal{D}(\mathbf{A}, \mathbf{TA}, r, \mathbf{T}r) = 1] - \mathbf{Pr}[\mathcal{D}(\mathbf{A}, \mathbf{TA}, r, u) = 1]|$$
$$\leq \mathbf{Adv}_{KLWE_{m,n-1,q}^{m_1}}^{\mathcal{A}_1} + \mathbf{Adv}_{KLWE_{m,n,q}^{m_1}}^{\mathcal{A}_1}$$

where $(\mathbf{A}, \mathbf{TA}) \leftarrow D_{KLWE_{m,n,q,\chi}^{m_1}}$, $r \overset{\$}{\leftarrow} \mathbb{Z}_q^m$ and $u \overset{\$}{\leftarrow} \mathbb{Z}_q^{m_1}$.

Proof. We can describe the formula in the lemma as:

$$|\mathbf{Pr}[\mathcal{D}(\mathbf{A}, \mathbf{TA}, r, \mathbf{T}r) = 1] - \mathbf{Pr}[\mathcal{D}(\mathbf{A}, \mathbf{TA}, r, u) = 1]|$$
$$\leq |\mathbf{Pr}[\mathcal{D}(\mathbf{A}, \mathbf{TA}, r, \mathbf{T}r) = 1] - \mathbf{Pr}[\mathcal{D}(\mathbf{A}, \mathbf{B}, r, u) = 1]|$$
$$+ |\mathbf{Pr}[\mathcal{D}(\mathbf{A}, \mathbf{B}, r, u) = 1] - \mathbf{Pr}[\mathcal{D}(\mathbf{A}, \mathbf{TA}, r, u) = 1]|$$
$$= |\mathbf{Pr}[\mathcal{D}([\mathbf{A}|r], \mathbf{T}[\mathbf{A}|r]) = 1] - \mathbf{Pr}[\mathcal{D}([\mathbf{A}|r], [\mathbf{B}|u]) = 1]|$$
$$+ |\mathbf{Pr}[\mathcal{D}(\mathbf{A}, \mathbf{B}, r, u) = 1] - \mathbf{Pr}[\mathcal{D}(\mathbf{A}, \mathbf{TA}, r, u) = 1]|$$

where $(\mathbf{A}, \mathbf{TA}) \leftarrow D_{\mathrm{KLWE}_{m,n,q,\chi}^{m_1}}$, $\mathbf{B} \overset{\$}{\leftarrow} \mathcal{U}(\mathbb{Z}_q^{m_1 \times (m-n)})$, and $r \overset{\$}{\leftarrow} \mathbb{Z}_q^m$, $u \overset{\$}{\leftarrow} \mathbb{Z}_q^{m_1}$. Based on the $\mathrm{KLWE}_{m,n-1,q}^{m_1}$ problem, we have

$$|\mathbf{Pr}[\mathcal{D}([\mathbf{A}|r], \mathbf{T}[\mathbf{A}|r]) = 1] - \mathbf{Pr}[\mathcal{D}([\mathbf{A}|r], [\mathbf{B}|u]) = 1]|$$
$$= |\mathbf{Pr}[\mathcal{D}(\mathbf{A}', \mathbf{TA}') = 1] - \mathbf{Pr}[\mathcal{D}(\mathbf{A}', \mathbf{B}') = 1]| \leq \mathbf{Adv}_{\mathrm{KLWE}_{m,n-1,q}^{m_1}}^{\mathcal{A}_1}$$

where $(\mathbf{A}', \mathbf{TA}') \leftarrow D_{\mathrm{KLWE}_{m,n-1,q,\chi}^{m_1}}$, $\mathbf{B}' \overset{\$}{\leftarrow} \mathcal{U}(\mathbb{Z}_q^{m_1 \times (m-(n-1))})$, and $r \overset{\$}{\leftarrow} \mathbb{Z}_q^m$, $u \overset{\$}{\leftarrow} \mathbb{Z}_q^{m_1}$.

Based on the $\text{KLWE}_{m,n,q}^{m_1}$ problem, r and u are independent with $(\mathbf{A}, \mathbf{TA})$, then we have

$$|\mathbf{Pr}\left[\mathcal{D}(\mathbf{A}, \mathbf{B}, r, u) = 1\right] - \mathbf{Pr}\left[\mathcal{D}(\mathbf{A}, \mathbf{TA}, r, u) = 1\right]|$$
$$= |\mathbf{Pr}\left[\mathcal{D}(\mathbf{A}, \mathbf{B}) = 1\right] - \mathbf{Pr}\left[\mathcal{D}(\mathbf{A}, \mathbf{TA}) = 1\right]| \leq \mathbf{Adv}_{\text{KLWE}_{m,n,q}^{m_1}}^{\mathcal{A}_1}$$

where $(\mathbf{A}, \mathbf{TA}) \leftarrow \mathcal{D}_{\text{KLWE}_{m,n,q,\chi}^{m_1}}$, $\mathbf{B} \xleftarrow{\$} \mathcal{U}(\mathbb{Z}_q^{m_1 \times (m-n)})$, and $r \xleftarrow{\$} \mathbb{Z}_q^m$, $u \xleftarrow{\$} \mathbb{Z}_q^{m_1}$.
Combining the above results, so we have

$$|\mathbf{Pr}\left[\mathcal{D}(\mathbf{A}, \mathbf{TA}, r, \mathbf{T}r) = 1\right] - \mathbf{Pr}\left[\mathcal{D}(\mathbf{A}, \mathbf{TA}, r, u) = 1\right]|$$
$$\leq \mathbf{Adv}_{\text{KLWE}_{m,n-1,q}^{m_1}}^{\mathcal{A}_1} + \mathbf{Adv}_{\text{KLWE}_{m,n,q}^{m_1}}^{\mathcal{A}_1}$$

$\square$

3.2   Extended-(knapsack) LWE

Next we review extended-LWE, which was firstly appeared in [13], and then [2] gave a tight reduction to it from the standard LWE problem. Let m, n, q be integers and $\chi = \bar{\Psi}_\alpha$ be an error distribution over $\mathbb{Z}_q$, $\varphi = D_{\mathbb{Z},r}$, then the distribution of extended-LWE is defined as

$$D_{\text{ELWE}_{m,n,q,\chi}} \stackrel{def}{=} ((\mathbf{A}, \mathbf{A}s + e, z, \langle z, e \rangle) | \mathbf{A} \xleftarrow{\$} \mathbb{Z}_q^{m \times n}, s \xleftarrow{\$} \mathbb{Z}_q^n, z \leftarrow \varphi^m, e \leftarrow \chi^m)$$

and the distribution of m_1-fold extended-LWE is defined as

$$D_{\text{ELWE}_{m,n,q,\chi}^{m_1}} \stackrel{def}{=} ((\mathbf{A}, \mathbf{AS} + \mathbf{E}, z, z^t\mathbf{E}) | \mathbf{A} \xleftarrow{\$} \mathbb{Z}_q^{m \times n}, \mathbf{S} \xleftarrow{\$} \mathbb{Z}_q^{n \times m_1}, z \leftarrow \varphi^m, \mathbf{E} \leftarrow \chi^{m \times m_1})$$

We say extended-LWE is also hard even if some information about the error e is leaked by $(z, \langle z, e \rangle)$. Let $\mathcal{B}, \mathcal{B}_1$ be any PPT adversaries attacking the extended-LWE problem and the m_1-fold extended-LWE problem respectively, then the advantages of $\mathcal{B}, \mathcal{B}_1$ are defined as

$$\mathbf{Adv}_{\text{ELWE}_{m,n,q}}^{\mathcal{B}} \stackrel{def}{=} |\mathbf{Pr}[\mathcal{B}(\mathbf{A}, \mathbf{A}s + e, z, \langle z, e \rangle) = 1] - \mathbf{Pr}[\mathcal{B}(\mathbf{A}, \mathbf{B}, z, \langle z, e \rangle) = 1]|$$

where $(\mathbf{A}, \mathbf{A}s + e, z, \langle z, e \rangle) \leftarrow D_{\text{ELWE}_{m,n,q,\chi}}$, and $\mathbf{B} \xleftarrow{\$} \mathcal{U}(\mathbb{Z}_q^m)$. And

$$\mathbf{Adv}_{\text{ELWE}_{m,n,q}^{m_1}}^{\mathcal{B}_1} \stackrel{def}{=} |\mathbf{Pr}[\mathcal{B}_1(\mathbf{A}, \mathbf{AS} + \mathbf{E}, z, z^t\mathbf{E}) = 1] - \mathbf{Pr}[\mathcal{B}_1(\mathbf{A}, \mathbf{C}, z, z^t\mathbf{E}) = 1]|$$

where $(\mathbf{A}, \mathbf{AS} + \mathbf{E}, z, z^t\mathbf{E}) \leftarrow D_{\text{ELWE}_{m,n,q,\chi}^{m_1}}$, and $\mathbf{C} \xleftarrow{\$} \mathcal{U}(\mathbb{Z}_q^{m \times m_1})$.
We can use the knapsack LWE instead of the standard LWE and redefine new distributions of extended knapsack LWE and m_1-fold extended knapsack LWE as follows:

$$D_{\text{EKLWE}_{m,n,q,\chi}} \stackrel{def}{=} ((\mathbf{H}, \mathbf{H}e, z, \langle z, e \rangle) | \mathbf{H} \xleftarrow{\$} \mathbb{Z}_q^{(m-n) \times m}, z \leftarrow \varphi^m, e \leftarrow \chi^m)$$

and

$$D_{\text{EKLWE}^{m_1}_{m,n,q,\chi}} \overset{def}{=} ((\mathbf{H}, \mathbf{HE}, \mathbf{z}, \mathbf{z}^t\mathbf{E}) | \mathbf{H} \overset{\$}{\leftarrow} \mathbb{Z}_q^{(m-n)\times m}, \mathbf{z} \leftarrow \varphi^m, \mathbf{E} \leftarrow \chi^{m\times m_1})$$

Similarly, the advantages of any PPT adversaries $\mathcal{C}, \mathcal{C}_1$ attacking the extended knapsack LWE and m_1-fold extended knapsack LWE are defined as

$$\mathbf{Adv}^{\mathcal{C}}_{\text{EKLWE}_{m,n,q}} \overset{\mathcal{D}}{=} |\mathbf{Pr}[\mathcal{C}(\mathbf{H}, \mathbf{He}, \mathbf{z}, \langle \mathbf{z}, \mathbf{e} \rangle) = 1] - \mathbf{Pr}[\mathcal{C}(\mathbf{H}, \mathbf{B}, \mathbf{z}, \langle \mathbf{z}, \mathbf{e} \rangle) = 1]|$$

where $(\mathbf{H}, \mathbf{He}, \mathbf{z}, \langle \mathbf{z}, \mathbf{e} \rangle) \leftarrow D_{\text{EKLWE}_{m,n,q,\chi}}$, and $\mathbf{B} \overset{\$}{\leftarrow} \mathcal{U}(\mathbb{Z}_q^{(m-n)})$, and

$$\mathbf{Adv}^{\mathcal{C}_1}_{\text{EKLWE}^{m_1}_{m,n,q}} \overset{\mathcal{D}}{=} |\mathbf{Pr}[\mathcal{C}_1(\mathbf{H}, \mathbf{HE}, \mathbf{z}, \mathbf{z}^t\mathbf{E}) = 1] - \mathbf{Pr}[\mathcal{C}_1(\mathbf{H}, \mathbf{C}, \mathbf{z}, \mathbf{z}^t\mathbf{E}) = 1]|$$

where $(\mathbf{H}, \mathbf{HE}, \mathbf{z}, \mathbf{z}^t\mathbf{E}) \leftarrow D_{\text{EKLWE}^{m_1}_{m,n,q,\chi}}$, and $\mathbf{C} \overset{\$}{\leftarrow} \mathcal{U}(\mathbb{Z}_q^{(m-n)\times m_1})$.

Theorem 1 ([2]). *Let $m \geq n + \omega(\log n), n$ be some integers, q be a prime and $\chi = D_{\mathbb{Z},\alpha q}$ be an error distribution over $\mathbb{Z}_q$, let $\varphi = D_{\mathbb{Z},r}$ be another error distribution over $\mathbb{Z}_q$. If $1/\alpha \geq r \cdot \sqrt{m+n}$, then there exists an PPT adversary $\mathcal{S}$ such that for any adversary $\mathcal{B}$,*

$$\mathbf{Adv}^{\mathcal{S}}_{LWE_{m,n,q}} \geq \frac{1}{2q-1} \cdot \mathbf{Adv}^{\mathcal{B}}_{ELWE_{m,n,q}}$$

In the following paper, in order to achieve our requirement, we set $r = \alpha q$ and the choice of parameter in our scheme will satisfy the condition in this theorem.

According to the hybrid argument, we can get a reduction from $\text{ELWE}_{m,n,q,\chi}$ to m_1-fold extended-LWE and m_1-fold extended knapsack LWE that reduces the advantage by a factor of m_1.

Lemma 3. *For any PPT adversary $\mathcal{B}_1$ attacking $\text{ELWE}^{m_1}_{m,n,q}$, then there exists a PPT adversary $\mathcal{C}_1$ attacking $\text{EKLWE}^{m_1}_{m,n,q}$ and a PPT adversary $\mathcal{B}$ attacking $ELWE_{m,n,q}$ such that*

$$\mathbf{Adv}^{\mathcal{C}_1}_{EKLWE^{m_1}_{m,n,q}} = \mathbf{Adv}^{\mathcal{B}_1}_{ELWE^{m_1}_{m,n,q}} \leq m_1 \cdot \mathbf{Adv}^{\mathcal{B}}_{ELWE_{m,n,q}}$$

3.3   The Gadget Matrix G

In this section, we will recall a parity-check matrix $\mathbf{G}$ used in [12], where $\mathbf{G}$ is defined as:

$$\mathbf{G}^t = \mathbf{I_n} \otimes \mathbf{g}^t \in \mathbb{Z}^{n\times nk}, k = \lceil \log q \rceil$$

where $\mathbf{g}^t = (1, 2, 2^2, \dots, 2^{k-1}) \in \mathbb{Z}^k$ is a special vector, $\mathbf{I_n} \in \mathbb{Z}^{n\times n}$ is the identity matrix and $\otimes$ denotes the tensor product.

For the lattice $\Lambda^\perp(\mathbf{g}^t)$, we have a good basis $\mathbf{S}$ as follows, and then $\mathbf{B} = \mathbf{I_n} \otimes \mathbf{S} \in \mathbb{Z}^{nk\times nk}$ is the basis of $\Lambda^\perp(\mathbf{G}^t)$, that is,

$$\mathbf{S} := \begin{bmatrix} 2 & & & & q_0 \\ -1 & 2 & & & q_1 \\ & & \ddots & & \vdots \\ & & & 2 & q_{k-2} \\ & & & -1 & q_{k-1} \end{bmatrix} \in \mathbb{Z}^{k\times k}, \mathbf{B} := \mathbf{I} \otimes \mathbf{S} = \begin{bmatrix} \mathbf{S} & & & \\ & \mathbf{S} & & \\ & & \ddots & \\ & & & \mathbf{S} \\ & & & \mathbf{S} \end{bmatrix} \in \mathbb{Z}^{nk\times nk}$$

where $q_0, \ldots, q_{k-1} \in \{0,1\}^k$ is the binary decomposition of $q = \Sigma_i(2^i \cdot q_i)$.
There are some properties of this gadget matrix $\mathbf{G}$ proposed in [12]:

- **Short Basis:** $\mathbf{B}$ is the basis of lattice $\Lambda^\perp(\mathbf{G^t})$ with $\|\tilde{\mathbf{B}}\| \leq \sqrt{5}$ and $\|\mathbf{B}\| \leq \{\sqrt{5}, \sqrt{k}\}$.
- **Inverting simply:** The function $\mathbf{g_G}(\boldsymbol{s}, \boldsymbol{e}) = \mathbf{G}\boldsymbol{s} + \boldsymbol{e} \bmod q$ can be inverted in quasi-linear time $O(n \cdot log^c n)$ for any $\boldsymbol{s} \in \mathbb{Z}_q^n$ and $\boldsymbol{e} \leftarrow \chi^m$ such that $\boldsymbol{e} \in \mathcal{P}_{1/2}(q \cdot \mathbf{B}^{-t})$ or $\boldsymbol{e} \in \mathcal{P}_{1/2}(q \cdot \tilde{\mathbf{B}}^{-t})$. In this paper, we will denote the algorithm of inverting $\mathbf{g_G}(\boldsymbol{s}, \boldsymbol{e}) = \mathbf{G}\boldsymbol{s} + \boldsymbol{e}$ by an oracle $\mathcal{O}(\mathbf{G})$.

Besides, we will introduce some facts about the Gaussian random matrices:

Lemma 4 ([18]). *Let $\mathbf{T} \in \mathbb{Z}^{n \times m}$ be a Gaussian random matrix with parameter s, There exists an universal constant $\mathbf{C} > 0$ such that for any $t \geq 0$, the largest singular value $s_1(\mathbf{T}) \leq \mathbf{C} \cdot s(\sqrt{m} + \sqrt{n} + t)$ except with probability at most $2e^{-\pi t^2}$.*

4   Description of Our Scheme

In our scheme, we will adopt the similar double trapdoor generator that proposed in [10]. For the tag, we will use the ring $\mathcal{R}$ and the function h that defined in [12], where $\mathcal{R} = \mathbb{Z}_q[x]/(f(x))$ for some monic degree-n polynomial $f(x) = x^n + f_{n-1}x^{n-1} + \ldots + f_0 \in \mathbb{Z}[x]$ which is irreducible module every prime factor that dividing q, and h is an injective ring homomorphism that maps any $\tau \in \mathcal{R}$ to an invertible matrix $h(\tau)$. We also utilize a large set $\mathcal{T} = \{\tau_0, \ldots, \tau_l\} \subset \mathcal{R}$ that satisfies the "unit differences" property: for any $i \neq j$, $\tau_i - \tau_j \in \mathcal{T}$, and therefore $h(\tau_i) - h(\tau_j) = h(\tau_i - \tau_j)$ is invertible according to the property of function h.

4.1   Our Scheme

The algorithm of our scheme is described as follows:

Parameter: Let m, n be integers, q is a prime and p is relatively prime to q, and $m_1 = nk, k = \lceil \log q \rceil$.
KeyGen$(1^k) \to (sk, pk)$: The algorithm samples $\mathbf{T}_0, \mathbf{T}_1 \leftarrow \bar{\Psi}_\alpha^{m_1 \times m}$ and $\mathbf{A} \xleftarrow{\$} \mathbb{Z}_q^{m \times (m-n)}$ and let $\mathbf{B}_0 := \mathbf{T}_0\mathbf{A} - \mathbf{G}h(\tau_0)$, $\mathbf{B}_1 := \mathbf{T}_1\mathbf{A} - \mathbf{G}h(\tau_1)$ and ek=$(\mathbf{A}, \mathbf{B}_0, \mathbf{B}_1)$. Then the pair of (sk, pk) is defined as follows:

$$\begin{cases} sk := (\tau_0, \mathbf{T}_0) \\ pk := (\mathbf{A}, \mathbf{B}_0, \mathbf{B}_1) \end{cases}$$

Enc$(pk, \tau, \boldsymbol{m}) \to \mathbf{CT} = (\boldsymbol{c}, \boldsymbol{c}_0, \boldsymbol{c}_1)$: The algorithm picks:

$$\boldsymbol{m} \xleftarrow{\$} \{0, \ldots, p-1\}^{m_1}, \boldsymbol{e}_1 \leftarrow \bar{\Psi}_\alpha^m, \mathbf{T}_0', \mathbf{T}_1' \leftarrow \bar{\Psi}_\alpha^{m_1 \times m}, \boldsymbol{s} \xleftarrow{\$} \mathbb{Z}_q^{m-n}$$

and defines:

$$\begin{cases} \boldsymbol{c} := \mathbf{A}\boldsymbol{s} + p\boldsymbol{e}_1 & \mod q \in \mathbb{Z}_q^m \\ \boldsymbol{c}_0 := (\mathbf{G}h(\tau) + \mathbf{B}_0)\boldsymbol{s} + \mathbf{T}_0' \cdot p\boldsymbol{e}_1 + \boldsymbol{m} \mod q \in \mathbb{Z}_q^{m_1} \\ \boldsymbol{c}_1 := (\mathbf{G}h(\tau) + \mathbf{B}_1)\boldsymbol{s} + \mathbf{T}_1' \cdot p\boldsymbol{e}_1 + \boldsymbol{m} \mod q \in \mathbb{Z}_q^{m_1} \end{cases}$$

The algorithm outputs $\mathbf{CT} = (\boldsymbol{c}, \boldsymbol{c}_0, \boldsymbol{c}_1)$.

$\mathbf{Dec}(sk, \tau, \mathbf{CT}) \to (\boldsymbol{m} \ or \ \perp)$: The algorithm runs as follows:

1. Parse $sk = (\tau_0, \mathbf{T}_0)$, $\mathbf{CT} = (\boldsymbol{c}, \boldsymbol{c}_0, \boldsymbol{c}_1)$.
2. Compute $\boldsymbol{c}_0 - \mathbf{T}_0\boldsymbol{c} = \mathbf{G}h(\tau - \tau_0)\boldsymbol{s} + (\mathbf{T}_0' - \mathbf{T}_0) \cdot p\boldsymbol{e}_1 + \boldsymbol{m}$.
3. Then recover $h(\tau - \tau_0)\boldsymbol{s}$ using $\mathcal{O}(\mathbf{G})$, and compute $\boldsymbol{s} = h^{-1}(\tau - \tau_0)h(\tau - \tau_0)\boldsymbol{s}$ from $h(\tau - \tau_0)\boldsymbol{s}$.
4. Compute $\boldsymbol{m} = \boldsymbol{c}_0 - \mathbf{T}_0\boldsymbol{c} - \mathbf{G}h(\tau - \tau_0)\boldsymbol{s} \mod p$.
5. Testify whether $\boldsymbol{c}_1 - (\mathbf{G}h(\tau) + \mathbf{B}_1)\boldsymbol{s} = \boldsymbol{m} \mod p$. If not, output $\perp$.
6. If anyone of the conditions in the following is matched, the algorithm outputs $\perp$.
   (a) not every entry of $\boldsymbol{c} - \mathbf{A}\boldsymbol{s}$ is a multiple of p or $|\frac{\boldsymbol{c} - \mathbf{A}\boldsymbol{s}}{p}| \geq \alpha q\sqrt{m}$;
   (b) not every entry of $\boldsymbol{c}_0 - (\mathbf{G}h(\tau) \mid \mathbf{B}_u)\boldsymbol{s} - \boldsymbol{m}$ is a multiple of p or any entry of $|\frac{\boldsymbol{c}_0 - (\mathbf{G}h(\tau) + \mathbf{B}_0)\boldsymbol{s} - \boldsymbol{m}}{p}|$ is more than $(\alpha q)^2\sqrt{m} \cdot w(\log m)$;
   (c) not every entry of $\boldsymbol{c}_1 - (\mathbf{G}h(\tau) + \mathbf{B}_1)\boldsymbol{s} - \boldsymbol{m}$ is a multiple of p or any entry of $|\frac{\boldsymbol{c}_1 - (\mathbf{G}h(\tau) + \mathbf{B}_1)\boldsymbol{s} - \boldsymbol{m}}{p}|$ is more than $(\alpha q)^2\sqrt{m} \cdot w(\log m)$;
7. Otherwise, the algorithm outputs $\boldsymbol{m}$.

4.1.1   The Parameter Setting and Correctness of Scheme

In this section, we will describe two sets of parameters in the following Table 2: $p = 2$ and $p = O(n)$. For the different p, the effects on the error rate and the choice of the modulus q are also different.

Table 2. Parameter for different p

Parameter	p	q is a prime	m	$1/\alpha$
Case 1	$p = 2$	$q = O(n^2) \cdot w(\log n)$	$m = 2n$	$O(n^{1.5})w(\log n)$
Case 2	$p = O(n)$	$q = O(n^3) \cdot w(\log n)$	$m = 2n$	$O(n^{2.5})w(\log n)$

Theorem 2 (Correctness). *With the above two sets of parameter, the decryption algorithm* $\mathbf{Dec}(sk, \tau, \mathbf{CT})$ *will output $\boldsymbol{m}$ with overwhelming probability over* $\mathbf{CT} \leftarrow \mathbf{Enc}(pk, \tau, \boldsymbol{m})$ *by the choice of* (sk, pk) *and for all* $\tau \in \mathcal{T}, \boldsymbol{m} \xleftarrow{\$} \{0, \ldots, p - 1\}^{m_1}$.

Proof. In this scheme, firstly, the oracle $\mathcal{O}(\mathbf{G})$ can restore $\boldsymbol{s}$ correctly.

For $\mathbf{T}_0', \mathbf{T}_0 \leftarrow \bar{\varPsi}_\alpha^{m_1 \times m}$, by the Lemma 4 we have $S_1(\mathbf{T}_0' - \mathbf{T}_0) \leq O(\alpha \cdot q \cdot (\sqrt{m_1} + \sqrt{m}))$ except with probability $2^{-\Omega(n)}$. For $\boldsymbol{e}_1 \leftarrow \bar{\varPsi}_\alpha^m$, then we have

$$\|(\mathbf{T}_0' - \mathbf{T}_0) \cdot p\boldsymbol{e}_1\| \leq pS_1(\mathbf{T}_0' - \mathbf{T}_0) \cdot \|\boldsymbol{e}_1\| \leq O(\alpha^2 \cdot pq^2 \cdot (\sqrt{m_1 m} + m))$$
$$= O(\alpha^2 \cdot pq^2 \cdot n)w(\sqrt{\log q})$$

And by the definition of $\mathbf{B}$, for all i, $\|b_i^t m\| \leq kp$. Because $\|\mathbf{B}\| \leq \{\sqrt{5}, \sqrt{k}\}$, then we have

$$\|b_i^t((\mathbf{T}_0' - \mathbf{T}_0) \cdot pe_1 + m)\| \leq \|b_i^t((\mathbf{T}_0' - \mathbf{T}_0) \cdot pe_1)\| + \|b_i^t m\|$$
$$\leq \|\mathbf{B}\| \cdot \|(\mathbf{T}_0' - \mathbf{T}_0) \cdot pe_1\| + kp$$
$$\leq O(\alpha^2 \cdot pq^2 \cdot n) \cdot \omega(\log q)$$

In the case of $p = 2$, for large enough $1/\alpha = O(n^{1.5}) \cdot \omega(\log n)$, then $1/\alpha^2 = O(nq) \cdot \omega(\log n)$. We have

$$(\mathbf{T}_0' - \mathbf{T}_0) \cdot 2e_1 + m \in \mathcal{P}_{1/2}(q \cdot \mathbf{B}^{-t}) \tag{1}$$

In the case of $p = O(n)$, for large enough $1/\alpha = O(n^{2.5}) \cdot \omega(\log n)$, then $1/\alpha^2 = O(n^5) \cdot \omega(log^2 n) = O(n^2 q) \cdot \omega(\log n)$. We also have

$$(\mathbf{T}_0' - \mathbf{T}_0) \cdot pe_1 + m \in \mathcal{P}_{1/2}(q \cdot \mathbf{B}^{-t}) \tag{2}$$

Secondly, it needs to reconstruct the message m from $(\mathbf{T}_0' - \mathbf{T}_0) \cdot pe_1 + m = c_0 - \mathbf{T}_0 c - \mathbf{G}h(\tau - \tau_0)s$. If $(\mathbf{T}_0' - \mathbf{T}_0) \cdot pe_1 + m \in \mathcal{P}_{1/2}(q \cdot \mathbf{B}^{-t})$, then every entry of $(\mathbf{T}_0' - \mathbf{T}_0) \cdot pe_1 + m$ is less than $\frac{q}{2}$. Therefore, we can recover m correctly from $(\mathbf{T}_0' - \mathbf{T}_0) \cdot pe_1 + m \bmod p$. And the decryption algorithm $\mathbf{Dec}(sk, \tau, \mathbf{CT})$ outputs m with overwhelming probability as desired.

$\square$

Remark. For Theorem 1, we set $r = \alpha q$. With the above parameter settings, when $p = 2$, $1/\alpha^2 = O(n^3) \cdot \omega(log^2 n) \geq O(n^2) \cdot \omega(\log n) \cdot \sqrt{3n} = q\sqrt{m+n}$; when $p = O(n)$, $1/\alpha^2 = O(n^5) \cdot \omega(log^2 n) \geq O(n^3) \cdot \omega(\log n) \cdot \sqrt{3n} = q\sqrt{m+n}$. In both cases, we have $1/\alpha \geq r \cdot \sqrt{m+n}$ that meets the requirement of Theorem 1.

4.1.2 Proof of CCA Security

Theorem 3 (CCA Security). *Under the LWE assumption, this scheme is secure against the selective-tag weak CCA adversaries. In particular, the advantage for any adversary $\mathcal{A}$ is*

$$\mathbf{Adv}_{TBE,\mathcal{A}}^{stw-cca} \leq 4\mathbf{Adv}_{KLWE_{m,n,q}^{m_1}}^{\mathcal{A}_1} + 6\mathbf{Adv}_{EKLWE_{m,n,q}^{m_1}}^{\mathcal{C}_1} + \mathbf{Adv}_{LWE_{n,m,q,\chi}}^{\mathcal{S}_1}$$
$$+ 2\mathbf{Adv}_{KLWE_{m,n-1,q}^{m_1}}^{\mathcal{A}_1} + negl(n)$$

Proof. Let $\mathcal{A}$ be the adversary attacking this TBE scheme. In the following, we will start from the actual attack to the scheme, which denotes by **Game 0**, and then modify it step by step in order to obtain a game in which the advantage of adversary $\mathcal{A}$ is zero. In each game, the simulator $\mathcal{S}$ either changes the construction of public key and the challenge ciphertext $\mathbf{CT}^* = (c^*, c_0^*, c_1^*)$, or decides which secret key is used to answer the decryption queries. For the adversary $\mathcal{A}$, it cannot distinguish the difference between every step under the LWE assumption. The goal of the simulator $\mathcal{S}$ is to use the ability of adversary $\mathcal{A}$ to solve the hard problem, e.g., the LWE problem. The advantage of the game is defined as

$$\mathbf{Adv}_{TBE,\mathcal{A}}^{stw-cca} = |\mathbf{Pr}[b = b'] - \frac{1}{2}|$$

where b is randomly chosen from $\{0, 1\}$ and b' is the output of adversary $\mathcal{A}$. For every game, we use $\mathbf{Pr}\left[\mathcal{A}[\mathbf{Game\ i}] = 1\right]$ denotes $\mathbf{Pr}[b = b']$.

Before generating the challenge ciphertext, the simulator computes a part of ciphertext which are independent with the message in the stage of initializing.

$$c^* := \mathbf{A}s^* + pe_1^*,\ c_0' := (\mathbf{G}h(\tau^*) + \mathbf{B}_0)s^* + \mathbf{T}_0^* pe_1^*,\ c_1' := (\mathbf{G}h(\tau^*) + \mathbf{B}_1)s^* + \mathbf{T}_1^* pe_1^*$$

And the challenge ciphertext $\mathbf{CT}^*$ is

$$c^* := \mathbf{A}s^* + pe_1^*,\ c_0^* := c_0' + m_b,\ c_1^* := c_1' + m_b$$

We will describe these games concretely.

$\boxed{\textbf{Game 0:}}$ This is the actual TBE selective-tag weak CCA security game between the adversary $\mathcal{A}$ and the simulator $\mathcal{S}$. So we have

$$\mathbf{Adv}_{\mathrm{TBE},\mathcal{A}}^{stw-cca} = \left| \mathbf{Pr}\left[\mathcal{A}[\mathbf{Game\ 0}] = 1\right] - \frac{1}{2} \right|$$

$\boxed{\textbf{Game 1:}}$ In this game, we change the construction of public key $\mathbf{B}_1$ and the c_1^* item of the challenge ciphertext $\mathbf{CT}^*$, but the adversary $\mathcal{A}$ cannot distinguish **Game 1** from **Game 0** if the $\mathrm{ELWE}_{m,n,q}^{m_1}$ and $\mathrm{EKLWE}_{m,n,q}^{m_1}$ problems are hard.

During generating the public key, the simulator $\mathcal{S}$ replaces the hidden tag of $\mathbf{T}_1$ from 0 to τ^*, that is, $\mathbf{B}_1 = \mathbf{T}_1 \mathbf{A} - \mathbf{G}h(\tau^*)$. Simultaneously, the simulator uses $\mathbf{T}_1 pe_1^*$ instead of $\mathbf{T}_1^* pe_1^*$ in computing c_1^* and obtains $c_1^* := \mathbf{T}_1 \cdot c^* + m_b$, the other part of $\mathbf{CT}*$ is the same with those in **Game 0**.

In particular, after replacing $\mathbf{B}_1 = \mathbf{T}_1 \mathbf{A}$ with $\mathbf{B}_1 = \mathbf{T}_1 \mathbf{A} - \mathbf{G}h(\tau^*)$, then the simulator cannot use this trapdoor $\mathbf{T}_1$ to answer the decryption queries and to replace $\mathbf{T}_1^* pe_1^*$. During the process of initializing, however, if the simulator can simultaneously replace the $\mathbf{T}_1 \mathbf{A}$ and $\mathbf{T}_1^* pe^*$ by $\mathbf{T}_1 \mathbf{A} - \mathbf{G}h(\tau^*)$ and $\mathbf{T}_1 pe^*$, then he can figure out this problem. And this truth is based on the following lemma.

Lemma 5. Let $a \xleftarrow{\$} \mathbb{Z}_q^{m \times (m-n)}$, $e \leftarrow \bar{\Psi}_\alpha^m$ and $\mathbf{T}, \mathbf{T}_1 \leftarrow \bar{\Psi}_\alpha^{m_1 \times m}$ are independent, let $\mathcal{A}$ be any PPT adversary, then we have

$$|\mathbf{Pr}[\mathcal{A}(\mathbf{A}, \mathbf{T}_1 \mathbf{A}, e, \mathbf{T}e) = 1] - \mathbf{Pr}[\mathcal{A}(\mathbf{A}, \mathbf{T}_1 \mathbf{A} - \mathbf{G}h(\tau), e, \mathbf{T}_1 e) = 1]|$$
$$\leq \mathbf{Adv}_{\mathrm{KLWE}_{m,n,q}^{m_1}}^{\mathcal{A}_1} + 3 \cdot \mathbf{Adv}_{\mathrm{EKLWE}_{m,n,q}^{m_1}}^{\mathcal{C}_1}$$

Proof. Firstly, because $\mathbf{T}, \mathbf{T}_1$ are independent, $(e, \mathbf{T}e)$ can not provide any information about $\mathbf{T}_1$. For $\mathbf{B} \xleftarrow{\$} \mathbb{Z}_q^{m_1 \times (m-n)}$, we have

$$|\mathbf{Pr}[\mathcal{A}(\mathbf{A}, \mathbf{T}_1 \mathbf{A}, e, \mathbf{T}e) = 1] - \mathbf{Pr}[\mathcal{A}(\mathbf{A}, \mathbf{B}, e, \mathbf{T}e) = 1]|$$
$$= |\mathbf{Pr}[\mathcal{A}(\mathbf{A}, \mathbf{T}_1 \mathbf{A}) = 1] - \mathbf{Pr}[\mathcal{A}(\mathbf{A}, \mathbf{B}) = 1]| \leq \mathbf{Adv}_{\mathrm{KLWE}_{m,n,q}^{m_1}}^{\mathcal{A}_1}$$

Secondly, by the hardness of extended-knapsack LWE, we have

$$|\mathbf{Pr}[\mathcal{A}(\mathbf{A}, \mathbf{T}_1 \mathbf{A}, e, \mathbf{T}e) = 1] - \mathbf{Pr}[\mathcal{A}(\mathbf{A}, \mathbf{T}_1 \mathbf{A}, e, \mathbf{T}_1 e) = 1]|$$
$$\leq |\mathbf{Pr}[\mathcal{A}(\mathbf{A}, \mathbf{T}_1 \mathbf{A}, e, \mathbf{T}e) = 1] - \mathbf{Pr}[\mathcal{A}(\mathbf{A}, \mathbf{B}, e, \mathbf{T}e) = 1]|$$
$$+ |\mathbf{Pr}[\mathcal{A}(\mathbf{A}, \mathbf{B}, e, \mathbf{T}e) = 1] - \mathbf{Pr}[\mathcal{A}(\mathbf{A}, \mathbf{T}\mathbf{A}, e, \mathbf{T}e) = 1]|$$
$$+ |\mathbf{Pr}[\mathcal{A}(\mathbf{A}, \mathbf{T}\mathbf{A}, e, \mathbf{T}e) = 1] - \mathbf{Pr}[\mathcal{A}(\mathbf{A}, \mathbf{T}_1 \mathbf{A}, e, \mathbf{T}_1 e) = 1]|$$
$$\leq \mathbf{Adv}_{\mathrm{KLWE}_{m,n,q}^{m_1}}^{\mathcal{A}_1} + \mathbf{Adv}_{\mathrm{EKLWE}_{m,n,q}^{m_1}}^{\mathcal{C}_1}$$

Where $|\mathbf{Pr}[\mathcal{A}(\mathbf{A}, \mathbf{TA}, e, \mathbf{T}e) = 1] - \mathbf{Pr}[\mathcal{A}(\mathbf{A}, \mathbf{T}_1\mathbf{A}, e, \mathbf{T}_1 e) = 1]| = 0$, because $(\mathbf{A}, \mathbf{TA}, e, \mathbf{T}e)$ and $(\mathbf{A}, \mathbf{T}_1\mathbf{A}, e, \mathbf{T}_1 e)$ are the same distribution and $\mathbf{T}, \mathbf{T}_1$ are independent.

Finally, with the above results we have

$$|\mathbf{Pr}[\mathcal{A}(\mathbf{A}, \mathbf{T}_1\mathbf{A}, e, \mathbf{T}e) = 1] - \mathbf{Pr}[\mathcal{A}(\mathbf{A}, \mathbf{T}_1\mathbf{A} - \mathbf{G}h(\tau^*), e, \mathbf{T}_1 e) = 1]|$$
$$\leq |\mathbf{Pr}[\mathcal{A}(\mathbf{A}, \mathbf{T}_1\mathbf{A}, e, \mathbf{T}e) = 1] - \mathbf{Pr}[\mathcal{A}(\mathbf{A}, \mathbf{B}, e, \mathbf{T}_1 e) = 1]|$$
$$+ |\mathbf{Pr}[\mathcal{A}(\mathbf{A}, \mathbf{B} - \mathbf{G}h(\tau^*), e, \mathbf{T}_1 e) = 1] - \mathbf{Pr}[\mathcal{A}(\mathbf{A}, \mathbf{T}_1\mathbf{A} - \mathbf{G}h(\tau^*), e, \mathbf{T}_1 e) = 1]|$$
$$\leq |\mathbf{Pr}[\mathcal{A}(\mathbf{A}, \mathbf{T}_1\mathbf{A}, e, \mathbf{T}e) = 1] - \mathbf{Pr}[\mathcal{A}(\mathbf{A}, \mathbf{T}_1\mathbf{A}, e, \mathbf{T}_1 e) = 1]|$$
$$+ |\mathbf{Pr}[\mathcal{A}(\mathbf{A}, \mathbf{T}_1\mathbf{A}, e, \mathbf{T}_1 e) = 1]| - \mathbf{Pr}[\mathcal{A}(\mathbf{A}, \mathbf{B}, e, \mathbf{T}_1 e) = 1]|$$
$$+ |\mathbf{Pr}[\mathcal{A}(\mathbf{A}, \mathbf{B} - \mathbf{G}h(\tau^*), e, \mathbf{T}_1 e) = 1] - \mathbf{Pr}[\mathcal{A}(\mathbf{A}, \mathbf{T}_1\mathbf{A} - \mathbf{G}h(\tau^*), e, \mathbf{T}_1 e) = 1]|$$
$$\leq \mathbf{Adv}^{\mathcal{A}_1}_{\mathrm{KLWE}^{m_1}_{m,n,q}} + 3 \cdot \mathbf{Adv}^{\mathcal{C}_1}_{\mathrm{EKLWE}^{m_1}_{m,n,q}}$$

where

$$|\mathbf{Pr}[\mathcal{A}(\mathbf{A}, \mathbf{B} - \mathbf{G}h(\tau^*), e, \mathbf{T}_1 e) = 1] - \mathbf{Pr}[\mathcal{A}(\mathbf{A}, \mathbf{T}_1\mathbf{A} - \mathbf{G}h(\tau^*), e, \mathbf{T}_1 e) = 1]|$$
$$\leq |\mathbf{Pr}[\mathcal{A}(\mathbf{A}, \mathbf{B}, e, \mathbf{T}_1 e) = 1] - \mathbf{Pr}[\mathcal{A}(\mathbf{A}, \mathbf{T}_1\mathbf{A}, e, \mathbf{T}_1 e) = 1]|$$

□

Then we can describe an Algorithm $\mathcal{B}$ which simulates **Game 1** or **Game 0** with different inputs.

Algorithm $\mathcal{B}$		
Input: $(\mathbf{A}, \mathbf{B}_1^*, e^*, \mathbf{T}_1' e^*)$		
Initialize:	Compute:	Output (sk, pk):
$\mathbf{B}_0 = \mathbf{T}_0\mathbf{A},$	$c^* := \mathbf{A}s^* + pe_1^*$	$sk = (0, \mathbf{T}_0)$
$ek = (\mathbf{A}, \mathbf{B}_0, \mathbf{B}_1^*)$	$c_0' := (\mathbf{G}h(\tau^*) + \mathbf{B}_0)s^* + \mathbf{T}_0^* pe_1^*$	$pk = ek$
$s^* \leftarrow \mathbb{Z}_q^n, \mathbf{T}_0^* \leftarrow \bar{\Psi}_\alpha^{m_1 \times m}$	$c_1' := (\mathbf{G}h(\tau^*) + \mathbf{B}_1)s^* + \mathbf{T}_1^* pe_1^*$	
Dec Query$(\tau, \mathbf{CT})$:	**Ciphertext $\mathbf{CT^*} = (c^*, c_0^*, c_1^*)$**	
If $\tau = \tau^*$, Output $\perp$;	Choose $b \xleftarrow{\$} \{0, 1\}$, for challenge$(m_0, m_1)$	
Otherwise,	$c^* := \mathbf{A}s^* + pe_1^*$	
Output $\mathbf{Dec}(sk, \tau, \mathbf{CT})$	$c_0^* := c_0' + m_b = (\mathbf{G}h(\tau^*) + \mathbf{B}_0)s^* + \mathbf{T}_0^* pe_1^* + m_b$	
	$c_1^* := c_1' + m_b = (\mathbf{G}h(\tau^*) + \mathbf{B}_1)s^* + \mathbf{T}_1^* pe_1^* + m_b$	

When the Algorithm $\mathcal{B}$ inputs with a tuple $(\mathbf{A}, \mathbf{T}_1\mathbf{A}, e_1^*, \mathbf{T}_1^* e_1^*)$, then it simulates the **Game 0**. When $\mathcal{B}$ inputs with a tuple $(\mathbf{A}, \mathbf{T}_1\mathbf{A} - \mathbf{G}h(\tau^*), e_1^*, \mathbf{T}_1 e_1^*)$, then it simulates the **Game 1**. If the adversary $\mathcal{A}$ can distinguish the two games, then there exists an algorithm $\mathcal{A}'$ can discriminate the two tuples.

From the Lemma 5, the difference between **Game 1** and **Game 0** is

$$|\mathbf{Pr}[\mathcal{A}[\textbf{Game } 0] = 1] - \mathbf{Pr}[\mathcal{A}[\textbf{Game } 1] = 1]|$$
$$\leq |\mathbf{Pr}[\mathcal{A}'(\mathbf{A}, \mathbf{T}_1\mathbf{A}, e_1^*, \mathbf{T}_1^* e_1^*) = 1] - \mathbf{Pr}[\mathcal{A}'(\mathbf{A}, \mathbf{T}_1\mathbf{A} - \mathbf{G}h(\tau^*), e_1^*, \mathbf{T}_1 e_1^*) = 1]|$$
$$\leq \mathbf{Adv}^{\mathcal{A}_1}_{\mathrm{KLWE}^{m_1}_{m,n,q}} + 3 \cdot \mathbf{Adv}^{\mathcal{C}_1}_{\mathrm{EKLWE}^{m_1}_{m,n,q}}$$

Game 2: In this game, we will show that the simulator uses different trapdoor from **Game 1** to answer the decryption queries, but introduce only $negl(n)$ statistical difference with **Game 1**.

When given a ciphertext by adversary $\mathcal{A}$, the simulator $\mathcal{S}$ uses the trapdoor $\mathbf{T}_1$ instead of $\mathbf{T}_0$ to answer the decryption queries and outputs the correct plaintext. What we need is that the adversary $\mathcal{A}$ can not discriminate between two behaviors issued by two trapdoors.

By the following lemma, we have that the **Dec** algorithm can output the same distribution with the different trapdoors with overwhelming probability.

Lemma 6. *Given* $\mathbf{CT} = (\boldsymbol{c}, \boldsymbol{c}_0, \boldsymbol{c}_1) \leftarrow \mathbf{Enc}(pk, \tau, \boldsymbol{m})$, *and* $sk_0 = (\tau_0, \mathbf{T_0})$, $sk_1 = (\tau_1, \mathbf{T_1})$, *the decryption algorithm* $\mathbf{Dec}(sk_0, \tau, \mathbf{CT})$ *and* $\mathbf{Dec}(sk_1, \tau, \mathbf{CT})$ *will output the same distribution with overwhelming probability over the choice of public and secret keys.*

Proof. Given $\mathbf{CT} = (\boldsymbol{c}, \boldsymbol{c}_0, \boldsymbol{c}_1) \leftarrow \mathbf{Enc}(pk, \tau, \boldsymbol{m})$, we need to recover the $\boldsymbol{m}$ under two conditions in order to testify the correctness of the two decryption oracles. The decryption algorithm $\mathbf{Dec}(sk_0, \tau, \mathbf{CT})$ will get $\boldsymbol{m}$ from $(\boldsymbol{c}, \boldsymbol{c}_0)$, then we have

$$\boldsymbol{c}_0 - \mathbf{T}_0\boldsymbol{c} = \mathbf{G}h(\tau - \tau_0)\boldsymbol{s} + (\mathbf{T}'_0 - \mathbf{T}_0) \cdot p\boldsymbol{e}_1 + \boldsymbol{m}$$

From the expression above, we will reconstruct $\boldsymbol{s}$ if $(\mathbf{T}'_0 - \mathbf{T}_0) \cdot p\boldsymbol{e}_1 + \boldsymbol{m} \in \mathcal{P}_{1/2}(q \cdot \mathbf{B}^{-t})$. Then we need to check the consistency of plaintext by recovering the $\boldsymbol{m}$ from $\boldsymbol{c}_1$ with the $\boldsymbol{s}$, that is,

$$\boldsymbol{m} = \boldsymbol{c}_1 - (\mathbf{G}h(\tau) + \mathbf{B}_1)\boldsymbol{s} \bmod p$$

According to the above decryption, we have $\mathbf{Dec}(sk_1, \tau, \mathbf{CT})$ to get

$$\boldsymbol{c}_1 - \mathbf{T}_1\boldsymbol{c} = \mathbf{G}h(\tau - \tau_1)\boldsymbol{s} + (\mathbf{T}'_1 - \mathbf{T}_1) \cdot p\boldsymbol{e}_1 + \boldsymbol{m}$$

The oracle $\mathcal{O}(\mathbf{G})$ will reconstruct the same $\boldsymbol{s}$ if $(\mathbf{T}'_1 - \mathbf{T}_1) \cdot p\boldsymbol{e}_1 + \boldsymbol{m} \in \mathcal{P}_{1/2}(q \cdot \mathbf{B}^{-t})$. Because for the $\mathbf{Dec}(sk_0, \tau, \mathbf{CT})$, the condition $(\mathbf{T}'_1 - \mathbf{T}_1) \cdot p\boldsymbol{e}_1 + \boldsymbol{m} \in \mathcal{P}_{1/2}(q \cdot \mathbf{B}^{-t})$ is matched, and $\mathbf{T}_0, \mathbf{T}'_0, \mathbf{T}_1, \mathbf{T}'_1$ are sampled from the same distribution, then we have

$$\mathbf{Pr}_{\mathbf{T}'_1}[\forall \boldsymbol{e}_1, |\boldsymbol{e}_1| \leq \alpha q\sqrt{m}, \boldsymbol{m} \in \{0,1\}^{m_1} : (\mathbf{T}'_1 - \mathbf{T}_1) \cdot p\boldsymbol{e_1} + \boldsymbol{m} \in \mathcal{P}_{1/2}(q \cdot \mathbf{B}^{-t})]$$
$$\geq 1 - negl(n)$$

If $\mathcal{O}(\mathbf{G})$ will reconstruct the same $\boldsymbol{s}$, then the $\mathbf{Dec}(sk_1, \tau, \mathbf{CT})$ will output the same $\boldsymbol{m}$. So we complete the proof. □

From the above lemma, the adversary cannot figure out which trapdoor is used to answer the decryption queries, otherwise he can distinguish **Game 2** from **Game 1**. And the difference between two games is

$$|\mathbf{Pr}\left[\mathcal{A}[\textbf{Game 1}] = 1\right] - \mathbf{Pr}\left[\mathcal{A}[\textbf{Game 2}] = 1\right]| \leq negl(n)$$

Game 3: In this game, we will change the construction of public key $\mathbf{B_0}$ and the c_0^* item of challenge ciphertext $\mathbf{CT}^*$, but the adversary $\mathcal{A}$ cannot distinguish **Game 3** from **Game 2** if the $\mathrm{ELWE}_{m,n,q}^{m_1}$ and $\mathrm{EKLWE}_{m,n,q}^{m_1}$ problems are hard.

During generating the public key and c_0^*, the simulator $\mathcal{S}$ replaces the hidden tag of $\mathbf{T_0}$ from 0 to τ^*, and uses $\mathbf{T_0}pe_1^*$ instead of $\mathbf{T_0^*}pe_1^*$. Then $\mathbf{B_1} = \mathbf{T_1 A} - \mathbf{G}h(\tau^*)$ and $c_0^* := \mathbf{T_0} \cdot \mathbf{c}^* + \mathbf{m_b}$, the other part of $\mathbf{CT}^*$ is the same with those in **Game 2**.

The advantage of the adversary $\mathcal{A}$ distinguishing two games is analogous to the advantage of $\mathcal{A}$ distinguishing **Game 1** from **Game 0**, then we have

$$|\mathbf{Pr}\left[\mathcal{A}[\mathbf{Game\ 2}] = 1\right] - \mathbf{Pr}\left[\mathcal{A}[\mathbf{Game\ 3}] = 1\right]| \leq \mathbf{Adv}_{\mathrm{KLWE}_{m,n,q}^{m_1}}^{\mathcal{A}_1} + 3 \cdot \mathbf{Adv}_{\mathrm{EKLWE}_{m,n,q}^{m_1}}^{\mathcal{C}_1}$$

Game 4: In this game, we show that the ciphertext $\mathbf{c}^*$ can be replaced by a random value over $\mathbb{Z}_q^m$, but the adversary $\mathcal{A}$ cannot distinguish **Game 4** from **Game 3** under the standard LWE assumption.

During generating the $\mathbf{c}^* = \mathbf{A}s^* + pe_1^*$ item of $\mathbf{CT}^*$, the simulator chooses a random value over $\mathbb{Z}_q^m$. Then simulators generates $c_0^* := \mathbf{T_0} \cdot \mathbf{c}^* + \mathbf{m_b}$, $c_1^* := \mathbf{T_1} \cdot \mathbf{c}^* + \mathbf{m_b}$ with $\mathbf{T_0}, \mathbf{T_1}$. Then we have

$$|\mathbf{Pr}\left[\mathcal{A}[\mathbf{Game\ 3}] = 1\right] - \mathbf{Pr}\left[\mathcal{A}[\mathbf{Game\ 4}] = 1\right]| \leq \mathbf{Adv}_{\mathrm{LWE}_{n,m,q,\chi}}^{\mathcal{S}_1}$$

Game 5: In this game, we change c_0' item during initializing, but the adversary $\mathcal{A}$ cannot distinguish **Game 5** from **Game 4** under the KLWE problem..

In the initialization, the simulator replaces the $c_0' = \mathbf{T_0} \cdot \mathbf{c}^*$ in **Game 4** with a random value from $\mathcal{U}(\mathbb{Z}_q^{m_1})$. The rest part is as same as **Game 4**. In this game, the trapdoor $\mathbf{T_1}$ is used to answer decryption answers. Then we describe an Algorithm $\mathcal{C}$ that simulates **Game 4** and **Game 5** with different inputs.

Algorithm $\mathcal{C}$		
Input:$(\mathbf{A}, \mathbf{B_0^*}, \mathbf{c}^*, c_0')$		
Initialize:	Compute:	Output: (sk, pk)
$\mathbf{B_1^*} = \mathbf{T_1 A} - \mathbf{G}h(\tau^*)$	$c_1' := \mathbf{T_1}\mathbf{c}^*$	$sk = (\tau^*, \mathbf{T_1})$
ek=$(\mathbf{A}, \mathbf{B_0^*}, \mathbf{B_1^*})$		$pk = ek$
Dec Query$(\tau, \mathbf{CT})$:	**Ciphertext** $\mathbf{CT}^*$=$(\mathbf{c}^*, c_0^*, c_1^*)$	
If $\tau = \tau^*$, Output $\perp$;	Choose $b \xleftarrow{\$} \{0,1\}$, for challenge$(\mathbf{m_0}, \mathbf{m_1})$	
Otherwise,	$\mathbf{c}^*$	
Output **Dec**$(sk, \tau, \mathbf{CT})$	$c_0^* := c_0' + \mathbf{m_b}$	
	$c_1^* := c_1' + \mathbf{m_b} = \mathbf{T_1}\mathbf{c}^* + \mathbf{m_b}$	

When the Algorithm $\mathcal{C}$ inputs with a tuple $(\mathbf{A}, \mathbf{T_0 A} - \mathbf{G}h(\tau^*), \mathbf{c}^*, \mathbf{T_0}\mathbf{c}^*)$, then it simulates the **Game 4**. When the algorithm inputs with a tuple $(\mathbf{A}, \mathbf{T_0 A} - \mathbf{G}h(\tau^*), \mathbf{c}^*, \mathbf{u})$, then it simulates the **Game 5**. If the adversary $\mathcal{A}$ can distinguish the two games, then there exists an Algorithm $\mathcal{B}$ can discriminate the two tuples. Then the difference between **Game 4** and **Game 5** is

$$|\mathbf{Pr}\left[\mathcal{A}[\mathbf{Game\ 4}] = 1\right] - \mathbf{Pr}\left[\mathcal{A}[\mathbf{Game\ 5}] = 1\right]| \leq$$
$$|\mathbf{Pr}\left[\mathcal{D}(\mathbf{A}, \mathbf{T_0 A} - \mathbf{G}h(\tau^*), \mathbf{c}^*, \mathbf{T_0}\mathbf{c}^*) = 1\right] - \mathbf{Pr}\left[\mathcal{D}(\mathbf{A}, \mathbf{T_0 A} - \mathbf{G}h(\tau^*), \mathbf{c}^*, \mathbf{u}) = 1\right]|$$

From the Lemma 2, we have

$$|\mathbf{Pr}\left[\mathcal{D}(\mathbf{A}, \mathbf{T}_0\mathbf{A}, \boldsymbol{c}^*, \mathbf{T}_0\boldsymbol{c}^*) = 1\right] - \mathbf{Pr}\left[\mathcal{D}(\mathbf{A}, \mathbf{T}_0\mathbf{A}, \boldsymbol{c}^*, \boldsymbol{u}) = 1\right]|$$
$$\leq \mathbf{Adv}_{\mathrm{KLWE}_{m,n-1,q}^{m_1}}^{\mathcal{A}_1} + \mathbf{Adv}_{\mathrm{KLWE}_{m,n,q}^{m_1}}^{\mathcal{A}_1}$$

where $\boldsymbol{c}^* \xleftarrow{\$} \mathbb{Z}_q^m$ and $\boldsymbol{u} \xleftarrow{\$} \mathbb{Z}_q^{m_1}$. Because $\mathbf{G}h(\tau^*)$ is independent with $\mathbf{A}, \mathbf{T}_0\mathbf{A}$, then we have

$$|\mathbf{Pr}\left[\mathcal{D}(\mathbf{A}, \mathbf{T}_0\mathbf{A} - \mathbf{G}h(\tau^*), \boldsymbol{c}^*, \mathbf{T}_0\boldsymbol{c}^*) = 1\right] - \mathbf{Pr}\left[\mathcal{D}(\mathbf{A}, \mathbf{T}_0\mathbf{A} - \mathbf{G}h(\tau^*), \boldsymbol{c}^*, \boldsymbol{u}) = 1\right]|$$
$$\leq \mathbf{Adv}_{\mathrm{KLWE}_{m,n-1,q}^{m_1}}^{\mathcal{A}_1} + \mathbf{Adv}_{\mathrm{KLWE}_{m,n,q}^{m_1}}^{\mathcal{A}_1}$$

With the above result, if the adversary $\mathcal{A}$ can distinguish **Game 5** from **Game 4**, then there exists an PPT algorithm can solve the knapsack LWE problem. Therefore, we have

$$|\mathbf{Pr}\left[\mathcal{A}[\mathbf{Game\ 4}] = 1\right] - \mathbf{Pr}\left[\mathcal{A}[\mathbf{Game\ 5}] = 1\right]| \leq \mathbf{Adv}_{\mathrm{KLWE}_{m,n-1,q}^{m_1}}^{\mathcal{A}_1} + \mathbf{Adv}_{\mathrm{KLWE}_{m,n,q}^{m_1}}^{\mathcal{A}_1}$$

Game 6: In this game, we will show that the simulator uses different trapdoor from **Game 5** to answer the decryption queries, but introduces only $negl(n)$ statistical difference with **Game 5**.

During the process of answering the decryption queries, the simulator $\mathcal{S}$ uses the trapdoor $\mathbf{T_0}$ instead of $\mathbf{T_1}$ to answer the decryption queries and output the correct plaintext. And the analysis is analogous to the **Game 2**.

From the advantage between **Game 1** and **Game 2**, we have

$$|\mathbf{Pr}\left[\mathcal{A}[\mathbf{Game\ 5}] = 1\right] - \mathbf{Pr}\left[\mathcal{A}[\mathbf{Game\ 6}] = 1\right]| \leq negl(n)$$

Game 7: In this game, we change $\boldsymbol{c}_1'$ item during initializing, but the adversary $\mathcal{A}$ cannot distinguish **Game 7** from **Game 6** under the KLWE problem.

In the step of initializing, the simulator $\mathcal{S}$ replaces the $\boldsymbol{c}_1'$ with a random value chosen from $\mathcal{U}(\mathbb{Z}_q^{m_1})$. The rest part is as same as **Game 6**. And the analysis is analogous to the **Game 5**. Then we have

$$|\mathbf{Pr}\left[\mathcal{A}[\mathbf{Game\ 6}] = 1\right] - \mathbf{Pr}\left[\mathcal{A}[\mathbf{Game\ 7}] = 1\right]| \leq \mathbf{Adv}_{\mathrm{KLWE}_{m,n-1,q}^{m_1}}^{\mathcal{A}_1} + \mathbf{Adv}_{\mathrm{KLWE}_{m,n,q}^{m_1}}^{\mathcal{A}_1}$$

Game 8: In this game, we show that the advantage of the adversary $\mathcal{A}$ will be $1/2$, that is, $\mathbf{Pr}\left[\mathcal{A}[\mathbf{Game\ 8}] = 1\right] = \frac{1}{2}$, because the challenge ciphertext is independent with the message.

We observe that $\boldsymbol{c}_0', \boldsymbol{c}_1'$ are both randomly chosen from $\mathcal{U}(\mathbb{Z}_q^{m_1})$ in **Game 7**, so the items $\boldsymbol{c}_0^* = \boldsymbol{c}_0' + \boldsymbol{m}_b$, $\boldsymbol{c}_1^* = \boldsymbol{c}_1' + \boldsymbol{m}_b$ are according to the distribution $\mathcal{U}(\mathbb{Z}_q^{m_1})$. In the process of generating the challenge ciphertext, the simulator $\mathcal{S}$ can chooses the $\boldsymbol{c}_0^*, \boldsymbol{c}_1^*$ randomly from $\mathcal{U}(\mathbb{Z}_q^{m_1})$ to replace those in **Game 7**. And the difference between **Game 7** and **Game 8** is

$$|\mathbf{Pr}\left[\mathcal{A}[\mathbf{Game\ 7}] = 1\right] - \mathbf{Pr}\left[\mathcal{A}[\mathbf{Game\ 8}] = 1\right]| \leq negl(n)$$

Because c_0^*, c_1^* are according to $\mathcal{U}(\mathbb{Z}_q^{m_1})$, the message is completely hidden by them and the challenge ciphertext in **Game 8** is independent with the message. Therefore, we have

$$Pr\left[\mathcal{A}[\textbf{Game 8}] = 1\right] = \frac{1}{2}$$

By summing up all the above probabilities, we complete the proof.   □

5   Conclusion

We construct an efficient chosen ciphertext secure PKE scheme under the LWE assumption and our scheme can decrease the message expansion rate to $O(1)$. In our work, we hide the message by the knapsack LWE assumption to decrease the size of ciphertext and increase the message space to reduce the message expansion rate.

References

1. Agrawal, S., Boneh, D., Boyen, X.: Efficient lattice (H)IBE in the standard model. In: Gilbert, H. (ed.) EUROCRYPT 2010. LNCS, vol. 6110, pp. 553–572. Springer, Heidelberg (2010)
2. Alperin-Sheriff, J., Peikert, C.: Circular and KDM security for identity-based encryption. In: Fischlin, M., Buchmann, J., Manulis, M. (eds.) PKC 2012. LNCS, vol. 7293, pp. 334–352. Springer, Heidelberg (2012)
3. Boneh, D., Canetti, R., Halevi, S., Katz, J.: Chosen-ciphertext security from identity-based encryption. SIAM J. Comput. **36**(5), 1301–1328 (2006)
4. Boneh, D., Katz, J.: Improved efficiency for CCA-secure cryptosystems built using identity-based encryption. In: Menezes, A. (ed.) CT-RSA 2005. LNCS, vol. 3376, pp. 87–103. Springer, Heidelberg (2005)
5. Brakerski, Z., Gentry, C., Vaikuntanathan, V.: (Leveled) fully homomorphic encryption without bootstrapping. In: Proceedings of the 3rd Innovations in Theoretical Computer Science Conference, ITCS 2012, pp. 309–325. ACM, New York (2012)
6. Brakerski, Z., Langlois, A., Peikert, C., Regev, O., Stehlé, D.: Classical hardness of learning with errors. In: STOC, pp. 575–584 (2013)
7. Canetti, R., Halevi, S., Katz, J.: Chosen-ciphertext security from identity-based encryption. In: Cachin, C., Camenisch, J.L. (eds.) EUROCRYPT 2004. LNCS, vol. 3027, pp. 207–222. Springer, Heidelberg (2004)
8. Håstad, J., Impagliazzo, R., Levin, L.A., Luby, M.: A pseudorandom generator from any one-way function. SIAM J. Comput. **28**(4), 1364–1396 (1999)
9. Kiltz, E.: Chosen-ciphertext security from tag-based encryption. In: Halevi, S., Rabin, T. (eds.) TCC 2006. LNCS, vol. 3876, pp. 581–600. Springer, Heidelberg (2006)
10. Kiltz, E., Masny, D., Pietrzak, K.: Simple chosen-ciphertext security from low-noise LPN. In: Krawczyk, H. (ed.) PKC 2014. LNCS, vol. 8383, pp. 1–18. Springer, Heidelberg (2014)
11. Micciancio, D., Mol, P.: Pseudorandom knapsacks and the sample complexity of LWE search-to-decision reductions. In: Rogaway, P. (ed.) CRYPTO 2011. LNCS, vol. 6841, pp. 465–484. Springer, Heidelberg (2011)

12. Micciancio, D., Peikert, C.: Trapdoors for lattices: simpler, tighter, faster, smaller. In: Pointcheval, D., Johansson, T. (eds.) EUROCRYPT 2012. LNCS, vol. 7237, pp. 700–718. Springer, Heidelberg (2012)
13. O'Neill, A., Peikert, C., Waters, B.: Bi-deniable public-key encryption. In: Rogaway, P. (ed.) CRYPTO 2011. LNCS, vol. 6841, pp. 525–542. Springer, Heidelberg (2011)
14. Peikert, C.: Public-key cryptosystems from the worst-case shortest vector problem: extended abstract. In: STOC, pp. 333–342 (2009)
15. Peikert, C., Waters, B.: Lossy trapdoor functions and their applications. In; STOC, pp. 187–196 (2008)
16. Regev, O.: On lattices, learning with errors, random linear codes, and cryptography. In: STOC, pp. 84–93 (2005)
17. Rosen, A., Segev, G.: Chosen-ciphertext security via correlated products. In: Reingold, O. (ed.) TCC 2009. LNCS, vol. 5444, pp. 419–436. Springer, Heidelberg (2009)
18. Vershynin, R.: Introduction to the non-asymptotic analysis of random matrices (2010)

Secret Sharing

Almost Optimum Secret Sharing Schemes with Cheating Detection for Random Bit Strings

Hidetaka Hoshino and Satoshi Obana[✉]

Hosei University, Tokyo, Japan
15t0015@cis.k.hosei.ac.jp, obana@hosei.ac.jp

Abstract. Cheating detectable secret sharing is a secret sharing scheme with an extra property to detect forged shares in reconstructing a secret. Such a property is indispensable when we have to store shares in possibly malicious environment (e.g., cloud storage.) Because of its importance in the real world applications, cheating detectable secret sharing is actively studied, and many efficient schemes have been presented so far. However, interestingly, no optimum scheme is known when the secret is an element of finite field of characteristic two. Since $\mathbb{F}_{2^N}$ is the most natural representation of a bit string, an efficient scheme supporting $\mathbb{F}_{2^N}$ is highly desired.

In this paper, we present cheating detectable secret sharing schemes which is secure when the secret is an element of $\mathbb{F}_{2^N}$. The size of share of the proposed scheme is almost optimum in the sense that the bit length of the share meets the lower bound with equality. Moreover, the proposed schemes are applicable to any linear secret schemes.

Keywords: Secret sharing · Cheating detection · Bit string

1 Introduction

Secret sharing scheme is a cryptographic primitive to guarantee only a qualified set of users can recover the secret and a non-qualified set of users obtain no information about the secret. Secret sharing scheme was first presented by Shamir [9] and Blakley [2] independently, and is now one of the most important building block in constructing various cryptographic protocols.

Cheating detection in secret sharing was first considered by Tompa and Woll [10]. They pointed out that in Shamir's k-out-of-n threshold scheme, even a malicious user can make the reconstruction algorithm reconstruct an invalid secret by submitting a forged share, and only the malicious user can obtain the correct secret. They also presented a scheme which detects share forgery by at most $n-1$ colluding users.

Followed by the seminal work by Tompa and Woll, the area of cheating detectable secret sharing has been extensively studied, and the lower bound of size of share [5,8] and many efficient schemes have been presented so far [1,4,6–8].

This work was supported by JSPS KAKENHI Grant Number 15K00193.

K. Tanaka and Y. Suga (Eds.): IWSEC 2015, LNCS 9241, pp. 213–222, 2015.
DOI: 10.1007/978-3-319-22425-1_13

Historically, two different models are considered for cheating detectable secret sharing. The first model assumes that cheaters know the secret in forging their shares. We call this model the CDV model. On the other hand, the second model assumes cheaters do not know the secret in forging their shares. We call this model the OKS model. The scheme secure under CDV model possesses such a merit that the successful cheating probability of the scheme does not depend on the probability distribution of the secret. On the other hand, when we can assume the secret is uniformly distributed, the merit of the scheme secure under the OKS model is in its smaller size of share. Namely, let $|\mathcal{S}|$ and ϵ be the size (cardinality) of the secret and the successful cheating probability, respectively, then the lower bound of sizes of share for the CDV model and the OKS model are given by $(|\mathcal{S}| - 1)/\epsilon^2 + 1$ and $(|\mathcal{S}| - 1)/\epsilon + 1$, respectively.

In this paper, we present schemes with the following properties:

- the scheme is secure under the OKS model when a secret s is an element of a finite field of characteristic two (i.e., $s \in \mathbb{F}_{2^N}$),
- bit length of the share meets the lower bound with equality, that is, the following equation holds where $|\mathcal{V}_i|$ denotes the size (cardinality) of a share

$$\lceil \log_2 |\mathcal{V}_i| \rceil = \left\lceil \log_2 \left(\frac{|\mathcal{S}| - 1}{\epsilon} + 1 \right) \right\rceil .$$

The first property is important from the practical point of view since $\mathbb{F}_{2^N}$ is the most natural representation of a bit string, and a bit string is the most natural representation of data processed in the computer network. Interestingly and surprisingly, there is no existing scheme which possesses both of above properties even though constructions of scheme secure under the OKS model have been well studied. The proposed schemes possess such desired properties that they are applicable to any any linear secret schemes.

The rest of paper is organized as follows. In Sect. 2, we formally define the model of cheating detectable secret sharing schemes, and review the related work. In Sect. 3, we presents our proposed schemes. Concluding remarks and future work is described in Sect. 4.

2   Preliminaries

2.1   Secret Sharing Scheme

Secret sharing scheme is constructed by n users $\mathcal{P} = P_1, \cdots, P_n$ and dealer D. The set of users who are authorized to reconstruct a secret is characterized by an access structure $\Gamma \subseteq 2^{\mathcal{P}}$, that is, the set of users $\hat{\mathcal{P}}$ is authorized to reconstruct the secret if and only if $\hat{\mathcal{P}} \in \Gamma$. Secret sharing scheme consists of two algorithms, ShareGen and Reconst. Share generation algorithm ShareGen takes a secret $s \in \mathcal{S}$ as input and outputs a list of shares $(v_1, \cdots, v_n)$ where share v_i is distributed to user P_i. Secret reconstruction algorithm Reconst takes list of shares as input and outputs a secret $s \in \mathcal{S}$.

When a secret sharing scheme satisfies following conditions, the scheme is called *perfect*.

- if $\{P_{i_1}, \cdots, P_{i_k}\} \in \Gamma$ then $\Pr[\mathsf{Reconst}(v_{i_1}, \cdots .v_{i_k})] = 1$
- if $\{P_{i_1}, \cdots, P_{i_k}\} \notin \Gamma$ then $\Pr[S = s | V_{i_1} = v_{i_1}, \cdots, V_{i_k} = v_{i_k}] = \Pr[S = s]$ for any $s \in \mathcal{S}$

here, S and V_i denote random variables defined over the set of secret and share of user P_i, respectively, and the probabilities are taken over the random tape of ShareGen.

2.2 Cheating Detectable Secret Sharing Schemes

Cheating detectable secret sharing schemes was first presented by Tompa and Woll to prevent share forgery by cheaters.

As an ordinary secret sharing scheme, cheating detectable secret sharing scheme consists of ShareGen and Reconst. A share generation algorithm ShareGen of secure schemes is the same as that of an ordinary secret sharing scheme. A secret reconstruction algorithm Reconst is slightly changed. When cheating is not detected, the algorithm Reconst outputs the reconstructed secret $\hat{s}$. By contrast, when cheating is detected, Reconst outputs the special symbol $\bot$. This symbol $\bot$ indicates that at least one user submits a forged share to Reconst. The security of cheating detectable secret sharing schemes are defined through the following game for any (k, n) threshold secret sharing schemes $\mathbf{SS} = (\mathsf{ShareGen}, \mathsf{Reconst})$ and for any Turing machine $\mathcal{A} = (\mathcal{A}_1, \mathcal{A}_2)$ representing cheaters.

Game$(\mathbf{SS}, \mathcal{A})$
$\quad s \leftarrow \mathcal{S}; \quad$ // according to the probability distribution over $\mathcal{S}$.
$\quad (v_1, \ldots, v_n) \leftarrow \mathsf{ShareGen}(s);$
$\quad (i_1, \ldots, i_{k-1}) \leftarrow \mathcal{A}_1(X);$
$\quad$ // set $X = s$ for the CDV model, $X = \emptyset$ for the OKS model.
$\quad (v'_{i_1}, \ldots, v'_{i_{k-1}}, i_k) \leftarrow \mathcal{A}_2(v_{i_1}, \ldots, v_{i_{k-1}}, X);$

The advantage of cheaters is expressed as $Adv(\mathbf{SS}, \mathcal{A}) = \Pr[s' \in \mathcal{S} \wedge s' \neq s]$, where s' is an output of $\mathsf{Reconst}(v'_{i_1}, \cdots, v'_{i_{k-1}}, v_{i_k})$. Here, the probability is taken over the distribution of $\mathcal{S}$, and over the random tapes of ShareGen and $\mathcal{A}$.

Definition 1. *A (k, n) threshold secret sharing scheme $\mathbf{SS}$ is called a (k, n, ϵ)-secure secret sharing scheme if $Adv(\mathbf{SS}, \mathcal{A}) \leq \epsilon$ for any adversary $\mathcal{A}$.*

In [8], Ogata, Kurosawa and Stinson showed the lower bound of size of share for (k, n, ϵ)-secure secret sharing scheme under the OKS model as follows.

Proposition 1. *For any (k, n, ϵ)-secure secret sharing scheme secure under the OKS model, the following inequality holds:*

$$|\mathcal{V}_i| \leq \frac{|\mathcal{S}| - 1}{\epsilon} + 1$$

Ogata *et al.* also presented a scheme which meets the above lower bound with equality. However, parameters supported by the scheme are very much limited.

Our motivation is to construct schemes which meet the above lower bound with equality when a secret is uniformly distributed over $\mathbb{F}_{2^n}$.

2.3   Existing Schemes Secure Under the OKS Model

In this section, we review the known cheating detectable secret sharing schemes secure under the OKS model.

All schemes described in this section are constructed based on a canonical technique to construct a cheating detectable secret sharing scheme secure under the OKS model. The canonical technique is very simple and described as follows. In share generation phase, ShareGen generates two types of shares. The first one is shares for the secret s to be shared, and the second one is shares for *check digit* $A(s)$. In secret reconstruction phase a secret $\hat{s}$ and a check digit $\hat{a}$ are reconstructed from shares input to Reconst. Then Reconst checks whether $\hat{s}$ and $\hat{a}$ are consistent by verifying $\hat{a} = A(\hat{s})$. If inconsistency between $\hat{s}$ and $\hat{a}$ is detected, Reconst concludes that at most one user submits an invalid share. Here, the check digit function A is a deterministic function, and the construction of A determine the properties (i.e., security, efficiency) of the scheme. Therefore, how to design the check digit function is key to construct *good* cheating detectable secret sharing schemes.

Cabello et al. proposed a cheating detectable secret sharing scheme in the OKS model using the squaring function $A(s) = s^2$ as a check digit function [4]. The properties of the scheme is summarized by the following proposition.

Proposition 2. *[4] If the secret is uniformly distributed then there exists a (k, n, ϵ)-secure secret sharing scheme under the OKS model with the following parameters where p is an odd prime:*

$$|\mathcal{S}| = p^n, \quad \epsilon = 1/p^n, \quad |\mathcal{V}_i| = p^{2n}.$$

The scheme is almost optimum in the sense that the size of share satisfies $|\mathcal{V}_i| = |\mathcal{S}|/\epsilon$, and is at most one bit longer than the lower bound. However, the scheme does not support $\mathbb{F}_{2^n}$ as a secret. More precisely, the successful cheating probability becomes 1 when we apply the scheme for the secret $s \in \mathbb{F}_{2^n}$. We note that they also proposed generalization of the scheme which enable us to flexibly determine the successful cheating probability ϵ.

In [7], Ogata and Araki proposed a cheating detectable secret sharing scheme which guarantees security even when the secret is an element of $\mathbb{F}_{2^n}$ under the OKS model. To guarantee security even when $s \in \mathbb{F}_{2^n}$, the scheme employs a cubic function $A(s) = s^3$ as a check digit function. The properties of the scheme is summarized by the following proposition.

Proposition 3. *[7] If the secret is uniformly distributed then there exists a (k, n, ϵ)-secure secret sharing scheme under the OKS model with the following parameters where p is a prime such that $p \neq 3$:*

$$|\mathcal{S}| = p^n, \quad \epsilon = 2/p^n, \quad |\mathcal{V}_i| = p^{2n}.$$

A size of share of the scheme is slightly longer than the lower bound. More precisely, size of share of the scheme is 1 bit longer than lower bound in OKS model. This is not a problem in practice, though, the possibility to construct a optimum scheme is of interest from the theoretical point of view. A generalized scheme to enable flexible choice of parameters is also presented for this scheme.

In IWSEC 2014, Obana and Tsuchida presented a cheating detectable secret sharing scheme which can deal with a secret of an arbitrary finite field [6]. The check digit function employed in the scheme is $A(s) = s^2 + s^3$, and is simple combination of check digit functions employed in [4] and [7]. The properties of the scheme is summarized by the following proposition.

Proposition 4. *[6] If the secret is uniformly distributed then there exists a (k, n, ϵ)-secure secret sharing scheme under the OKS model with the following parameters where p is an arbitrary prime:*

$$|\mathcal{S}| = p^n, \quad \epsilon = 2/p^n, \quad |\mathcal{V}_i| = p^{2n} .$$

As the scheme in [7], the size of share of the scheme is 1 bit longer than lower bound in OKS model.

3   Proposed Schemes

In this section, we present two (k, n, ϵ_{OKS})-secure secret sharing schemes for uniformly distributed bit strings under in the OKS model. The sizes of share of proposed schemes are almost optimum, that is, the bit lengths of the share achieve the lower bound with equality.

The basic idea of propose scheme is to employ bit decomposition technique. Let the secret s be a bit string, then two substrings s_1 and s_2 are generated by decomposing s (e.g., $101110 \rightarrow 101, 110$ when secret is 101110.) Therefore, the proposed scheme requires the secret s to be an even bit string. We employ $A(x_1, x_2) = x_1 \cdot x_2$ as the check digit function in the proposed scheme where $\cdot$ denotes multiplication over $\mathbb{F}_{2^n}$. The share generation algorithm ShareGen and share reconstruction algorithm Reconst are described as follows.

ShareGen
Input   : secret $s \in \mathbb{F}_{2^{2n}}$
Output: list of shares $(v_1, \cdots, v_n)$

1. Generate a random polynomial $f_s(x) \in \mathbb{F}_{2^{2n}}[X]$ of degree $k - 1$ such that $f_s(0) = s$.
2. Generate two bit strings s_1 and s_2 by decomposing s.
3. Generate a check digit $a = A(s_1, s_2) = s_1 \cdot s_2$.
4. Generate a random polynomial $f_a(x) \in \mathbb{F}_{2^n}[X]$ of degree $k - 1$ such that $f_a(0) = a$.
5. Compute $v_i = (f_x(i), f_a(i))$ and output $(v_1, \cdots, v_n)$

Reconst
Input   : List of shares $(v_{i_1}, \cdots, v_{i_m})$ where $m \geq k$
Output: Reconstructed secret $\hat{s}$ or $\perp$

1. Reconstruct two polynomials $\hat{f}_s(x)$ and $\hat{f}_a(x)$ from $v_{i_1}, \cdots, v_{i_m}$ using Lagrange interpolation.
2. If $\deg(\hat{f}_s) > k - 1$ or $\deg(\hat{f}_a) > k - 1$ holds, output $\perp$.
3. Compute $\hat{s} = (\hat{s}_1, \hat{s}_2) = \hat{f}_s(0)$ and $\hat{a} = \hat{f}_a(0)$.
4. Output $\hat{s}$ if $\hat{a} = \hat{s}_1 \cdot \hat{s}_2$ holds. Otherwise Reconst outputs $\perp$.

The following theorem holds for the above scheme.

Theorem 1. *If the secret s is uniformly distributed over $\mathbb{F}_{2^{2n}}$ then the proposed scheme is (k, n, ϵ)-secure secret sharing schemes in the OKS model. The parameters of the scheme are $|\mathcal{S}| = 2^{2n}$ and $|\mathcal{V}_i| = |\mathcal{S}|/\epsilon = 2^{3n}$, and the successful cheating probability ϵ is $1/2^n$. The bit length of share satisfies the lower bound.*

Proof. Without loss of generality, we can assume $P_1 \cdots P_k$ submit their shares to Reconst. We consider a case that $k - 1$ cheaters $P_2, \cdots, P_k$ try to fool an authorized user P_1. In the reconstruction, cheaters submits forge shares $(v_{s,i} + \delta_{s,i}, v_{a,i} + \delta_{a,i})(2 \leq i \leq k)$ and P_1 submits a correct share $v_i = (v_{1,s}, v_{1,a})$. Then, reconstructed secret $\hat{s}$ and check digit $\hat{a}$ is computed using Lagrange interpolation as follows.

$$
\hat{s} = \left(\sum_{i=1}^{k-1} \prod_{j=1, j \neq i}^{k} \frac{-j}{i-j}(v_{s,i} + \delta_{s,i}) \right) + \prod_{j=1}^{k-1} \frac{-j}{k-j} v_{s,k}
$$

$$
= \left(\sum_{i=1}^{k} \prod_{j=1, j \neq i}^{k} \frac{-j}{i-j} v_{s,i} \right) + \left(\sum_{i=1}^{k-1} \prod_{j=1, j \neq i}^{k} \frac{-j}{i-j} \delta_{s,i} \right)
$$

$$
= s + \delta_s
$$

The check digit $\hat{a}$ is calculated in the same way. Let $\hat{s}_1$ and $\hat{s}_2$ be substrings generated by decomposing $\hat{s}$. We describe $\hat{s}_1 = s_1 + \delta_1$, $\hat{s}_2 = s_2 + \delta_2$ and $\hat{a} = a + \delta_a$ where δ_1 and δ_2 are the difference between $\hat{s}$ and original secret s, and δ_a is the difference between $\hat{a}$ and original check digit a. Here, the values δ_1, δ_2 and δ_a are arbitrary controlled by cheaters. We note that $(\delta_1, \delta_2) \neq (0, 0)$ holds since if both δ_1 and δ_2 are equal to 0, reconstructed secret $\hat{s}$ equals to the original secret s. We evaluate successful cheating probability ϵ of cheaters $P_2, \cdots, P_k$. From the definition of Reconst, we detects the cheating by verifying consistency of $\hat{s}$ and $\hat{a}$. Therefore, $P_2, \cdots, P_k$ succeed to fool P_1 if the following equation holds.

$$
\hat{a} = \hat{s}_1 \cdot \hat{s}_2
$$

Since $\hat{s}_1 = s_1 + \delta_1$. $\hat{s}_2 = s_2 + \delta_2$ and $\hat{a} = a + \delta_a$ hold, above equation is rewritten as follows.

$$
a + \delta_a = (s_1 + \delta_1)(s_2 + \delta_2)
$$

The above equation is further transformed into the following equation.

$$
\delta_2 s_1 + \delta_1 s_2 + \delta_1 \delta_2 - \delta_a = 0
$$

For fixed δ_1, δ_2 and δ_a, there are 2^n solutions (s_1, s_2) satisfying the above equation. On the other hand, there are 2^{2n} possible combinations of s_1 and s_2 since (s_1, s_2) are uniformly distributed over $\mathbb{F}_{2^n} \times \mathbb{F}_{2^n}$. Therefore, the successful cheating probability for this scheme ϵ becomes $2^n / 2^{2n} = 1/2^n$.

Next, we show the bit length of share of the proposed scheme is optimum. Since $|\mathcal{V}_i| = 2^{2n} \cdot 2^n$ holds in the proposed scheme, it is easy to see that bit length of the share $\lceil \log |\mathcal{V}_i| \rceil$ satisfies $\lceil \log |\mathcal{V}_i| \rceil = 3n$. Now we estimate the lower bound of bit length of the share. Since lower bound of size of share in the OKS model is given by $|\mathcal{V}_{OKS}| = (|\mathcal{S}| - 1)/\epsilon + 1$, the lower bound of bit length of share in the OKS model is given by $\lceil \log |\mathcal{V}_{OKS}| \rceil$. Therefore, the lower bound of bit length of share with parameters $|\mathcal{S}| = 2^{2n}$ and $\epsilon = 1/2^n$ is evaluated as follows:

$$
\begin{aligned}
\lceil \log |\mathcal{V}_{OKS}| \rceil &= \left\lceil \log \left(\frac{2^{2n} - 1}{2^{-n}} + 1 \right) \right\rceil \\
&= \left\lceil \log \left(2^{3n} - 2^n + 1 \right) \right\rceil \\
&= \left\lceil \log \left(2^{3n} \left(1 - \frac{1}{2^{2n}} + \frac{1}{2^{3n}} \right) \right) \right\rceil \\
&= \left\lceil \log 2^{3n} + \log \left(1 - \frac{1}{2^{2n}} + \frac{1}{2^{3n}} \right) \right\rceil \\
&= \left\lceil 3n + \log \left(1 - \frac{1}{2^{2n}} + \frac{1}{2^{3n}} \right) \right\rceil \\
&= 3n
\end{aligned}
$$

The last equality holds since the following inequality holds:

$$
-1 < \log \left(1 - \frac{1}{2^{2n}} + \frac{1}{2^{3n}} \right) < 0.
$$

Therefore, we see that the bit length of the proposed scheme meets the lower bound with equality. $\qquad \square$

3.1 Generalization

In this section, we give the generalization of the first scheme.

In the first scheme, $\epsilon = 1/\sqrt{|\mathcal{S}|}$ holds. Therefore, when the secret is 1024 bits, successful cheating probability ϵ becomes $1/2^{512}$. However, in the usual settings, $\epsilon = 1/2^{512}$ is too small. Therefore, we want a scheme in which the security level is adequate (e.g., $\epsilon = 1/2^{40}$) and the size of share is much smaller than the first scheme. The first scheme can be generalized to be suitable for such a situation. The parameters of generalized scheme are $|\mathcal{S}| = 2^{2n}$, $\epsilon = 2^{n-i}$ and $|\mathcal{V}_i| = 2^{3n-i}$ where i is an arbitrary integer such that $0 \le i < n$. The generalization is realized by slightly changing the function $f_a(s_1, s_2) = s_1 \cdot s_2$ into $f'_a(s_1, s_2) = \phi(s_1 \cdot s_2)$, where ϕ is a surjective linear mapping $\phi : \mathbb{F}_{2^n} \to \mathbb{F}_{2^{n-i}}$. We should note that such a technique is very popular, and is employed in various schemes [4,7].

The share generation algorithm ShareGen and share reconstruction algorithm Reconst are very similar to the first scheme.

ShareGen

Input : secret $s \in \mathbb{F}_{2^{2n}}$

Output: List of shares $(v_1, \cdots, v_n)$

1. Generate a random polynomial $f_s(x) \in \mathbb{F}_{2^{2n}}[X]$ of degree $k-1$ such that $f_s(0) = s$.
2. Generate two bit strings s_1 and s_2 by decomposing s.
3. Generate a check digit $a = f'(s_1, s_2) = \phi(s_1 \cdot s_2)$.
4. Generate a random polynomial $f_a(x) \in \mathbb{F}_{2^{n-i}}[X]$ of degree $k-1$ such that $f'_a(0) = a$.
5. Compute $v_i = (f_x(i), f'_a(i))$ and output $(v_1, \cdots, v_n)$

Reconst

Input : List of shares $(v_{i_1}, \cdots, v_{i_m})$ where $m \geq k$

Output: Reconstructed secret $\hat{s}$ or $\perp$

1. Reconstruct two polynomials $\hat{f}_s(x)$ and $\hat{f}'_a(x)$ from $v_{i_1}, \cdots, v_{i_m}$ using Lagrange interpolation.
2. If $\deg(\hat{f}_s) > k-1$ or $\deg(\hat{f}'_a) > k-1$ holds, output $\perp$.
3. Compute $\hat{s} = (\hat{s}_1, \hat{s}_2) = \hat{f}_s(0)$ and $\hat{a} = \hat{f}'_a(0)$.
4. Output $\hat{s}$ if $\hat{a} = \phi(\hat{s}_1 \cdot \hat{s}_2)$ holds. Otherwise Reconst outputs $\perp$.

The following theorem holds for the generalized scheme.

Theorem 2. *When secret s is uniformly distributed over $\mathbb{F}_{2^{2n}}$, the proposed scheme is (k, n, ϵ)-secure secret sharing schemes in the OKS model. The parameters of the scheme are $|\mathcal{S}| = 2^{2n}$ and $|\mathcal{V}_i| = |\mathcal{S}|/\epsilon = 2^{3n-i}$, and the successful cheating probability ϵ is $1/2^{n-i}$. The bit length of share satisfies the lower bound.*

Proof. First, we will prove the security of the generalized scheme. Let the situation be same as the first scheme then reconstructed values become as follows.

- $\hat{s} = (\hat{s}_1, \hat{s}_2) = (s_1 + \delta_1, s_2 + \delta_2)$: a value reconstructed from shares of the secret.
- $\hat{a} = a + \delta_a$: a value reconstructed from shares of the check digit.

Here, δ_1, δ_2 and δ_a are difference between reconstructed values and original values, and are arbitrarily controlled by cheaters $P_2, \cdots, P_k$. Cheaters succeed in cheating if the following equation hold.

$$a + \delta_a = \phi((s_1 + \delta_1)(s_2 + \delta_2))$$

The above equation is equivalent to the following equation.

$$\phi(s_1 s_2) + \delta_a = \phi((s_1 + \delta_1)(s_2 + \delta_2))$$

The above equation is transformed as follows by the linearity of ϕ.

$$\phi(\delta_2 s_1) + \phi(\delta_1 s_2) + \phi(\delta_1 \delta_2) - \delta_a = 0$$

For fixed δ_1, δ_2 and δ_2, there are $2^n \cdot 2^i$ solutions (s_1, s_2) satisfying the above equation. On the other hand, there are 2^{2n} possible combinations of s_1 and s_2

since (s_1, s_2) are uniformly distributed over $\mathbb{F}_{2^n} \times \mathbb{F}_{2^n}$. Therefore, the successful cheating probability for this scheme ϵ becomes $(2^n \cdot 2^i)/2^{2n} = 1/2^{n-i}$.

Next, we show the bit length of share of the proposed scheme is optimum. Since $|\mathcal{V}_i| = 2^{2n} \cdot 2^{n-i}$ in the proposed scheme, it is easy to see that bit length of the share $\lceil \log |\mathcal{V}_i| \rceil$ satisfies $\lceil \log |\mathcal{V}_i| \rceil = 3n - i$. Now we estimate the lower bound of bit length of the share. Since lower bound of size of share in the OKS model is given by $|\mathcal{V}_{\mathsf{OKS}}| = (|\mathcal{S}| - 1)/\epsilon + 1$, the lower bound of bit length of share in the OKS model is given by $\log\lceil |\mathcal{V}_{\mathsf{OKS}}| \rceil$. Therefore, the lower bound of bit length of share with parameters $|\mathcal{S}| = 2^{2n}$ and $\epsilon = 1/2^{n-i}$ is evaluated as follows:

$$
\begin{aligned}
\lceil \log |\mathcal{V}_{\mathsf{OKS}}| \rceil &= \left\lceil \log \left(\frac{2^{2n} - 1}{2^{-n+i}} + 1 \right) \right\rceil \\
&= \left\lceil \log \left(2^{3n-i} - 2^{n-i} + 1 \right) \right\rceil \\
&= \left\lceil \log \left(2^{3n-i} \left(1 - \frac{1}{2^{2n-i}} + \frac{1}{2^{3n-i}} \right) \right) \right\rceil \\
&= \left\lceil \log 2^{3n-i} + \log \left(1 - \frac{1}{2^{2n-i}} + \frac{1}{2^{3n-i}} \right) \right\rceil \\
&= \left\lceil 3n - i + \log \left(1 - \frac{1}{2^{2n-i}} + \frac{1}{2^{3n-i}} \right) \right\rceil \\
&= 3n - i
\end{aligned}
$$

The last equality holds since the following inequality holds:

$$
-1 < \log \left(1 - \frac{1}{2^{2n-i}} + \frac{1}{2^{3n-i}} \right) < 0 .
$$

Therefore, we see that the bit length of the proposed scheme meets the lower bound with equality. $\square$

We should note that the above proof does not rely on any property which is specific to $\mathbb{F}_{2^n}$. In fact, the same proof works even if we replace $\mathbb{F}_{2^{2n}}$ with $\mathbb{F}_{p^{2n}}$ for any prime power p (in this case we also need to replace ϕ to be $\phi : \mathbb{F}_{p^n} \to \mathbb{F}_{p^{n-i}}$.)

4   Concluding Remarks

In this paper, we propose cheating detectable secret sharing schemes which guaranty security for random bit strings under the OKS model. The proposed schemes

Table 1. The properties known schemes

Scheme	Size of share	Supported secret	Condition on ϵ
Proposed	$|\mathcal{S}|/\epsilon$ (optimum)	$\mathbb{F}_{p^2}$ for any prime power p	$\epsilon \geq 1/\sqrt{|\mathcal{S}|}$
Cabello et al.[4]	$|\mathcal{S}|/\epsilon$ (optimum)	Arbitrary finite field except $\mathbb{F}_{2^n}$	$\epsilon \geq 1/|\mathcal{S}|$
Ogata et al.[7]	$2|\mathcal{S}|/\epsilon$	Arbitrary finite field except $\mathbb{F}_{3^n}$	$\epsilon \geq 2/|\mathcal{S}|$
Obana et al.[6]	$2|\mathcal{S}|/\epsilon$	Arbitrary finite field	$\epsilon = 2/|\mathcal{S}|$

detect forged share when at most $k-1$ cheaters try to fool honest user. Let $|\mathcal{S}|$ be the size of share and ϵ be the successful cheating probability of cheaters, then size of share $|\mathcal{V}_i|$ of generalized scheme becomes $|\mathcal{V}_i| = |\mathcal{S}|/\epsilon$, which is almost optimum in the sense that the bit length of share meets the lower bound with equality.

Table 1 compares properties of existing schemes secure under the OKS model to the proposed scheme. We will see that the proposed scheme is the first scheme which satisfy both (1) the scheme is secure when the secret is an element of $\mathbb{F}_{2^N}$, and (2) the bit length of the share meets the lower bound with equality. On the other hand, the drawback of the proposed scheme is that only the limited values ϵ are chosen as the successful cheating probability compared to the other schemes. To construct an optimum scheme satisfying the above property (1) and (2) and supporting $\epsilon < 1/\sqrt{|\mathcal{S}|}$ is our future challenge.

References

1. Araki, T., Ogata, W.: A simple and efficient secret sharing scheme secure against cheating. IEICE Trans. Fundam. **E94–A**(6), 1338–1345 (2011)
2. Blakley, G.R.: Safeguarding cryptographic keys. In: Proceedings AFIPS 1979, National Computer Conference, vol. 48, pp. 313–137 (1979)
3. Brickell, E.F., Stinson, D.R.: The detection of cheaters in threshold schemes. SIAM J. Discret. Math. **4**(4), 502–510 (1991)
4. Cabello, S., Padró, C., Sáez, G.: Secret sharing schemes with detection of cheaters for a general access structure. Des. Codes Crypt. **25**, 175–188 (2002)
5. Carpentieri, M., De Santis, A., Vaccaro, U.: Size of shares and probability of cheating in threshold schemes. In: Helleseth, T. (ed.) EUROCRYPT 1993. LNCS, vol. 765, pp. 118–125. Springer, Heidelberg (1994)
6. Obana, S., Tsuchida, K.: Cheating detectable secret sharing schemes supporting an arbitrary finite field. In: Yoshida, M., Mouri, K. (eds.) IWSEC 2014. LNCS, vol. 8639, pp. 88–97. Springer, Heidelberg (2014)
7. Ogata, W., Araki, T.: Cheating detectable secret sharing schemes for random bit strings. IEICE TRANS. FUNDAM. **E96–A**(11), 2230–2234 (2013)
8. Ogata, W., Kurosawa, K., Stinson, D.R.: Optimum secret sharing scheme secure against cheating. SIAM J. Discret. Math. **20**(1), 79–95 (2006)
9. Shamir, A.: How to share a secret. Commun. ACM **22**(11), 612–613 (1979)
10. Tompa, M., Woll, H.: How to share a secret with cheaters. J. Cryptol. **1**(3), 133–138 (1989)

Privacy-Preserving and Anonymity

k-Anonymous Microdata Release via Post Randomisation Method

Dai Ikarashi, Ryo Kikuchi$^{(\boxtimes)}$, Koji Chida, and Katsumi Takahashi

NTT Secure Platform Laboratories, Tokyo, Japan
{ikarashi.dai,kikuchi.ryo,chida.koji,takahashi.katsumi}@lab.ntt.co.jp

Abstract. The problem of the release of anonymized microdata is an important topic in the fields of statistical disclosure control (SDC) and privacy preserving data publishing (PPDP), and yet it remains sufficiently unsolved. In these research fields, k-anonymity has been widely studied as an anonymity notion for mainly deterministic anonymization algorithms, and some probabilistic relaxations have been developed. However, they are not sufficient due to their limitations, i.e., being weaker than or incomparable to the original k-anonymity, or requiring strong parametric assumptions. In this paper, we propose Pk-anonymity, a new probabilistic k-anonymity. It is proven that Pk-anonymity is a mathematical extension of k-anonymity rather than a relaxation, and requires no parametric assumptions. These properties have a significant meaning in the viewpoint that it enables us to compare privacy levels of probabilistic microdata release algorithms with deterministic ones. We then apply Pk-anonymity to the post randomization method (PRAM), which is an SDC algorithm based on randomization. PRAM is proven to satisfy Pk-anonymity in a controlled way, i.e., one can control PRAM's parameter so that Pk-anonymity is satisfied.

Keywords: Post randomization method (PRAM) · k-anonymity · PPDP

1 Introduction

Releasing microdata while preserving privacy has been widely studied in the fields of statistical disclosure control (SDC) and privacy preserving data publishing (PPDP). Microdata has significant value, especially for data analysts who wish to conduct various type of analyses involving the viewing of whole data and determining what type of analysis they should conduct.

The most common privacy notion for microdata release is k-*anonymity* proposed by Samarati and Sweeney [11,13]. It means that "no one can narrow down a person's record to k records." This semantics is quite simple and intuitive. Therefore, many studies have been conducted on k-anonymity, and many relevant privacy notions such as ℓ-diversity [8], have also been proposed. Among these relevant studies, applying k-anonymity to probabilistic algorithms is a significant research direction. Samarati [10] first employed generalization and suppression for satisfying k-anonymity and its enforcement. Subsequent studies have

© Springer International Publishing Switzerland 2015
K. Tanaka and Y. Suga (Eds.): IWSEC 2015, LNCS 9241, pp. 225–241, 2015.
DOI: 10.1007/978-3-319-22425-1_14

followed the direction and most k-anonymization algorithms deterministically generalize or partition microdata. However, there are probabilistic SDC methods such as random swapping, random sampling, and post randomization method (PRAM) [5]. How are these probabilistic algorithms related to k-anonymity?

Regarding random swapping, for example, Soria-Comas and Domingo-Ferrer answered the above question by relaxing k-anonymity to a probabilistic k-anonymity, which means that "no one can *correctly link* a person to a record with a higher probability than $1/k$" [12]. Intuitively, this semantics seems to be very close to that of the original k-anonymity. However, its precise relation to k-anonymity has not been argued, and we still cannot definitely say that an algorithm satisfying their probabilistic k-anonymity also is k-anonymous.

PRAM was proposed by Kooiman et al. in 1997. It changes data into other random data according to the probability on a *transition probability matrix*. Agrawal et al. also developed privacy preserving OLAP (Online Analytical Processing) [3] by retention-replacement perturbation, which is an instantiation of PRAM. For many years, PRAM's privacy was not clarified; however, PRAM has been recently proven to satisfy ε-*differential privacy (DP)* [7], which is another privacy notion that has attracted a great deal of attention recently.

1.1   Motivations

After the proposal, ε-DP has been widely researched and is now known to be strong privacy notion. However, especially in the PPDP field, k-anonymity is as important as ε-DP, although it takes only re-identification into consideration and several papers showed the limitation of k-anonymity [6,8]. This notion is very simple and intuitive; therefore, the enormous number of techniques has been invented, and as a result, k-anonymity has already spread among the businesspeople, doctors, etc., who are conscious about privacy, not only among the researchers. From the viewpoint of practice, it is a great merit that people recognize and understand the notion.

PRAM has several good features, and we believe that it is one of promising candidates for PPDP. The anonymization step in PRAM is performed by a record-wise fashion so anonymizing data in parallel is easy, and we can extend PRAM to a local perturbation, i.e., an individual anonymizes his/her data before sending them to the central server. In addition, PRAM does not needs generalization, so we can obtain anonymized data with fine granularity and perform a fine-grained analysis on them.

Although PRAM has these features and was proposed [5] before when the methods satisfying k-anonymization [13] and satisfying ε-DP [4] were proposed, it has been studied less than other approaches in the area of PPDP. Most popular methods for PPDP are evaluated in the context of k-anonymity. However, PRAM is a probabilistic method, so it cannot be evaluated in the context of k-anonymity. This means that no one can compare PRAM with other methods for PPDP in the same measure.

From the above circumstances, our aim of the paper is twofold. First, we extend *k*-anonymity for probabilistic methods (not only PRAM). Second, we evaluate how strongly PRAM preserves privacy in the context of *k*-anonymity.

1.2 Contributions

Our contributions are the following two points.

Extending k-Anonymity for Probabilistic Methods. We propose *Pk*-anonymity, which has the following advantages compared to current probabilistic *k*-anonymity notions.

1. It is formally defined and sufficient to prove that it is a rigorous extension of the original *k*-anonymity. Specifically, we prove that *k*-anonymity and *Pk*-anonymity are totally equivalent if an anonymization algorithm is deterministic, in other words, if the algorithm is in the extent of conventional *k*-anonymization. We claim that one can consider a set of microdata anonymized by using a probabilistic algorithm as *k*-anonymous if it is *Pk*-anonymous.

2. *Pk*-anonymity never causes failure of anonymization. Some current probabilistic *k*-anonymity notions are defined as "satisfaction of *k*-anonymity with certain probability." Unlike these notions, *Pk*-anonymity always casts a definite level of re-identification hardness to the adversary while it is defined via the theory of probability.

3. It is non-parametric; that is, no assumption on the distribution of raw microdata is necessary. Furthermore, it does not require any raw microdata to evaluate *k*.

Applying Pk-Anonymity to PRAM. *Pk*-anonymity on PRAM is analyzed. The value of *k* is derived with no parametric assumption. Furthermore, an algorithm for PRAM satisfying *Pk*-anonymity with any value of *k* is given.

1.3 Related Work

On Probabilistic k-Anonymity Notions. There are many studies on *k*-anonymity, and it has many supplemental privacy notions such as ℓ-diversity and *t*-closeness [6]. There have also been several studies that are relevant to the probability.

Wong et al. proposed (α, k)-anonymity [14]. Roughly speaking, (α, k)-anonymity states that (the original) *k*-anonymity is satisfied with probability α. Lodha and Thomas proposed $(1 - \beta, k)$-anonymity. This is a relaxation from *k*-anonymity in a sample to that in a population. These two notions are essentially based on the original *k*-anonymity and are relaxations that allow failures of anonymization in a certain probability. *Pk*-anonymity is fully probabilistically defined and never causes failure of anonymization.

Aggarwal proposed a probabilistic *k*-anonymity [1]. Their goal was the same as with *Pk*-anonymity; however, it requires a parametric assumption that the distribution of raw microdata is a parallel translation of randomized microdata, and this seems to be rarely satisfied since a randomized distribution is generally flatter than the prior distribution.

Soria-Comas and Domingo-Ferrer also proposed their probabilistic k-anonymity [12]. They applied it to random swapping and micro-aggregation. The semantics of their anonymity is "no one can *correctly link* a person to a record with a higher probability than $1/k$" and Pk-anonymity is stronger. Unfortunately, further comparison is difficult since we could not find a sufficiently formal version of the definition.

On Privacy Measures Applicable to PRAM. Aggarwal and Agrawal proposed a privacy measure based on conditional differential entropy [2]. This measure requires both raw and randomized data to be evaluated, unlike Pk-anonymity.

Agrawal et al. proposed (s, ρ_1, ρ_2) Privacy Breach [3], which is based on probability and applicable to retention-replacement perturbation. In contrast to k-anonymity, it does not take into account background knowledge concerning raw data, that is, concerning quasi-identifier attributes.

Rebollo-Monedero et al. [9] proposed a t-closeness-like privacy criterion and a distortion criterion which are applicable to randomization, and showed that PRAM can meet these criteria. Their work was aimed at clarifying the privacy-distortion trade-off problem via information theory, in the area of attribute estimation. Therefore, they did not mention whether PRAM can satisfy a well known privacy notion such as k-anonymity.

2 Preliminaries

2.1 Publishing Scenario

There are two scenarios of microdata release using randomization. One is the setting in which a database administrator gathers individuals' records and randomizes them. The other is that in which individuals randomize their own records.

PRAM is not only applicable to the former but also applicable to the latter [3] in contrast, k-anonymity can only be applied to the former. Thus, our Pk-anonymity is applicable to both scenarios via PRAM. More precisely, we show that PRAM's parameter satisfying Pk-anonymity can be determined by only the expected record count and number of possible attribute values (in other words, metadata of attributes).

2.2 Notation

On the notation of mathematical formula, $\mathbb{R}$ denotes the space of the real numbers and $\circ$ denotes the composition of functions.

We treat a table-formed database as both private and released data. Since the record count is revealed at the same time that the data are released in an ordinary microdata release, we assume that the record count is public and static in theory.

Basically, we use the following notations.

- T: the set of any private tables
- τ, T: a private table as an instance/random variable
- T': the set of any released tables
- τ', T': a released table as an instance/random variable
- R, R': the sets of all records in a private/released table
- V, V': the sets of any values in a private/released table
- A: a transition probability matrix in PRAM
- f_X: the probability function of X where X is a random variable

In the discussion of multi-attributes, we also use the following notations.

- $\mathcal{A}, \mathcal{A}'$: the sets of attributes in a private/released table
- $\mathcal{V}_a, \mathcal{V}'_{a'}$ where $a \in \mathcal{A}$ and $a' \in \mathcal{A}'$: the sets of values in a private/released table, i.e., $\mathcal{V} = \prod_{a \in \mathcal{A}} \mathcal{V}_a$ and $\mathcal{V} = \prod_{a' \in \mathcal{A}'} \mathcal{V}'_a$ hold where $\prod$ means the direct product.
- A_a where $a \in \mathcal{A}$: transition probability matrix of each attribute

We consider a table $\tau \in T$ (or, $\tau' \in T'$) as a map from R to V (or, R' to V'). More formally, we define τ (or, τ') as follows.

Definition 1 (*Tables*). *Let a record set R and a value set V be finite sets. Then, the following map τ is called a table on (R, V).*

$$\tau : R \to V,$$

When we discuss a multi-attribute table, V is represented as $\prod_{a \in \mathcal{A}} \mathcal{V}_a$, where an attribute set $\mathcal{A}$ is a finite set, each $\mathcal{V}_a$ is also a finite set for any $a \in \mathcal{A}$, and $\prod$ means the direct product.

2.3 PRAM

PRAM [5] is a privacy preserving method for microdata release. It changes data according to a transition probability matrix, A. The transition probability matrix consists of probabilities in which each value in a private table will be changed into other specific (or the same) values. $A_{u,v}$ denotes the probability $u \in V$ is changed into $v \in V'$.

PRAM is a quite general method. Invariant PRAM [5], retention-replacement perturbation [3], etc., are known as instantiations of it. Specifically, retention-replacement perturbation is simple and convenient.

Retention-Replacement Perturbation. In retention-replacement perturbation, individuals probabilistically replace their data with random data according to *retention probability ρ*. First, data are retained with ρ, and if the data are not retained, they will be replaced with a uniformly random value chosen from the attribute domain. For example, for an attribute "sex", when $\rho = 0.5$, "male" is retained with $1/2$ probability, and with the remaining $1/2$ probability, it is replaced with a uniformly random value, namely, a value "female" and a value "male", which is the same as the original, both with $1/2 \times 1/2 = 1/4$ probability. Eventually, the probability that "male" changes into "female" is $1/4$, and the probability that it does not change is $3/4$. The lower the retention probability, the higher privacy is preserved. On the contrary, the lower the probability, the lower utility. These probabilities form the following transition probability matrix.

$$\begin{bmatrix} 0.75 & 0.25 \\ 0.25 & 0.75 \end{bmatrix}$$

Generally, the transition probability matrix A_a of an attribute a is written as

$$(A_a)_{v_a, v'_a} = \begin{cases} \rho_a + \dfrac{(1 - \rho_a)}{|\mathcal{V}_a|} & \text{if } v_a = v'_a \\ \dfrac{(1 - \rho_a)}{|\mathcal{V}_a|} & \text{otherwise} \end{cases}$$

where for any $v \in \mathcal{V}$ and $a \subset \mathcal{A}$, $v_a \subset \mathcal{V}_a$ is an element of $v \in \mathcal{V}$ corresponding to a, and ρ_a is the retention probability corresponding to a.

2.4   k-Anonymity

The k-anonymity [11,13] is a privacy notion that is applicable to table-formed databases and defined as "for all database records, there are at least k records whose values are the same", in other words, "no one can narrow down a person's record to less than k records."

Using the notations in Sect. 2.2, we represent the definition of k-anonymity [13] as follows.

Definition 2 (*k-Anonymity*). *For a positive integer k, a released table $\tau' \in \mathcal{T}'$ is said to satisfy k-anonymity (or to be k-anonymous), if and only if it satisfies the following condition.*

For any $r' \in \mathcal{R}'$, there are k or more $\hat{r}'$'s such that $\hat{r}' \in \mathcal{R}'$ and $\tau'(r') = \tau'(\hat{r}')$.

A released table τ' in the above definition represents all columns corresponding to quasi-identifier attributes of an anonymized table.

However, the definition in [13] is problematic; i.e., there are some tables that satisfy k-anonymity but do not achieve its aim. For example, a table generated by copying all a private table's records k times satisfies k-anonymity but it is obviously not safe. Therefore, we assume $|\mathcal{R}| = |\mathcal{R}'|$ to strengthen the above definition in the discussion of k-anonymity in this paper.

2.5   Anonymization and Privacy Mechanisms

We define anonymization and privacy mechanisms separately to discuss them formally. First we define anonymization.

Definition 3 (*Anonymization*). *Let $\mathcal{R}$, $\mathcal{R}'$, $\mathcal{V}$ and $\mathcal{V}'$ be finite sets, $\mathcal{T}$ and $\mathcal{T}'$ be the sets of all tables on $(\mathcal{R}, \mathcal{V})$ and $(\mathcal{R}', \mathcal{V}')$, respectively, and let π be a map $\pi : \mathcal{R} \to \mathcal{R}'$.*

Then, for any $\tau \in \mathcal{T}$ and $\tau' \in \mathcal{T}'$, a map $\delta : \mathcal{T} \to (\mathcal{R} \to \mathcal{V}')$ is called anonymization with π from τ to τ' if and only if they satisfy

$$\delta(\tau) = \tau' \circ \pi, \tag{1}$$

where the notation $\mathcal{X} \to \mathcal{Y}$ denotes the set of all maps from $\mathcal{X}$ to $\mathcal{Y}$ for any set $\mathcal{X}$ and $\mathcal{Y}$.

Anonymization δ represents an anonymization algorithm such as perturbation, k-anonymization, etc. A map π represents an anonymous communication channel, the shuffling function, or another component which hides the order of records in τ. In this paper, we adopt the uniformly random permutation as π.[1]

Privacy mechanisms involve not only δ but also π, $\mathcal{R}$, $\mathcal{R}'$, $\mathcal{V}$ and $\mathcal{V}'$, and random variables are brought to extend the above definitions to probabilistic ones. Random variables corresponding to τ, τ', π, and δ are denoted by T, T', Π, and Δ, respectively. We assume T, Π, and Δ are mutually independent as probabilistic events, while T' is dependent on the other three random variables.

Definition 4 (*Privacy Mechanisms*). *Let $\mathcal{R}$, $\mathcal{R}'$, $\mathcal{V}$, $\mathcal{V}'$, $\mathcal{T}$, and $\mathcal{T}'$ be the same as Definition 3, and let T, T', Π, and Δ be random variables on $\mathcal{T}$, $\mathcal{T}'$, $\mathcal{R} \to \mathcal{R}'$, and $\mathcal{T} \to (\mathcal{R} \to \mathcal{V}')$, respectively, such that T, Π, and Δ are mutually independent as probabilistic events, where the notation $\mathcal{X} \to \mathcal{Y}$ denotes the set of all maps from $\mathcal{X}$ to $\mathcal{Y}$ for any set $\mathcal{X}$ and $\mathcal{Y}$.*

Then, the 6-tuple $(\mathcal{R}, \mathcal{V}, \mathcal{R}', \mathcal{V}', \Pi, \Delta)$ is called a privacy mechanism from T to T' if and only if they satisfy the following equation.

$$\Delta(T) = T' \circ \Pi$$

3 Pk-**Anonymity**

k-anonymity is one of the most popular privacy notion and essential for many subsequent notions. However, it cannot be directly applied to a randomized table. Imagine that one randomizes all records' values uniformly randomly. The randomized table does not satisfy k-anonymity because it may have a unique record, while an adversary cannot identify anyone's record.

Therefore, we need a new definition of k-anonymity applicable to randomization. Intuitively, Pk-anonymity guarantees that "no one estimates which person the record came from with more than $1/k$ probability." It is represented as *all* the probability of linkage[2] between a record in the private table and one in released table is bounded by $1/k$.

---

[1] A map π is essential for anonymization. For example, if the first record in the private table is to be the first record in the released table, identification is trivial.

[2] We take into account an adversary's incorrect presumption since the probability of linkage is bounded whichever the linkage is correct or not.

3.1   Background Knowledge of Adversary

In the definition of k-anonymity, there is no adversary, and this definition is described as a simple condition to be satisfied in a table. This is convenient for measuring k-anonymity. At the same time, however, it makes the meaning of privacy unclear.

Therefore, there is an adversary in our model of Pk-anonymity. The probability of linkage is varied according to the background knowledge of the adversary. In the Pk-anonymity model, an adversary's background knowledge is represented as a probabilistic function f_T[3] on the private table. Pk-anonymity requires the privacy mechanism of that the probability of linkage is bounded by $1/k$ for *all* f_T. It means that we deal with an adversary who has arbitrary knowledge about the private table: The adversary might know the private table itself and incorrect private tables.

We note that even if in the extreme case where the adversary knows the private table itself, Pk-anonymity can be satisfied by using the randomness in the privacy mechanisms. Of course, we assume that the adversary knows the released table, the anonymization algorithm, and parameters used in the system in addition to the background knowledge.

3.2   Definition of Pk-Anonymity

We now define our new anonymity, Pk-anonymity, and Pk-anonymization which is a privacy mechanism that always satisfies Pk-anonymity.

First, we define an attack by an adversary, which is represented as an estimation by the following probability, where τ' is a released table, Π is a uniformly random injective map from $\mathcal{R}$ to $\mathcal{R}'$, $r \in \mathcal{R}$, $r' \in \mathcal{R}'$ and $\Delta(T) = T' \circ \Pi$.

$$\Pr[\Pi(r) = r' | T' = \tau']$$

The term $\mathcal{R}$ represents a set of individuals, and $\mathcal{R}'$ represents a set of record IDs (not necessarily explicit IDs. In anonymized microdata, it maybe just a location in storage.). The Π's randomness represents that "an adversary has no knowledge of the linkage between individuals and the records in τ'." Taken together, the above probability represents $\Pr[$"a person r's record in τ' is r'"$]$ from the standpoint of an adversary who saw τ'. We denote the above probability as $\mathcal{E}(f_T, \tau', r, r')$.

Next, we define Pk-anonymity and Pk-anonymization.

Definition 5 (*Pk-Anonymity*). *Let $\mathcal{R}$, $\mathcal{V}$, $\mathcal{R}'$, and $\mathcal{V}'$ be finite sets, and Π and Δ be random variables on $\mathcal{R} \to \mathcal{R}'$ and $\mathcal{T} \to (\mathcal{R} \to \mathcal{V}')$, respectively, where $\mathcal{T}$ denotes the set of tables on $(\mathcal{R}, \mathcal{V})$ and the notation $\mathcal{X} \to \mathcal{Y}$ denotes the set of all maps from $\mathcal{X}$ to $\mathcal{Y}$ for any set $\mathcal{X}$ and $\mathcal{Y}$. Furthermore, let $\tilde{\Delta}$ denote a 6-tuple $(\mathcal{R}, \mathcal{V}, \mathcal{R}', \mathcal{V}', \Pi, \Delta)$.*

---

[3] It means that the adversary knows that the private table is τ_1 with probability x_1, τ_2 with x_2 and so on. It is not a distribution of values in a specific table, but the distribution on the space of all tables.

Then, for any real number $k \geq 1$ and a table τ' on $(\mathcal{R}', \mathcal{V}')$, a pair $(\tilde{\Delta}, \tau')$ is said to satisfy Pk-anonymity (or to be Pk-anonymous) if and only if for any random variables T of tables on $(\mathcal{R}, \mathcal{V})$ and T' of tables on $(\mathcal{R}', \mathcal{V}')$ such that $\tilde{\Delta}$ is a privacy mechanism from T to T', any record $r \in \mathcal{R}$ of the private table T and any record $r' \in \mathcal{R}'$ of the released table τ', the following equation is satisfied.

$$\Pr[\Pi(r) = r' | T' = \tau'] \leq \frac{1}{k}$$

Definition 6 (Pk-Anonymization Algorithms). *Let $\mathcal{R}$, $\mathcal{V}$, $\mathcal{R}'$, $\mathcal{V}'$, Π, Δ, and $\tilde{\Delta}$ be the same as Definition 5, and let T' denote the set of all tables on $(\mathcal{R}', \mathcal{V}')$.*

Then, for any real number $k \geq 1$, $\tilde{\Delta}$ is said to be a Pk-anonymization if and only if $(\tilde{\Delta}, \tau')$ satisfies Pk-anonymity for any released table $\tau' \in T'$ such that there exists a private table τ on $(\mathcal{R}, \mathcal{V})$ which satisfies $\Pr[\Delta(\tau) = \tau' \circ \Pi] \neq 0$.

In other sections, we treat only Δ within 6-tuple of a privacy mechanism $(\mathcal{R}, \mathcal{V}, \mathcal{R}', \mathcal{V}', \Pi, \Delta) = \tilde{\Delta}$; thus, we do not differentiate $\tilde{\Delta}$ and Δ.

Pk-anonymity's direct meaning is "no one estimates which person the record came from with more than $1/k$ probability". Intuitively, it seems to be similar to "no one can narrow down a person's record to less than k records", which is an intuitive concept of *k*-anonymity. This intuitive similarity can also be confirmed mathematically. Furthermore, as far as deterministic anonymization algorithms, such as *k*-anonymization algorithms, are concerned, two anonymity notions can be shown to be equivalent to each other. Therefore, we say *k*-anonymity is satisfied in a randomized table if *Pk*-anonymity is satisfied in the table.

Theorem 1. *For any positive integer k, privacy mechanism Δ, and released table τ', the following relation holds if Δ is deterministic, i.e., for any $\tau \in T$, there exists unique anonymized table $\hat{\tau}$ and $\Delta(\tau) = \hat{\tau}$.*

   τ' is k-anonymous $\Leftrightarrow$ (Δ, τ') is Pk-anonymous

This theorem represents equality of *Pk*-anonymity and *k*-anonymity under the consideration of deterministic anonymization algorithms, which are the applicable field of *k*-anonymity. Therefore, *Pk*-anonymity is deemed as an extension of *k*-anonymity.

This theorem is shown with the following two lemmas.

Lemma 1. *For any positive integer k, if a released table τ' is k-anonymous, then (Δ, τ') is Pk-anonymous for any privacy mechanism Δ.*

Lemma 2. *For any real number $t \geq 1$, positive number k such that $k \leq t$, any deterministic privacy mechanism Δ, and released table τ', if (Δ, τ') is Pt-anonymous, then τ' is k-anonymous.*

Roughly, Lemma 1 states that "$k \Rightarrow Pk$ always", and Lemma 2 states that "$Pk \Rightarrow k$ if an anonymization algorithm is deterministic".

(*Proof of Lemma* 1). First, we use notation $\sharp_{\tau'}(v')$ as $|\tau'^{-1}(\{v'\})|$, and say r' is k-anonymous in τ' if $\sharp_{\tau'}(\tau'(r')) \geq k$. Then, k-anonymity of τ' is represented as "r' is k-anonymous in τ' for any $r' \in \mathcal{R}'$."

We show that an estimation probability, $\mathcal{E}(f_T, \tau', r, r')$, is equal to or less than $1/k$. For any adversary $f_T : \mathcal{T} \to \mathbb{R}$, any $r \in \mathcal{R}$ and any $r' \in \mathcal{R}'$, the following equations hold.

$$\mathcal{E}(f_T, \tau', r, r') = \Pr[\Pi(r) = r' | T' = \tau']$$

$$= \frac{\Pr[\Pi(r) = r' \wedge T' = \tau']}{\Pr[T' = \tau']} = \frac{\Pr[\Pi(r) = r' \wedge \Delta(T) = \tau' \circ \Pi]}{\Pr[\Delta(T) = \tau' \circ \Pi]}$$

(from Eq. 1)

$$= \frac{\sum_{\substack{\delta:\mathcal{T}\to(\mathcal{R}\to\mathcal{V}') \\ \tau\in\mathcal{T}}} f_\Delta(\delta) f_T(\tau) \Pr[\Pi(r) = r' \wedge \delta(\tau) = \tau' \circ \Pi]}{\sum_{\substack{\delta:\mathcal{T}\to(\mathcal{R}\to\mathcal{V}') \\ \tau\in\mathcal{T}}} f_\Delta(\delta) f_T(\tau) \Pr[\delta(\tau) = \tau' \circ \Pi]}$$

(since T, Δ, and Π are independent of each other)

We define two propositions $\Phi(\delta, \tau)$ and $\hat{\Phi}(\delta, \tau)$ as

$$\Phi(\delta, \tau) = [\text{There exists } \hat{\pi} : \mathcal{R} \to \mathcal{R}' \text{ such that } \delta(\tau) = \tau' \circ \hat{\pi}]$$

$$\hat{\Phi}(\delta, \tau) = [\Phi(\delta, \tau) \text{ and } (\delta(\tau))(r) = \tau'(r')]$$

respectively. Since Π is a uniformly random permutation, the following equations hold.

$$\Pr[\delta(\tau) = \tau' \circ \Pi] = \begin{cases} \dfrac{\prod_{v'\in\text{Im}(\tau')} \sharp_{\tau'}(v')!}{|\mathcal{R}|!} & \text{(if } \Phi \text{ holds)} \\ 0 & \text{(otherwise)} \end{cases}$$

$$\Pr[\Pi(r) = r' \wedge \delta(\tau) = \tau' \circ \Pi]$$

$$= \begin{cases} \dfrac{(\sharp_{\tau'}(\tau'(r')) - 1)! \prod_{v'\in\text{Im}(\tau')\setminus\{\tau'(r')\}} \sharp_{\tau'}(v')!}{|\mathcal{R}|!} & \text{(if } \hat{\Phi} \text{ holds)} \\ 0 & \text{(otherwise)} \end{cases}$$

$$= \begin{cases} \dfrac{\prod_{v'\in\text{Im}(\tau')} \sharp_{\tau'}(v')!}{\sharp_{\tau'}(\tau'(r'))|\mathcal{R}|!} & \text{(if } \hat{\Phi} \text{ holds)} \\ 0 & \text{(otherwise)} \end{cases}$$

Therefore, the primary equation $\Pr[\Pi(r) = r' | T' = \tau']$ is transformed as

$$\frac{\sum_{\hat{\Phi}(\delta,\tau)} f_\Delta(\delta) f_T(\tau) \dfrac{\prod_{v'\in\text{Im}(\tau')} \sharp_{\tau'}(\tau'(s'))!}{\sharp_{\tau'}(\tau'(r'))|\mathcal{R}|!}}{\sum_{\Phi(\delta,\tau)} f_\Delta(\delta) f_T(\tau) \dfrac{\prod_{v'\in\text{Im}(\tau')} \sharp_{\tau'}(\tau'(s'))!}{|\mathcal{R}|!}} \leq \frac{\sum_{\Phi(\delta,\tau))} f_\Delta(\delta) f_T(\tau) \dfrac{\prod_{v'\in\text{Im}(\tau')} \sharp_{\tau'}(\tau'(s'))!}{\sharp_{\tau'}(\tau'(r'))|\mathcal{R}|!}}{\sum_{\Phi(\delta,\tau)} f_\Delta(\delta) f_T(\tau) \dfrac{\prod_{v'\in\text{Im}(\tau')} \sharp_{\tau'}(\tau'(s'))!}{|\mathcal{R}|!}}$$

(since $\hat{\Phi} \Rightarrow \Phi$)

$$= \frac{1}{\sharp_{\tau'}(\tau'(r'))} \leq \frac{1}{k}.$$

(from k-anonymity)□

(*Proof of Lemma 2*). Next, we prove the opposite direction. In the proof we use and show the following contraposition.

> For any privacy mechanism Δ, if τ' is not k-anonymous, then (Δ, τ') is also not Pt-anonymous.

We consider adversary's background knowledge, f_T, satisfying $f_T(\tau) = 1$. Let $r' \in \mathcal{R}'$ be a record that is not k-anonymous in τ' and that satisfies $r \in \pi^{-1}(r')$. As in the proof of Lemma 1, the following equation holds.

$$\mathcal{E}(f_T, \tau', r, r') = \frac{\displaystyle\sum_{\hat{\Phi}(\delta, \tau)} f_\Delta(\delta) f_T(\tau) \frac{\prod_{v' \in \mathrm{Im}(\tau')} \sharp_{\tau'}(v')!}{\sharp_{\tau'}(\tau'(r')) |\mathcal{R}|!}}{\displaystyle\sum_{\Phi(\delta, \tau)} f_\Delta(\delta) f_T(\tau) \frac{\prod_{v' \in \mathrm{Im}(\tau')} \sharp_{\tau'}(v')!}{|\mathcal{R}|!}}$$

Since Δ is deterministic and $f_T(\tau) = 1$, we transform the above equation as follows.

$$\frac{\displaystyle\sum_{\hat{\Phi}(\delta, \tau)} f_\Delta(\delta) f_T(\tau) \frac{\prod_{v' \in \mathrm{Im}(\tau')} \sharp_{\tau'}(v')!}{\sharp_{\tau'}(\tau'(r')) |\mathcal{R}|!}}{\displaystyle\sum_{\Phi(\delta, \tau)} f_\Delta(\delta) f_T(\tau) \frac{\prod_{v' \in \mathrm{Im}(\tau')} \sharp_{\tau'}(v')!}{|\mathcal{R}|!}} = \frac{\frac{\prod_{v' \in \mathrm{Im}(\tau')} \sharp_{\tau'}(v')!}{\sharp_{\tau'}(\tau'(r')) |\mathcal{R}|!}}{\frac{\prod_{v' \in \mathrm{Im}(\tau')} \sharp_{\tau'}(v')!}{|\mathcal{R}|!}} = \frac{1}{\sharp_{\tau'}(\tau'(r'))}.$$

We assume r' is not k-anonymous; therefore, $\dfrac{1}{\sharp_{\tau'}(\tau'(r'))} \gtrless \dfrac{1}{k}.$ □

Through Theorem 1, we have seen that Pk-anonymity is an exact mathematical extension of k-anonymity. Moreover, the intuitive meaning of k-anonymity, "no one can narrow down a person's record to less than k records" is applicable from the following viewpoint. Under a privacy mechanism Δ and a certain released table τ', an adversary's estimation $\mathcal{E}(f_T, \tau', r, r')$ is $1/k$ or less for any $r \in \mathcal{R}$ and $r' \in \mathcal{R}'$, when (Δ, k) is Pk-anonymous. Then by definition, for any $k - 1$ records $\{r'_i\}_{0 \leq i < k-1}$ in τ', the following relation holds.

$$\sum_{0 \leq i < k-1} \mathcal{E}(f_T, \tau', r, r'_i) \leq 1 - \frac{1}{k} \lessgtr 1$$

This relation means that when one has chosen $k - 1$ records from τ', there is always $1/k$ probability that r is not in these $k - 1$ records in τ'. This precisely means that "no one can narrow down a person's record to less than k records."

4   Applying Pk-Anonymity to PRAM

In this section, we apply Pk-anonymity to PRAM. First, we show a theorem on general PRAM for calculating k. Next, we describe a more concrete formula on retention-replacement perturbation introduced in Sect. 2.3. Without loss of generality, we consider the case where there is one attribute.

Theorem 2 (*Pk-Anonymity on PRAM*). *A PRAM mechanism whose transition probability matrix is A is a Pk-anonymization if and only if k is described as follows.*

$$k \leq 1 + (|\mathcal{R}| - 1) \min_{\substack{u,v \in \mathcal{V} \\ u',v' \in \mathcal{V}'}} \frac{A_{u,v'} A_{v,u'}}{A_{u,u'} A_{v,v'}}$$

Note that this theorem shows the tight bound of k.

(*Proof of Theorem 2*). This theorem is shown by evaluating the maximum probability of estimation $\mathcal{E}(f_T, \tau', r, r')$ on $r \in \mathcal{R}, r' \in \mathcal{R}', \tau' \in \mathcal{T}'$ and adversary's background knowledge $f_T : \mathcal{T} \to \mathbb{R}$. When $\mathcal{E}(f_T, \tau', r, r')$ takes the maximum? Let's consider the case represented as Table 1 where there are four records and one attribute. Intuitively, if $A_{v,v'}$ and $A_{u,u'}$ are big, and $A_{u,v'}$ and $A_{v,u'}$ are small, then $\mathcal{E}(f_T, \tau', r_0, r_0')$ is big since v is likely to be not u' but v' and u is likely to be not v' but u'. In fact, we can show Table 1 is the worst case if Table 1 satisfying the following.

Table 1. An example of the private/released table.

Private table τ			Released table τ'	
Record $\mathcal{R}$	Attribute $\mathcal{V}$		Record $\mathcal{R}'$	Attribute $\mathcal{V}'$
r_0	v		r_0'	v'
r_1	u		r_1'	u'
r_2	u		r_2'	u'
r_3	u		r_3'	u'

- v and u are hard to be transformed to the other: $\dfrac{A_{\bar{u},\bar{v}'} A_{\bar{v},\bar{u}'}}{A_{\bar{u},\bar{u}'} A_{\bar{v},\bar{v}'}}$ takes the minimum in $\bar{v} \in \mathcal{V}, \bar{u} \in \mathcal{V}, \bar{v}' \in \mathcal{V}'$ and $\bar{u}' \in \mathcal{V}'$ when $\bar{v} = v, \bar{u} = u, \bar{v}' = v'$ and $\bar{u}' = u'$.

- The adversary knows all about the private table, i.e., $f_T(\tau) = \begin{cases} 1 & \text{if } \tau = \tau' \\ 0 & \text{otherwise} \end{cases}$

Formally, we show this theorem by evaluating the maximum of $\mathcal{E}(f_T, \tau', r, r')$ on $r \in \mathcal{R}, r' \in \mathcal{R}', \tau' \in \mathcal{T}'$ and $f_T : \mathcal{T} \to \mathbb{R}$. Similar to the proof of Lemma 1, the following equation holds.

$$\mathcal{E}(f_T, \tau', r, r') = \Pr[\Pi(r) = r' | T' = \tau'] = \frac{\Pr[\Delta(T) = \tau' \circ \Pi \wedge \Pi(r) = r']}{\Pr[\Delta(T) = \tau' \circ \Pi]}$$

(from Eq. 1)

$$= \frac{\sum_{\tau \in \mathcal{T}} f_T(\tau) \Pr[\Delta(\tau) = \tau' \circ \Pi \wedge \Pi(r) = r']}{\sum_{\tau \in \mathcal{T}} f_T(\tau) \Pr[\Delta(\tau) = \tau' \circ \Pi]}$$

Next we show that which f_T maximizes the above estimation probability.

Lemma 3. *Let $\mathbb{R}^{n+}$ be the set of non-zero n-dim vectors whose elements are non-negative real numbers without 0. Then for any vector $a, b \in \mathbb{R}^{n+}$, the maximum of*

$$g(x) \stackrel{\text{def}}{=} \frac{b \cdot x}{a \cdot x} \left(= \frac{\sum_{i<n} b_i x_i}{\sum_{i<n} a_i x_i} \right)$$

on a variable x on $\mathbb{R}^{n+}$ is $\max_{i<n} \frac{b_i}{a_i}$ where $\cdot$ denotes the inner product, and x satisfies

$$\text{for any } i < n \text{ such that } \frac{b_i}{a_i} \neq \max_{i<n} \frac{b_i}{a_i}, x_i = 0.$$

(Proof of Lemma 3). Let $y_i = a_i x_i$ for $i < n$, $\sum_{i<n} y_i = \alpha$, and x^* be an n-dim vector satisfying $x_i = 0$ if $\frac{b_i}{a_i} \neq \max_{i<n} \frac{b_i}{a_i}$ for any $i < n$.
Then,

$$g(x) = \frac{\sum_{i<n} b_i x_i}{\sum_{i<n} a_i x_i} = \frac{\sum_{i<n} \frac{b_i}{a_i} y_i}{\sum_{i<n} y_i} = \frac{1}{\alpha} \sum_{i<n} \frac{b_i}{a_i} y_i \leq \frac{1}{\alpha} \left(\max_{j<n} \frac{b_j}{a_j} \right) \sum_{i<n} y_i = \max_{i<n} \frac{b_i}{a_i}$$

and

$$g(x^*) = \max_{i<n} \frac{b_i}{a_i}.$$

□ (Lemma 3)

From the above lemma, when $\mathcal{E}(f_T, \tau', r, r')$ takes the maximum, f_T makes the following formula maximum,

$$\frac{\Pr[\Delta(\tau) = \tau' \circ \Pi \wedge \Pi(r) = r')]}{\Pr[\Delta(\tau) = \tau' \circ \Pi]} \tag{2}$$

and the maximum of Formula (2) is equal to that of $\mathcal{E}(f_T, \tau', r, r')$.

Since Π is a uniformly random permutation, Formula (2) is transformed as follows.

$$\text{Formula(2)} = \frac{\frac{1}{|\mathcal{R}|!} \sum_{\pi(r) = r'} \Pr[\Delta(\tau) = \tau' \circ \pi]}{\frac{1}{|\mathcal{R}|!} \sum_{\pi} \Pr[\Delta(\tau) = \tau' \circ \pi]}$$

$$= \frac{\sum_{\pi(r) = r'} \Pr[\Delta(\tau) = \tau' \circ \pi]}{\sum_{\pi} \Pr[\Delta(\tau) = \tau' \circ \pi]} = \frac{\sum_{\pi(r) = r'} \prod_{s \in \mathcal{R}} \Pr[(\Delta(\tau))(s) = \tau'(\pi(s))]}{\sum_{\pi} \prod_{s \in \mathcal{R}} \Pr[(\Delta(\tau))(s) = \tau'(\pi(s))]}$$

(since Δ is independent from each record)

Let a matrix $A^{\tau,\tau'}$ be

$$A^{\tau,\tau'}_{s,s'} \stackrel{\text{def}}{=} \Pr[(\Delta(\tau))(s) = \tau'(s')]$$

for any $s \in \mathcal{R}$, $s' \in \mathcal{R}'$. Then, the above formula is represented as follows.

$$F(A^{\tau,\tau'}) \stackrel{\text{def}}{=} \frac{\sum_{\pi(r)=r'} \prod_{s \in \mathcal{R}} A^{\tau,\tau'}_{s,\pi(s)}}{\sum_{\pi} \prod_{s \in \mathcal{R}} A^{\tau,\tau'}_{s,\pi(s)}}$$

We would rather find the minimum of the reciprocal than the maximum of $F(A^{\tau,\tau'})$ itself. In the case of $|\mathcal{R}| \geq 2$, the reciprocal is transformed as follows.

$$\frac{1}{F(A^{\tau,\tau'})} = \frac{\sum_{\pi} \prod_{s \in \mathcal{R}} A^{\tau,\tau'}_{s,\pi(s)}}{\sum_{\pi(r)=r'} \prod_{s \in \mathcal{R}} A^{\tau,\tau'}_{s,\pi(s)}} = 1 + \frac{\sum_{\pi(r)\neq r'} \prod_{s \in \mathcal{R}} A^{\tau,\tau'}_{s,\pi(s)}}{\sum_{\pi(r)=r'} \prod_{s \in \mathcal{R}} A^{\tau,\tau'}_{s,\pi(s)}}$$

$$-\frac{\sum_{\substack{t\neq r \\ t'\neq r'}} A^{\tau,\tau'}_{t,r'} A^{\tau,\tau'}_{r,t'} \sum_{\substack{\pi(t)=r' \\ \pi(r)=t'}} \prod_{s\neq t,r} A^{\tau,\tau'}_{s,\pi(s)} + A^{\tau,\tau'}_{r,r'} \sum_{\pi(r)=r'} \prod_{s\neq r} A^{\tau,\tau'}_{s,\pi(s)}}{A^{\tau,\tau'}_{r,r'} \sum_{\pi(r)=r'} \prod_{s\neq r} A^{\tau,\tau'}_{s,\pi(s)}}$$

$$= 1 + \frac{\sum_{\substack{t\neq r \\ t'\neq r'}} A^{\tau,\tau'}_{t,r'} A^{\tau,\tau'}_{r,t'} \sum_{\substack{\pi(t)=r' \\ \pi(r)=t'}} \prod_{s\neq t,r} A^{\tau,\tau'}_{s,\pi(s)}}{A^{\tau,\tau'}_{r,r'} \sum_{\pi(r)=r'} \prod_{s\neq r} A^{\tau,\tau'}_{s,\pi(s)}}$$

$$= 1 + \frac{\sum_{\substack{t\neq r \\ t'\neq r'}} A^{\tau,\tau'}_{t,r'} A^{\tau,\tau'}_{r,t'} \sum_{\substack{\pi(r)=r' \\ \pi(t)=t'}} \prod_{s\neq t,r} A^{\tau,\tau'}_{s,\pi(s)}}{A^{\tau,\tau'}_{r,r'} \sum_{\pi(r)=r'} \prod_{s\neq r} A^{\tau,\tau'}_{s,\pi(s)}}$$

$$= 1 + \frac{\sum_{t\neq r} A^{\tau,\tau'}_{t,r'} A^{\tau,\tau'}_{r,\pi(t)} \sum_{\pi(r)=r'} \frac{\prod_{s\neq r} A^{\tau,\tau'}_{s,\pi(s)}}{A^{\tau,\tau'}_{t,\pi(t)}}}{A^{\tau,\tau'}_{r,r'} \sum_{\pi(r)=r'} \prod_{s\neq r} A^{\tau,\tau'}_{s,\pi(s)}}$$

$$= 1 + \frac{\sum_{\pi(r)=r'} \sum_{t\neq r} \frac{A^{\tau,\tau'}_{t,r'} A^{\tau,\tau'}_{r,\pi(t)}}{A^{\tau,\tau'}_{t,\pi(t)}} \prod_{s\neq r} A^{\tau,\tau'}_{s,\pi(s)}}{\sum_{\pi(r)=r'} A^{\tau,\tau'}_{r,r'} \prod_{s\neq r} A^{\tau,\tau'}_{s,\pi(s)}}$$

We show the following lemma.

Lemma 4. Let g_i and h_i be $g_i, h_i : \mathbb{R}^{\mathcal{I}} \to \mathbb{R}$ for any index $i \in \mathcal{I}$, where $\mathcal{I}$ is a set of indices. If some $x \in \mathbb{R}^{\mathcal{I}}$ and $z \in \mathbb{R}$ satisfy $\frac{h_i(x)}{g_i(x)} = \min_{x' \in \mathbb{R}^{\mathcal{I}}} \frac{h_i(x')}{g_i(x')} = z$ for any $i \in \mathcal{I}$, then the following equation is satisfied.

$$\min_{x' \in \mathbb{R}^{\mathcal{I}}} \frac{\sum_{i \in \mathcal{I}} h_i(x')}{\sum_{i \in \mathcal{I}} g_i(x')} = \frac{\sum_{i \in \mathcal{I}} h_i(x)}{\sum_{i \in \mathcal{I}} g_i(x)}$$

(*Proof of Lemma 4*). From the assumption of the lemma, $h_i(x') \geq z g_i(x')$ hold for all $i \in \mathcal{I}$ and any $x' \in \mathbb{R}$. Therefore,

$$\frac{\sum_{i \in \mathcal{I}} h_i(x')}{\sum_{i \in \mathcal{I}} g_i(x')} \geq z,$$

then

$$\min_{x' \in \mathbb{R}^n} \frac{\sum_{i \in \mathcal{I}} h_i(x')}{\sum_{i \in \mathcal{I}} g_i(x')} = \frac{\sum_{i \in \mathcal{I}} h_i(x)}{\sum_{i \in \mathcal{I}} g_i(x)}$$

holds.   □ (Lemma 4)

Let $h_\pi(A^{\tau,\tau'})$ and $g_\pi(A^{\tau,\tau'})$ be

$$h_\pi(A^{\tau,\tau'}) = \sum_{t \neq r} \frac{A^{\tau,\tau'}_{t,r'} A^{\tau,\tau'}_{r,\pi(t)}}{A^{\tau,\tau'}_{t,\pi(t)}} \prod_{s \neq r} A^{\tau,\tau'}_{s,\pi(s)}, \qquad g_\pi(A^{\tau,\tau'}) = A^{\tau,\tau'}_{r,r'} \prod_{s \neq r} A^{\tau,\tau'}_{s,\pi(s)}$$

for any $\pi : \mathcal{R} \to \mathcal{R}'$. Thanks to Lemma 4, it is sufficient to consider $\dfrac{h_\pi(A^{\tau,\tau'})}{g_\pi(A^{\tau,\tau'})}$

only. Because it is transformed into $\dfrac{1}{A^{\tau,\tau'}_{r,r'}} \sum_{t \neq r} \dfrac{A^{\tau,\tau'}_{t,r'} A^{\tau,\tau'}_{r,\pi(t)}}{A^{\tau,\tau'}_{t,\pi(t)}}$, it takes the mini-

mum for any $\pi : \mathcal{R} \to \mathcal{R}'$ when τ and τ' are as follows.

There exists $v \in \mathcal{V}$ and $v' \in \mathcal{V}'$ such that $\dfrac{A_{v,\tau'(r')} A_{\tau(r),v'}}{A_{\tau(r),\tau'(r')} A_{v,v'}} =$

$\min\limits_{\substack{u,v \in \mathcal{V} \\ u',v' \in \mathcal{V}'}} \dfrac{A_{u,v'} A_{v,u'}}{A_{u,u'} A_{v,v'}}$, $\tau(s) = v$ for any $s \neq r$ and $\tau'(s') = v'$ for any $s' \neq r'$.

Since k is to be the reciprocal of the maximum of $F(A^{\tau,\tau'})$, k is found to be the following value.

$$k = 1 + (|\mathcal{R}| - 1) \min_{\substack{u,v \in \mathcal{V} \\ u',v' \in \mathcal{V}'}} \frac{A_{u,v'} A_{v,u'}}{A_{u,u'} A_{v,v'}}$$

It is easy to confirm that the above equation also holds when $|\mathcal{R}| = 1$. In this case, since only one π exists (denoted as $\hat{\pi}$), k equals 1 as follows.

$$k = \frac{1}{F(A^{\tau,\tau'})} = \frac{\sum_\pi \prod_{s \in \mathcal{R}} A^{\tau,\tau'}_{s,\pi(s)}}{\sum_{\pi(r)=r'} \prod_{s \in \mathcal{R}} A^{\tau,\tau'}_{s,\pi(s)}} = \frac{\prod_{s \in \mathcal{R}} A^{\tau,\tau'}_{s,\hat{\pi}(s)}}{\prod_{s \in \mathcal{R}} A^{\tau,\tau'}_{s,\hat{\pi}(s)}}$$

$$= 1 = 1 + (|\mathcal{R}| - 1) \min_{\substack{u,v \in \mathcal{V} \\ u',v' \in \mathcal{V}'}} \frac{A_{u,v'} A_{v,u'}}{A_{u,u'} A_{v,v'}}$$

(since $|\mathcal{R}| = 1$)□

We describe the multi-attribute version of Theorem 2.

Corollary 1. *A PRAM mechanism whose transition probability matrices are A_a for each attribute a is a Pk-anonymization when k is described as follows,*

$$k = 1 + (|\mathcal{R}| - 1) \prod_{a \in \mathcal{A}} \min_{\substack{u,v \in \mathcal{V} \\ u',v' \in \mathcal{V}'}} \frac{(A_a)_{u,v'} (A_a)_{v,u'}}{(A_a)_{u,u'} (A_a)_{v,v'}}$$

The following corollary is applicable to retention-replacement perturbation.

Corollary 2. *Retention-replacement perturbation whose retention probabilities are ρ_a for each attribute $a \in \mathcal{A}$, is a Pk-anonymization when k is described as follows,*

$$k = 1 + (|\mathcal{R}| - 1)(\prod_{a \in \mathcal{A}} \frac{1 - \rho_a}{1 + (|\mathcal{V}_a| - 1)\rho_a})^2 \tag{3}$$

Using Theorem 2, k is easily calculated with the record count $|\mathcal{R}|$ and transition probability matrix A. Regarding retention-replacement perturbation, A is determined independently with the instance of private data, k is calculated with the record count $|\mathcal{R}|$ and the numbers of attribute values $|\mathcal{V}_a|$ and retention probabilities ρ_a only, for each attribute a.

Conversely, ρ_a are also calculated in retention-replacement perturbation. By letting all ρ_a be the same ρ over all attributes, the equation is transformed as follows. Since k monotonically decreases on $0 \leq \rho \leq 1$ from Eq. (3), ρ is easily and uniquely solved using, for example, the bisection method for any k, $|\mathcal{R}|$ and $|\mathcal{V}_a|$(Algorithm 1). Note that k is allowed to be a real number, for example, $k = 1.5$.

Algorithm 1. determining ρ in retention-replacement perturbation from k
input: $k \in \mathbb{R}(k \geq 1)$, $|\mathcal{R}| \in \mathbb{N}$, $|\mathcal{V}_a|$ for each attribute
output: retention probability ρ

1: Set $\rho_0 = 1/2$.
2: Run the bisection method with ρ's initial value ρ_0 with respect to k using Eq. 3 and output the converged ρ.

For example, to ensuring $P100$-anonymity on $100,000$ records of data, ρ is calculated as roughly 0.303, where there are three attributes, sex, age from 20's to 60's, and 10-leveled annual income.

When the record count is uncertain since the data are to be collected thereafter, it is sufficient to use the expected record count. Even when the record count does not reach the expected value, Pk-anonymity is still satisfied for the following reason. When each record in table τ' is anonymous due to an anonymous communication channel, it can be said that only a part of table τ' is visible in the state in which τ' is being collected. An estimation in such a situation is equivalent to that from the algorithm that ignores the absent records. From Proposition 1, the algorithm cannot derive $\mathcal{E}(f_T, \tau', r, r') \ngeq 1/k$ if it is correct.

5 Conclusions

In the field of anonymized microdata release, we mainly presented the following two theories. We first proposed an anonymity notion, Pk-anonymity, which is an extension of k-anonymity to randomized microdata. We then applied Pk-anonymity to PRAM.

The contributions of the paper are

- an anonymity notion called Pk-anonymity,
- proofs that Pk-anonymity is an exact mathematical extension of k-anonymity,
- a formula for calculating k on PRAM,
- algorithms for determining the parameter of the retention-replacement perturbation according to k.

References

1. Aggarwal, C.C.: On unifying privacy and uncertain data models. In: Alonso, G., Blakeley, J.A., Chen, A.L.P.: (eds.) ICDE, pp. 386–395. IEEE (2008)
2. Agrawal, D., Aggarwal, C.C.: On the design and quantification of privacy preserving data mining algorithms. In: Buneman, P. (ed.) PODS. ACM (2001)
3. Agrawal, R., Srikant, R., Thomas, D.: Privacy preserving OLAP. In: Özcan, F. (ed.) SIGMOD Conference, pp. 251–262. ACM (2005)
4. Dwork, C.: Differential privacy. In: Bugliesi, M., Preneel, B., Sassone, V., Wegener, I. (eds.) ICALP 2006. LNCS, vol. 4052, pp. 1–12. Springer, Heidelberg (2006)
5. Kooiman, P., Willenborg, L., Gouweleeuw, J.: PRAM: a method for disclosure limitation of microdata. Reseach report no. 90 (1997)
6. Li, N., Li, T., Venkatasubramanian, S.: t-closeness: privacy beyond k-anonymity and l-diversity. In: Chirkova, R., Dogac, A., Özsu, M.T., Sellis, T.K. (eds.) ICDE, pp. 106–115. IEEE (2007)
7. Lin, B.-R., Wang, Y., Rane, S.: A framework for privacy preserving statistical analysis on distributed databases. In: WIFS, pp. 61–66. IEEE (2012)
8. Machanavajjhala, A., Kifer, D., Gehrke, J., Venkitasubramaniam, M.: l-diversity: privacy beyond k-anonymity. In: Liu, L., Reuter, A., Whang, K.-Y., Zhang, J. (eds.) ICDE, p. 24. IEEE Computer Society (2006)
9. Rebollo-Monedero, D., Forné, J., Domingo-Ferrer, J.: From t-closeness-like privacy to postrandomization via information theory. IEEE Trans. Knowl. Data Eng. **22**(11), 1623–1636 (2010)
10. Samarati, P.: Protecting respondents' identities in microdata release. IEEE Trans. Knowl. Data Eng. **13**(6), 1010–1027 (2001)
11. Samarati, P., Sweeney, L.: Generalizing data to provide anonymity when disclosing information (abstract). In: Mendelzon, A.O, Paredaens, J. (eds.) PODS, p. 188. ACM Press (1998)
12. Soria-Comas, J., Domingo-Ferrer, J.: Probabilistic k-anonymity through microaggregation and data swapping. In: FUZZ-IEEE, pp. 1–8. IEEE (2012)
13. Sweeney, L.: k-anonymity: a model for protecting privacy. Int. J. Uncertainty Fuzziness Knowl.-Based Syst. **10**(5), 557–570 (2002)
14. Wong, R.C.-W., Li, J., Fu, A.W.-C., Wang, K.: (alpha, k)-anonymity: an enhanced k-anonymity model for privacy preserving data publishing. In: Eliassi-Rad, T., Ungar, L.H., Craven, M., Gunopulos, D. (eds.) KDD, pp. 754–759. ACM (2006)

On Limitations and Alternatives of Privacy-Preserving Cryptographic Protocols for Genomic Data

Tadanori Teruya[1]([✉]), Koji Nuida[1,2], Kana Shimizu[3], and Goichiro Hanaoka[1]

[1] Information Technology Research Institute, National Institute of Advanced Industrial Science and Technology, AIST Tokyo Waterfront Bio-IT Research Building, 2-4-7 Aomi, Koto-ku, Tokyo, Japan
tadanori.teruya@aist.go.jp
[2] JST PRESTO, Tokyo, Japan
[3] Biotechnology Research Institute for Drug Discovery, National Institute of Advanced Industrial Science and Technology, AIST Tokyo Waterfront Bio-IT Research Building, 2-4-7 Aomi, Koto-ku, Tokyo, Japan

Abstract. The human genome can identify an individual and determine the individual's biological characteristics, and hence has to be securely protected in order to prevent privacy issues. In this paper we point out, however, that current standard privacy-preserving cryptographic protocols may be insufficient to protect genome privacy. This is mainly due to typical characteristics of genome information; it is immutable, and an individual's genome has correlations to those of the individual's progeny. Then, as an alternative, we propose to protect genome privacy by cryptographic protocols with *everlasting security*, which provides an appropriate mixture of computational and information-theoretic security. We construct a concrete example of a protocol with everlasting security, and discuss its practical efficiency.

1 Introduction

Recent progress in DNA sequencing technology has been a driving force of large-scale personal genome projects where various genome-wide association studies are conducted for discovering new biological knowledge such as identification of disease factors [2]. Although personal genome analysis has a high potential to expedite both biological and biomedical research, the data produced by those projects have not been fully utilized, because information that includes donors' privacy is difficult to handle. In most cases, a personal genome has been kept at the site where it was sequenced and only a limited number of researchers are allowed to access the database storing it. This conservative approach has a small risk of information leakage, however, it limits opportunities for many researchers to work on valuable databases, and it also hinders analyses on an enriched dataset consisting of different databases, which is unfortunate, since the performance of statistical methods generally increases according to sample size.

© Springer International Publishing Switzerland 2015
K. Tanaka and Y. Suga (Eds.): IWSEC 2015, LNCS 9241, pp. 242–261, 2015.
DOI: 10.1007/978-3-319-22425-1_15

In order to overcome the dilemma between privacy and utility, several groups [3,4,14,18] recently attempted to use a technology called privacy-preserving data mining (PPDM) which allows data mining results to be obtained without accessing individual data records [1]. Among various techniques used in PPDM, cryptography plays a crucial role, and generally, the entire security achieved by a PPDM method deeply depends on each cryptosystem, which is used as a building block of the method.

Personal genome enables ones to identify an individual. On the other hand, association between genotypic and phenotypic information has not been fully uncovered yet, which makes it difficult to estimate how serious a privacy breach would be if donors' genomes are leaked. Therefore, personal genome should, with no doubt, be protected in a highly secure manner. To the best of our knowledge, most of the conventional studies to cryptographically protect personal genome used public-key cryptography just as in the cases of other sensitive information, e.g., bank accounts communicated in an e-commerce, and security requirements that arise from typical characteristics of genomic data, have not been discussed well.

The main aim of this paper is to point out that, *straightforward uses of public-key cryptography are practically not sufficient for protecting genome privacy.* Namely, the genome of one person remains (in principle) unchanged throughout the person's lifetime, and is furthermore partly inherited by the person's descendants. Due to this, cryptography for the purpose should be able to provide security during an unbounded or unpredictably long period of time, which is in fact not realistic in the currently standard, practical framework of handling public-key cryptosystems. See Sect. 2.2 for a detailed argument. Then, as an alternative solution, we propose to use cryptographic protocols with *everlasting security* [11,23], which is, roughly speaking, a security model involving a formalization of the notion of "compromise of cryptographic primitives".

In our proposal, the one-time pad is used in order to achieve everlasting security, which is stronger than the usual computational security. It seems to trivially achieve useful functionalities by assuming the availability of pre-shared keys. However, the availability of pre-shared keys does *not* imply the general information-theoretically secure computation immediately. For example, the impossibility results are proved by [17,23], which showed that the quantum channel (which is one of mechanisms for the information-theoretically secure key sharing) cannot imply an information-theoretically secure oblivious transfer. Therefore, our proposal and its concept in this paper are never trivial even in the presence of pre-shared keys.

To overcome the impossibilities, we examine security requirements based on our careful consideration and observation described in Sect. 3. More precisely, we classify the secrets into two necessary security levels, namely, information-theoretic security and computational security. In our observation, as explained later, genomic privacy and related secret information are to be protected by the former security to provide long-term security, on the other hand, the secrets such as the intellectual property, for which short-term security is sufficient, may be protected by the latter security. Because of the security requirements different from [17,23], we can overcome the impossibility results in [17,23].

We evaluate the possibility and the efficiency of our solution by constructing a concrete example of such a protocol and analyzing its execution costs (including necessary but a significantly reduced amount of pre-shared keys).

The rest of the paper is organized as follows. We describe a detailed problem setting and the appropriate cryptographic security in order to provide long-term security for genome privacy in Sect. 2. In Sect. 3, we discuss and consider a cryptographic security model for genome privacy protection, and then define a security notion guaranteeing the appropriate security. Preliminaries for our protocol construction are described in Sect. 4. We describe the construction of our proposal in Sect. 5, and a discussion of our proposal is in Sect. 6. The conclusion is given in Sect. 7.

2   Protecting Genome Privacy By Cryptography

2.1   Example: Privacy-Preserving Data Mining for Allele Frequency

For the purpose of discussion, we suppose the following possible scenario in medical research based on genomic data: To develop a model for contraction risk of a disease, a researcher may refer to the allele frequencies of a large number of donors (for example, the 100th positions consist of 40 %, 50 %, 3 % and 7 % of A, T, G and C, respectively) in a database owned by a stakeholder.

Here we emphasize that, for the client-side security, some researcher would want to hide from anyone, even the database holder, which information in the database one is interested in, since the target information itself may be closely related to one's intellectual properties. On the other hand, for the server-side security, one may feel that the information on each individual donor is certainly protected, since the allele frequencies are just statistical information calculated from donors' data. However, recent studies have revealed risks of learnable information on individual donors even from statistical genomic information [6,12]. This suggests the requirements for protecting such databases more securely.

Various cryptographic protocols have been proposed for the purpose of privacy-preserving data mining (PPDM) [3,4,14,18]. Therefore, a reasonable solution to enhance the security in the above-mentioned scenario would be to apply such cryptographic tools. The oblivious transfer is one of the popular building-block cryptographic protocols for PPDM.

2.2   Limitation of Computationally Secure Cryptographic Tools

In most of the major cryptographic protocols for PPDM, the security provided by the protocols is computational security. Since the necessary computation time to break the protocols depends on *the sizes (bit-lengths) of cryptographic keys* used in the protocols, we must select sufficiently long key lengths to guarantee a sufficient security level.

It is observed in [16] that the appropriate key length depends on the following four factors (quoted from [16]):

1. Life span: The expected time the information needs to be protected.
2. Security margin: An acceptable degree of infeasibility of a successful attack.
3. Computing environment: The expected change in computational resources available to attackers.
4. Cryptanalysis: The expected developments in cryptanalysis. .

We note that, in fact, the current recommendations for standard key lengths by NIST [20] and ECRYPT II [7] are decided for setting the effective time of the security to be around 10–20 years. One of the reasons for such recommendations is that too long keys make the cryptographic protocols unnecessarily inefficient. On the other hand, another and more noteworthy reason is that, long-standing measurements and future predictions of the above-mentioned factors, especially Computing environment and Cryptanalysis, are quite difficult. A typical example of the difficulty is as follows: It has been known that practical development of quantum computers will efficiently weaken and break many current standard cryptosystems [9, 22], and most of the researchers would think that the situation will not be realized in a near future, but there seem to exist no common opinions by the experts on whether practical quantum computers will become available in the next 50 years or not.

Here is a limitation of current standard cryptographic tools for protecting genome privacy. Despite that the genome of a person is immutable, the only 20-year security, which is currently guaranteed by those cryptographic tools, is much shorter than the average length of human life. Moreover, the heredity of the genome makes the situation worse; a person must protect an individual's genome not only for oneself, but also for the individual's progeny, which yields further longer life spans required for the cryptographic tools. By the observation, we conclude that *it is quite difficult to protect genome privacy by using the standard cryptography with computational security*.

2.3   Everlasting Security

From the argument above, one may think that cryptographic tools with *information-theoretic security* seem to be appropriate for protecting genome privacy, since it is never compromised even by the passage of long time. However, the information-theoretic security also has disadvantages. In general, cryptographic tools to provide information-theoretic security tend to be very inefficient, e.g., a large amount of cryptographic keys or unrealistic assumptions are required. This issue may be more crucial in the applications to genomic information, since the sizes of genomes themselves are very large and it would yield a further large total amount of cryptographic keys in the protocol. Furthermore, the known techniques (the noisy channel models, the quantum key distribution, the bounded storage model, the limited access model, and the proactive secret sharing) to provide information-theoretic security are impractical [5] (note that the information-theoretically secure key exchange of a huge amount of pre-shared keys is also needed to construct the proactive secret sharing). The secure multiparty computations based on the linear secret sharing are also one of methodologies achieving information-theoretic security. However, the security is proven on the strong assumptions that secret information is divided into shares; divided shares are distributed and transmitted to each party on authenticated secure channels; parties do not collude or are not corrupted up to a certain number of them.

In order to reduce the problem, here we propose to use cryptographic protocols which provide *everlasting security* [11,23]. This is a kind of blend of computational security and information-theoretic security. As mentioned above, ordinary cryptographic protocols with computational security are likely to become easily breakable after a long time, e.g., 20 years. The adversarial model of everlasting security captures this situation as follows. First, during the execution of a protocol, the computational power of each party is supposed to be bounded by probabilistic polynomial-time, which is the standard model in cryptographic study. Then, after the protocol execution, each party is supposed to get unbounded computational power, which enables the party to break every computationally secure cryptographic tools used in the protocol. We emphasize that, despite that an adversary has unbounded computational power after the protocol, one cannot perform forgery or man-in-the-middle attacks for the communications in the protocol since the protocol has already finished. We can expect that the use of everlasting security instead of information-theoretic security would indeed reduce the total cost of the protocol, since some appropriate part of the protocol is protected by only computational security.

3   An Example of Everlasting Security

From now, we discuss and evaluate the possibility and the effectiveness of everlasting security for protecting genome privacy. For the purpose, we consider a concrete example of cryptographic protocols, namely oblivious transfer protocols, applied to the scenario described in Sect. 2.1. The oblivious transfer provides security for the client's query to the database, as well as a kind of access control by the server for the records of the database.

3.1   Observation for the Desired Security

For applications of everlasting security to the oblivious transfer protocols, we note that, it is technically very difficult to protect every data in the protocol by the strongest security level. By the reason, we first evaluate the necessary security level for each data to be protected, and appropriately classify the whole information into one for which relatively short-term security is sufficient, and the other, such as the genomic data, for which long-term (or information-theoretically secure) protection is required. Then we protect each of them by computational security or information-theoretic security depending on the above-mentioned classification. In other words, in this work, by relaxing the security level for some parts of the information appropriately, we overcome the above-mentioned difficulty of construction of the protocols.

Table 1. Summary of desired terms of security for the information against the parties, where "short-term" and "long-term" indicate that the information listed in the header at the same column should be protected in short-term and long-term, respectively, against the party listed in the party column at the same row, and "owner" means that the information in the header at the same column is possessed by the party listed in the party column at the same row

Party	Database	Query	Authentication code
Server S	(Owner)	Short-term	Short-term
Client C	Long-term	(Owner)	Short-term
Adversary A	Long-term	Long-term	Short-term

In the privacy-preserving database search for genomic allele frequency based on the oblivious transfer, it is desirable that the database possessed by a server S is protected for a long time against any third-party adversary A. On the other hand, since the query by a client C is supposed to be motivated from an interest in the database, it is reasonable to observe that the client's query and the content of the database would have some correlation. Therefore, the query by C should also be protected for a long time against A.

For the client C, since information gained from the database yields one's benefits, it is reasonable to assume that one has an incentive for illegal database query, which is not valid formatted query expected by S, to gain more information than the allowable extent, therefore such illegal queries should be prevented. In other words, C should be modeled as a malicious party that may not follow the specification of protocol.

In contrast, since the queries by the client C would be related to intellectual property rights rather than privacy-related genomic data possessed by C, relatively short-term security would suffice for protection of these data against the server S. On the other hand, since S is usually a company or a public organization, which is watched by the society, it would be reasonable to assume that the possibility of illegal actions by S is small. In other words, we model S as a semi-honest party.

Finally, in order to achieve everlasting security, confidentiality and integrity protection are needed. More precisely, the former is guaranteed by the information-theoretically secure manner, but the latter is guaranteed by the computationally secure manner. This is because the adversary A and the client C can attack actively but they are computationally bounded during the protocol execution, and after the protocol execution, their power are unlimited and they can recover all the input values from output values of all the computationally secure cryptographic tools. Therefore, information-theoretic confidentiality is needed to protect secrets after the protocol execution, and computational integrity protection is sufficient to prevent active attacks by A and C during the protocol execution. Additionally, the man-in-the-middle attack by A must be prevented during the protocol execution. We use one-time pad and then message authentication code (MAC), to guarantee information-theoretic confidentiality, and computational integrity protection, and also to prevent the man-in-the-middle attack. The summary of the observations in this section is shown at Table 1.

3.2 Security Definition

Based on the argument above and the existing behavior model of parties in the everlasting security setting [23], we formalize the security notion for our oblivious transfer protocol (see Sect. 4 for some terminology).

First of all, we introduce an execution and communication model of our oblivious transfer protocol. In our model, the protocol consists of two phases called *the key sharing phase* and *the transfer phase*. In the key sharing phase, the client C and the server S exchange and share the keys and public parameters of the protocol, and the adversary A does nothing. In the transfer phase, C and S jointly execute the protocol, and without loss of generality, we suppose that each communication from C or S is always intercepted by A and then resent (with or without modification) to the other party. Based on our observation in the previous section, we model C and A to be computationally bounded malicious parties during the protocol execution and computationally unbounded semi-honest parties afterwards, and S to be a computationally bounded semi-honest party during the protocol execution and a computationally unbounded semi-honest party afterwards. We note that any coalition of C or S with the A does not cause security issues. Since C and S already have more information than A and have the same computational power as A, they get no advantages by colluding with A. The reason of introducing A is to capture the differences between everlasting security and computational security, especially the behaviors of unbounded parties after the protocol execution. Note that our security formalization, we define the random transcripts as random sampled queries and replies from the query space and the reply space, and distinguish the legal query and the illegal query. Since they depend on the description of the protocol, their definitions are described in our proposal at Sect. 5. Now, we define a notion of everlasting security for the database possessed by S in our model:

Definition 1 (Everlasting Security for Database). *For an oblivious transfer protocol in our model, we say that the protocol provides* everlasting security for the database, *if the following conditions are satisfied:*

- *For any adversary A, if the client C makes a legal query and the modification of the communication between C and the server S in the transfer phase is performed by A using a probabilistic polynomial-time algorithm, then all the transcripts obtained by A in the transfer phase are statistically indistinguishable from random transcripts.*
- *When C makes a legal query for an index I, the transcripts generated by S, except those for plaintext X_I associated to the content of the database at the index I, are statistically indistinguishable for C from random transcripts.*
- *When C makes an illegal query by using a probabilistic polynomial-time algorithm, either the protocol aborts before replying to C, or the transcripts generated by S are statistically indistinguishable from random transcripts.*

– *When C makes a legal query, for S, the transcripts generated by C are computationally indistinguishable from random transcripts.*

4   Preliminaries for Protocol Construction

In this section, we introduce notations, security notions, and cryptographic primitives.

Mathematical Notation. We denote a security parameter by $\kappa \in \mathbb{N}$, and we say that $\mathsf{negl} : \mathbb{N} \to \mathbb{R}$ is a negligible function if for all polynomial function f, there exists $n \in \mathbb{N}$, such that for all integers $n' > n$, $\mathsf{negl}(n') < 1/f(n')$ holds. Let $n \in \mathbb{N}$, then the set of integers from 0 to $n - 1$ is denoted by $[n] := \{0, 1, \ldots, n - 1\}$, and an n-element set is denoted by $\{a_i\}_{i \in [n]} := \{a_0, a_1, \ldots, a_{n-1}\}$. A n-component vector is denoted by $(a_i)_{i \in [n]} := (a_0, a_1, \ldots, a_{n-1})$. We also denote a vector by a bold symbol as $\boldsymbol{a}$ if the length and the components of $\boldsymbol{a}$ can be ignored. We denote an operation, assigning the value of a formula b to a variable or variables in a formula a, by $a \leftarrow b$ (e.g., let $\boldsymbol{v} = (0, 1, 2)$ be a vector, then $(a, b, c) \leftarrow \boldsymbol{v}$ means $a = 0$, $b = 1$, and $c = 2$ after this operation), and we also denote a sequence of n operations $(a_i)_{i \in [n]} \leftarrow (f(b_i))_{i \in [n]}$ by $(a_i \leftarrow f(b_i))_{i \in [n]}$ in short. Let n, m be two positive integers and let $\{x_i\}_{i \in [nm]}$ be an nm-element set. An $n \times m$ matrix, which has an $(i + 1)$th row and $(j + 1)$th column component x_{im+j} for $0 \le i < n$ and $0 \le j < m$, is denoted by $\mathrm{Mat}(x_{im+j})_{i \in [n], j \in [m]}$.

Parties. In our proposed protocols, there are three parties; a client, a server, and an adversary, and these are denoted by $\mathcal{C}$, $\mathcal{S}$, $\mathcal{A}$, respectively. The server $\mathcal{S}$ has a N-entry database $\mathrm{DB} = (X_i)_{i \in [N]}$. Since the number of allele of the human genome is about 3.2 billion, $N = 2^{32}$ (about 4.3 billion) is supposed. The client $\mathcal{C}$ wants to know Ith entry record X_I of the database. The adversary $\mathcal{A}$ will eavesdrop, forge, and manipulate a communication between $\mathcal{S}$ and $\mathcal{C}$ to get the information of the database maliciously. As mentioned in Sect. 3.2, we model behaviors of $\mathcal{C}$ and $\mathcal{A}$ to be computationally bounded malicious parties during the protocol execution and computationally unbounded semi-honest parties afterwards, and $\mathcal{S}$ to be a computationally bounded semi-honest party during the protocol execution and a computationally unbounded semi-honest party afterwards.

Indistinguishability. Let Ω be a sample space, and let $(X_i)_{i \in \mathbb{N}}$, $(Y_i)_{i \in \mathbb{N}}$ be two sequences of random variables. We say that $(X_i)_{i \in \mathbb{N}}$ and $(Y_i)_{i \in \mathbb{N}}$ are *statistically indistinguishable* if $\frac{1}{2} \sum_{\omega \in \Omega} \left| \Pr[X_i = \omega] - \Pr[Y_i = \omega] \right| < \mathsf{negl}(i)$ is held for a negligible function negl and for all $i \in \mathbb{N}$. We say that $(X_i)_{i \in \mathbb{N}}$ and $(Y_i)_{i \in \mathbb{N}}$ are *computationally indistinguishable* if for any probabilistic polynomial-time algorithm D, there exists a negligible function $\mathsf{negl}(i)$ such that, we have $\left| \Pr[D(i, X_i) = 1] - \Pr[D(i, Y_i) = 1] \right| < \mathsf{negl}(i)$ for all $i \in \mathbb{N}$. Note that if two sequences are statistically indistinguishable, then an unbounded adversary (in other words, an adversary having infinite computational power) cannot distinguish them. Therefore, one can prove the information-theoretic security by proving the statistical indistinguishability between a ciphertext of a secret and a random sequence.

Random Number Generator. In this paper, we discriminate true and pseudo random number generators because generated sequences have different entropy. Let A be a finite set, then let $\mathsf{PRG}(A)$ be a pseudo random number generator whose output is distributed on A, and let $\mathsf{TRG}(A)$ be a true random number generator whose output is distributed on A uniformly. Note that PRG generates a sequence of values from a short random sequence, and generated sequence is computationally indistinguishable with a uniformly distributed random sequence.

Message Authentication Code. Our proposed protocol uses a *message authentication code (MAC)* to guarantee the integrity of transcripts between two parties who have a pre-shared secret key.

Definition 2 (Message Authentication Code). *The message authentication code is defined by the following two functions* $(\mathsf{MAC}, \mathsf{MACVer})$:

MAC: *This function takes as input a secret key k and a plaintext $m \in \{0,1\}^*$, and output an authentication code t.*
MACVer: *This function takes as input a secret key k, a plaintext $m \in \{0,1\}^*$, and an authentication code t, then it outputs 1 if t is a valid authentication code of k and m, 0 otherwise.*

A notion of security called existential unforgeability against adaptive chosen message attack (EUF-ACMA) is introduced in [13,19], and this security notion achieves the strongest security of MACs. We also define the advantage of an adversary $\mathcal{A}$ against the EUF-ACMA of a MAC scheme on a given security parameter κ by the success probability of breaking EUF-ACMA security and it is denoted by $\mathsf{Adv}_{\mathcal{A}}^{\mathrm{MAC}}(\kappa)$. Then we say that a MAC scheme is EUF-ACMA secure if $\mathsf{Adv}_{\mathcal{A}}^{\mathrm{MAC}}(\kappa) \leq \mathsf{negl}(\kappa)$ for any κ, where negl is a negligible function depending on the $\mathcal{A}$. An EUF-ACMA secure MAC scheme is proposed [13,19].

Elliptic-Curve Lifted-ElGamal Encryption. Our proposed protocols need additively homomorphic encryption. We use the elliptic-curve lifted-ElGamal encryption (ECLEEnc) [8,10]. The ECLEEnc achieves a notion of security called indistinguishability against chosen plaintext attack (IND-CPA) under the decisional Diffie–Hellman (DDH) assumption. The IND-CPA guarantees computational indistinguishability between a ciphertext of a plaintext and a ciphertext of a random value, and it means no information is leaked from any ciphertexts.

Definition 3 (Elliptic-Curve Lifted-ElGamal Encryption). *The elliptic-curve lifted-ElGamal encryption is defined by the following three functions* $(\mathsf{KeyGen}, \mathsf{Enc}, \mathsf{Dec})$:

KeyGen: *This function takes as input a security parameter κ, and outputs a pair of secret key and public key* (sk, pk). *The public key and secret key are defined as* pk $= (\mathbb{G}, g, h)$ *and* sk $= (x)$, *respectively, where $\mathbb{G}$ is a cyclic group of prime order r, g is a generator of $\mathbb{G}$, x is an element of $\mathbb{Z}_r$, and $h \in \mathbb{G}$ is defined as $h = g^x$.*

Enc: *This function takes as input a public key* pk, *a plaintext $m \in \mathbb{Z}_r$, and a random value $s \in \mathbb{Z}_r$, and outputs* $\mathsf{Enc}(\mathrm{pk}, m; s) = (g^s, g^m h^s)$. *We call the first and the second component of a ciphertext $(g^s, g^m h^s)$ the random term and the plaintext term, respectively. The ciphertext space C is $\mathbb{G} \times \mathbb{G}$, and its length is denoted by ℓ_C, and the message space is denoted by $\mathsf{M} \subseteq \mathbb{Z}_r$. The length of M is denoted by ℓ_M, and ℓ_M is usually quite small for κ ($\ell_M = O(\log \kappa)$).*

Dec: *This function takes as input a public key* pk, *a secret key* sk, *and a ciphertext $c = (g_0, g_1)$, then it computes the discrete logarithm $\log_g(g_1/g_0^x)$. If $c = (g^s, g^m h^s)$, where $s, m \in \mathbb{Z}_r$, and the computation of discrete logarithm is succeeded, then* $\mathsf{Dec}(\mathrm{sk}, \mathrm{pk}, c) = m$. *Otherwise,* $\mathsf{Dec}(\mathrm{sk}, \mathrm{pk}, c) = \perp$.

In this paper, we assume (by assuming the DDH assumption) that a ciphertext generated by an ECLEEnc scheme is computationally indistinguishable from a ciphertext of a random value.

In general, the computation of discrete logarithm is hard. This is reason of why the length of the plaintext space of ECLEEnc should be sufficiently small. Note that the ECLEEnc scheme enables us to easily decide whether $c \in \mathsf{C}$ for checking the validity of ciphertexts.

We denote the homomorphic addition of two ciphertexts by $(g^{s_0}, g^{m_0} h^{s_0}) \cdot (g^{s_1}, g^{m_1} h^{s_1}) := (g^{s_0+s_1}, g^{m_0+m_1} h^{s_0+s_1})$. Let $\ell \in \mathbb{N}$, then the inner product of a vector of ciphertexts $\boldsymbol{c} = (c_i)_{i \in [\ell]}$ and a vector of plaintexts $\boldsymbol{m} = (m_i)_{i \in [\ell]}$ is denoted by $\boldsymbol{c} \cdot \boldsymbol{m}^\mathsf{T} := \prod_{i \in [\ell]} c_i^{m_i}$, and we define $\boldsymbol{m} \cdot \boldsymbol{c}^\mathsf{T} := \boldsymbol{c} \cdot \boldsymbol{m}^\mathsf{T}$.

In order to regard a ciphertext of an ECLEEnc scheme as a sequence of plaintexts for the same scheme, in this paper we suppose that we have a one-to-one and efficiently computable encoding function $\phi_{\mathsf{C},\mathsf{M}} : \mathsf{C} \to \mathsf{M}^{\ell_D}$, where ℓ_D is the smallest positive integer such that $\ell_C \leq \ell_D \ell_M$.

Computationally-Private Information Retrieval. A computationally-private information retrieval (cPIR) [15,21] used in this paper is the following two-party protocol utilizing the additively homomorphic encryption: (1) A client $\mathcal{C}$ generates his key pair of the additively homomorphic encryption, and then publishes its public key. Then, $\mathcal{C}$ generates encrypted query Q of an index I by using the key pair, and then $\mathcal{C}$ sends Q to a server $\mathcal{S}$. (2) $\mathcal{S}$ generates encrypted reply R of received Q and a database by using $\mathcal{C}$'s public key, and then $\mathcal{S}$ sends R to $\mathcal{C}$. (3) $\mathcal{C}$ decrypts R, and then $\mathcal{C}$ obtain the Ith entry X_I of the database. The cPIR enables $\mathcal{C}$ to see the database without revealing $\mathcal{C}$'s query.

Let ℓ_V and d be two positive integers such that $N \leq \ell_V^d$, where N is the number of records in a database. In a variant of cPIR proposed in [21], the encrypted query Q consists of d vectors, and each vector consists of ℓ_V ciphertexts of an additively homomorphic encryption scheme. The encrypted reply R also

---

Algorithm 1. cPIRQuery: Encrypted query generation of cPIR

Input: A public key pk, the number of components ℓ_V of each vector, the number of vectors d, and an index I.

Output: An encrypted query Q of I.

1: **for** $i = 0$ to $d - 1$ **do**
2:    $q_i \leftarrow \lfloor I/\ell_V \rfloor$; $r_i \leftarrow I \bmod \ell_V$; $I \leftarrow q_i$.
3:    $c_i \leftarrow \left(\mathsf{Enc}\left(\mathrm{pk}, b_j; \mathsf{PRG}(\mathbb{Z}_r)\right) \right)_{j \in [\ell_V]}$, where $b_{r_i} = 1$ and $b_j = 0$ for $j \neq r_i$.
4: **end for**
5: **return** $Q \leftarrow \left((c_i)_{i \in [d]} \right)$.

Algorithm 2. cPIRReply: Encrypted reply generation in cPIR

Input: A public key pk, an encrypted query $Q = (c_i)_{i \in [d]}$, and elements $(X_i)_{i \in [N]}$ of the database.

Output: An encrypted reply R.

1: $M_1 \leftarrow \mathrm{Mat}(X_{i\ell_V + j})_{i \in [\ell_V^{d-1}], j \in [\ell_V]}$; $(c_{1,i})_{i \in [\ell_V^{d-1}]} \leftarrow M_1 \cdot c_0^\mathsf{T}$.

2: $c_1' \leftarrow \left(c_{1,i} \cdot \mathsf{Enc}\left(\mathrm{pk}, 0; \mathsf{TRG}(\mathbb{Z}_r)\right) \right)_{i \in [\ell_V^{d-1}]}$.

3: **for** $k = 2$ to d **do**

4:    $(c_{k-1,i})_{i \in [\ell_V^{d-k+1}]} \leftarrow c_{k-1}'$; $\left((m_{k-1,i,j})_{j \in [\ell_D]} \right) \leftarrow \phi_{\mathsf{C,M}}(c_{k-1,i}))_{i \in [\ell_V^{d-k+1}]}$.

5:    $M_k \leftarrow \mathrm{Mat}(m_{k-1, r_k \ell_V + j, q_k})_{i \in [\ell_D^{k-1} \ell_V^{d-k}], j \in [\ell_V]}$, where $q_k = \lfloor i/\ell_V^{d-k} \rfloor$, $r_k = i \bmod \ell_V^{d-k}$.

6:    $(c_{k,i})_{i \in [\ell_D^{k-1} \ell_V^{d-k}]} \leftarrow M_k \cdot c_{k-1}^\mathsf{T}$.

7:    $c_k' \leftarrow \left((c_{k,i}) \cdot \mathsf{Enc}\left(\mathrm{pk}, 0; \mathsf{TRG}(\mathbb{Z}_r)\right) \right)_{i \in [\ell_D^{k-1} \ell_V^{d-k}]}$.   (Re-randomization by TRG)

8: **end for**
9: **return** $R \leftarrow (c_d')$.

---

consists of ciphertexts which contain encryption of X_I. Typically, cPIR does not care the confidentiality of the database. Note that the variant of cPIR in [21] can be executed as oblivious transfer, namely hiding records of the database except for X_I. Hereinafter, suppose that $N = \ell_V^d$ for the notational convenience. Algorithms of generating encrypted query cPIRQuery, encrypted reply cPIRReply, and decryption cPIRDec of the variant of cPIR are described at Algorithm 1, 2, and 3, respectively.

Special Honest Verifier Zero-Knowledge Unit Vector Proof of Knowledge. In this paper, we utilize a protocol of special honest verifier zero-knowledge proof of knowledge (SHVZKPoK) for unit vector defined over the ECLEEnc, which is proposed in [24]. We briefly introduce the procedure and security property of this protocol, we refer the reader to [24] for the details.

Let $n \in \mathbb{N}$, a statement x of the protocol is a vector of n ciphertexts $\left(\mathsf{Enc}(\mathrm{pk}, b_i; r_i) \right)_{i \in [n]}$ that its each component is the ciphertext of each component of a unit vector, i.e., $\left(\bigwedge_{i \in [n]} (b_i = 0 \vee b_i = 1) \right) \wedge \left(1 = \sum_{i \in [n]} b_i \right)$, and a witness w of the x is i such that $b_i = 1$. The protocol proposed in [24] is the

Algorithm 3. cPIRDec: Decryption of cPIR

Input: A secret key sk, a public key pk, positive integers ℓ_D and d, and an encrypted
    reply $R = (c_{1,i})_{i \in [\ell_D^{d-1}]}$.

Output: A record X.

1: **for** $k = 1$ **to** $d - 1$ **do**

2:    $\left(c_{k+1,i} \leftarrow \phi_{\mathsf{C,M}}^{-1} \left(\left(\mathsf{Dec}(\mathrm{sk}, \mathrm{pk}, c_{k,i+j\ell_D^{d-k-1}}) \right)_{j \in [\ell_D]} \right) \right)_{i \in [\ell_D^{d-k-1}]}$.

3: **end for**

4: **return** $X \leftarrow \mathsf{Dec}(\mathrm{sk}, \mathrm{pk}, c_{d,0})$.

---

following two-party protocol between a probabilistic polynomial-time prover $\mathcal{P}$
and a probabilistic polynomial-time honest verifier $\mathcal{V}$. $\mathcal{P}$ takes as input a witness
w, $\mathcal{P}$ and $\mathcal{V}$ take as common input a statement x for the w, and then: (1) $\mathcal{P}$
generates a value a (also called a commitment), and then $\mathcal{P}$ sends a to $\mathcal{V}$. (2) $\mathcal{V}$
picks a random value e (also called a challenge), and then $\mathcal{V}$ sends e to $\mathcal{P}$. (3) $\mathcal{P}$
generates z, then $\mathcal{P}$ sends z to $\mathcal{V}$, and then $\mathcal{V}$ output 1 or 0.

Note that the protocol holds the following security properties:

Completeness: When $(x, w) \in R$, and $\mathcal{P}$ and $\mathcal{V}$ follow the protocol correctly,
    then $\mathcal{V}$ always outputs 1, and we call this situation $\mathcal{V}$ is convinced by $\mathcal{P}$.

Soundness: For all witness w, when $(x, w) \notin R$, the probability of $\mathcal{V}$ outputs 1
    is negligible.

Zero-Knowledge: There exists a polynomial time algorithm (also called sim-
    ulator) which takes as input a statement x and a challenge e, and output
    (a, e, z) such that its distribution is identical to a transcript of the protocol
    between $\mathcal{P}$ and $\mathcal{V}$ when $\mathcal{V}$ outputs 1.

More precisely, in our proposed protocol below, we use an enhanced version of SHVZKPoK, where each of the transcripts is protected by using one-
time pad and then authenticated by using MAC. As a result, though the
original SHVZKPoK has only computational zero-knowledge property, any
third-party adversary $\mathcal{A}$ can learn no information from the transcripts of
the enhanced SHVZKPoK even if $\mathcal{A}$ has unbounded computational power.
We will denote the verification result of the enhanced SHVZKPoK by
$\mathsf{UVProof}(\mathrm{pk}, k_{\mathsf{MAC}}, \mathsf{UVProofKeys}, c) \in \{0, 1\}$, where c is a vector of ciphertexts
corresponding to a public key pk, k_{MAC} is a MAC key and UVProofKeys is one-
time-pad keys for the transcript (and note that UVProof stands for "unit vector
proof").

5   Proposed Oblivious Transfer Protocol

5.1   Overview of the Construction

In the construction of our proposal, to achieve the security in Definition 1, we
use the one-time pad together with a true random number generator TRG and a

MAC. Owing to this, we assume that the client $\mathcal{C}$ and the server $\mathcal{S}$ can share the keys for one-time pad via an authenticated secure communication channel in the proposed protocol. We note that the one-time pad seems to be two-party secret sharing to guarantee information-theoretic security, since transcripts seems to be true random values without pre-shared one-time-pad keys.

Our proposed protocol uses the ECLEEnc scheme to computationally protect the client $\mathcal{C}$'s query against the server $\mathcal{S}$, use MACs to prevent forgery of communication and the man-in-the-middle attacks by the adversary $\mathcal{A}$. A briefly saying, the query and its reply of the proposed protocol is one-time padded query and reply of cPIR (explained in Sect. 4), in order to reduce the length of required random numbers generated by TRG. It means that the legal query of our protocol is defined as vectors of ciphertexts, and each vector is an encrypted unit vector by ECLEEnc and one-time pad. More precise definition is explained next. In our proposal, $\mathcal{S}$ uses the SHVZKPoK for unit vector implemented on the ECLEEnc scheme [24] with the pre-shared keys in order to verify that a received $\mathcal{C}$'s query.

In fact, it is practically not easy to generate true random numbers and to share the keys for the one-time pad via secure channels. Due to this, in this paper, we regard the sizes of these true random numbers and the one-time-pad keys as the execution costs of our proposed protocol.

5.2   Construction

Our proposed protocol consists of the following three steps; Setup, ShareKey, and Transfer. The Setup, ShareKey steps correspond to the key sharing phase, and the Transfer step corresponds to the transfer phase, in the sense of the definition of our model explained in Sect. 3.2. The main purpose of Setup step is the exchange and sharing the keys and public parameters of computationally secure cryptographic primitives, and their keys are reusable in the multiple invocation of proposed protocol. On the other hand, one-time-pad keys are shared in the ShareKey step, they cannot be reused. Thus, every invocation of our protocol, it always needs an independent execution of the ShareKey step.

The procedures of Setup, ShareKey, and Transfer steps of our proposal are defined as follows:

Setup *Step.* (1) The client $\mathcal{C}$ generates a secret key and public key pair (sk, pk) by executing KeyGen(κ), and publishes pk. (2) $\mathcal{C}$ and the server $\mathcal{S}$ share a secret key $k_{\mathsf{MAC}} \in \{0,1\}^{\ell_{\mathsf{MAC}}}$ for the MAC. (3) $\mathcal{C}$ and $\mathcal{S}$ jointly determine the length ℓ_{V} of unit vectors used for queries in cPIR and the number d of these vectors.

ShareKey *Step.* $\mathcal{C}$ and $\mathcal{S}$ share, via an information-theoretically secure authenticated secure channel, $d\ell_{\mathsf{V}}$ elements $\left((k_{i,j})_{j\in[\ell_{\mathsf{V}}]}\right)_{i\in[d]}$ of $\mathbb{Z}_r$ and $k' \in \{0,1\}^{\ell_{\mathsf{M}}}$,

and $(\mathsf{UVProofKeys}_\ell)_{\ell\in[d]} := \left(\left(k_{\ell,i}^\rho, \left(k_{\ell,i,j}^z, (k_{\ell,i,t}^{(j)})_{t\in\{1,2\}}\right)_{j\in[2]}\right)_{i\in[\ell_{\mathsf{V}}]}, k_\ell^\rho, k_\ell^Z \right)_{\ell\in[d]}$

as one-time-pad keys, and, for all $\ell \in [d]$, $\mathsf{UVProofKeys}_\ell$ consists of $7\ell_{\mathsf{V}} + 2$ elements of $\mathbb{Z}_r$. As mentioned above, the keys shared in this step cannot be reused.

Transfer *Step*. (1) $\mathcal{C}$ generates an encrypted query Q by executing Algorithm 5, and sends Q to $\mathcal{S}$. (2) After receiving Q from $\mathcal{C}$, $\mathcal{S}$ generates an encrypted reply R by executing Algorithm 6, and sends R to $\mathcal{C}$. Here, the UVProof in the line 3 means an execution of Algorithm 4, which is an SHVZKPoK for a unit vector [24] combined with the one-time pad and the MAC, with $\mathcal{C}$ as the prover $\mathcal{P}$ and $\mathcal{S}$ as the verifier $\mathcal{V}$. (3) After receiving $R = (R_{\mathsf{cPIR}}, \delta)$ from $\mathcal{S}$, $\mathcal{C}$ first verifies the validity of the identification entity δ. If it is valid, then $\mathcal{C}$ decrypts R by computing $\mathsf{cPIRDec}(\mathsf{sk}, \mathsf{pk}, \ell_D, d, R_{\mathsf{cPIR}}) \oplus k'$ and obtains X as the result. Otherwise, $\mathcal{C}$ outputs the abort symbol $\perp$.

Now we define the legal and the illegal query precisely. The legal query consists of vectors $\left((c_{i,j})_{j \in [\ell_\mathsf{V}]}\right)_{i \in [d]}$ and an authentication code of these vectors satisfying the following statements, for all $i \in [d]$:

$$\left(\bigwedge_{j \in [\ell_\mathsf{V}]} \left(b_{i,j} \in \{0,1\} \wedge m_{i,j} = k_{i,j} + b_{i,j} \wedge r_{i,j} \in \mathbb{Z}_r \wedge c_{i,j} = \mathsf{Enc}(\mathsf{pk}, m_{i,j}; r_{i,j})\right)\right)$$

$$\wedge \sum_{j \in [\ell_\mathsf{V}]} b_{i,j} = 1, \quad (1)$$

where pk is the public key of ECLEEnc, and the values $k_{i,j}$ for all i, j are the pre-shared one-time-pad keys. The illegal query does not hold Equation (1). In our protocol, $\mathcal{S}$ uses the MAC scheme and the SHVZKPoK for unit vector to verify that a received query is legal.

Note that if $\mathcal{C}$ or $\mathcal{S}$ outputs $\perp$, then they halt the protocol immediately, and also note that the one-time-pad keys are used only one time, i.e., these are thrown away regardless of success or abort.

5.3   Security Analysis

Theorem 4. *Assume that an EUF-ACMA MAC exists, the DDH assumption holds, and the algorithm* ShareKey *can be correctly executed. Then the oblivious transfer protocol constructed in Sect. 5.2 provides everlasting security for the database.*

Proof. First of all, we consider the case that, either the client $\mathcal{C}$ is malicious and makes an illegal query, or ($\mathcal{C}$ is semi-honest and) the adversary $\mathcal{A}$ modifies a communication between $\mathcal{C}$ and the server $\mathcal{S}$ in the transfer phase. By the EUF-ACMA property of the MAC and the soundness of the UVProof, the protocol aborts by detecting the attack (hence the malicious party obtains no information) except for a negligible probability. This implies that, the conditional distribution of the transcripts during the protocol, conditioned on the case that the protocol does not abort, is statistically close to the distribution of the transcripts for the case without any illegal query by $\mathcal{C}$ and any modification of communication by $\mathcal{A}$. Therefore, it suffices to prove the conditions in Definition 1 for the case that $\mathcal{C}$ is semi-honest and $\mathcal{A}$ does not modify the communication.

For the security against the server $\mathcal{S}$, we note that any communication from the client $\mathcal{C}$ to $\mathcal{S}$ in the transfer phase is either a ciphertext by ECLEEnc

Algorithm 4. UVProof: Special honest verifier zero-knowledge unit vector proof [24] with the one-time pad and the MAC

Additional Keys: A MAC key k_{MAC}, and one-time-pad keys UVProofKeys $= \left(\left(k_i^\rho, (k_{i,j}^z, (k_{i,t}^{(j)})_{t \in \{1,2\}})_{j \in [2]} \right)_{i \in [n]}, k^\rho, k^Z \right)$.

Common Input: A public key pk.

Prover $\mathcal{P}$'s Private Input: $(b_i)_{i \in [n]}$, $(r_i)_{i \in [n]}$, where $b_i \in \{0,1\}$, $r_i \in \mathbb{Z}_r$ for all $i \in [n]$.

Statement: $\left((e_{i,1}, e_{i,2})\right)_{i \in [n]} = \left((g^{r_i}, g^{b_i} h^{r_i})\right)_{i \in [n]}$.

Step 1, $\mathcal{P}$ computes the followings:

1: **for** $i = 0$ **to** $n-1$ **do**

2:     $\overline{b}_i \leftarrow 1 - b_i$; $s_i, \rho_{i,\overline{b}_i}, z_{i,\overline{b}_i} \leftarrow \mathsf{PRG}(\mathbb{Z}_r)$.

3:     $a_{i,1}^{(b_i)} \leftarrow g^{s_i}$; $a_{i,2}^{(b_i)} \leftarrow h^{s_i}$.

4:     $a_{i,1}^{(\overline{b}_i)} \leftarrow g^{z_{i,\overline{b}_i}} e_{i,1}^{-\rho_{i,\overline{b}_i}}$; $a_{i,2}^{(\overline{b}_i)} \leftarrow h^{z_{i,\overline{b}_i}} \cdot (e_{i,2} \cdot g^{-\overline{b}_i})^{-\rho_{i,\overline{b}_i}}$.

5: **end for**

6: $(a_{i,k}^{(j)} \leftarrow a_{i,k}^{(j)} \cdot g^{k_{i,k}^{(j)}})_{i \in [n], j \in [2], k \in \{1,2\}}$; $s \leftarrow \mathsf{PRG}(\mathbb{Z}_r)$; $(A_1, A_2) \leftarrow (g^s, h^s)$; $m_a \leftarrow \left((a_{i,k}^{(j)})_{i \in [n], j \in [2], k \in \{1,2\}}, A_1, A_2\right)$; $\alpha \leftarrow \mathsf{MAC}(k_{\mathsf{MAC}}, m_a)$.

7: $\mathcal{P}$ sends m_a and α to $\mathcal{V}$.

Step 2, $\mathcal{V}$ computes the followings:

8: **if** $\mathsf{MACVer}(k_{\mathsf{MAC}}, m_a, \alpha) = 0$ **then return** $\perp$.

9: $\rho \leftarrow \mathsf{PRG}(\mathbb{Z}_r^*) + k^\rho$; $\beta \leftarrow \mathsf{MAC}(k_{\mathsf{MAC}}, \rho)$.

10: $\mathcal{V}$ sends ρ and β to $\mathcal{P}$.

Step 3, $\mathcal{P}$ computes the followings:

11: **if** $\mathsf{MACVer}(k_{\mathsf{MAC}}, \rho, \beta) = 0$ **then return** $\perp$.

12: **for** $i = 0$ **to** $n-1$ **do**

13:     $\rho_{i,b_i} \leftarrow \rho - \rho_{i,\overline{b}_i} - k^\rho$; $z_{i,b_i} \leftarrow r_i \rho_{i,b_i} + s_i$.

14: **end for**

15: $Z \leftarrow \rho \cdot \sum_{i \in [n]} r_i + s + k^Z$; $\left((\rho_{i,0}, (z_{i,j})_{j \in [2]}) \leftarrow (\rho_{i,0} + k_i^\rho, (z_{i,j} + k_{i,j}^z)_{j \in [2]})\right)_{i \in [n]}$; $m_z \leftarrow \left((\rho_{i,0}, z_{i,0}, z_{i,1})_{i \in [n]}, Z\right)$; $\gamma \leftarrow \mathsf{MAC}(k_{\mathsf{MAC}}, m_z)$.

16: $\mathcal{P}$ sends m_z and γ to $\mathcal{V}$.

Verification: $\mathcal{V}$ verifies as follows,

17: **return** 1 if the following statement is satisfied, 0 otherwise:

$$\left(\bigwedge_{i \in [n]} \left(\bigwedge_{t \in \{1,2\}} e_{i,t}^{\rho_{i,0} - k_i^\rho} a_{i,t}^{(0)} g^{-k_{i,t}^{(0)}} = g^{z_{i,0} - k_{i,0}^z}\right)\right.$$

$$\wedge \; e_{i,1}^{\rho_{i,1}} a_{i,1}^{(1)} g^{-k_{i,1}^{(1)}} = g^{z_{i,1} - k_{i,1}^z} \wedge (e_{i,2}/g)^{\rho_{i,1}} a_{i,2}^{(1)} g^{-k_{i,2}^{(1)}} = h^{z_{i,1} - k_{i,1}^z})$$

$$\wedge \; E_1^{\rho - k^\rho} A_1 = g^{Z - k^Z} \wedge (E_2/g)^{\rho - k^\rho} A_2 = h^{Z - k^Z}$$

$$\wedge \; \mathsf{MACVer}(k_{\mathsf{MAC}}, m_z, \gamma) = 1,$$

where $E_1 \leftarrow \prod_{i \in [n]} e_{i,1}$, $E_2 \leftarrow \prod_{i \in [n]} e_{i,2}$, and $\rho_{i,1} \leftarrow \rho - k^\rho - \rho_{i,0} - k_i^\rho$ for all $i \in [n]$.

Algorithm 5. Query generation in the cPIR-based method

Input: A public key pk, a MAC key k_{MAC}, the length of each encrypted unit vector ℓ_V, the number of encrypted unit vectors d in an encrypted query, one-time-pad keys $\big((k_{i,j})_{j\in[\ell_V]}\big)_{i\in[d]}$ for the encrypted query, an index I specified by the client $\mathcal{C}$.

Output: The encrypted query Q of I.

1: $\Big(\big((c_{i,j})_{j\in[\ell_V]}\big)_{i\in[d]}\Big) \leftarrow \mathsf{cPIRQuery}(\text{pk}, \ell_V, d, I)$.

2: $Q_{\mathsf{cPIR}} \leftarrow \Big(\big((c_{i,j} \cdot \mathsf{Enc}(\text{pk}, k_{i,j}; 0))_{j\in[\ell_V]}\big)_{i\in[d]}\Big)$.

3: **return** $Q \leftarrow \big(Q_{\mathsf{cPIR}}, \mathsf{MAC}(k_{\mathsf{MAC}}, Q_{\mathsf{cPIR}})\big)$.

Algorithm 6. Reply generation in the cPIR-based method

Input: A public key pk, a MAC key k_{MAC}, one-time-pad keys $(\mathsf{UVProofKeys}_\ell)_{\ell\in[d]}$, $\big((k_{i,j})_{j\in[\ell_V]}\big)_{i\in[d]}$, and k', elements of DB $(X_i)_{i\in[N]}$, and an encrypted query $Q = (Q_{\mathsf{cPIR}}, \sigma)$.

Output: An encrypted reply R or the abort symbol $\perp$.

1: **if** $\mathsf{MACVer}(k_{\mathsf{MAC}}, Q_{\mathsf{cPIR}}, \sigma) = 0$ **then return** $\perp$.

2: $\Big(\big((c'_{i,j})_{j\in[\ell_V]}\big)_{i\in[d]}\Big) \leftarrow Q_{\mathsf{cPIR}};\ (c_i)_{i\in[d]} \leftarrow \Big(\big(c'_{i,j} \cdot \mathsf{Enc}(\text{pk}, -k_{i,j}; 0)\big)_{j\in[\ell_V]}\Big)_{i\in[d]}$.

3: **if** $\bigvee_{i\in[d]}\big(\mathsf{UVProof}(\text{pk}, k_{\mathsf{MAC}}, \mathsf{UVProofKeys}_i, c_i) = 0\big)$ **then return** $\perp$.

4: $R_{\mathsf{cPIR}} \leftarrow \mathsf{cPIRReply}\Big(\text{pk}, \big((c_i)_{i\in[d]}\big), (k' \oplus X_i)_{i\in[N]}\Big)$.

5: **return** $R \leftarrow \big(R_{\mathsf{cPIR}}, \mathsf{MAC}(k_{\mathsf{MAC}}, R_{\mathsf{cPIR}})\big)$.

---

or a transcript of UVProof. Therefore, the (computational) security against $\mathcal{S}$ is implied by the security of ECLEEnc and the zero-knowledge property of UVProof.

On the other hand, the required security against the (semi-honest) client $\mathcal{C}$ is implied by the security property of the underlying cPIR protocol (see [15, 21] for the detail) as well as the fact that the re-randomization performed by the reply generation in the cPIR protocol uses a true random number (see Algorithm 2).

Finally, for the security against the adversary $\mathcal{A}$, note that, by the line 2 in Algorithm 5, the plaintext for a ciphertext $c_{i,j}$ (temporarily denoted by $m_{i,j}$) is replaced with $m_{i,j} + k_{i,j}$ by the homomorphic addition. On the other hand, by the line 4 in Algorithm 6, the target content X_I of the database is replaced with $X_I + k'$. Since $k_{i,j}$ and k' are not known by $\mathcal{A}$, the additions by $k_{i,j}$ and k' behave as one-time pad, therefore the resulting ciphertexts in the transcripts are statistically close to random ciphertexts from the view of $\mathcal{A}$ (note that, as mentioned in the previous paragraph, the information on the other contents of the database is already statistically removed from the reply to the client $\mathcal{C}$ due to the property of the underlying cPIR protocol). This completes the proof of Theorem 4.   □

5.4 Execution Cost

In the proposed protocol, the required one-time-pad keys for each protocol execution consist of $d(8\ell_V + 2)$ elements of $\mathbb{Z}_r$ and a binary sequence of length ℓ_M, therefore the total length is $\ell_r d(8\ell_V + 2) + \ell_M$ bit. On the other hand, when the server generates the encrypted reply, the re-randomization of $(\ell_V^d - \ell_D^d)/(\ell_V - \ell_D)$ ciphertexts is required in total. Therefore, the length of the required random numbers is $\ell_r(\ell_V^d - \ell_D^d)/(\ell_V - \ell_D)$ bit.

Now, we evaluate the actual length of one-time-pad keys. Fix the security parameter $\kappa = 128$, it does not imply huge key lengths, but sufficient for the current security level. Then the several parameters of ECLEEnc are determined as $\ell_r = 256$ and $\ell_C = 514$ (Y-coordinate compression is used). Then we define the small plaintext space $\mathsf{M} := [0, 2^{16} - 1]$, and then we have $\ell_M = 16$ and $\ell_D = \left\lceil \frac{\ell_C}{\ell_M} \right\rceil = 33$. Next, we set the parameter for cPIR as $d = 2$ in order to reduce the execution cost for the $\mathcal{S}$, then we have $\ell_V = 2^{16}$ since $N = \ell_V^d = 2^{32}$. Therefore the length of the one-time-pad keys is $\ell_r d(8\ell_V + 2) + \ell_M = 256 \times 2 \times (8 \times 2^{16} + 2) + 16 \approx 2.684 \times 10^8$ bit, and the length of the true random numbers is $\ell_r(\ell_V^d - \ell_D^d)/(\ell_V - \ell_D) = 256 \times (2^{16 \times 2} - 33^2)/(2^{16} - 33) \approx 1.68 \times 10^7$ bit. As a result, the lengths of one-time-pad keys shared by the client and the server and the true random numbers generated by the server to protect the database is significantly small in comparing to the number of allele of the human genome $\approx 3.2 \times 10^9$.

6 Discussion

Here we give some discussions on our results in this paper. First, our proposed oblivious transfer protocol uses internally some one-time pads in order to achieve stronger security than the usual computational security. One may then suspect that every task of secure communication, including oblivious transfer, becomes trivial by assuming the availability of pre-shared keys. However, it is known that pre-shared keys do *not* imply information-theoretically secure oblivious transfer [17,23]. Therefore, our task in this paper is never trivial even in the presence of pre-shared keys.

We also emphasize that, the size of pre-shared keys in our protocol is significantly reduced in comparison to a naive solution where every information appearing in an oblivious transfer protocol is protected by an *independent* one-time pad. For example, in our protocol, a single one-time-pad key is reused to encrypt *every* entry in a database. Intuitively, this was made possible by compressing the communicated information by using the functionality of additively homomorphic encryption. If we instead use somewhat or fully homomorphic encryption, it is expected that the richer functionality would enable further compression of the communicated information and therefore the size of pre-shared keys would be further reduced, while it enlarges the size of each ciphertext communicated in the protocol, and this concept is summarized as follows:

$$\mathsf{OneTimePad}\left(\mathsf{Compress}\left(\mathsf{Analyze}\left(\mathsf{Enc}(m_0),\mathsf{Enc}(m_1),\dots\right)\right),k\right) =$$

$$\mathsf{Enc}\left(\mathsf{OneTimePad}\left(\mathsf{Compress}\left(\mathsf{Analyze}(m_0,m_1,\dots)\right),k\right)\right). \quad (2)$$

An valuation of the trade-off between the cost in pre-shared keys and the communication complexity will be a future research topic.

The main concern for our proposed solution would be how to share the one-time-pad keys between the client and the server. For the problem, we note that the personal genome to be protected by the protocol is highly sensitive and somewhat professional objects, in comparison to more familiar data in ubiquitous encrypted communications on the Internet (e.g., an e-commerce). As a consequence, it seems reasonable to assume that our protocol will be for special purposes and mainly utilized between not fully trusted but somewhat authorized parties (such as a hospital and a distinguished researcher) rather than the anonymous users on the Internet. Owing to this, we expect that the initial cost of sharing the (shortened) keys (at the time of registration, for example) is still allowable in practical applications.

7 Conclusion

In this paper, we discussed the appropriate cryptographic security for protecting genome privacy. The conclusion is that the computational security of the standard public-key cryptosystems might not work. Indeed, genome protection requires long-term security due to its biological characteristics, which is far beyond the short-term (e.g., 20 years) security achievable by the current framework for selecting appropriate cryptographic key lengths. On the other hand, information-theoretic security is never compromised and therefore can provide long-term security, but the existing information-theoretically secure cryptographic protocols are known to be impractical [5].

To overcome the problem, we proposed a design principle to construct cryptographic protocols for genome privacy, where the most sensitive information such as genomic data are protected by information-theoretic security, while the other data such as the authentication information are protected by computational security in order to reduce the total execution cost. Technically, our protocol provides everlasting security, which provides an appropriate blend of long-term security and short-term security. Our proposed design principle was explained by focusing on a specific scenario described in Sect. 2, namely privacy-preserving data mining for the allele frequency database, observing the appropriate security level for each data type in the protocol, and then showing a concrete construction of an oblivious transfer protocol to provide the security (by combining well-known cryptographic primitives such as message authentication codes, elliptic-curve lifted-ElGamal encryption, one-time pad, computationally-private information retrieval, and zero-knowledge proofs).

Finally, we note that, our intention in this paper is not to claim that cryptography is the silver bullet to realize genome privacy protection. For example, in a situation of shopping on the Internet, the secure socket layer (SSL) and the transport layer security (TLS) need the public-key certification and its trust management framework. The discussion of the appropriate cryptographic security for protecting genome privacy in this paper focused on the theoretical aspect of cryptography. On the other hand, at the same time, we also addressed the value of reasonable adjustment of required security levels to overcome the impracticality of information-theoretic security. We believe that our discussion here is helpful for future constructions and analyses of practical systems for genome privacy protection.

Acknowledgements. The authors thank the members of Shin-Akarui-Angou-Benkyo-Kai for their precious comments, especially Yusuke Sakai for his valuable advice about the zero-knowledge proofs, Yohei Watanabe for discussion about the information-theoretically secure cryptographic protocols, and Jacob Schuldt for his valuable comments and helpful advice on our manuscript. The authors also thank Masao Nagasaki and Soichi Ogishima for valuable comments on our work. The authors had a fruitful discussion with Toshiaki Katayama and Kiyoshi Asai. The authors thank the anonymous reviewers of IWSEC 2015 for their valuable discussions and comments. This study was supported by the Japan-Finland Cooperative Scientific Research Program of Japan Science and Technology Agency (JST).

References

1. Agrawal, R., Srikant, R.: Privacy-preserving data mining. In: ACM SIGMOD, pp. 439–450 (2000)
2. Ayday, E., Cristofaro, E.D., Hubaux, J., Tsudik, G.: The chills and thrills of whole genome sequencing. Computer **99**(PrePrints), 1 (2013)
3. Ayday, E., Raisaro, J.L., Hengartner, U., Molyneaux, A., Hubaux, J.-P.: Privacy-preserving processing of raw genomic data. In: Garcia-Alfaro, J., Lioudakis, G., Cuppens-Boulahia, N., Foley, S., Fitzgerald, W.M. (eds.) DPM 2013 and SETOP 2013. LNCS, vol. 8247, pp. 133–147. Springer, Heidelberg (2014)
4. Baldi, P., Baronio, R., Cristofaro, E.D., Gasti, P., Tsudik, G.: Countering GAT-TACA: efficient and secure testing of fully-sequenced human genomes. In: ACM CCS, pp. 691–702 (2011)
5. Braun, J., Buchmann, J., Mullan, C., Wiesmaier, A.: Long term confidentiality: a survey. Des. Codes Cryptograph. **71**(3), 459–478 (2014)
6. Braun, R., Rowe, W., Schaefer, C., Zhang, J., Buetow, K.: Needles in the haystack: identifying individuals present in pooled genomic data. PLoS Genet. **5**(10), e1000668 (2009)
7. ECRYPT II: yearly report on algorithms and keysize (2011–2012), September 2012
8. ElGamal, T.: A public key cryptosystem and a signature scheme based on discrete logarithms. IEEE Trans. Inf. Theor. **31**(4), 469–472 (1985)
9. Grover, L.K.: A fast quantum mechanical algorithm for database search. In: STOC, pp. 212–219 (1996)
10. Hankerson, D., Menezes, A.J., Vanstone, S.: Guide to Elliptic Curve Cryptography. Springer-Verlag New York Inc, Secaucus (2004)

11. Harnik, D., Naor, M.: On everlasting security in the *hybrid* bounded storage model. In: Bugliesi, M., Preneel, B., Sassone, V., Wegener, I. (eds.) ICALP 2006. LNCS, vol. 4052, pp. 192–203. Springer, Heidelberg (2006)

12. Homer, N., Szelinger, S., Redman, M., Duggan, D., Tembe, W., Muehling, J., Pearson, J.V., Stephan, D.A., Nelson, S.F., Craig, D.W.: Resolving individuals contributing trace amounts of DNA to highly complex mixtures using high-density SNP genotyping microarrays. PLoS Genet. **4**(8), e1000167 (2008)

13. Iwata, T., Kurosawa, K.: OMAC: one-key CBC MAC. In: Johansson, T. (ed.) FSE 2003. LNCS, vol. 2887, pp. 129–153. Springer, Heidelberg (2003)

14. Jha, S., Kruger, L., Shmatikov, V.: Towards practical privacy for genomic computation. In: IEEE Symposium on Security and Privacy, pp. 216–230 (2008)

15. Kushilevitz, E., Ostrovsky, R.: Replication is NOT needed: SINGLE database, computationally-private information retrieval. In: FOCS, pp. 364–373 (1997)

16. Lenstra, A.K., Verheul, E.R.: Selecting cryptographic key sizes. J. Cryptol. **14**(4), 255–293 (2001)

17. Mayers, D.: Unconditionally secure quantum bit commitment is impossible. Phys. Rev. Lett. **78**, 3414–3417 (1997)

18. Naveed, M., Agrawal, S., Prabhakaran, M., Wang, X., Ayday, E., Hubaux, J., Gunter, C.A.: Controlled functional encryption. In: ACM CCS, pp. 1280–1291 (2014)

19. NIST: Special publication 800–38B, recommendation for block cipher modes of operation: The CMAC mode for authentication, May 2005

20. NIST: Special publication 800–57, recommendation for key management - part 1: General (revision 3), July 2012

21. Ostrovsky, R., Skeith III, W.E.: A survey of single-database private information retrieval: techniques and applications. In: Okamoto, T., Wang, X. (eds.) PKC 2007. LNCS, vol. 4450, pp. 393–411. Springer, Heidelberg (2007)

22. Shor, P.W.: Polynomial-time algorithms for prime factorization and discrete logarithms on a quantum computer. SIAM J. Comput. **26**(5), 1484–1509 (1997)

23. Unruh, D.: Everlasting multi-party computation. In: Canetti, R., Garay, J.A. (eds.) CRYPTO 2013, Part II. LNCS, vol. 8043, pp. 380–397. Springer, Heidelberg (2013)

24. Zhang, B., Lipmaa, H., Wang, C., Ren, K.: Practical fully simulatable oblivious transfer with sublinear communication. In: Sadeghi, A.-R. (ed.) FC 2013. LNCS, vol. 7859, pp. 78–95. Springer, Heidelberg (2013)

Anonymous Credential System with Efficient Proofs for Monotone Formulas on Attributes

Shahidatul Sadiah[1(✉)], Toru Nakanishi[1], and Nobuo Funabiki[2]

[1] Department of Information Engineering, Hiroshima University,
Higashi-hiroshima, Hiroshima, Japan
{D152447,t-nakanishi}@hiroshima-u.ac.jp
[2] Department of Communication Network Engineering,
Okayama University, Okayama, Japan
funabiki@okayama-u.ac.jp

Abstract. An anonymous credential system allows a user to convince a service provider anonymously that he/she owns certified attributes. Previously, a system to prove AND and OR relations simultaneously by CNF formulas was proposed. To achieve a constant-size proof of the formula, this system adopts an accumulator that compresses multiple attributes into a single value. However, this system has a problem: the proof generation requires a large computational time in case of lots of OR literals in the formula. One of the example formulas consists of lots of birthdate attributes to prove age. This greatly increases the public parameters correspondent to attributes, which causes a large delay in the accumulator computation due to multiplications of lots of parameters. In this paper, we propose an anonymous credential system with constant-size proofs for *monotone formulas* on attributes, in order to obtain more efficiency in the proof generation. The monotone formula is a logic formula that contains any combination of AND and OR relations. Our approach to prove the monotone formula is that the accumulator is extended to be adapted to the tree expressing the monotone formula. Since the use of monotone formulas increases the expression capability of the attribute proof, the number of public parameters multiplied in the accumulator is greatly decreased, which impacts the reduction of the proof generation time.

1 Introduction

1.1 Backgrounds

In Web services, user authentications are required to protect a malicious access. In conventional ID-based authentications, the privacy problem may occur, since Service Provider (SP) can trace the user's ID, grasps user's service history, and might use it to attempt malicious activities. On the other hand, from the SP's point of view, the authentication using the user's attributes such as a gender, an occupation, and an age is more advantageous for commercial values. Thus, an attribute-based authentication with a strong privacy protection is in demand,

© Springer International Publishing Switzerland 2015
K. Tanaka and Y. Suga (Eds.): IWSEC 2015, LNCS 9241, pp. 262–278, 2015.
DOI: 10.1007/978-3-319-22425-1_16

where users can disclose the minimal amount of personal information necessary for the service instead of his/her ID.

For the demand, in [5,7,9,11], *anonymous credential systems* were proposed, where a user can anonymously convince an SP about the possession of specified attributes. There exist three entities in the anonymous credential system: an issuer, users, and an SP. The issuer issues a certificate to the user in advance, where the certificate ensures the attributes of the user. Then, the user can prove the certified attributes to the SP. In the authentication, the SP requests the user to prove his/her certified attributes and their relation. For example, when accessing an alcohol-related company's website, most companies would ask for both nationality and birthdate attributes (to prove the age) during the authentication, since different countries have different legal drinking age. Thus, the user needs to prove the AND relation of his/her nationality and age. In general, complex relations on attributes can be expressed by logic formulas. The AND relation is used when proving the possession of all the multiple attributes. The OR relation represents the possession of one of the multiple attributes. In the authentication, a zero-knowledge type of proof allows the user to hide any other information beside the satisfaction of the relations.

1.2   Previous Works

In [5,11], anonymous credential systems were proposed, where the proof in the authentication has the constant size for the number of attributes. However, simple AND or OR relations on attributes are only available. In [9], a system with the constant size proofs is proposed, where the inner-product on attributes can be proved (Thus, CNF and DNF formulas are also available). However, in this system, the proof generation needs exponentiations depending on the number of literals in the OR relations, which causes a large delay in case of formulas with lots of OR literals.

In [3], an anonymous credential system with the constant size proofs was proposed, where a user can prove any CNF formulas on attributes. In this system, the proof generation is more efficient than [9], since it needs only multiplications depending the number of OR literals. However, this system still suffers from the inefficiency in case of much more OR literals, due to the less expression capability of CNF formulas. The typical example is to prove the age using birthdate attributes. To achieve the constant-size proof, this system utilizes an *accumulator* that compresses multiple attributes into a single value. In the compression, the multiplications are needed. Consider the above example in accessing an alcohol related website. An example of CNF formula is $(Japan \vee Iceland \vee \ldots) \wedge (1914, December31 \vee \cdots \vee 1994, February12)$, where each birthdate is encoded to one attribute value such as "1914, December31". The accumulator requires that all public parameters assigned to the attribute values of OR literals in the formula are multiplied. For the above example, the number of attributes for the birthdates between year 1914 to 1994, i.e., 80 years, is approximately $80 \times 365 = 29,200$. The multiplications causes a large delay in the authentication.

1.3   Our Contributions

In this paper, we propose an anonymous credential system with constant-size proofs for *monotone formulas* on attributes, in order to obtain more efficiency in the proof generation. The monotone formula is a logic formula that contains any combination of AND and OR relations. In the system, the birthdate attribute is composed of the year, the month, and the day, and one birthdate is expressed as $(year \land month \land day)$ in the logic formula. Thus, we replace the CNF formula with a monotone formula. For the above example, the monotone formula is ($Japan \lor Iceland \lor \ldots) \land (1914 \lor \ldots \lor (1994 \land (January \lor (February \land (1st \lor \ldots \lor 12th)))))$). Using this type of formula, the number of public parameters multiplied in the accumulator is decreased to approximately 100, i.e., the sum of the number of attributes for 80 years, 2 months, and 12 days, which greatly impacts the reduction of authentication time.

Our approach to prove the monotone formula is that the accumulator is extended to be adapted to the tree structure to express the monotone formula. In the tree, leaves indicate the attributes and internal nodes are the AND or OR relations. For instance, consider the following proved monotone formula:

$$((a_1 \land a_2) \lor a_3) \land (a_4 \lor a_5) \land a_6).$$

As the preparation of the accumulator, a tag assignment is executed as follows. At the root, a series of tags $c_1, \ldots, c_4$ are generated. Then, these tags are divided and assigned to the leaves. The same tags are distributed to the children on an OR relation, while different tags are distributed to the children on an AND relation. The tag assignment result for the above formula is

$$((a_1^{c_1} \land a_2^{c_2}) \lor a_3^{c_1,c_2}) \land (a_4^{c_3} \lor a_5^{c_3}) \land a_6^{c_4}),$$

where the superscript in each attribute means the assigned tags. In this assignment, the attributes of the user satisfy the formula if and only if the tags for the attributes are exactly the same as the initial tags. For example, when a user with satisfying attributes a_3, a_5, a_6, the assigned tags are $\{(c_1, c_2), c_3, c_4\}$, which compose the initial tags. In the verification of the accumulator, it is checked using a pairing relation, which is extended from that of [3].

2   Preliminaries

In this section, we show the cryptographic tool and proof system used as building blocks of our anonymous credential system.

2.1   Bilinear Maps

Our scheme utilizes bilinear groups and bilinear map.

1. $\mathbb{G}$ and $\mathbb{G}_T$ are multiplicative cyclic groups of prime order p.
2. g is a randomly chosen generator of $\mathbb{G}$.
3. e is a computable bilinear map, $e : \mathbb{G} \times \mathbb{G} \to \mathbb{G}_T$ with the following properties:
   - Bilinearity: for all $u, v \in \mathbb{G}$, and $a, b \in \mathbb{Z}, e(u^a, v^b) = e(u, v)^{ab}$.
   - Non-degeneracy: $e(g, g) \neq 1_{\mathbb{G}_T}$ where $1_{\mathbb{G}_T}$ is an identity element of $\mathbb{G}_T$.

2.2 Complexity Assumptions

As in the previous system [3], the security of our proposed system is based on DLIN (Decision LINear) assumption [4], the n-DHE (DH Exponent) assumption [6] and the q-SFP (Simultaneous Flexible Pairing) assumption [2].

Definition 1 (DLIN Assumption). For all PPT algorithm $\mathcal{A}$, the probability

$$Pr\left[\mathcal{A}(g, g^a, g^b, g^{ac}, g^{bd}, g^{c+d}) = 1\right] - Pr\left[\mathcal{A}(g, g^a, g^b, g^{ac}, g^{bd}, g^z) = 1\right]$$

is negligible, where $g \in_R \mathbb{G}$ and $a, b, c, d, z, \in_R \mathbb{Z}_p$.

Definition 2 (n-DHE Assumption). For all PPT algorithm $\mathcal{A}$, the probability

$$Pr\left[\mathcal{A}(g, g^a, \ldots, g^{a^n}, g^{a^{n+2}}, \ldots, g^{a^{2n}}) = g^{a^{n+1}}\right]$$

is negligible, where $g \in_R \mathbb{G}$ and $a \in_R \mathbb{Z}_p$.

Definition 3 (q-SFP Assumption). For all PPT algorithm $\mathcal{A}$, the probability

$$Pr\left[\begin{array}{l} \mathcal{A}\left(g_z, h_z, g_r, h_r, a, \tilde{a}, b, \tilde{b}, \{(z_j, r_j, s_j, t_j, u_j, v_j, w_j)\}_{j=1}^q\right) \\ = (z^*, r^*, s^*, t^*, u^*, v^*, w^*) \in \mathbb{G}_1{}^7 \wedge e(a, \tilde{a}) \\ = e(g_z, z^*)e(g_r, r^*)e(s^*, t^*) \wedge e(b, \tilde{b}) \\ = e(h_z, z^*)e(h_r, u^*)e(v^*, w^*) \wedge z^* \neq 1_{\mathbb{G}_1} \wedge z^* \neq z_j \text{ for all } 1 \leq j \leq q \end{array}\right]$$

is negligible, where $(g_z, h_z, g_r, h_r, a, \tilde{a}, b, \tilde{b}) \in \mathcal{G}^8$ and all tuples $\{(z_j, r_j, s_j, t_j, u_j, v_j, w_j)\}_{j=1}^q)$ satisfy the above relations.

2.3 AHO Structure-Preserving Signatures

In previous system, the AHO signature scheme [1,2] is utilized for the structure-preserving signatures, since the knowledge of the signature can be proved by Groth-Sahai proofs. Using the AHO scheme, multiple group elements are signed to obtain a constant-size signature. As in the previous construction, a single group element is signed, and thus we described in the case of single message to be signed.

AHOKeyGen: Select bilinear groups $\mathbb{G}, \mathbb{G}_T$ with a prime order p and a bilinear map e. Select $g, G_r, H_r \in \mathbb{G}$ and $\mu_z, \nu_z, \mu, \nu, \alpha_a, \alpha_b \in_R \mathbb{Z}_p$. Compute $G_z = G_r^{\mu_z}, H_z = H_r^{\nu_z}, G = G_r^{\mu}, H = H_r^{\nu}, A = e(G_r, g^{\alpha_a}), B = e(H_r, g^{\alpha_b})$. Output the public key as $pk = (\mathbb{G}, \mathbb{G}_T, p, e, g, G_r, H_r, G_z, H_z, G, H, A, B)$, and the secret key as $sk = (\alpha_a, \alpha_b, \mu_z, \nu_z, \mu, \nu)$.

AHOSign: Given message $M \in \mathbb{G}$ together with sk, choose $\beta, \epsilon, \eta, \iota, \kappa \in_R \mathbb{Z}_p$, and compute $\theta_1 = g^\beta$, and

$$\theta_2 = g^{\epsilon - \mu_z \beta} M^{-\mu}, \quad \theta_3 = G_r^\eta, \quad \theta_4 = g^{\alpha_a - \epsilon/\eta},$$

$$\theta_5 = g^{\iota - \nu_z \beta} M^{-\nu}, \quad \theta_6 = H_r^\kappa, \quad \theta_7 = g^{\alpha_b - \iota/\kappa}.$$

Output the signature $\sigma = (\theta_1, \ldots, \theta_7)$.

AHOVerify: Given the message M and the signature $\sigma = (\theta_1, \ldots, \theta_7)$, accept these if following equations are hold:

$$A = e(G_z, \theta_1) \cdot e(G_r, \theta_2) \cdot e(\theta_3, \theta_4) \cdot e(G, M),$$
$$B = e(H_z, \theta_1) \cdot e(H_r, \theta_5) \cdot e(\theta_6, \theta_7) \cdot e(H, M).$$

This signature is existentially unforgeably against chosen-message attacks under the q-SFP assumption [2].

Using the re-randomization algorithm, this signature can be publicly randomized to obtain another signature $(\theta_1', \ldots, \theta_7')$ on the same message. As a result, in the following Groth -Sahai proof, $(\theta_i')_{i=3,4,6,7}$ can be safely revealed, while $(\theta_i')_{i=1,2,5}$ have to be committed as mentioned in [10].

2.4   Groth-Sahai (GS) Proof

To prove the secret knowledge in relations of the bilinear maps, we utilize Groth-Sahai (GS) proofs [8]. As in [3], we adopt the instantiation based on DLIN assumption. For the bilinear groups, the proof system needs a common reference string $(\boldsymbol{f}_1, \boldsymbol{f}_2, \boldsymbol{f}_3) \in \mathbb{G}^3$ for $\boldsymbol{f}_1 = (f_1, 1, g)$, $\boldsymbol{f}_2 = (1, f_2, g)$ for some $f_1, f_2 \in \mathbb{G}$. The commitment to an element $X \in \mathbb{G}$ is computed as $\boldsymbol{C} = (1, 1, X) \cdot \boldsymbol{f}_1^r \cdot \boldsymbol{f}_2^s \cdot \boldsymbol{f}_3^t$ for $r, s, t \in_R \mathbb{Z}_p^*$. In case of the CRS setting for perfectly sound proofs, $\boldsymbol{f}_3 = \boldsymbol{f}_1^{\xi_1} \cdot \boldsymbol{f}_2^{\xi_2}$ for $\xi_1, \xi_2 \in_R \mathbb{Z}_p^*$. Then, the commitment $\boldsymbol{C} = (f_1^{r+\xi_1 t}, f_2^{s+\xi_2 t}, X g^{r+s+t(\xi_1+\xi_2)})$ is the linear encryption in [4]. On the other hand, in the setting of the witness indistinguishability, $\boldsymbol{f}_1, \boldsymbol{f}_2, \boldsymbol{f}_3$ are linealy independent, and thus $\boldsymbol{C}$ is perfectly hidden. The DLIN assumption implies the indistinguishability of the CRS. To prove that the committed variables in the pairing relations, the prover prepares the commitments, and replaces the variables in the pairing relations by the commitments. The GS Proof allows us to prove the set of pairing product equations:

$$\prod_{i=1}^n e(A_i, X_i) \cdot \prod_{i=1}^n \prod_{j=1}^n e(X_i, X_j)^{a_{ij}} = t$$

for variables $X_1, \ldots, X_n \in \mathbb{G}$ and constants $A_1, \ldots, A_n \in \mathbb{G}, a_{ij} \in \mathbb{Z}_p, t \in \mathbb{G}_T$.

3   Accumulator to Verify Monotone Formulas

Before constructing our improved anonymous credential system, we show an accumulator to verify monotone formulas as the key primitive. The construction of the accumulator is based on the previous work [3], but it newly needs a tag assignment algorithm where tags are assigned in leaf nodes (attributes) in the binary tree expression of the given monotone formula.

3.1   Tag Assignment Algorithm

In this tag assignment algorithm, the input is a monotone formula $\mathcal{M}$ on attributes. The formula $\mathcal{M}$ is represented by a binary tree, where any intermediate node is either AND or OR, and the leaf nodes are attributes. Each node is indexed by an integer i. The output of the algorithm is a non-negative integer T showing the maximum of tag indexes, and the sequence of tag indexes, $\mathcal{S}_i = \{\tilde{t}_i, \tilde{t}_i + 1, \ldots, \tilde{t}'_i\}$ to each attribute i in $\mathcal{M}$.

From the tree of given monotone formula $\mathcal{M}$, consider a *minimum satisfaction tree*, where, in any intermediate OR node, one child node is needed in minimum for satisfying the OR node remains, but another redundant child node (and the descendant subtree) is removed. Note that, in any intermediate AND node, both child nodes remain because the child nodes are needed for the satisfaction of $\mathcal{M}$. Hereafter, we define a predicate $\mathcal{MS}(\tilde{U}, \mathcal{M})$ on the subset of attributes $\tilde{U}$ and the monotone formula $\mathcal{M}$, where $\mathcal{MS}(\tilde{U}, \mathcal{M}) = 1$ if $\tilde{U}$ consists of attributes in the minimum satisfaction tree of $\mathcal{M}$, and otherwise $\mathcal{MS}(\tilde{U}, \mathcal{M}) = 0$. We call set $\tilde{U}$ s.t. $\mathcal{MS}(\tilde{U}, \mathcal{M}) = 1$ a *minimum attribute set* of $\mathcal{M}$.

The goal of this algorithm is that, for the minimum attribute set $\tilde{U}$ of $\mathcal{M}$, the set of $\mathcal{S}_i$ for all $i \in \tilde{U}$ is a partition of the initial $\mathcal{S}_{root}$, which is used for verifying the monotone formula in the accumulator.

In the algorithm, at first, set the initial set $\mathcal{S}_{root} = \{1, \ldots, T\}$ at the root node with $T = s_{root} + 1$, where s_{root} is the total number of AND nodes in the tree of the monotone formula. While traversing the tree from the root, for an AND node, the indexes $\mathcal{S}_i$ that are assigned to this node are separated and distributed to each child. For the distribution, we also need auxiliary sequences $\mathcal{A}_i$ which holds a separation points of the sequence. The assignment process is executed as follows:

Preparation: Traverse the tree to find parameter s_i for each node i showing the number of AND nodes in the subtree rooted by node i. Then, set $T = s_{root} + 1$, $\mathcal{S}_{root} = \{1, \ldots, T\}$, and $\mathcal{A}_{root} = \{1, \ldots, T - 1\}$ at the root node.

Assignment: Traverse the tree once again from root to the leaves with these following steps. Here, let i be the current node, let i_0 be the left child of node i, and let i_1 be the right child.

  (S1) If the current node i has no child, skip.

  (S2) If the current node has child nodes, check either the current node is AND or OR.

   (S2-1) If the current node is OR,

   (i) Divide $\mathcal{A}_i = \{t_i, t_i+1, \ldots, t'_i\}$ into $\mathcal{A}_{i_0} = \{t_i, t_i+1, \ldots, t_i+s_{i_0}-1\}$ for the left child i_0, and $\mathcal{A}_{i_1} = \{t_i + s_{i_0}, t_i + s_{i_0} + 1, \ldots, t'_i\}$ for the right child i_1, where $t'_i = t_i + s_{i_0} + s_{i_1} - 1$.

   (ii) Set the exact same $\mathcal{S}_i$ to the child nodes as $\mathcal{S}_{i_0}$ and $\mathcal{S}_{i_1}$.

   (S2-2) If the current node is AND,

   (i) Divide $\mathcal{A}_i = \{t_i, t_i+1, \ldots, t'_i\}$ into $\mathcal{A}_{i_0} = \{t_i, t_i+1, \ldots, t_i+s_{i_0}-1\}$ for the left child i_0, and $\mathcal{A}_{i_1} = \{t_i + s_{i_0} + 1, t_i + s_{i_0} + 2, \ldots, t'_i\}$ for the right child i_1, where $t'_i = t_i + s_{i_0} + s_{i_1}$. The index $t_i + s_{i_0}$ is consumed for the separation at the current AND node.

   (ii) Divide $\mathcal{S}_i$ into $\mathcal{S}_{i_0} = \{\tilde{t}_i, \tilde{t}_i + 1, \ldots, \tilde{t}_i + s_{i_0}\}$ for the left child i_0 and $\mathcal{S}_{i_1} = \{\tilde{t}_i + s_{i_0} + 1, \tilde{t}_i + s_{i_0} + 2, \ldots, \tilde{t}'_i\}$ for the right child i_1 where $\tilde{t}'_i = \tilde{t}_i + s_{i_0} + s_{i_1} + 1$.

  (S4) Move to the next traversed node and set it as the current node.

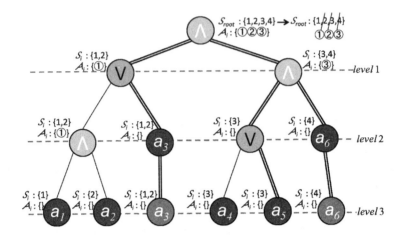

Fig. 1. Tag assignment in tree of monotone formula

We explain the algorithm using an example of monotone formula $((a_1 \wedge a_2) \vee a_3) \wedge (a_4 \vee a_5) \wedge a_6)$ whose tree is shown in Fig. 1. In Fig. 1, $\mathcal{S}_{root} = \{1, 2, 3, 4\}$ and $\mathcal{A}_{root} = \{①, ②, ③\}$). The dividing point t in $\mathcal{A}_i$ means that $\mathcal{S}_i = \{\ldots, t, \ldots\}$ is separable to $\mathcal{S}_{i_0} = \{\ldots, t\}$ and $\mathcal{S}_{i_1} = \{t+1, \ldots\}$. When traversing the AND node, one dividing point in $\mathcal{A}_i$ is consumed, and $\mathcal{S}_i$ is divided into the left and right child nodes. The remaining dividing points are also distributed to the child nodes such that the descendant AND nodes can consume the dividing points. In Fig. 1, the root node is an AND node and it consumes dividing point ②, thus $\mathcal{S}_{i_0} = \{1, 2\}$ and $\mathcal{S}_{i_1} = \{3, 4\}$ are assigned to left child i_0 and right child i_1, respectively. The dividing points $\mathcal{A}_i = \{①, ②, ③\}$ are distributed to $\mathcal{A}_{i_0} = \{①\}$ and $\mathcal{A}_{i_1} = \{③\}$, where ② is consumed in the root.

When the current node i is an OR node, none of the dividing point is consumed and thus both child nodes receive the same $\mathcal{S}_i$. The dividing points are distributed to the child nodes such that the descendant AND nodes can consume the dividing points. In the OR node of level 1 (Fig. 1), the dividing point ① is distributed to the left child with the AND descendant.

In Fig. 1, the subtree connected with double lines branch is the minimum satisfaction tree where $\tilde{U} = \{a_3, a_5, a_6\}$ s.t. $\mathcal{MS}(\tilde{U}, \mathcal{M}) = 1$. The set of $\mathcal{S}_i$ for all $i \in \tilde{U}$, i.e., $\{\{1, 2\}, \{3\}, \{4\}\}$ is a partition of $\mathcal{S}_{root} = \{1, 2, 3, 4\}$.

3.2 Correctness of Tag Assignment Algorithm

We have the following theorem of the correctness of the tag assignment algorithm.

Theorem 1. *For $\mathcal{S}_i$ and $\mathcal{S}_{root}$ outputted in the tag assignment algorithm on input $\mathcal{M}$, for any set of attributes $\tilde{U}$, if and only if $\mathcal{MS}(\tilde{U}, \mathcal{M}) = 1$, $\{\mathcal{S}_i | i \in \tilde{U}\}$ is a partition of $\mathcal{S}_{root}$, i.e., $\bigcup_{i \in \tilde{U}} \mathcal{S}_i = \mathcal{S}_{root}$ and $\mathcal{S}_i$'s are mutually disjoint.*

Although the proof will be shown in the full version, the proof sketch is shown here. At first, let's consider the proof that, if $\mathcal{MS}(\tilde{U}, \mathcal{M}) = 1$, $\{\mathcal{S}_i | i \in \tilde{U}\}$ is a partition of $\mathcal{S}_{root}$. In the algorithm, since each s_i is correctly computed, each AND node can always consume a dividing point in $\mathcal{A}_i$. Note that, at each level in the minimum satisfaction tree, the set of all $\mathcal{S}_i$ remains a partition of $\mathcal{S}_{root}$. This is because $\mathcal{S}_i$ in an AND node is partitioned to two children and $\mathcal{S}_i$ in an OR node is inherited to the child in the minimum satisfaction. As in Fig. 1, a leaf node at a level is replaced by an virtual intermediate node connected to the virtual leaf node at the bottom level via the virtual intermediate nodes. Then, at the bottom level, the set of all $\mathcal{S}_i$ remains a partition of $\mathcal{S}_{root}$.

Next, consider the reverse proof that, if $\mathcal{MS}(\tilde{U}, \mathcal{M}) \neq 1$, $\{\mathcal{S}_i | i \in \tilde{U}\}$ is not a partition of $\mathcal{S}_{root}$. Consider the subtree of $\mathcal{M}$ with only the paths from the root to all leafs in $\tilde{U}$. We assume that $\tilde{U}$ is not empty, since it means that $\{\mathcal{S}_i | i \in \tilde{U}\}$ is also empty. In this case, since the subtree is not the satisfaction tree, there is an AND node i^* with a child i_0 connected to a leaf in $\tilde{U}$ and the other child i_1 that is not connected to the leafs in $\tilde{U}$. In the AND node i^*, tag indexes are divided to $\mathcal{S}_{i_0} = \{t_1, \ldots, t\}$ and $\mathcal{S}_{i_1} = \{t+1, \ldots, t_2\}$ at a dividing point t (or $\mathcal{S}_{i_1} = \{t_1, \ldots, t\}$ and $\mathcal{S}_{i_0} = \{t+1, \ldots, t_2\}$, but this case is omitted since it is similarly discussed). Then, tag indexes $\{t+1, \ldots, t_2\}$ are not assigned to leaf attributes in $\tilde{U}$ that are descendants of the AND node i^*. On the other hand, from ancestors of the AND node i^*, using other paths that do not including the AND node i^*, the tag indexes $\{t+1, \ldots, t_2\}$ may be distributed to leaf attributes in $\tilde{U}$. However, this is not true. This is because tag indexes are not divided at t in other AND nodes, which means that any leaf except the descendants of i^* cannot obtain $\{t+1, \ldots, t'\}$ for any t'. Therefore, $\{\mathcal{S}_i | i \in \tilde{U}\}$ is not a partition of $\mathcal{S}_{root}$.

3.3 Accumulator to Verify Monotone Formulas

The following accumulator is used to verify a monotone formula where the formula is compressed into one single value to obtain a constant-size proof. In the accumulator, each attribute is indexed by an integer in $\{1, \ldots, n\}$. Let $\mathcal{M}_\mathcal{A}$ be the set of attribute indexes in $\mathcal{M}$. In a monotone formula, the same attribute may be included twice or more, such as $(Japanese \land student) \lor (Japanese \land professor)$. In this paper, for simplicity, the same attributes are indexed by different indexes (e.g., the first $Japanese$ is indexed by 1 and the second one is indexed by 2). Thus, we assume that the attribute indexes in $\mathcal{M}$ are all different. Here, let U be the set of user's attributes (indexes) and set $\tilde{U} = U \cap \mathcal{M}_\mathcal{A}$ showing the user's attributes included in the monotone formula. We assume that $|U|$ is bounded by the upper bound η.

AccSetup: This is the algorithm to output the public parameters. Select bilinear group $\mathbb{G}$ with a prime order p and a bilinear map e. Select $\tilde{g} \in_R \mathbb{G}$. Select $\gamma \in_R \mathbb{Z}_p$, and compute and publish $p, \mathbb{G}, e, \tilde{g}, \tilde{g}_1 = \tilde{g}^{\gamma^1}, \ldots, \tilde{g}_n = \tilde{g}^{\gamma^n}, \tilde{g}_{n+2} = \tilde{g}^{\gamma^{n+2}}, \ldots, \tilde{g}_{2n} = \tilde{g}^{\gamma^{2n}}$ and $z = e(\tilde{g}, \tilde{g})^{\gamma^{n+1}}$ as the public parameters.

AccGen: This is the algorithm to compute the accumulator given the public parameters and a monotone formula $\mathcal{M}$. For $\mathcal{M}$, run the tag assignment algorithm. For $i \in \mathcal{M}_\mathcal{A}$, let $\mathcal{S}_i$ be tag indexes assigned to each attribute i for the maximum index T, as the outputs of the tag assignment algorithm. Set a tag value as $c_t = (\eta + 1)^{t-1}$ for all $1 \leq t \leq T$. The accumulator of $\mathcal{M}$ is computed as $acc_\mathcal{M} = \prod_{i \in \mathcal{M}_\mathcal{A}} \tilde{g}_{n+1-i}^{\sum_{t \in \mathcal{S}_i} c_t}$.

AccWitGen: This is the algorithm to compute the witness of the minimum satisfaction of $\mathcal{M}$ by $\tilde{U}$ (i.e., $\mathcal{MS}(\tilde{U}, \mathcal{M}) = 1$), given the public parameters and $U, \mathcal{M}, \mathcal{S}_1, \ldots, \mathcal{S}_{|\mathcal{M}_\mathcal{A}|}, c_1, \ldots, c_T$. The witness is computed as
$$W = \prod_{j \in U} \prod_{i \in \mathcal{M}_\mathcal{A}, i \neq j} \tilde{g}_{n+1-i+j}^{\sum_{t \in \mathcal{S}_i} c_t}.$$

AccVerify: This is the algorithm to verify the minimum satisfaction of $\mathcal{M}$ by $\tilde{U}$, given the public parameters and $acc_\mathcal{M}, U, W, c_1, \ldots, c_T$. Set $u = c_1 + \ldots + c_T$, accept if,
$$\frac{e(\prod_{i \in U} \tilde{g}_i, acc_\mathcal{M})}{e(\tilde{g}, W)} = z^u.$$

3.4 Correctness and Security of Accumulator

We can show that the proposed accumulator is correct and secure, as follows. We show only the proof sketches, which are similar to the underlying accumulator [3].

Theorem 2. *If each algorithm is correctly performed for U and $\mathcal{M}$ s.t. $\mathcal{MS}(\tilde{U}, \mathcal{M}) = 1$ for $\tilde{U} = U \cap \mathcal{M}_\mathcal{A}$, **AccVerify** accepts $acc_\mathcal{M}, U, W$ outputted by the algorithms.*

Proof Sketch. By substituting $acc_\mathcal{M}$ and W to the verification equation, the left hand is equal to
$$= \frac{e(\prod_{j \in U} \tilde{g}_i, \prod_{i \in \mathcal{M}_\mathcal{A}} \tilde{g}_{n+1-i}^{\sum_{t \in S(i)} c_t})}{e(\tilde{g}, \prod_{j \in U} \prod_{i \in \mathcal{M}_\mathcal{A}, i \neq j} \tilde{g}_{n+1-i+j}^{\sum_{t \in S(i)} c_t})} = \frac{e(\tilde{g}, \prod_{j \in U} \prod_{i \in \mathcal{M}_\mathcal{A}} \tilde{g}_{n+1-i+j}^{\sum_{t \in S(i)} c_t})}{e(\tilde{g}, \prod_{j \in U} \prod_{i \in \mathcal{M}_\mathcal{A}, i \neq j} \tilde{g}_{n+1-i+j}^{\sum_{t \in S(i)} c_t})}$$
$$= e(\tilde{g}, \prod_{i \in \tilde{U}} \tilde{g}_{n+1}^{\sum_{t \in S(i)} c_t}) = e(\tilde{g}, \tilde{g}_{n+1}^{\sum_{i \in \tilde{U}} \sum_{t \in S(i)} c_t}).$$

This is equal to $z^u = e(\tilde{g}, \tilde{g}_{n+1})^{c_1 + \cdots + c_T}$, since $\{\mathcal{S}_i | i \in \tilde{U}\}$ is a partition of $\mathcal{S}_{root}$ due to Theorem 1.

Theorem 3. *Under the n-DHE assumption, on inputs of the public parameters correctly outputted by **AccSetup**, any adversary cannot output $\tilde{U}, \mathcal{M}, W$ s.t. **AccVerify** accepts $\tilde{U}^*, acc_\mathcal{M}, W$, but $\mathcal{MS}(\tilde{U}^*, \mathcal{M}) = 0$.*

Proof Sketch. Assume that such an adversary exists. Let $\tilde{g}_{n+1} = \tilde{g}^{\gamma^{n+1}}$. By substituting $acc_\mathcal{M}$ to the calculation in the verification equation, we can obtain

$$\frac{e(\prod_{i\in\tilde{U}}\tilde{g}_i,\prod_{i\in\mathcal{M}_\mathcal{A}}\tilde{g}_{n+1-i}^{\sum_{t\in\mathcal{S}_i}c_t})}{e(\tilde{g},W)} = z^u = e(\tilde{g},\tilde{g}_{n+1})^u,$$

$$e(\tilde{g}, \prod_{j\in\tilde{U}^*}\prod_{i\in\mathcal{M}_\mathcal{A}}\tilde{g}_{n+1-i+j}^{\sum_{t\in\mathcal{S}_i}c_t}) = e(\tilde{g}, W\tilde{g}_{n+1}^u).$$

Thus, we have

$$\prod_{j\in\tilde{U}^*}\prod_{i\in\mathcal{M}_\mathcal{A}}\tilde{g}_{n+1-i+j}^{\sum_{t\in\mathcal{S}_i}c_t} = W\tilde{g}_{n+1}^u$$

Here, let $\tilde{U}^* = U^* \cap \mathcal{M}_\mathcal{A}$, and, for all $1 \le t \le T$, let λ_t be the number of $\mathcal{S}_i$ s.t. $i \in \tilde{U}^*$ and $t \in \mathcal{S}_i$. Then, we have

$$\prod_{j\in\tilde{U}^*}\prod_{i\in\mathcal{M}_\mathcal{A},i\ne j}\tilde{g}_{n+1-i+j}^{\sum_{t\in\mathcal{S}_i}c_t} \cdot \prod_{1\le t\le T}\tilde{g}_{n+1}^{\lambda_t c_t} = W\tilde{g}_{n+1}^u$$

$$\prod_{j\in\tilde{U}^*}\prod_{i\in\mathcal{M}_\mathcal{A},i\ne j}\tilde{g}_{n+1-i+j}^{\sum_{t\in\mathcal{S}_i}c_t} = W\tilde{g}_{n+1}^{u-\sum_{1\le t\le T}\lambda_t c_t} \tag{1}$$

Set

$$\Delta = u - \sum_{1\le t\le T}\lambda_t c_t = \sum_{1\le t\le T}(1-\lambda_t)c_t.$$

Eq. (1) means that, if $\Delta \ne 0 \pmod p$, we can compute $\tilde{g}_{n+1}$ from the other $\tilde{g}_1,\ldots,\tilde{g}_{n+1},\tilde{g}_{n+2},\ldots,\tilde{g}_{2n}$. Then, we can prove $\Delta \ne 0 \pmod p$ for $u = c_1 + \ldots + c_T$ in the similar way to [3].

4 Syntax and Security Model of Anonymous Credential System

The syntax and security model of anonymous credential system for monotone formulas can be defined similarly to the previous work [3]. Here, we show the syntax and the informal security model. The formal model is shown in Appendix A.

4.1 Syntax

The attribute value is indexed by an integer from $\{1,\ldots,n\}$, where n is the total number of attribute values. All attribute values in all attribute types are indexed by using the universal set $\{1,\ldots,n\}$.

The anonymous credential system consists of the following algorithms:

IssuerKeyGen: The inputs of this algorithm are n,η, where η is the maximum number of users' attributes. The outputs are issuer's public key *ipk* and issuer's secret key *isk*.

CertObtain: This is an interactive protocol between a probabilistic algorithm
CertObtain-$\mathcal{U}$ for the user and a probabilistic algorithm **CertObtain-$\mathcal{I}$**
for an issuer, where the issuer issues the certificate including the attributes
to the user. **CertObtain-$\mathcal{U}$**, on input ipk and $U \subset \{1, \ldots, n\}$ that is indexes
corresponding to the attributes of the user, outputs the certificate $cert$ ensur-
ing the attributes of the user. On the other hand, **CertObtain-$\mathcal{I}$** is given
ipk, isk as inputs.

ProofGen: This probabilistic algorithm, on inputs ipk, U, $cert$, $\mathcal{M}$ that is the
monotone formula on attributes to be proved, outputs the proof σ.

Verify: This is a deterministic algorithm for verification. The input is ipk, a
proof σ, and the formula $\mathcal{M}$. Then the output is 'valid' if the attributes in
U satisfy $\mathcal{M}$, or 'invalid' otherwise.

4.2   Security Model

The security model consists of misauthentication resistance and anonymity.
The misauthentication resistance requirement captures the soundness of the
attribute proof. This means that an adversary $\mathcal{A}$ cannot try to forge a proof
for a monotone formula, where the attributes of any user corrupted by $\mathcal{A}$ do not
satisfy the formula. The anonymity requirement captures the anonymity and
unlinkability of proofs, as in the group signatures. The formal definitions are in
Appendix A.

5   Proposed Anonymous Credential System

5.1   Construction Overview

Basically, the construction of the proposed system is similar to the previous
system [3], as follows. The certificate of attributes is an AHO signature, where
the user's attributes in set U are unified to $P_U = \prod_{i \in U} \tilde{g}_i$ and embedded for
the accumulator verification. Together with the accumulated value $acc_{\mathcal{M}}$, P_U
are applied in the verification equation of accumulator to authenticate the user.
In the authentication, the user proves that the attributes in U satisfies the given
monotone formula $\mathcal{M}$, using the verification equation of the accumulator in
Sect. 3.3. To conceal any information beyond the satisfaction, we utilize the GS
proofs for the pairing equations in the verifications of the AHO signature and
the accumulator.

However, $U \cap \mathcal{M}_{\mathcal{A}}$ satisfies the formula $\mathcal{M}$, but might not be a minimum
attribute set due to redundant attributes included in U, which causes a fail-
ure in the verification. Thus, the user must derive a subset $\tilde{U}$ from U and use
$\tilde{U}$ such that $\tilde{U}$ is a minimum attribute set in the verification equation. However, it is normally infeasible for the user to derive a certificate of $P_{\tilde{U}}$ from the
issued certificate of P_U. Hence, we have the following approach: Consider all
possible candidates $U_k (1 \leq k \leq K)$ of subsets of U such that a U_k satisfying
$\mathcal{MS}(U_k, \mathcal{M}) = 1$ exists for any monotone formula $\mathcal{M}$. Then, the user is issued

a number of certificates for all U_k. In the authentication, for a given monotone formula $\mathcal{M}$, the user selects a certificate for a $U_k = \tilde{U}$ such that $\tilde{U}$ is the minimum attribute set, and can prove the correctness of the certified attributes of the user using the verification of AHO signatures.

5.2 Construction

IssuerKeyGen. It is given n that is the total number of attributes and η is the maximum number of user's attributes.

1. Select bilinear groups $\mathbb{G}, \mathbb{G}_T$ with the same order p and the bilinear map e, and $\tilde{g} \in_R \mathbb{G}$.
2. Generate public parameters of the accumulator: Select $\gamma \in_R \mathbb{Z}_p$, and compute
$$pk_{acc} = (\tilde{g}_1 = \tilde{g}^{\gamma^1}, \ldots, \tilde{g}_n = \tilde{g}^{\gamma^n}, \tilde{g}_{n+2} = \tilde{g}^{\gamma^{n+2}}, \ldots, \tilde{g}_{2n} = \tilde{g}^{\gamma^{2n}}, z = (\tilde{g}, \tilde{g})^{\gamma^{n+1}}).$$
3. Generate a key pair for the AHO signatures:

$$pk_{\mathrm{AHO}} = (G_r, H_r, G_z, H_z, G, H, A, B)$$
$$sk_{\mathrm{AHO}} = (\alpha_a, \alpha_b, \mu_z, \nu_z, \mu, \nu)$$

4. Generate a CRS for the GS NIWI proof: Select $\boldsymbol{f} = (\boldsymbol{f}_1, \boldsymbol{f}_2, \boldsymbol{f}_3) \in \mathbb{G}^3$ where $\boldsymbol{f}_1 = (f_1, 1, g), \boldsymbol{f}_2 = (1, f_2, g), \boldsymbol{f}_3 = \boldsymbol{f}_1^{\varepsilon_1} \cdot \boldsymbol{f}_2^{\varepsilon_2}$ for $f_1, f_2 \in_R \mathbb{G}$ and $\varepsilon_1, \varepsilon_2 \in_R \mathbb{Z}_p^*$.
5. Output the issuer public key $ipk = (p, \mathbb{G}, \mathbb{G}_T, e, \tilde{g}, pk_{acc}, pk_{\mathrm{AHO}}, \boldsymbol{f})$, and the issuer secret key $isk = sk_{\mathrm{AHO}}$.

CertObtain. In this protocol, the common inputs are ipk, isk, and the user's attribute set U, and the issuer's input is isk.

1. As all possible subsets of set U, the issuer prepares U_k for all $1 \leq k \leq K$ where K is the total number of subsets.
2. Using sk_{AHO}, the issuer generates each AHO signatures on $P_k = \prod_{i \in U_k} \tilde{g}_i$ as σ_k for all $1 \leq k \leq K$ and then send them to the user.
3. The user outputs the obtained signatures $cert = \{(\sigma_k)_{1 \leq k \leq K}\}$, as the certificates.

ProofGen. The inputs are $ipk, U, cert$ and the monotone formula $\mathcal{M}$. Define $\tilde{U} \subseteq U$ is the minimum attribute set selected by the user to satisfy the formula $\mathcal{M}$.

1. Run the tag assignment algorithm. For each attribute i in $\mathcal{M}$, a series of tags $\mathcal{S}_i$ is assigned, where the tags are $c_1, \ldots, c_T$.
2. User calculates $P_{\tilde{U}} = \prod_{i \in \tilde{U}} \tilde{g}_i$.
3. User selects certificate σ_k w.r.t. $P_{\tilde{U}}$ s.t. $\tilde{U} = U_k$ for some U_k from $cert$.
4. Compute the accumulator:

$$acc_{\mathcal{M}} = \prod_{i \in \mathcal{M}_{\mathcal{A}}} \tilde{g}_{n+1-i}^{\sum_{t \in \mathcal{S}(i)} c_t}.$$

5. Compute the witness that $\tilde{U}$ satisfies $\mathcal{M}$ for $acc_{\mathcal{M}}$:

$$W = \prod_{j \in \tilde{U}} \prod_{i \in \mathcal{M}_A, i \neq j} \tilde{g}_{n+1-i+j}^{\sum_{t \in S(i)} c_t}$$

and sets $u = c_1 + \ldots + c_T$.

6. Compute GS commitments $com_{P_{\tilde{U}}}, com_W$ to $P_{\tilde{U}}, W$. Then re-randomize the AHO signature σ_k to obtain $\sigma'_k = \{\theta'_1, \ldots, \theta'_7\}$, and compute GS commitments $\{com_{\theta'_i}\}_{i \in \{1,2,5\}}$.

7. Generate the GS proofs $\{\pi_i\}_{i=1}^3$ s.t.

$$z^u = e(P_{\tilde{U}}, acc_{\mathcal{M}}) \cdot e(g, W)^{-1}, \tag{2}$$

$$A \cdot e(\theta'_3, \theta'_4)^{-1} = e(G_z, \theta'_1) \cdot e(G_r, \theta'_2) \cdot e(G, P_{\tilde{U}}), \tag{3}$$

$$B \cdot e(\theta'_6, \theta'_7)^{-1} = e(H_z, \theta'_1) \cdot e(H_r, \theta'_5) \cdot e(H, P_{\tilde{U}}) \tag{4}$$

8. Output $\sigma = (\{\theta'_i\}_{i=3,4,6,7}), com_{P_{\tilde{U}}}, com_W, \{com_{\theta'_i}\}_{i=1,2,5}, \{\pi_i\}_{i=1}^3$.
The Eq. (2) shows the verification relation of accumulator:

$$\frac{e(P_{\tilde{U}}, acc_{\mathcal{M}})}{e(g, W)} = z^u,$$

where $P_{\tilde{U}} = \prod_{i \in \tilde{U}} \tilde{g}_i$. The Eqs. (3),(4) show the knowledge of the AHO signature of $P_{\tilde{U}}$.

Verify. The inputs are ipk, the proof σ, and the proved formula $\mathcal{M}$.
1. Compute the accumulator $acc_{\mathcal{M}}$, as in **ProofGen.**
2. Accept σ, if the verifications of all GS proofs are successful.

5.3   Security

We can prove the following security of our construction.

Theorem 4. *The proposed system satisfies the misauthentication resistance under the security of the AHO signatures and the accumulator.*

Theorem 5. *The proposed system satisfies the anonymity under the DLIN assumption.*

These theorems can be proved similarly to [3], which will be shown in the full paper. As for the proof of the misauthentication resistance, using the adversary against the misauthentication resistance game, we can construct the adversary for AHO signatures in case that the certificate of P_{U^*} has not been issued for P_{U^*} used in the accumulator verification of the forged proof of formula $\mathcal{M}^*$, or the adversary for the accumulator in case that the certificate of P_{U^*} has been issued (the latter case means that $\mathcal{MS}(U^*, \mathcal{M}^*) = 0$ but the verification accepts U^*). On the other hand, the anonymity holds, since the real GS proof in **ProofGen** is indistinguishable to the case of the witness indistinguishable setting.

6   Efficiency Consideration

In this section, we show the efficiency of our proposed system using an example of monotone formula for accessing alcohol related websites. In the formula, four categories of attributes are used; the country name, the birth-year, the birth-month and the birth-day. The example of the monotone formula is as follows:

$(Japan \vee Iceland \vee \ldots) \wedge (1914 \vee \ldots \vee (1994 \wedge (January \vee (February \wedge (1st \vee \ldots \vee 12th)))))$.

This formula expresses that the prover's age is more than 20 years old a some point of time, using the birthdate. As shown in Introduction, the CNF type of formula is as follows:

$(Japan \vee Iceland \vee \ldots) \wedge (1914, December31 \vee \cdots \vee 1994, February12)$,

where each birthdate is encoded to one attribute value such as "1914, $December31$". Using these formulas, compare the efficiency between the previous system [3] for CNF formulas and the proposed system for monotone formulas.

Concentrate in the more frequently executed authentication protocol that consists of **ProofGen** and **Verify** algorithms. In both systems, **ProofGen** mainly consists of computations of $acc_\mathcal{M}$ and W and GS proof generations. **Verify** consists of the computation of $acc_\mathcal{M}$ and GS proof verifications. The GS proof generations and verifications need constant computations for the size of formulas (The number of GS proofs are reduced and thus the costs are also reduced from the previous system to the proposed system). Thus, concentrate on the computational cost of W, since the cost depends on the size of the formulas and the cost of $acc_\mathcal{M}$ is smaller.

Here, we review the computations of W of the previous system and the proposed system. In the previous system,

$$W = \prod_{j \in U} \prod_{1 \le l \le L} (\prod_{j \in V_l}^{i \ne j} g_{n+1-i+j})^{c_l},$$

where c_ℓ are similar tags, and V_ℓ includes the attribute literals of ℓ-th clause in the CNF formula. In this computation, by arranging the calculation order, the number of the exponentiations of c_ℓ can be reduced to L. On the other hand, the number of multiplications is approximately $|U| \cdot \sum_{1 \le \ell \le L} |V_\ell|$, which means that the cost is linear in the number of OR literals. Thus, as shown in Introduction, since the above example needs approximately 29,200 literals, the number of multiplications becomes very large.

In the proposed system,

$$W = \prod_{j \in \tilde{U}} \prod_{i \in \mathcal{M}_\mathcal{A}, i \ne j} \tilde{g}_{n+1-i+j}^{\sum_{t \in S_i} c_t}.$$

Also, we can arrange the calculation order to

$$W = \prod_{1 \le t \le T} (\prod_{j \in \tilde{U}} (\prod_{i \in S_{\mathcal{M}_\mathcal{A}}^{-1}(t), i \ne j} \tilde{g}_{n+1-i+j}))^{c_t},$$

where $S_{\mathcal{M}_{\mathcal{A}}}^{-1}(t)$ is the set of i such that attribute i is assigned to tag c_t. The number of exponentiations of c_t is approximately the number of AND relations. In the above example using lots of OR relations, the number of AND relations is only 3. The number of multiplications is at most $T \cdot |\tilde{U}| \cdot |\mathcal{M}_{\mathcal{A}}|$. As shown in Introduction, since the above monotone formula example needs approximately 100 literals, the number of multiplications is greatly reduced.

7   Conclusions

In this paper, we have proposed an anonymous credential system with constant-size proofs for monotone formulas on attributes. As the key primitive, we also have constructed an accumulator to verify the monotone formulas. The monotone formula is more expressive and compact than CNF formulas, and thus the proposed system can reduce the proof generation time, compared to the previous system for CNF formulas.

Our future work includes detailed comparisons to the previous system based on the implementations.

Acknowledgments. his work was partially supported by JSPS KAKENHI Grant Number 25330153.

A   Security Model of Anonymous Credential System

This security model is similar to the previos work [3].

A.1   Misauthentication Resistance

Consider the following misauthentication resistance game.

Misauthentication Resistance Game: The challenger runs **IssuerKeyGen**, and obtains ipk and isk. He provides $\mathcal{A}$ with ipk, and run $\mathcal{A}$. He sets CU with empty, where CU denotes the set of IDs of users corrupted by $\mathcal{A}$. In the run, $\mathcal{A}$ can query the challenger about the following issuing query:

C-Issuing: $\mathcal{A}$ can request the certificates on attribute set $U^{(i)}$ of user i. Then, $\mathcal{A}$ as the user executes **CertObtain** protocol with the challenger as the issuer.

Finally, $\mathcal{A}$ outputs a monotone formula $\mathcal{M}^*$, and a proof σ^*.

Then, $\mathcal{A}$ wins if

1. **Verify**$(ipk, \sigma^*, \mathcal{M}^*) = $ valid, and
2. for all i $\in CU$, $U^{(i)}$ does not satisfy $\mathcal{M}^*$.

Misauthentication resistance requires that for all PPT $\mathcal{A}$, the probability that $\mathcal{A}$ wins the misauthentication resistance game is negligible.

A.2 Anonymity

Consider the following anonymity game.

Anonymity Game: The challenger runs **IssuerKeyGen**, and obtains ipk, isk. He provides $\mathcal{A}$ with ipk, isk, and run $\mathcal{A}$. He sets HU with empty. In the run, $\mathcal{A}$ can query the challenger, as follows.

H-Issuing: $\mathcal{A}$ can request the certificates on attribute set $U^{(i)}$ of user i. Then, $\mathcal{A}$ as the issuer executes **CertObtain** protocol with the challenger as the user. The challenger adds this user to HU.

Proving: $\mathcal{A}$ can request the user i's proof on formula $\mathcal{M}$. Then, the challenger responds the proof on $\mathcal{M}$ of the user i, if the user is in HU.

During the run, as the challenge, $\mathcal{A}$ outputs a formula $\mathcal{M}$, and two users i_0 and i_1, such that both $U^{(i_0)}$ and $U^{(i_1)}$ satisfy $\mathcal{M}^*$. If $i_0 \in HU$ and $i_1 \in HU$, the challenger chooses $\phi \in_R \{0, 1\}$, and responds the proof on $\mathcal{M}^*$ of user i_ϕ. After that, similarly, $\mathcal{A}$ can make the queries.

Finally, $\mathcal{A}$ outputs a bit ϕ' indicating its guess of ϕ.

If $\phi' = \phi$, $\mathcal{A}$ wins. We define the advantage of $\mathcal{A}$ as $|\Pr[\phi' = \phi] - 1/2|$.

Anonymity requires that for all PPT $\mathcal{A}$, the advantage of $\mathcal{A}$ on the anonymity game is negligible.

References

1. Abe, M., Fuchsbauer, G., Groth, J., Haralambiev, K., Ohkubo, M.: Structure-preserving signatures and commitments to group elements. In: Rabin, T. (ed.) CRYPTO 2010. LNCS, vol. 6223, pp. 209–236. Springer, Heidelberg (2010)
2. Abe, M., Haralambiev, K., Ohkubo, M.: Signing on elements in bilinear groups for modular protocol design. Cryptology ePrint Archieve, Report 2010/133 (2010)
3. Begum, N., Nakanishi, T., Funabiki, N.: Efficient proofs for CNF formulas on attributes in pairing-based anonymous credential system. In: Kwon, T., Lee, M.-K., Kwon, D. (eds.) ICISC 2012. LNCS, vol. 7839, pp. 495–509. Springer, Heidelberg (2013)
4. Boneh, D., Boyen, X., Shacham, H.: Short group signatures. In: Franklin, M. (ed.) CRYPTO 2004. LNCS, vol. 3152, pp. 41–55. Springer, Heidelberg (2004)
5. Camenisch, J., Groß, T.: Efficient attributes for anonymous credentials. In: Proceedings of the ACM Conference on Computer and Communications Security 2008 (ACM-CCS 2008), pp. 345–356 (2008)
6. Camenisch, J., Kohlweiss, M., Soriente, C.: An accumulator based on bilinear maps and efficient revocation for anonymous credentials. In: Jarecki, S., Tsudik, G. (eds.) PKC 2009. LNCS, vol. 5443, pp. 481–500. Springer, Heidelberg (2009)
7. Camenisch, J.L., Lysyanskaya, A.: Dynamic accumulators and application to efficient revocation of anonymous credentials. In: Yung, M. (ed.) CRYPTO 2002. LNCS, vol. 2442, pp. 61–76. Springer, Heidelberg (2002)
8. Groth, J., Sahai, A.: Efficient non-interactive proof systems for bilinear groups. In: Smart, N.P. (ed.) EUROCRYPT 2008. LNCS, vol. 4965, pp. 415–432. Springer, Heidelberg (2008)

9. Izabachène, M., Libert, B., Vergnaud, D.: Block-wise P-signatures and non-interactive anonymous credentials with efficient attributes. In: Chen, L. (ed.) IMACC 2011. LNCS, vol. 7089, pp. 431–450. Springer, Heidelberg (2011)
10. Libert, B., Peters, T., Yung, M.: Scalable group signatures with revocation. In: Pointcheval, D., Johansson, T. (eds.) EUROCRYPT 2012. LNCS, vol. 7237, pp. 609–627. Springer, Heidelberg (2012)
11. Sudarsono, A., Nakanishi, T., Funabiki, N.: Efficient proofs of attributes in pairing-based anonymous credential system. In: Fischer-Hübner, S., Hopper, N. (eds.) PETS 2011. LNCS, vol. 6794, pp. 246–263. Springer, Heidelberg (2011)

Secure Protocol

Secure Multi-Party Computation Using Polarizing Cards

Kazumasa Shinagawa[1,2]([✉]), Takaaki Mizuki[3], Jacob Schuldt[2], Koji Nuida[2],
Naoki Kanayama[1], Takashi Nishide[1], Goichiro Hanaoka[2], and Eiji Okamoto[1]

[1] University of Tsukuba, Tsukuba, Japan
shinagawa@cipher.risk.tsukuba.ac.jp
[2] National Institute of Advanced Industrial Science and Technology (AIST),
Tokyo, Japan
[3] Tohoku University, Sendai, Japan

Abstract. It is known that, using just a deck of cards, an arbitrary number of parties with private inputs can securely compute the output of any function of their inputs. In 2009, Mizuki and Sone constructed a six-card COPY protocol, a four-card XOR protocol, and a six-card AND protocol, based on a commonly used encoding scheme in which each input bit is encoded using two cards. However, up until now, it has remained an open problem to construct a set of COPY, XOR, and AND protocols based on a two-cards-per-bit encoding scheme, which all can be implemented using only four cards. In this paper, we show that it is possible to construct four-card COPY, XOR, and AND protocols using polarizing plates as cards and a corresponding two-cards-per-bit encoding scheme. Our protocols are optimal in the setting of two-cards-per-bit encoding schemes since four cards are always required to encode the inputs. As applications of our protocols, we show constructions of optimal input-preserving XOR and AND protocols, which we combine to obtain optimal half-adder, full-adder, voting protocols, and more.

1 Introduction

1.1 Background

Secure Multi-Party Computation (MPC) enables an arbitrary number of parties with secret inputs to compute the output of a function without revealing their inputs. MPC protocols are usually implemented in a computer-based environment. On the other hand, there are various works constructing MPC protocols using only a deck of physical cards, referred to as *card-based protocols* [1–12]. Compared to computer-based protocols, a card-based protocol has several advantages; (1) It is easy to understand the correctness and security of protocols even for non-experts, (2) Since card-based protocols do not rely on computers, they can be performed without the use of electricity, (3) In contrast to online protocols where players are invisible to each other, there is a potentially high cost of acting maliciously in card-based protocols, (4) Playing a card-based protocol is a lot of fun!

© Springer International Publishing Switzerland 2015
K. Tanaka and Y. Suga (Eds.): IWSEC 2015, LNCS 9241, pp. 281–297, 2015.
DOI: 10.1007/978-3-319-22425-1_17

The *Five-Card Trick* proposed by den Boer [2] in 1989 is the first card-based protocol, which enables two parties to securely compute the AND ($\wedge$) function of their secret inputs using five cards (two $\clubsuit$s and three $\heartsuit$s). In [2] (as well as in previous works [1–12]), a binary input is encoded by two cards as follows: $\boxed{\clubsuit}\boxed{\heartsuit} = 0, \boxed{\heartsuit}\boxed{\clubsuit} = 1$. In the input phase, a party encodes his input as above, and places the cards face down in a public space (e.g. on a table). This pair of cards is called a *commitment*. In 1993, Crépeau and Kilian [1] achieved any functionality by constructing an eight-card COPY protocol, a 14-card XOR protocol, and a ten-card AND protocol. These protocols generate a commitment to the result, and hence allow secure sequential composition (the original AND protocol from [2] directly outputs the value $x \wedge y$). Note that, in contrast to electronic computations, it is non-trivial to perform a COPY operation, which takes a commitment to x as input, and outputs two commitments to x. In 2009, Mizuki and Sone [8] designed a six-card COPY protocol, a four-card XOR protocol, and a six-card AND protocol, applying a new shuffle called a *random bisection cut*. The four-card XOR protocol is optimal with respect to the above encoding scheme since four cards are required to represent the input commitments, i.e., it is impossible to construct an XOR protocol with less than four cards using a similar encoding scheme. On the other hand, it is unknown whether it is possible to design COPY and AND protocols using less than six cards in this setting. Note that in particular an efficient COPY protocol is important in the construction of more general protocols. Consider computing a given function $f(x_0, x_1, \cdots, x_{n-1})$ from commitments to $x_0, x_1, \cdots, x_{n-1} \in \{0, 1\}$. In many cases, it is required to preserve the input commitments since they might be needed in later stages of the protocol. To achieve this, the commitments are copied using the COPY protocol before evaluation of $f(x_0, x_1, \cdots, x_{n-1})$. If the commitments are copied in parallel, and the COPY protocol requires two additional cards besides the cards for storing the commitments, $2n$ additional cards are required in total.[1] Consequently, constructing an efficient COPY protocol is important not only from a theoretical viewpoint but also from a practical viewpoint.

Unfortunately, it seems impossible to construct a four-card COPY protocol for the above described encoding scheme because, to ensure correctness, it is necessary to open a face-down card, but the number of face-down cards is less than four after this opening. Due to this, it appears to be necessary to introduce a new type of cards or encoding scheme.

1.2   Our Contribution

In this paper, we show that it is possible to construct a four-card COPY protocol, a four-card XOR protocol, and a four-card AND protocol, by using *polarizing plates* as cards and a corresponding two-cards-per-bit encoding. The use of polarizing plates enables us to overcome the limitations of ordinary cards. As an additional technical contribution we introduce a new shuffle, a *diagonal flip shuffle*,

---

[1] If the commitments are copied one-by-one, only two additional cards are needed, but this is inefficient with respect to required computational time.

Table 1. Protocols using two-cards-per-bit encoding scheme

	Type of cards	# of cards	Optimality
○ COPY protocol			
Mizuki-Sone [8]	♣, ♡	6	?
Ours	Polarizing	4	✓
○ XOR protocol			
Mizuki-Sone [8]	♣, ♡	4	✓
Ours	Polarizing	4	✓
○ AND protocol			
Mizuki-Sone [8]	♣, ♡	6	?
Ours	Polarizing	4	✓
○ Input-preserving XOR protocol			
Nishida et al. [10]	♣, ♡	6	?
Ours	Polarizing	4	✓
○ Input-preserving AND protocol			
Nishida et al. [10]	♣, ♡	6	?
Ours	Polarizing	4	✓

which is used in our four-card AND protocol. Since four cards are always required to encode the inputs, our constructions are optimal in the setting of two-cards-per-bit encoding [2] (see Table 1). As applications of our protocols, we show constructions of optimal input-preserving XOR and AND protocols, which we combine to obtain an optimal half-adder protocol, an optimal full-adder protocol, an optimal voting protocol, and optimal protocols for arbitrary symmetric functions.

Polarizing Cards. Polarization of light is a physical phenomenon that transforms light waves oscillating in various directions into light waves oscillating in a certain direction. A polarizing plate is a material which has such a polarization property. If two polarizing plates with a same direction are superimposed, then light will pass through the plates which will appear to be transparent. On the other hand, if two plates with perpendicular directions are superimposed, no light will pass through the plates which will appear to be black. In addition, a polarizing plate has an important property, which we will refer to as a *direction indistinguishability*. Specifically, it is difficult to distinguish the direction of the polarization without superimposing the plate with another polarizing plate. A polarizing card used in this paper is a square polarizing plate with either vertical or horizontal direction. As far as we know, this is the first study of

---

[2] Note that the optimality is defined with respect to the used encoding scheme. Hence, if an encoding scheme using less than two cards per bit can be used, it might be possible to reduce the number of cards required by the protocols. In fact, in Sect. 6.1, we propose a one-card-per-bit encoding scheme for polarizing cards and corresponding efficient protocols. However, this encoding scheme requires the use of a *base card* (see Sect. 6.1 for a discussion of this approach).

MPC protocols using cards which are not an analogy of ordinary cards similar to playing cards. Let $\updownarrow$ / $\leftrightarrow$ denote a card with vertical/horizontal direction, respectively. We define a new encoding: To encode 0, we use a sequence of two cards with the same direction, i.e., $(\updownarrow, \updownarrow)$ or $(\leftrightarrow, \leftrightarrow)$. Likewise, to encode 1, we use a sequence of two cards with opposite directions, i.e., $(\updownarrow, \leftrightarrow)$ or $(\leftrightarrow, \updownarrow)$. This sequence is called a *commitment*. Note that it is impossible to distinguish a commitment to 0 and a commitment to 1 due to the direction indistinguishability property. For the sake of convenience, we assign a value 0 or 1 to a direction $\updownarrow$ or $\leftrightarrow$, respectively; we often regard a commitment to $a \in \{0, 1\}$ as a sequence of cards (x, y) using two bits $x, y \in \{0, 1\}$ with $x \oplus y = a$. (This is natural because the binary operation $\oplus$ corresponds to superimposing as easily imagined below.)

COPY Protocol. We will now explain the idea of our COPY protocol. Let x be a card with a secret direction and let r be a card with a random direction (in this regard, these directions are either $\updownarrow$ or $\leftrightarrow$). When x, r are superimposed, we only learn whether x, r have the same direction or not, i.e., the result of superimposing the cards is either *white* or *black*. If it is *white* then r is a copy of x, otherwise r rotated 90° is a copy of x. In its simplicity, this is the COPY protocol. To obtain a card with a random direction, we use a rotation shuffle. Roughly speaking, a rotation shuffle rotates a deck of cards a random number of times. As a result, we obtain an optimal COPY protocol using a rotation shuffle while the protocol from [8] needs two additional cards.

AND Protocol. We construct an optimal four-card AND protocol while the protocol from [8] requires six cards. Here, we explain the idea of our AND protocol. Since polarizing cards have two-sided symmetry, a sequence (x, y, z) and the "flipped and rotated" sequence $(\overline{z}, \overline{y}, \overline{x})$ are indistinguishable, where $\overline{x}$ denotes the 90° rotation of a card x. Here, the "flip and rotate" operation is equivalent to an operation that flips a deck of cards along the diagonal axis. Consequently, we obtain a new shuffle, a *diagonal flip shuffle*, that generates one of the two sequences (x, y, z) or $(\overline{z}, \overline{y}, \overline{x})$ with probability exactly $1/2$. This shuffle immediately gives us a four-card AND protocol. Given two commitments to x, y as inputs, and applying the diagonal flip shuffle to the input sequence, we can obtain a card r and one of the two sequence $(\overline{r}, r \oplus x, r \oplus y)$ or $(r \oplus \overline{y}, r \oplus \overline{x}, r)$, using only four cards (see more detail in Sect. 4.3). If the middle card of the sequence and the card r is a commitment to 1 then the right card (combined with r) is a commitment to $x \wedge y$, otherwise the left card (combined with r) is a commitment to $\overline{x \wedge y}$. Composing our optimal protocols, we can securely compute various functions with a small number of cards.

Input-Preserving XOR/AND Protocols. We construct an optimal four-card XOR protocol and an optimal four-card AND protocol, both with *input-preserving property*, which also output one of the input commitments. Based on these protocols and applying the techniques from [3,10], we obtain the construction of an optimal four-card half-adder protocol, an optimal six-card full-adder protocol, an optimal voting protocol, protocols for any k-variable functions using $2k + 4$ cards, and optimal protocols for any k-variable symmetric functions using $2k$ cards.

1.3 Related Works

In 1989, using two types of cards ($\clubsuit$, $\heartsuit$), den Boer [2] proposed a five-card AND protocol that reveals the output value at the end of the protocol. Most of subsequent works use the same type of cards as in [2]. In 2014, Mizuki-Shizuya [7] discussed the advantages and disadvantages of a one-card-per-bit encoding scheme that enables the construction of a three-card AND protocol, a two-card XOR protocol, and a three-card COPY protocol. A possible application of a similar encoding scheme to our polarizing cards is discussed in Sect. 6.1.

Based on elementary protocols, many applications are proposed [3, 10, 11]. In 2013, Mizuki, Asiedu, and Sone [3] constructed an eight-card half-adder protocol, a ten-card full-adder protocol, and a $(2\lceil \log n \rceil + 6)$-card voting protocol where n is the number of parties. In 2015, Nishida et al. [10] constructed a six-card AND protocol with input-preserving property, which also outputs one of the input commitments. They also show that it is possible to construct a six-card half-adder protocol and protocols for any k-variable functions using $2k + 6$ cards and any k-variable symmetric functions using $2k + 2$ cards [10]. The numbers of cards for such advanced protocols are also reduced due to our improvements of the elementary protocols mentioned above.

2 Polarizing Cards

In this section, we first propose a card called a polarizing card, which is a square polarizing plate. Next, we define three operations, permute, superimpose, and shuffle, that can be applied to a sequence of cards in a protocol. Furthermore, we propose two concrete shuffles, a *rotation shuffle* and a *diagonal flip shuffle*.

2.1 Polarizing Cards

Polarizing Cards. A *polarizing plate* is a material with a polarization property, for which it is difficult to distinguish the direction of the polarization without superimposing the plate with another polarizing plate (see also Sect. 1.2). We say that a polarizing plate is a *polarizing card* when it has two symmetries, 90° *rotational symmetry* and *two-sided symmetry*, i.e., the face of the plate is invariant even if it is rotated 90° or flipped (e.g. square polarizing plates as shown on the left in Fig. 1). A deck of polarizing cards is a number of cards such that any two cards either have a same direction or perpendicular directions. Without loss of generality, we assume that polarizing cards have a vertical or horizontal direction and we denote by $\mathcal{D} = \{\updownarrow, \leftrightarrow\}$ the set of polarizing cards. If two cards with the same direction are superimposed, then light can pass through the plates and they appear transparent. We say that this result is "*white*" (Fig. 1, center). On the other hand, if two plates with perpendicular directions are superimposed, no light can pass through the plates. We say that this result is "*black*" (Fig. 1, right). A k-sequence of polarizing cards is cards lying in the public space such that each of the cards is isolated and not superimposed on top of each other or any other polarization plates. We denote by $(a_0, a_1, \cdots, a_{k-1})$ a sequence of polarizing cards.

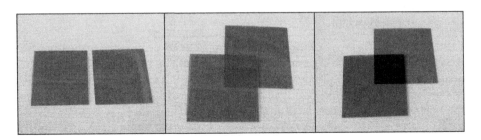

Fig. 1. Polarizing cards

New Encoding. How should we encode $x \in \{0, 1\}$ using polarizing cards? One possibility is to use the previous encoding scheme from [2], i.e., $(\updownarrow, \leftrightarrow) = 0$ and $(\leftrightarrow, \updownarrow) = 1$. However, it is difficult to encode and decode due to the direction indistinguishability. For this reason, we propose a new encoding scheme, which is explicitly based on the polarizing property of the cards. The new encoding Enc is a function that maps $x \in \{0, 1\}$ to a sequence of two cards. Specifically, to encode 0, we use a sequence of two cards with the same direction, which will result in white when superimposed, i.e., $\mathsf{Enc}(0) = \{(\updownarrow, \updownarrow), (\leftrightarrow, \leftrightarrow)\}$. Likewise, to encode 1, we use a sequence of two cards with opposite directions, which will result in black when superimposed, i.e., $\mathsf{Enc}(1) = \{(\updownarrow, \leftrightarrow) = (\leftrightarrow, \updownarrow)\}$.[3] We say that a sequence is a *commitment* to a when it is an element of $\mathsf{Enc}(a)$. Note that, due to the direction indistinguishability property, nobody can distinguish whether a commitment belongs to $\mathsf{Enc}(0)$ or $\mathsf{Enc}(1)$ without superimposing the used cards. From now on, in order to simplify the exposition, we denote $\mathcal{D}$ by $\{0, 1\}$ instead of $\{\updownarrow, \leftrightarrow\}$. Using this notation, a commitment to x can be represented as $(c, c \oplus x)$, and this is equivalent to $(c' \oplus x, c')$ where $c = c' \oplus x$.

Remark: Stacking Cards without Cover. When two polarizing cards x, y are superimposed, we learn whether $x = y$ or not. Therefore, stacking polarizing cards reveals partial information relevant to the secret value of the commitments. In order to avoid this, we use a cover which has a 90° rotational symmetric pattern when we stack two or more cards (Fig. 2). The cover is only used in a shuffle operation.

2.2   Operations

Permutation. In a previous work [6], the group of permutations for sequences of k cards is set to be the symmetric group S_k, which acts on each sequence by changing the order of cards. In this paper, we extend the definition of permutations for cards in order to deal with rotations of each card in a sequence (which is typical for our polarizing cards) as well as the re-ordering. We define permutation operations as follows.

---

[3] The other possibility is to use the encoding scheme such that $\mathsf{Enc}(0) = \{(\updownarrow, \leftrightarrow), (\leftrightarrow, \updownarrow)\}$ and $\mathsf{Enc}(1) = \{(\updownarrow, \updownarrow), (\leftrightarrow, \leftrightarrow)\}$. They are equivalent.

Fig. 2. Cover with polarizing cards

Definition 1. *Let* Σ_k *be the group of (extended) permutations generated by* σ_i *for* $i = 0, 1, \ldots, k - 2$ *and* ρ_i *for* $i = 0, 1, \ldots, k - 1$*, where* σ_i *and* ρ_i *denote the adjacent transposition and the fundamental rotation, respectively, defined as follows:*

$$\sigma_i(x_0, \cdots, x_i, x_{i+1}, \cdots, x_{k-1}) = (x_0, \cdots, x_{i+1}, x_i, \cdots, x_{k-1})$$
$$\rho_i(x_0, \cdots, x_i, \cdots, x_{k-1}) = (x_0, \cdots, \overline{x_i}, \cdots, x_{k-1}).$$

A permutation operation, *defined by* $\sigma \in \Sigma_k$*, takes as input a sequence* $s = (x_0, \ldots, x_{k-1})$*, and outputs* $\sigma(s)$*.*

Superimposing. We define superimposing operations as follows.

Definition 2. *We define a* superimposing operation *that takes as inputs* $s = (x_0, \cdots, x_{k-1})$ *and* i, j *where* $i, j \in \{0, \cdots, k - 1\}$ *are different indices* $(i \neq j)$*, and outputs "white" if* $x_i = x_j$ *or "black" if* $x_i \neq x_j$*.*

This is physically implemented by superimposing cards x_i and x_j. This operation is similar to an *opening operation* that is used in ordinary card-based protocols. However, a superimposing operation to x, y reveals only the direction difference between x, y, i.e., $x \oplus y$, while an opening operation reveals the value of card itself.

Shuffle. We define shuffle operations that play an important role in achieving perfect secrecy in card-based protocols. Firstly, we provide a general definition of shuffles that might contain a shuffle which has no easy physical implementation:

Definition 3. *Let* $\sigma \in \Sigma_k$ *be a permutation with order* ℓ*, i.e.,* $\sigma^\ell = id$ *where* id *is the identity mapping. We define a* shuffle operation *based on* σ *that takes* $s = (x_0, \cdots, x_{k-1})$ *as input, and outputs* $\sigma^r(s)$*, where* r *is uniformly chosen from* $\{0, \cdots, \ell - 1\}$*. The value* r *is assumed to be hidden from the parties executing the protocol.*

We show two implementations of a shuffle based on σ. The first one is simple. A party applies the permutation σ to s a random number of times until all of parties are satisfied. The random number is not known to either of them even the party who operates the shuffle. The other method does not require a party who takes responsibility of security. Let $P_0, \cdots, P_{n-1}$ be the parties participating in the protocol. P_0 uniformly chooses $r_0 \in \{0, \cdots, \ell-1\}$ where ℓ is the order of σ, operates permutation σ^{r_0} to s, and sends $\sigma^{r_0}(s)$ to P_1. This is repeated from P_0 to P_{n-1}. Finally, P_{n-1} places $\sigma^r(s)$ in a public space where $r = r_0 + \cdots + r_{n-1}$. Obviously, the value r is uniformly chosen from $\{0, \cdots, \ell-1\}$ and nobody knows it. In practice, the former method is usually used since it is easy to demonstrate. Therefore, it is important that a shuffle is easily implemented by hand. In this paper, we say that a shuffle has an *easy physical implementation* if it is achieved only using "one operation" that can be performed by hand.

We propose two shuffles, a *rotation shuffle* and a *diagonal flip shuffle*. These two shuffles have an easy physical implementation. The idea of a rotation shuffle is used in the protocols [7], but we define it more formally:

Definition 4. *Let $\rho \in \Sigma_k$ be a permutation, called a rotation operation, such that $\rho(x_0, \cdots, x_{k\,1}) = (\overline{x_0}, \cdots, \overline{x_{k-1}})$. A rotation shuffle Rotation($\cdot$) is defined as a shuffle based on ρ, i.e., Rotation($x_0, \cdots, x_{k-1}$) = $\rho^r(x_0, \cdots, x_{k-1})$ where r is uniformly chosen from $\{0, 1\}$, and it results in one of the two possibilities*

$$\text{Rotation}(x_0, \cdots, x_{k-1}) = \begin{cases} (x_0, \cdots, x_{k-1}) \\ (\overline{x_0}, \cdots, \overline{x_{k-1}}) \end{cases}$$

with probability exactly $1/2$.

The rotation operation ρ for $(x_0, \cdots, x_{k-1})$ is physically implemented by stacking the k cards (using a cover; see Sect. 2.1) and then rotating them $90°$ (Fig. 3).

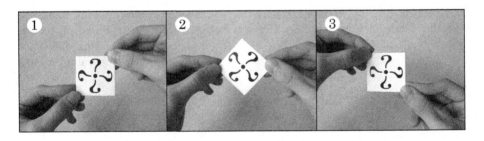

Fig. 3. An implementation of the rotation operation ρ

Next, we define a diagonal flip shuffle that is the main technical contribution in this paper:

Definition 5. *Let $\tau \in \Sigma_k$ be a permutation, called a diagonal flip operation, such that $\tau(x_0, x_1, \cdots, x_{k-1}) = (\overline{x_{k-1}}, \cdots, \overline{x_1}, \overline{x_0})$. A diagonal flip shuffle DiagFlip($\cdot$) is defined as a shuffle based on τ, i.e., DiagFlip($x_0, \cdots, x_{k-1}$) =*

$\tau^r(x_0, \cdots, x_{k-1})$ *where r is uniformly chosen from $\{0,1\}$, and it results in one of the two possibilities*

$$\mathsf{DiagFlip}(x_0, \cdots, x_{k-1}) = \begin{cases} (x_0, x_1, \cdots, x_{k-1}) \\ (\overline{x_{k-1}}, \cdots, \overline{x_1}, \overline{x_0}) \end{cases}$$

with probability exactly $1/2$.

The diagonal flip operation τ for $(x_0, \cdots, x_{k-1})$ is physically implemented by stacking the k cards (using two covers; see Sect. 2.1) and then flipping along the diagonal axis (Fig. 4).

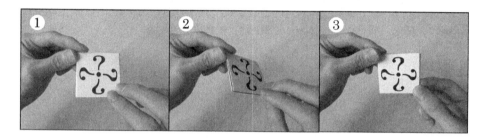

Fig. 4. An implementation of the diagonal flip operation τ

3   Multi-party Computation Using Polarizing Cards

3.1   Model

Now we are ready to define the notion of MPC protocols for polarizing cards. In contrast to standard MPC protocols, the number of parties is not essential in our protocols, as in most previous card-based protocols, since our protocols do not require private memories for parties, i.e., each of parties does not have any private information except his input information. Thus, our definition of a protocol is simply as follows.

Definition 6 (Protocol). *A k-card protocol Π is specified by a transition table with multiple rows where each row of the transition table contains a symbol corresponding to one of the following five operations:*

- *$\langle$Input$\rangle$: The protocol starts from here. This symbol only appears in the 0th row. Let s_0 be the initial sequence of k cards which encodes the input to the protocol[4] and contains any additional cards required by the protocol. Set the current sequence to s_0, and go to the next row.*
- *$\langle$Permutation, $\sigma\rangle$ where $\sigma \in \Sigma_k$: Apply the permutation σ to the current sequence, and go to the next row.*

---

[4] The alignment of the encoded input values must be specified by the protocol.

- $\langle$Superimposing$, (i,j), ind_0, ind_1\rangle$ where $i, j \in \{0, \cdots, k-1\}$, $i \neq j$, and $ind_0 \neq ind_1$: Let γ be the result of superimposing the i-th and the j-th cards. If γ is white, then go to the ind_0-th row, otherwise go to the ind_1-th row.
- $\langle$Shuffle$, \sigma\rangle$ where $\sigma \in \Sigma_k$: Apply the shuffle based on σ to the current sequence, and go to the next row.
- $\langle$Output$, S\rangle$ where $S \subset \{0, \cdots, k-1\}$: Set the final sequence to the current sequence. The protocol outputs the subsequence indexed by S, and terminates.

From now on, for the simplicity, we use $\langle$Shuffle$, name, S\rangle$ where the name is a shuffle name (e.g. Rotation or DiagFlip), and S is a subset of $\{0, \cdots, k-1\}$ that designates the cards which the shuffle should be applied to.

For a protocol Π, let $s_0(x_0, x_1, \cdots, x_{n-1})$ be the initial sequence corresponding to the input values [5] $(x_0, x_1, \cdots, x_{n-1}) \in \{0,1\}^n$. We say that a protocol Π computes $f(x_0, x_1, \cdots, x_{n-1})$ correctly if the following holds:

$$\forall x_0, x_1, \cdots, x_{n-1} \Big[\Pi(s_0(x_0, x_1, \cdots, x_{n-1})) \in \mathsf{Enc}(f(x_0, x_1, \cdots, x_{n-1})) \Big]$$

where $\Pi(s_0(x_0, x_1, \cdots, x_{n-1}))$ denotes the output sequence.

3.2 Security

We assume that all of parties are *honest-but-curious*, i.e., any parties do not deviate from the protocol (see Sect. 6.2, attacks outside our security model). Roughly speaking, security of a protocol is that any unbounded adversary can not learn the input information when he can see an execution of the protocol. However, due to the direction indistinguishability property, any adversary should be unable to obtain information regarding the direction of a sequence of cards. For this reason, what an adversary can learn essentially is results of superimposing.

For a protocol Π, its move depends only on the input values x and the random values generated in shuffles. We use Γ_Π to denote a random variable representing the values of superimposing; $\Gamma_\Pi = (\gamma_0, \cdots, \gamma_{m-1})$, where $\gamma_i \in \{white, black\}$, means that the i-th result of superimposing in the execution is γ_i.

Definition 7 (Security). *For a protocol Π, let X be a random variable taking over the input values $(x_0, x_1, \ldots, x_{n-1})$. We say that Π is secure if the following holds:*

$$H(X|\Gamma_\Pi) = H(X).$$

Corollary 1. *Let Π be a protocol. If Γ_Π is independent from the input values $(x_0, x_1, \ldots, x_{n-1})$ (e.g. each γ_i is either constant or random), then the protocol Π is secure.*

---

[5] Note that a vector of the input values is not equal to the initial sequence in general. For example, our XOR protocol (see Sect. 4.2) takes as an initial sequence $s_0 = (c_0, c_0 \oplus x, c_1, c_1 \oplus y)$. In the case, the vector of the input values is (x, y).

4   COPY, XOR, and AND Protocols

4.1   COPY Protocol

We construct a two-card COPY protocol that takes a card x, and outputs two copies of x. Using the protocol twice, we can immediately obtain a four-card COPY protocol for commitments. Note that a shuffle used in our COPY protocol does not require a cover since it is applied to a card (not a sequence).

0. ⟨Input⟩: Let s_0 be the initial sequence as follows.

$$s_0 = (x, c).$$

1. ⟨Shuffle, Rotation, $\{1\}$⟩: Apply a rotation shuffle to c:

$$s_1 = (x, \mathsf{Rotation}(c)) = (x, r)$$

where r is uniformly chosen from $\{0, 1\}$.
2. ⟨Superimposing, $(0, 1), 4, 3$⟩: Let γ be the result of superimposing x and r. If γ is *white*, i.e., $x = r$, then go to the 4th row, otherwise, $\overline{x} = r$, go to the 3rd row.

$$s_2 = \begin{cases} (x, x) & \text{if } \gamma \text{ is } white. \\ (x, \overline{x}) & \text{if } \gamma \text{ is } black. \end{cases}$$

3. ⟨Permutation, ρ_1⟩: ρ_1 is a fundamental rotation (See Definition 1). Rotate the right card by $90°$.

$$s_3 = \rho_1(s_2) = (x, x).$$

4. ⟨Output, $\{0, 1\}$⟩: Output the current sequence.

$$s_4 = (x, x), \text{the final and the output sequence.}$$

Security. A superimposing operation appears in the 2nd row and it only reveals $x \oplus r$. This is uniformly random since r is uniformly chosen from $\{0, 1\}$. Thus, our COPY protocol is secure.

4.2   XOR Protocol

Given two inputs $(c_0, c_0 \oplus x) \in \mathsf{Enc}(x)$ and $(c_1, c_1 \oplus y) \in \mathsf{Enc}(y)$, our optimal four-card XOR protocol proceeds as follows.

0. ⟨Input⟩: Let s_0 be the initial sequence as follows.

$$s_0 = (c_0, c_0 \oplus x, c_1, c_1 \oplus y).$$

1. ⟨Shuffle, Rotation, $\{0, 1\}$⟩: Apply a rotation shuffle to $(c_0, c_0 \oplus x)$:

$$s_1 = (\mathsf{Rotation}(c_0, c_0 \oplus x), c_1, c_1 \oplus y) = (r, r \oplus x, c_1, c_1 \oplus y)$$

where r is uniformly chosen from $\{0, 1\}$.

2. $\langle$Superimposing$, (0, 2), 4, 3)\rangle$: Let γ be the result of superimposing r and c_1. If γ is *white*, i.e., $r = c_1$, then go to the 4th row, otherwise, i.e., $\overline{r} = c_1$, go to the 3rd row.

$$s_2 = \begin{cases} (r, r \oplus x, r, r \oplus y) & \text{if } \gamma \text{ is } white. \\ (r, r \oplus x, \overline{r}, \overline{r} \oplus y) & \text{if } \gamma \text{ is } black. \end{cases}$$

3. $\langle$Permutation$, \rho_2 \circ \rho_3\rangle$: ρ_2 and ρ_3 are fundamental rotations (See Definition 1). Rotate $\overline{r}$ and $\overline{r} \oplus y$ by $90°$.

$$s_3 = \rho_2 \circ \rho_3(s_2) = (r, r \oplus x, r, r \oplus y).$$

4. $\langle$Output$, \{1, 3\}\rangle$: Output the sequence of the 1st and the 3rd cards. Note that the far-left one is the 0th card.

$$s_4 = (r, r \oplus x, r, r \oplus y), \text{the final sequence.}$$
$$\text{The output sequence is } (r \oplus x, r \oplus y).$$

Security. A superimposing operation appears in the 2nd row and only reveals $r \oplus c_1$. This is uniformly random since r is uniformly chosen from $\{0, 1\}$. Thus, our XOR protocol is secure.

4.3 AND Protocol

Given two inputs $(c_0, c_0 \oplus x) \in \mathsf{Enc}(x)$ and $(c_1, c_1 \oplus y) \in \mathsf{Enc}(y)$, our optimal four-card AND protocol proceeds as follows.

0. $\langle$Input$\rangle$: Let s_0 be the initial sequence as follows.

$$s_0 = (c_0, c_0 \oplus x, c_1, c_1 \oplus y).$$

1. Apply our XOR protocol to s_0. Let s_1 be the final sequence of the XOR protocol.
$$s_1 = (r, r \oplus x, r, r \oplus y)$$
where r is uniformly chosen from $\{0, 1\}$.

2. $\langle$Permutation$, \sigma_2 \circ \rho_0\rangle$: ρ_0 is a fundamental rotation and σ_2 is an adjacent transposition. Rotate the 0th card r by $90°$ and interchange the 2nd and the 3rd cards.
$$s_2 = \sigma_2 \circ \rho_0(s_1) = (\overline{r}, r \oplus x, r \oplus y, r).$$

3. $\langle$Shuffle$, \mathsf{DiagFlip}, \{0, 1, 2\}\rangle$: Apply a diagonal flip shuffle to $(\overline{r}, r \oplus x, r \oplus y)$:

$$s_3 = (\mathsf{DiagFlip}(\overline{r}, r \oplus x, r \oplus y), r) = \begin{cases} (\overline{r}, r \oplus x, r \oplus y, r) & \text{(A)} \\ (\overline{r \oplus y}, \overline{r \oplus x}, r, r) & \text{(B)}. \end{cases}$$

$$= \begin{cases} (\overline{r} \oplus (x \wedge y), r, r \oplus (\overline{x} \wedge y), r) & \text{when (A)} \wedge (x = 0) \text{ or (B)} \wedge (x = 1). \\ (\overline{r} \oplus (\overline{x} \wedge y), \overline{r}, r \oplus (x \wedge y), r) & \text{when (A)} \wedge (x = 1) \text{ or (B)} \wedge (x = 0). \end{cases}$$

where (A) and (B) are events which are outcomes of the shuffle.

4. $\langle$Superimposing, $(1,3), 5, 7)\rangle$: Let γ be the result of superimposing the 1st card and the 3rd card. If γ is *white*, i.e., the 1st card is r, then go to the 5th row, otherwise, i.e., the 1st card is $\overline{r}$, go to the 7th row.

$$s_4 = \begin{cases} (\overline{r} \oplus (x \wedge y), r, r \oplus (\overline{x} \wedge y), r) & \text{if } \gamma \text{ is } white. \\ (\overline{r} \oplus (\overline{x} \wedge y), \overline{r}, r \oplus (x \wedge y), r) & \text{if } \gamma \text{ is } black. \end{cases}$$

5. $\langle$Permutation, $\rho_0\rangle$: Rotate the 0th card by $90°$.

$$s_5 = \rho_0(s_4) = (r \oplus (x \wedge y), r, r \oplus (\overline{x} \wedge y), r).$$

6. $\langle$Output, $\{0,3\}\rangle$: This is the output phase when γ is *white*. The output is the sequence of the 0th and the 3rd cards.

$$s_6 = s_5 = (r \oplus (x \wedge y), r, r \oplus (\overline{x} \wedge y), r), \text{the final sequence.}$$
The output sequence is $(r \oplus (x \wedge y), r)$.

7. $\langle$Output, $\{2,3\}\rangle$: This is the output phase when γ is *black*. The output is the sequence of the 2nd and the 3rd cards.

$$s_7 = (\overline{r} \oplus (\overline{x} \wedge y), \overline{r}, r \oplus (x \wedge y), r), \text{the final sequence.}$$
The output sequence is $(r \oplus (x \wedge y), r)$.

Security. Superimposing operations appear in the 1st row and the 4th row. The result in the 1st row is uniformly random (see the security of XOR protocol 4.2). The superimposing in the 4th row results in $(r \oplus x) \oplus r$ or $(\overline{r \oplus x}) \oplus r$ with probability exactly $1/2$. This is uniformly random. Thus, our AND protocol is secure.

5 Applications

In this section, we construct an optimal four-card input-preserving XOR protocol and an optimal four-card input-preserving AND protocol, each of which also outputs one of the input commitments. Based on these protocols and applying the techniques from [3,10], we obtain various applications.

5.1 Input-Preserving Protocols

Input-Preserving XOR Protocol. We design an optimal four-card input-preserving XOR protocol. This protocol is used in our half-adder and our full-adder protocols as a subroutine. It takes two commitments to x, y as inputs, and outputs two commitments to $x, x \oplus y$ as follows. Let $s_0 = (c_0, c_0 \oplus x, c_1, c_1 \oplus y)$ be the initial sequence. Apply our XOR protocol to s_0 and let $s_1 = (r, r \oplus x, r, r \oplus y)$ be the final sequence. We obtain the output sequence $s_2 = (r, r \oplus x, r \oplus x, r \oplus y)$ by applying our COPY protocol.

Input-Preserving AND Protocol. We design an optimal four-card input-preserving AND protocol. This protocol is used in our half-adder and full-adder protocols as a subroutine. It takes two commitments to x, y as inputs, and outputs two commitments to $x \wedge y, x$ as follows. Let $s_0 = (c_0, c_0 \oplus x, c_1, c_1 \oplus y)$ be the initial sequence. Apply our AND protocol to $(c_1, c_1 \oplus y, c_0, c_0 \oplus x)$. We operate a permutation to the final sequence of the AND protocol, and obtain the sequence $s_1 = (r, r \oplus (y \wedge x), r, r \oplus (\overline{y} \wedge x))$. Using our COPY protocol, we can obtain the sequence $s_2 = (r, r \oplus (y \wedge x), r \oplus (y \wedge x), r \oplus (\overline{y} \wedge x))$. This can be simplified into $(c_2, c_2 \oplus (x \wedge y), c_3, c_3 \oplus x)$.

5.2 Applications

Half-Adder and Full-Adder Protocols. Based on our input-preserving XOR/ AND protocols, we obtain the construction of an optimal four-card half-adder protocol and an optimal six-card full-adder protocol. These constructions are the same as [3,10], while the half-adder and full-adder protocols from [3,10] require six cards and ten cards, respectively.

Voting Protocol. Secure voting is one of the most suitable applications for card-based protocols. Similarly to [3], using half-adder and full-adder protocols, we can construct a secure voting protocol with two candidates. Our voting protocol for n parties requires $2\lceil \log n \rceil + 2$ cards while the protocol from [3] requires $2\lceil \log n \rceil + 6$ cards. Here, our protocol is optimal since a voting protocol takes as inputs $2\lceil \log n \rceil$ cards and 2 cards for a voting result of $n - 1$ parties and an input of a party.

Protocols for Any Functions and Symmetric Functions. Composing our protocols, we can securely compute any functions. However, it is not known *how many cards a protocol needs*. Nishida et al. [10] calculated a good upper bound on the number of cards used in a protocol for an arbitrary function. Applying the techniques from [10], we obtain an upper bound for the number of cards used in a protocol of polarizing cards.[6] For an arbitrary boolean function $f(x_0, \cdots, x_{n-1})$, we obtain the construction of a protocol for $f(x_0, \cdots, x_{n-1})$ using $2n + 4$ cards, while the protocol from [10] uses $2n+6$ cards. For an arbitrary symmetric boolean function $f(x_0, \cdots, x_{n-1})$, we obtain the construction of an optimal protocol for $f(x_0, \cdots, x_{n-1})$ using $2n$ cards, while the protocol from [10] requires $2n + 2$ cards.

6 Discussions

In this section, we will discuss one-card-per-bit encoding schemes and tradeoffs they provide. We will furthermore discuss attacks outside our security model, such as injection attacks and polarizing glasses attacks, as well as countermeasures against them.

---

[6] Note that this is not an improvement for the work [10] since the upper bound in our setting and [10] is different.

6.1   One-Card-Per-Bit Scheme

Most of the previous works on card-based protocols use a *two-cards-per-bit encoding scheme*. On the other hand, using cards with a rotationally symmetric back side, Mizuki and Shizuya [7] proposed an encoding scheme using just a single card (*one-card-per-bit encoding scheme*). The cards with a rotationally symmetric back side [7] enable the construction of a three-card COPY protocol, a two-card XOR protocol, and a three-card AND protocol. However, the shuffle used in the AND protocol is not known to have an easy physical implementation, i.e., it requires two operations, a rotation and a permutation, at the same time. Similarly to [7], polarizing cards can be used in a one-card-per-bit encoding scheme but requires a "base card" that specifies a standard direction and is placed in the public space. The encoding works as follows: if a card has the same direction as the base card, then it is a commitment to 0, otherwise it is a commitment to 1. Applying our techniques, it is possible to construct a two-card COPY protocol, a two-card XOR protocol, and a three-card AND protocol.[7] (The details of the constructions will appear in the full version.) In contrast to [7], the diagonal flip shuffle used in the AND protocol has an easy physical implementation (Sect. 2.2). Moreover, the COPY protocol in our scheme only uses two cards while the scheme [7] requires three cards. However, our scheme requires the base card for the encoding. Therefore, there is a trade-off between the previous cards [7] and polarizing cards in one-card-per-bit encoding scheme.

6.2   Attacks from Outside of Our Model

Injection Attack. In the commonly used encoding scheme, i.e., $(\clubsuit, \heartsuit) = 0$ and $(\heartsuit, \clubsuit) = 1$, a malicious party might submit an invalid commitment like $(\clubsuit, \clubsuit)$ or $(\diamondsuit, \clubsuit)$. This immediately breaks the security of protocols since an invalid commitment behaves as a marker, and the malicious party might learn the randomness used in a shuffle. We say that such an attack is a *valid-card injection attack* when it uses valid cards such as $(\clubsuit, \clubsuit)$, and an *invalid-card injection attack* when it uses invalid cards such as $(\diamondsuit, \clubsuit)$. Mizuki and Shizuya [7] proposed a countermeasure which is capable of detecting an invalid commitment without revealing any information.

In our polarizing cards, a valid-card injection attack does not exist since two valid cards are always a commitment to either 0 or 1. In addition, there is a countermeasure against invalid-card injection attacks. Without loss of generality, we assume that invalid-card injection attacks are just using a polarizing card with a direction that is neither $\updownarrow$ or $\leftrightarrow$. Using a random valid card r and applying superimposing operation to x and r, we can detect that x is invalid. Therefore, our scheme has countermeasures against injection attacks.

---

[7] We do not include the base card in the number of cards since it is independent of the protocols and common among all of protocols.

Polarizing Glasses Attack. *Polarizing glasses* are glasses that use a material with the polarizing property as glass (Fig. 5). If a malicious party wears polarizing glasses, he can learn the direction of the polarizing cards, and thereby the secret values of the other parties. Since a person who wears polarizing glasses can detect an adversary with polarizing glasses, we can use polarizing glasses before execution of protocols. As another countermeasure, we can use a cover with two-sided and rotational symmetries. If a polarizing card is covered by a cover, adversaries can not learn the direction of the card. In that case, when two cards will be superimposed, it is necessary to apply the rotation shuffle to the cards, and then the covers are removed to superimpose. The adversary can not learn the secret values more than the result of superimposing.

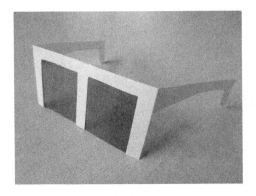

Fig. 5. A polarizing glasses

Acknowledgment. The authors would like to thank members of the study group "Shin-Akarui-Angou-Benkyou-Kai" for the valuable discussions and helpful comments, and thank the anonymous reviewers for their comments. This work was partially supported by JSPS KAKENHI Grant Numbers 26330001 and 26330151, Kurata Grant from The Kurata Memorial Hitachi Science and Technology Foundation, and JSPS A3 Foresight Program.

References

1. Crépeau, C., Kilian, J.: Discreet solitary games. In: Stinson, D.R. (ed.) CRYPTO 1993. LNCS, vol. 773, pp. 319–330. Springer, Heidelberg (1994)
2. den Boer, B.: More efficient match-making and satisfiability. In: Quisquater, J.-J., Vandewalle, J. (eds.) EUROCRYPT 1989. LNCS, vol. 434, pp. 208–217. Springer, Heidelberg (1990)
3. Mizuki, T., Asiedu, I.K., Sone, H.: Voting with a logarithmic number of cards. In: Mauri, G., Dennunzio, A., Manzoni, L., Porreca, A.E. (eds.) UCNC 2013. LNCS, vol. 7956, pp. 162–173. Springer, Heidelberg (2013)

4. Mizuki, T., Fumishige, U., Sone, H.: Securely computing XOR with 10 cards. Australas. J. Comb. **36**, 279–293 (2006)
5. Mizuki, T., Kumamoto, M., Sone, H.: The five-card trick can be done with four cards. In: Wang, X., Sako, K. (eds.) ASIACRYPT 2012. LNCS, vol. 7658, pp. 598–606. Springer, Heidelberg (2012)
6. Mizuki, T., Shizuya, H.: A formalization of card-based cryptographic protocols via abstract machine. Int. J. Inf. Sec. **13**, 15–23 (2014)
7. Mizuki, T., Shizuya, H.: Practical card-based cryptography. In: Ferro, A., Luccio, F., Widmayer, P. (eds.) FUN 2014. LNCS, vol. 8496, pp. 313–324. Springer, Heidelberg (2014)
8. Mizuki, T., Sone, H.: Six-card secure AND and four-card secure XOR. In: Deng, X., Hopcroft, J.E., Xue, J. (eds.) FAW 2009. LNCS, vol. 5598, pp. 358–369. Springer, Heidelberg (2009)
9. Niemi, V., Renvall, A.: Secure multiparty computations without computers. Theor. Comput. Sci. **191**(1–2), 173–183 (1998)
10. Nishida, T., Hayashi, Y., Mizuki, T., Sone, H.: Card-based protocols for any boolean function. In: Jain, R., Jain, S., Stephan, F. (eds.) TAMC 2015. LNCS, vol. 9076, pp. 110–121. Springer, Heidelberg (2015)
11. Nishida, T., Mizuki, T., Sone, H.: Securely computing the three-input majority function with eight cards. In: Dediu, A.-H., Martín-Vide, C., Truthe, B., Vega-Rodríguez, M.A. (eds.) TPNC 2013. LNCS, vol. 8273, pp. 193–204. Springer, Heidelberg (2013)
12. Stiglic, A.: Computations with a deck of cards. Theor. Comput. Sci. **259**(1–2), 671–678 (2001)

Systems Security

An Analysis Platform for the Information Security of In-Vehicle Networks Connected with External Networks

Takaya Ezaki[1], Tomohiro Date[1], and Hiroyuki Inoue[1,2(✉)]

[1] Graduate School of Information Sciences, Hiroshima City University,
3-4-1 Ozuka-Higashi, Asa-minami, Hiroshima 731-3194, Japan
{ezaki,date,hinoue}@v6.inet.info.hiroshima-cu.ac.jp
[2] National Institute of Information and Communications Technology,
4-2-1 Nukui-Kitamachi, Koganei, Tokyo 184-8795, Japan

Abstract. Most in-vehicle units, which are used for various services such as remote diagnosis, information gathering, telematics, car navigation with communications, etc., are connected to both an external network and an automotive network, referred to as an in-vehicle network. The information security of the in-vehicle network is an important concern. In this paper, we describe the analysis platform that we designed and developed for protection mechanisms to analyze and verify the information security of an in-vehicle network connected with external networks. The platform provides features to analyze messages of the in-vehicle network, to evaluate the effects of attacks on the in-vehicle network, and to verify protection mechanisms for the in-vehicle network. First, we developed a message analysis platform subsystem that analyzes messages of the in-vehicle network and performs send and receive processes in the in-vehicle network. Next, we designed an attack evaluation platform subsystem that evaluates attack mechanisms from servers or attackers on the Internet. We also developed a prototype (MoVIS) of the subsystem. We confirmed the effectiveness of the attack evaluation platform and MoVIS in an actual automobile, which was subjected to spoofing and DoS attacks to the in-vehicle network via the Internet. Finally, we discuss the requirements of the protection verification platform, which provides protection mechanisms and protection features based on the experimental results and the characteristics of the in-vehicle network.

Keywords: Automotive network · Information security · CAN · Spoofing · DoS

1 Introduction

Due to the advanced features and service diversification of automobiles and the rise of cellular networks with competitive pricing, automobiles are being equipped with always-on communications with wide-area networks such as the Internet. The information security of an automobile is an important concern [1]. The automobile connects external networks, including remote diagnostics, auto accident report systems, vehicle-to-vehicle communications, and infrastructure-to-vehicle communications.

© Springer International Publishing Switzerland 2015
K. Tanaka and Y. Suga (Eds.): IWSEC 2015, LNCS 9241, pp. 301–315, 2015.
DOI: 10.1007/978-3-319-22425-1_18

In particular, in-vehicle devices such as the car navigation system and the telematics device connecting to both the in-vehicle networks and the external networks are security risks. Such devices are often attached by the user and the automotive dealer after production of the automobile. Moreover, storage media such as mobile phones and USB memory are also connected to the automobile. So, the increasing number of malicious accesses and intrusion routes poses higher risks [2].

In this paper, we describe a platform that we designed and developed for analyzing and verifying the information security of an in-vehicle network. The most important purpose of this platform is that it introduces a protection mechanism for automobile information security. However, no general indicator of protection mechanisms of automobiles has existed to date, so we built a platform in stages. First, we analyze communication messages on the in-vehicle network, and then we examine the influences of attacks on the in-vehicle network, and finally we verify the protection mechanisms against the attacks. As an attack evaluation platform, we implemented an attack prototype system named MoVIS (Mobile vehicle Verification system for Information Security). MoVIS includes an in-vehicle device that is always connected to both the Internet and the Controller Area Network (CAN) [3], which is generally used as an in-vehicle network and connects the many electronic control units (ECUs) in an automobile. We confirmed that MoVIS can attack an in-vehicle network with spoofing and DoS attacks via the Internet to develop and establish the protection mechanism. Moreover, we discuss the requirements of the protection verification platform, which provides protection mechanisms based on the experimental results and the characteristics of the in-vehicle network.

2 Information Security of an Automobile

2.1 Security Issues of an Automobile Connected to External Networks

As the varieties of mobile communication media, personal mobile devices used by automotive users, and communication devices connected to the vehicle itself in-crease, the risk to automotive information security also increases [4]. As less well known examples, it is becoming common to use communications through the growing use of a probe car, which observes traffic situations according to the information of automotive status and motions, and remote inspecting or remote maintenance. In addition, more mobile devices or storage media of automotive users are directly connected to in-vehicle devices and wide-area networks such as the Internet. Researchers have pointed out the possibilities of malicious accesses or attacks by the connection of external networks or storage media [5].

Figure 1 shows a model of an automobile that can be connected to an outside network. The automobile is equipped with several tens of ECUs, such as engine control ECUs, transmission control ECUs, brake control ECUs, etc. The ECUs are connected with in-vehicle networks such as the CAN, which is widely used in automobiles. The CAN is connected to gateway devices, GPS navigation, or telematics devices. The diagnostic port equipped on an automobile, known as the on-board diagnostics OBD-II, includes the CAN bus signal and is accessible to the in-vehicle network. Because the

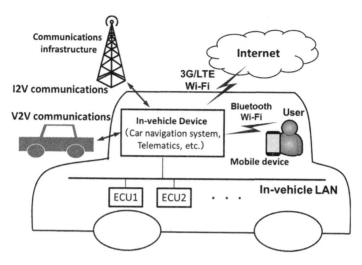

Fig. 1. System model

CAN protocol is very simple, numerous studies about their vulnerability have been conducted [6, 7].

2.2 Drawbacks of CAN

The CAN, which is assumed in this paper as the in-vehicle network, is a common bus system that communicates with other nodes in a broadcast manner. The CAN also communicates by using only the destination address, without the source address. In addition, the communication rate of the CAN is only 500 kbps, so the CAN is considered to be vulnerable to spoofing and DoS attacks because of its nature as a protocol.

2.3 Related Work

Attack Methods on In-Vehicle Networks. Related research has shown an in-vehicle network can be attacked from external networks (Bluetooth, FM radio, etc.) and has demonstrated the results. For example, attackers can use the connection point between in-vehicle networks and Bluetooth devices to attack an automobile [8]. Specifically, the attacker implements a Trojan application for a user's smartphone and sends attack messages to the in-vehicle network via the smartphone with background communication. In addition, an automobile can be attacked by obtaining the MAC address of the Bluetooth device by analyzing the communication traffic after pairing to the automotive Bluetooth device. As for other methods, researchers have shown that the number of attack routes using the connection points between in-vehicle networks and external networks is increasing [9, 10]. Moreover, one survey paper investigated recent trends with a focus on the CAN bus and automotive technology, and actually tested the powertrain, warning light, and air-bag. That paper also categorized four scenarios of attacks, based on existing Computer Emergency Response Team (CERT) classification criteria [11].

However, even though attacking methods of the automobile have been studied, few studies have investigated attacking methods connected to the Internet, and few platforms have been developed against such attacking methods.

Protection Mechanism for In-Vehicle Networks. The ECUs in an automobile manage and control the driving, so much damage would result if the ECUs were attacked with spoofing or taken over by other attackers; in reality, drivers' lives could be threatened. Since a lot of studies about attacking methods are now being conducted, studies about protection methods are also being carried out.

One proposed method extends the protocol of CAN by adding an authentication system [12–14]. Specifically, one study mentioned the requirements of the authentication protocol, the limitation of the CAN bus, issues caused by the transmission speed, and communications with multiple nodes [12]. It also suggested a protocol named CANAuth, which can authenticate with HMAC. By applying CAN+ for communications on the CAN bus, authentication on CAN would be possible. Another reference concerning authentication proposed a protocol based entirely on simple symmetric primitives, which make use of key splitting and MAC mixing [13].

In addition, research about the protection techniques for hardware has been done. One study introduced the hardware security module (HSM), which protects all communications taking place among in-vehicle ECUs based on a requirements engineering approach [15]. HSM protects the security of software by functioning as a trusted security anchor and limits the possibility of a hardware tampering attack by adapting effective tamper-protection measures. Plus, using highly optimized special circuitry instead of costly general purpose hardware makes it possible to reduce the security cost. In other studies, one research equipped "taint tracking tools" in a security framework for the conventional automobile communications system [16]. By observing the data flow in the control units with an architecture introducing data flow tracking, data could only be used for their intended purpose in applications, and the data could be monitored by ECUs when the code was executed.

However, despite these proposals, little progress has been made for the CAN protocol with security features, since it is difficult to improve ECUs and to make compatible products in the short term.

3   Development of the Analysis Platform

3.1   Outline

As mentioned in Sect. 2, it is necessary to introduce protection mechanisms for the information security of in-vehicle networks. However, since we did not have existing evaluations of the risk of attacks on in-vehicle networks, ECUs, and protection mechanisms, we designed and developed an analysis platform. The platform is designed as a generic platform based on models of the in-vehicle device connected with both external networks and in-vehicle networks. The platform is for a protection verification platform that adopts protection methods or implements protection mechanisms. The platform also provides an experimental environment for various kinds of attack and protection methods.

Figure 2 shows the overall structure of the analysis platform proposed in this paper. The platform is connected to a CAN bus as an in-vehicle network via the connection point of an automobile, and is also connected to the Internet with always-on communications with 3G/LTE communications, Wi-Fi, and Bluetooth. The protocol analysis feature communicates with the CAN bus, analyzes the functions of CAN messages between EUCs, and estimates the status of the automobile. The attack evaluation feature receives an order from the external network, and sends appropriate CAN messages to the automobile according to the order. A proper user is, for example, an automobile's owner or a organization authorized by the owner, such as the car dealer or the car insurance company. The attackers are on the Internet. The support server provides several services for the in-vehicle device, the user, and the attackers. The authentication feature authenticates whether a device connected to the connection point is proper.

To develop the platform, we designed and developed the following in this order: (1) message analysis platform, (2) attack evaluation platform, and (3) protection verification platform. These functions are described in detail in subsequent sections. In this study, we conducted experiments for (1) and (2), and then for (3), based on the knowledge we acquired through the experiments for (1) and (2), and we consider the protection methods and mechanisms here.

3.2 Message Analysis Platform

The message analysis platform communicates with ECUs in the in-vehicle network, verifies the protocol of the in-vehicle network, and analyzes the messages in the

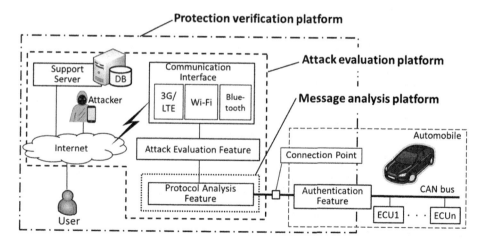

Fig. 2. Overall structure of proposed platform

network. The message analysis platform also observes messages and traffic, and obtains information about the supposed status of the automobile by capturing the messages. After developing the message analysis platform, we applied it to an actual car and conducted traffic and message analyses for CAN as a preliminary experiment. The actual car, which the author owns, is a hybrid vehicle that was first marketed in August 2013.

Experiment of CAN Messages Analysis. After the preliminary experiment using an actual car, we analyzed the CAN messages from the in-vehicle network. As the analysis method, we used the replay attack described below.

1. Capture the messages flowing on the CAN bus in the car for a certain time when the analysis target in the car is working properly for a certain action, and save the messages as a text file. For example, capture messages for three seconds when the windows are open.
2. Pick up the saved messages and send messages that are partly modified from the designated range to the CAN bus again.
3. Verify whether the analysis target operates with the expected action when sending back the messages, and if the expected action is observed, make the specified designated range more narrow and analyze it as a binary search tree.
4. Repeat (3), so that the last remaining message is the one related to the target's motion/action. If not found, return to (1) to capture it again and proceed as before.

In this procedure, by using the replay attack and analyzing the correlation between automotive action and observed messages, we carried out an analysis on CAN messages from the actual car. In addition, we classified the messages into the following three categories: messages of status, messages to perform some action, and messages of warning. The messages that could be analyzed are shown in Table 1.

Table 1. CAN messages to be analyzed

Category	Description	Remark
Status	Speed	km/h units
Status	RPM	rpm units
Status	Steering angle	Degree units
Status	Brake depression	
Status	Gear position	Reverse, neutral or drive (R, N, or D)
Action	Door lock open (1)	Valid in the ignition state OFF
Action	Door lock close (1)	
Action	Door lock open (2)	
Action	Door lock close (2)	
Action	Power window open	Any windows can be specified
Action	Power window close	
Action	Hazard flashing	Times can be specified
Warning	Seat belt warning	Turn on the warning light
Warning	Parking brake warning	Turn on the warning light
Warning	Engine oil pressure	Show warning light and message

The results of this analysis were obtained from the actual car we used, so we did not verify whether the same methods would be applicable to all kinds of automobiles. However, if we adopt this platform and follow the same procedures, it is possible that different types of automobiles could have the same results. Therefore, we should establish an automatic analysis method of messages for various kinds of automobiles and manufactures.

3.3   Attack Evaluation Platform

The attack evaluation platform uses servers that have databases on the Internet, manage the information, and provide APIs as well as a Web interface. It also has some functions to modify the messages for attacks by using the results from the message analysis platform. This time, we developed MoVIS, an attack prototype system that considers certain scenarios as the attack evaluation platform. Figure 3 shows the implementation of MoVIS. MoVIS consists of the in-vehicle device, which can be connected to both the in-vehicle network and the Internet, the support server on the Internet (MoVIS server), and the Web interface that supports the attacker. The in-vehicle device imitates a telematics device placed inside an automobile.

The scenarios we consider by using MoVIS are as follows.

1. Connect the in-vehicle device to the OBD-II port for diagnosis. Suppose the in-vehicle device is always connected to the Internet and accessible from an attacking program.
2. The in-vehicle device constantly sends information about the automobile to the support server, such as location information from the GPS, whether the ignition is on or off, etc.

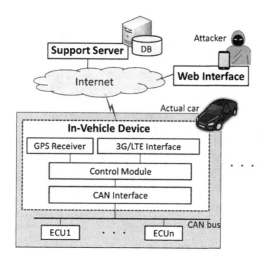

Fig. 3. Implementation of MoVIS

3. The attacker, using smartphones or the Web, issues specific orders to the specified in-vehicle device shown on the Web interface. For example, the orders are unlocking the doors, opening the designated window, and/or sending illegal data to the instrumental panel with some conditions.

In-Vehicle Device. The in-vehicle device is connected to the Internet by a telecommunication line such as 3G/LTE. If the in-vehicle device is a car navigation system or a telematics device, we assume that it also is connected to the CAN bus of the in-vehicle network. The inner program in the in-vehicle device captures and sends messages through the CAN bus, and enables functions that the attacker intends to alter. It also has the GPS location identification system and a function to constantly send notifications to the MoVIS server. The notifications contain information of the location information and the inner state of the automobile that can be acquired from the CAN bus.

Each in-vehicle device has its own identifier (ID), which has a MAC address, and a communication interface connected to the Internet, which has a global IP address. For the control module of the in-vehicle device, we used a small-scale Linux machine with the 3G communications interface board installed. For the CAN interface to the CAN bus, we used a microcomputer with the CAN interface installed. The microcomputer was connected via a USB serial communication to the Linux machine. Figure 4 shows a photo of the in-vehicle device, and Table 2 shows the specification of the computers. The liquid crystal display, shown in the front, shows information of the in-vehicle device, such as the IP address and status.

Fig. 4. Photo of the in-vehicle device

Table 2. Specification of computers in the in-vehicle device

(a) Control module	
Computer	Raspberry Pi B+ with Raspbian
CPU	ARM1176JZFS, 700 MHz
RAM	512 Mbyte
Storage	8 Gbyte (MicroSD card)
Interface	USB, PIO, I2C, 3G communication
(b) CAN interface	
Computer	Arduino UNO
CPU	ATmega328, 16 MHz
RAM	2 kbyte
Storage	32 kbyte
Interface	UART, CAN via SPI

MoVIS Server. The MoVIS server assists the attacker and the in-vehicle device, manages the information about the in-vehicle device, and offers APIs required for this system. Specifically, it has multiple functions to receive data constantly sent from several in-vehicle devices, record them in an internal database, manage the information, offer a user interface against an attacker, and send orders to a specific in-vehicle device based on the instructions. Plus, communication between the server and the in-vehicle device is accomplished with the global IP address that the in-vehicle device is assigned. Access to the server is authenticated with a password. The APIs offered by the server are as follows.

1. Search API
   Search for in-vehicle devices in which the information (ID, longitude, latitude, status, etc.) in the database is updated within 30 min. The API, which returns the results in xml format, is used when the Web interface displays information about the in-vehicle device location on a map.
2. Operation API
   Search for the designated ID of an in-vehicle device and obtain the IP address linked with the in-vehicle device. By using TCP, send the specified operation details to the in-vehicle device linked with the IP address. The API is used when the Web interface sends orders to the in-vehicle device.
3. Registration API
   Search to determine whether the designated ID exists. If the designated ID is not found, complete a new registration of the in-vehicle device. If found, update the in-vehicle ID, longitude, latitude, status, and updated time in the database. The API is used when the in-vehicle device notifies the MoVIS server of this information.

Web Interface. In the MoVIS scenario, we developed a Web user interface to enable operations through a smartphone or a PC. The Web interface enables an attacker to easily choose a target and give operation orders. The Web interface is shown in Fig. 5. The bigger markers on the Google map indicate the location and the ID of an in-vehicle device. The small marker indicates the location of the mobile device on the Web.

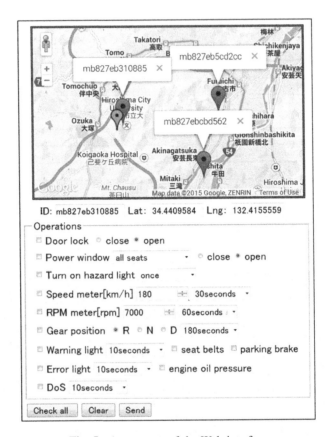

Fig. 5. Appearance of the Web interface

For an operation, the attacker first designates any in-vehicle device. Then, its ID and both longitude and latitude are shown on the bottom of the map. Next, the attacker specifies the orders of the operation to the automobile with the in-vehicle device, and the orders are sent to the device via the server.

4   Experiments with an Actual Car

We applied the same MoVIS system described above to the actual car we used in the preliminary experiment in Sect. 3.2. The results are stated below.

4.1   Acquisition of Automotive Status

We can obtain various kinds of information about the automotive status by using the messages in Table 1 from the in-vehicle device. For example, it is assumed that the automobile is in the running state by the gear status and the speed. It is also possible to identify the ignition state as ON or OFF by observing the traffic of all or specific CAN messages. From these, it is possible to estimate not only the automotive status but personal information, such as the driver's home address.

4.2 Spoofing Attack

By sending a message imitating an authentic one, which is sent from the engine control ECU or the transmission control ECU via the CAN bus, we verified the influences of the display on the instrument panel or the action on the automobile's equipment. Specifically, we observed the following actions and results.

1. Set the tachometer to an arbitrary value.
   The tachometer showed an unusual engine rotation value. The meter indicated that the values fluctuate rapidly because of the message contention between the authentic values and the arbitrary values.
2. Set the speedometer to an arbitrary value.
   The speedometer showed an unusual value for the speed. The meter indicated that the values fluctuate rapidly because of the message contention between the authentic values and the arbitrary values. Moreover, we confirmed significant interference to the odometer through spoofing attacks to the speedometer. Specifically, even though the automobile was not in motion at the parking gear position, the odometer increased when we sent the message of 180 km/h.
3. Change the gear position indicator.
   Regardless of whether an automobile is in motion, a spoofing attack is possible and we could change the indicator of the gear position. In the case of R (Reverse), despite the fact that the gear was not actually in reverse, the back buzzer beeped.
4. Turn on the warning lights for the seat belt or the parking brake.
   We confirmed that we could turn on the alarm light for the seat belt or the parking brake, or both. When the automobile was in motion, if we turned on the warning light for the parking brake, the alarm buzzer beeped as a side effect.
5. Turn on the hazard lights.
   We made the hazard lights blink by sending a message.
6. Unlock or lock the doors.
   Even when the key was not near the automobile and it was parked, we unlocked the doors with some kinds of the messages. Note that the message for unlock or lock the doors is different from the one when the ignition is on.
7. Open or close one of several windows.
   We opened or closed one of the windows. This action was possible when the automobile was parked and locked.

As for 1–4, the operation was successful when only the ignition was on. For 5–7, it worked even when the ignition was off. As an execution example, Fig. 6(a) and (b) show the display of the instrumental panel when we sent the data of 7000 rpm to the tachometer and 180 km/h to the speedometer.

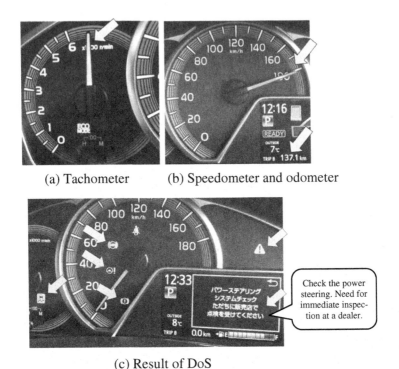

(a) Tachometer   (b) Speedometer and odometer

(c) Result of DoS

Fig. 6. Spoofing and DoS attacks to instrument panel

4.3 DoS Attack

A DoS attack was achieved by continuously sending the most prioritized messages to the CAN bus. By sending messages that had a CAN ID of 0 and the longest payloads, we observed a massive amount of errors that occurred when the message communications between ECUs were interrupted. The details are as follows.

1. The warning was displayed from the brake system, and the ABS function was inactivated.
2. The warning was displayed from the power steering system, and the power steering assistance was inactivated.
3. The warning was displayed from the hybrid system, and the message "check immediately at an automotive dealer" was displayed.
4. The message was displayed when the power was on, and the message was displayed only when the ignition was on.

Figure 6(c) shows examples of the instrumental panel during the DoS. When the DoS was conducted in the P (Parking) gear, we confirmed that the automobile never seemed to drive even when the gear was moved to the D (Drive) or R (Reverse) gears after the DoS messages finished; in this case, we finally needed to turn off the ignition and turn it back on to recover the correct functionalities. We expected that the DoS would affect the functionalities of an automobile widely, so we ignored the frequency and the importance of proper messages under the DoS attack.

4.4   Consideration Over Protection Mechanism

From the results of the message analysis platform and attack evaluation platform experiments, we demonstrated that operations via the Internet on the in-vehicle device including the CAN are possible. Therefore, we proved the existence of risk. It is necessary to introduce a protection mechanism and systems for the in-vehicle device. In this section, we apply our knowledge that we obtained from the experiments to the protection platform.

Protection Mechanism for the CAN Bus. As mentioned in Sect. 2, improving the CAN protocol or ECUs is proposed as one of the protection methods for automotive security. Adding an authentication system or encryption to the CAN protocol will be difficult to realize in the near future because of the drawbacks of the CAN mentioned in Sect. 2.2. A new Ethernet-based protocol that can replace CAN is being proposed and developed in Europe [17]. However, it is said that this new protocol will take several more years to deploy. In addition, improving the ECUs has some problems, such as its financial cost and difficulties in achieving compatibility with existing products. As the related studies and our experiment showed, attacks on the CAN bus or the ECUs are easily performed via the OBD-II port or a direct connection to the internal CAN bus. Introducing protection mechanisms for the CAN is an urgent need, and so we have to consider a protection system ready in the short term.

The connection point between the OBD-II port and car navigation or telematics device is also the connection point with the external networks. We propose a simple security gateway on the connection points, as shown in Fig. 7, for the short-term

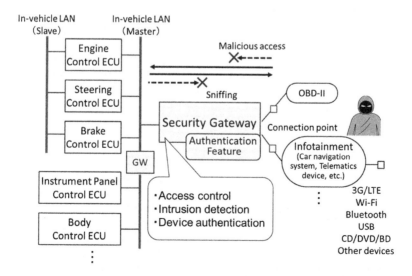

Fig. 7. Example of protection mechanism for the short term

solution. Introducing gateways on the connection point is the first phase of security enhancement. A gateway filters the messages on the CAN bus and works as an Internet Intrusion Prevention System (IPS). Specifically, for the connection point of the car navigation system, the gateway would pass only the message of the steering angle, and so collectively passes the messages required for action of the external device. For the OBD-II port, the gateway implements the following features: (1) passes the general messages that show the status of engines or bodies, (2) does not pass messages directly to an ECU, and (3) passes authentic messages from the device, as in the diagnostic tool used by automotive dealers with some kinds of authentication methods.

5 Conclusion

In this study, we designed an analysis platform for the information security of in-vehicle networks connected with external networks. To develop the protection verification platform for the in-vehicle network, we implemented the prototype of both the message analysis platform and the attack evaluation platform, and conducted experiments using an actual automobile. First, through the message analysis of CAN messages by using the actual automobile, we confirmed that we could analyze the messages that realized automotive functions. Next, we implemented MoVIS as an attack prototype system to verify the attacks from external networks. By applying MoVIS to the actual car and performing experiments, we verified that spoofing and DoS attacks on the in-vehicle network are possible and confirmed the effectiveness of the attack evaluation platform. The message analysis platform in Sect. 3.2 and the demonstration using MoVIS in Sects. 4.2 and 4.3 were conducted with an actual car on the market. Therefore, the demonstration results in this paper are obtained from the actual car we used, so we did not verify whether the same methods would be applicable to all kinds of automobiles. However, if we adopt this platform and follow the same procedures, it is possible that even different types of automobiles can give the same results. Finally, based on the results from the experiments on MoVIS and the characteristics of the CAN, we summarized the points for the protection mechanism and proposed a security gateway to be introduced in the short term. As future issues, we should establish an automatic analysis and protection method for various kinds of automobiles and manufactures.

References

1. McCarthy, C., Harnett, K., Carter, A.: Characterization of potential security threats in modern automobiles. Report No. DOT HS 812 074, National Highway Traffic Safety Administration, U.S. Department of Transportation, October 2014
2. Checkoway, S., McCoy, D., Kantor, B., Anderson, D., Shacham, H., Savage, S., Koscher, K., Czeskis, A., Roesner, F., Kohno, T.: Comprehensive experimental analyses of automotive attack surfaces. In: Proceedings of the 20th USENIX Conference on Security, pp. 77–92, August 2011

3. International Organization for Standardization: Road vehicles, controller area network (CAN), Part 1: Data link layer and physical signaling. ISO IS11898-1 (2003)
4. Kleberger, P., Olovsson, T., Jonsson, E.: Security aspects of the in-vehicle network in the connected car. In: Intelligent Vehicles Symposium (IV), pp. 528–533, June 2011
5. Duri, S., Gruteser, M., Liu, X., Moskowitz, P., Perez, R., Singh, M., Tang, J.: Framework for security and privacy in automotive telematics. In: Proceedings of the 2nd International Workshop on Mobile Commerce (WMC 2002), pp. 25–32, September 2002
6. Lemke, K., Paar, C., Wolf, M.: Embedded Security in Cars: Securing Current and Future Automotive IT Applications. Springer, Heidelberg (2005). ISBN 978-3540283843
7. Wolf, M., Weimerskirch, A., Paar, C.: Security in automotive bus systems. In: Proceedings of the Workshop on Embedded Security in Cars (ESCAR 2004) (2004)
8. Checkoway, S., et al.: Comprehensive experimental analyses of automotive attack surfaces. In: USENIX Security Symposium, August 2011
9. Studnia, I., Nicomette, V., Alata, E., Deswarte, Y., Kaâniche, M., Laarouchi, Y.: Survey on security threats and protection mechanisms in embedded automotive networks. In: The 43rd Annual IEEE/IFIP Conference on Dependable Systems and Networks (DSN 2013), pp. 1–12, June 2013
10. Koscher, K., Czeskis, A., Roesner, F., Patel, S., Kohno, T., Checkoway, S., McCoy, D., Kantor, B., Anderson, D., Shacham, H., Savage, S.: Experimental security analysis of a modern automobile. In: Proceedings of the 2010 IEEE Symposium on Security and Privacy, pp. 447–462, May 2010
11. Hoppe, T., Kiltz, S., Dittmann, J.: Security threats to automotive can networks – practical examples and selected short-term countermeasures. In: Harrison, M.D., Sujan, M.-A. (eds.) SAFECOMP 2008. LNCS, vol. 5219, pp. 235–248. Springer, Heidelberg (2008)
12. Herrewege, A., Singelee, D., Verbauwhede, I.: CANAuth-a simple, backward compatible broadcast authentication protocol for CAN bus. In: 9th Embedded Security in Cars Conference, September 2011
13. Groza, B., Murvay, S., van Herrewege, A., Verbauwhede, I.: LiBrA-CAN: a lightweight broadcast authentication protocol for controller area networks. In: Pieprzyk, J., Sadeghi, A.-R., Manulis, M. (eds.) CANS 2012. LNCS, vol. 7712, pp. 185–200. Springer, Heidelberg (2012)
14. Oka, D.K., Furue, T., Bayer, S., Vuillaume, C.: Analysis of performing secure remote vehicle diagnostics. In: Computer Security Symposium (CSS 2014), pp. 643–650, October 2014
15. Wolf, M., Gendrullis, T.: Design, implementation, and evaluation of a vehicular hardware security module. In: Kim, H. (ed.) ICISC 2011. LNCS, vol. 7259, pp. 302–318. Springer, Heidelberg (2012)
16. Schweppe, H., Roudier, Y.: Security and privacy for in-vehicle networks. In: 2012 IEEE 1st International Workshop on Vehicular Communications, Sensing, and Computing (VCSC), pp. 12–17. IEEE, June 2012
17. Hank, P., Suermann, T., Müller, S.: Automotive Ethernet, a holistic approach for a next generation in-vehicle networking standard. In: Meyer, G. (ed.) Advanced Microsystems for Automotive Applications 2012, pp. 79–89. Springer, Heidelberg (2012)

Beyond Scale: An Efficient Framework for Evaluating Web Access Control Policies in the Era of Big Data

Tong Liu and Yazhe Wang[(✉)]

State Key Laboratory of Information Security, Institute of Information Engineering,
Chinese Academy of Sciences, Beijing, China
wangyazhe@iie.ac.cn

Abstract. Recently explosive growth in data and the rapid development of communicate technologies bring us into the era of big data. As an important security strategy, access control calls for more highly efficient methods for its evaluation. However, most of the traditional work has already met the bottleneck. Although there've been some methods focusing on the performance of the evaluation engine, these methods are mostly either not in the setting of big data, or there are many limitations when they are deployed in practice. In this paper, we propose a novel framework based on a two-level structure employing multiple clustering techniques. Before building the framework, we propose some ideas for attributes' preprocessing. Then we obtain the two-level structure by a two-stage clustering. In first stage, we make a coarse-grained clustering in quality, and in second stage, we make a fine-grained clustering in quantity. Finally, we can obtain a further improvement by set-operations. In experiments, the framework is applied to some prevailing evaluation engines using large dataset, and the results show that our approach can get great improvements for all involved engines.

Keywords: XACML · Access Control · Performance · Clustering

1 Introduction

In modern network environments, the increasing openness and the enhanced regional interconnection bring us into the era of big data. The number of users and resources are dramatically increasing, which directly leads to more complicated relationships in access control. Given that, Attributed-Based Access Control (ABAC) [17,18] is proposed to make a fine-grained access control. In this paper, we mainly focus on XACML [1], which is a typical language for ABAC and has become the *de facto* standard for specifying and enforcing access control policies for web applications. But with the scale of policies trending to larger [6,39] and the relationships more complicated, the performance becomes a key issue while making a decision in near real time by the Policy Decision Point (PDP).

In previous work, there've been some PDP implementations proposed. For instance, Sun XACML [3], which is the first and most popular engine, searches

© Springer International Publishing Switzerland 2015
K. Tanaka and Y. Suga (Eds.): IWSEC 2015, LNCS 9241, pp. 316–334, 2015.
DOI: 10.1007/978-3-319-22425-1_19

the applicable policies with a brute force method and will encounter a bottleneck when the policies grow to large. Another famous engine is Enterprise XACML [4], which is featured by its result cache and its target indexing, but in fact it could not cut down the policy-scale very effectively. Regard to XEngine [5], which represents the *state-of-the-art* for XACML PDP, it converts the policies to numerical ranges with a flat structure to achieve a high efficiency, but so many limitations determine it's not very suitable for big data. Most engines [4,5,9,29, 36] use traditional ways (data structure, cache or index) for optimization design, instead there are very few methods proposed using the techniques of other fields.

- **Motivations.** It's not difficult to observe that although there're so many entities (subject or resource) in big data, there're many similarities between them. By exploiting this key property, we can get similar entities together by clustering [8,20–25], subsequently we can obtain corresponding sets of *approximate-applicable-policies* in small scale for each cluster, by which efficiently complete the evaluation. As far as we know, [6] is the first work to categorize the policies by clustering, opening up the new thinking in optimization for access control. [6] is an identity-based work without caring about the inner-relation of attributes, however, in real policies especially for open environment, it's not very often to directly make constraints by certain identities. [6] is focusing on identity, which leads to a coarse-grained access control and has many difficulties to implement. To enhance the usability and scalability, we propose a novel framework based on a two-level structure. The attributes are divided into *Categorical* and *Numerical* ones [7], based on which, we do a two-stage clustering. In first stage, we do the clustering for *Virtual-Entities* and in second stage we cluster the real entities for each first-stage-cluster. Subsequently, *approximate-applicable* policies in small scale are assigned to each cluster, then we can choose suitable clusters for a certain request and achieve high performance. Before evaluation, we further cut down the scale of policies by a set-optimization.

- **Overview.** Fig. 1 is the overview of the framework, which is based on the typical PCIM structure [35]. It consists of PEP, PDP, Preprocess-Module and Maintenance-Module. In PDP, we extend the original evaluation engine, adding an Intermediary-Service and a Set-Optimization Module. The two-stage clustering is conducted by Preprocess-Module, then the generated *mapping-tables* are submitted to Set-Optimization Module. When a request is coming, the Intermediary-Service will intercept it and do some analysis, then respond to the Set-Optimization Module. The Set-Optimization Module will retrieve the *mapping-tables* according to the response, get the corresponding policies and further cut down the policy-scale by the set intersection. Finally, the final policies in small scale will be submitted to evaluation engine. During that process, there may be some events triggering the related modules in Maintenance-Module. Figure 1 shows that our framework can treat the original evaluation engine as a black box with just a little modification, which provide the universality. Some abbreviations (eg. FSC, SSC) in Fig. 1 can be neglected temporarily, which will be introduced in the following sections.

In this work, the main contributions consist of:

- We propose a novel framework based on a two-level structure aiming at large-scale policies, by which we can achieve the fine-grained access control.
- We do a two-stage clustering to gather the entities and policies, when evaluating, the framework can cut down the policy-scale by a large margin.
- We apply the set operations to our work to further cut down the policy-scale.

The rest of the paper is organized as follows: In Sect. 2, we introduce some background information about XACML and propose some techniques to process the attributes. Section 3 describes the two-stage clustering in details. The experimental results are shown in Sect. 4. Then, Sect. 5 overviews the related work. Finally, in Sect. 6, we conclude this paper.

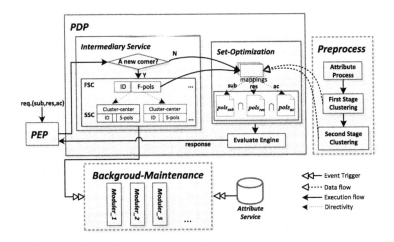

Fig. 1. Overview

2   Preliminaries

2.1   Overview of XACML

XACML [1] policies are organized in the form of hierarchy, which contains *Policy-Sets*, *Policies* and *Rules*. At the root of all XACML policies is a policy or policyset. A policyset consists of some other policysets or a series of policies, and a policy consists of some rules. Each of them has a *target* used to make a decision and the returned decision is either specified in *Effect* property if it's a rule, or by combining its children's decisions if it's a policyset or a policy [36], this process is recursive. The rule may contain a *Condition* to further refine the applicability of the rule. The *target* is a compound boolean expression over *Subject*(sub), *Resource*(res), *Action*(ac) and *Environment*(env), with a structure as formula (1), where X is the attributes involved in *Target*, and $Match_k$ is the match function.

$$Target(X) = \bigwedge_{i \in \{sub,\, res,\, ac,\, env\}} \left[\bigvee_j \left(\bigwedge_k Match_k \right) \right] \qquad (1)$$

$$
\begin{aligned}
Pred_{sub} &= [Role = FinStaff. \land Age \leq 25 \land Region = Gnv.] \lor \\
&\quad [Role = Stu. \land Credits \leq 50 \land Major = CS.] \\
Pred_{res} &= [FileType = Doc.]\ ,\quad Pred_{ac} = [ActionID = Read] \\
Target(X) &= Pred_{sub} \land Pred_{res} \land Pred_{ac}
\end{aligned}
\qquad (2)
$$

Appendix A (Fig. 6(a)) is an example of XACML policy. The Target of Rule1 is represented as formula (2), where some of the attribute-values are marked by abbreviations. When evaluating, for a request, if $T_{r^*} \land Cond_{r^*} = True$, then r^* is said *Applicable* to the request, otherwise *NotApplicable*, where r^* is a rule, T is Target and $Cond$ is Condition. Likewise, for a policy– pol^*, if $\exists r^* \in pol^*$ s.t. $T_{pol^*} \land T_{r^*} \land Cond_{r^*} = True$ and the *combining algorithm* is executed validly, then pol^* is said *Applicable* to the request.

2.2   Processing Attributes

In big data, there'll be great many attributes involved to achieve the fine-grained access control. To make sure our framework feasible and robust, some processing on attributes should be done. Easy to observe that the attributes can be mainly divided into two kinds: *Categorical-Attribute(CatAttr)* and *Numerical-Attribute(NumAttr)*. *CatAttr* is generally in string type, its value is pre-designed and discretely limited, such as File-Type, Role, etc. *NumAttr* can be in many forms such as integer, real, etc. Its value can vary in a large range, usually continuous type or infinite discrete type (eg. Age, FileSize). To some extent, *CatAttrs* and *NumAttrs* represent the qualitative and quantitative characteristics respectively.

Fig. 2. User Hierarchy

2.2.1   Attribute Selecting

Usually, there are many *CatAttrs* for an entity, but there're so limited attributes that will be involved in evaluation. Hence, some dominating *CatAttrs* need be selected. We use formula (3) to measure the *Significance* for every *CatAttr*, where f_{attr} is the total frequency of a certain $CatAttr^*$ appearing in the policy p, f_{or} is the total frequency that $CatAttr^*$ is in an atom predicate of a *disjunctive sequence* in p, and Set_{pol} is the set of all policies. For example, in Fig. 6(a), for the attribute[1] *Role*, $f_{attr} = 4$, $f_{or} = 1$ (in Rule r2). By (3), we traverse all the policies and count $\nabla_{CatAttr^*}$ for every $CatAttr$, if $\nabla_{CatAttr^*} > \xi$, then $CatAttr^*$ is selected. Thereinto, α and β are parameters, ξ is the threshold set by admin. About f_{or}, we'll talk more in Sect. 3.2.

$$\nabla_{CatAttr^*} = \alpha\Big[\sum_{p \in Set_{pol}} f_{attr}(p)\Big] + \beta\Big[\sum_{p \in Set_{pol}} f_{or}(p)\Big] \tag{3}$$

2.2.2 Attribute Similarity

The *hierarchy structure* in [7] is employed to estimate the similarity between the *CatAttr-values*. It's easy for administrators to build the tree for a certain *CatAttr*, within which every node represents a possible attribute value. The similarity is defined in formula (4), where $SPath(v_i, v_j)$ denotes the length of the shortest path between two nodes v_i and v_j, and H is the height of the hierarchy, in Fig. 2, $SPath(Student, professor) = 4$, $H = 3$.

$$S(v_i, v_j) = 1 - \frac{SPath(v_i, v_j)}{2H} \tag{4}$$

$$S_c(C_i, C_j) = S_c(C_j, C_i) = \begin{cases} \dfrac{\sum\limits_{v \in C_i} \sum\limits_{\hat{v} \in C_j} S(v, \hat{v})}{|C_i||C_j|} & C_i \neq C_j \\ 1 & C_i = C_j \end{cases} \tag{5}$$

$$S_c(C_i \cup C_j, C_k) = \frac{S_c(C_i, C_k)\,|C_i| + S_c(C_j, C_k)\,|C_j|}{|C_i \cup C_j|} \tag{6}$$

We call $C_\triangleright$ a *clique iff* $C_\triangleright \subseteq Set^\diamond_{CatValue}$, where $Set^\diamond_{CatValue}$ is the universal set of all the values for a certain $CatAttr^\diamond$. Formula (4) is the basic form at *element-level* and can be easily extended to *clique-level* by formula (5).

2.2.3 Attribute Compressing

When the values belong to one *CatAttr* grow large, the combinations of different *CatAttrs*, which is the *Virtual-Entity* behind, will also be large and it'll make the first stage clustering too costly. However, the values of a *CatAttr* can usually be categorized. In Fig. 2, intuitively, we can summarize three kinds of attributes to cover all the values, they are Student, Faculty and Staff. To generalize the attribute-compressing, we have two considerations: (a) The inherent affiliation for the values isn't expected to be broken, so only the values belong to a same parent should be combined, for example, PostDoc won't be combined with Business-Staff unless all the values are compressed into a single one. (b) It's reasonably believed that those value nodes with a larger depth are of greater demand to be compressed. According to these concerns, we use a variant of similarity to control the priority for the value nodes to be compressed. By formula (7), two points are ensured: Firstly, only the S^* between the clique and its direct parent

---

[1] Please distinguish the *attribute* and *attribute-value* in this paper.

can obtain an additional benefit, which conforms to the consideration (a); Secondly, the maximal depth is chosen as the additional benefit to get the priority for the deeper ones, which conforms to the consideration (b).

$$S^*(C_i, C_j) = S_c(C_i, C_j) + \begin{cases} max\{C_i.depth, C_j.depth\} & if \ SPath(C_i, C_j) = 1 \\ 0 & other \end{cases}$$

(7)

Algorithm 1. Compress Attributes

Input: $Attribute\_Hierarchy\_Tree$, n–the number of *cliques* wanted after compressing
Output: The compressed attribute cliques
1: //Creating a Clique for each node
2: **for each** $node_i \in Attribute\_Hierarchy\_Tree$ **do**
3:     $C_i \leftarrow \{node_i\}$
4: **end for**
5: //Initialize the S^*
6: **for each** $C_i \in Attribute\_Hierarchy\_Tree$ **do**
7:     **for each** $C_j \in Attribute\_Hierarchy\_Tree$ **do**
8:         **if** $SPath(C_i, C_j) = 1$ **then**
9:             $S^*(C_i, C_j) := S_c(C_i, C_j) + \max\{C_i.depth, C_j.depth\}$;
10:        **else**
11:            $S^*(C_i, C_j) := S_c(C_i, C_j)$;
12:        **end if**
13:    **end for**
14: **end for**
15: $cur := |Attribute\_Hierarchy\_Tree| - n$;
16: //Compressing
17: **while** $cur > 0$ **do**
18:    pick up a couple of Cliques with the maximal $S^*(C_i, C_j)$;
19:    $C_i \leftarrow C_i \bigcup C_j$; //assume C_i is the father Clique
20:    refresh the S^* and S_c ;
21:    $cur := cur - 1$;
22: **end while**
23: **return** The left cliques after compressing;

We propose a greedy method in Algorithm 1, it seems like the *hierarchical-clustering* [21,24] but also different. At first, we treat each value node as a single *clique* and initialize the S^* for every *clique-pair* in $Attribute\_Hierarchy\_Tree$, then compress each pair based on the priority S^* step by step. The refreshing for S^* and S_c can be done by formula (5) and (6).

3   Two-Stage Clustering

In this section, we do a two-stage clustering based on the selected *CatAttrs* and the *NumAttrs* respectively. Since subjects and resources are both composed of attributes and have a lot in common, we only introduce the process for subjects.

Definition 1 (Virtual-Entity). *Let $Cat_1, Cat_2, ..., Cat_n$ be the n kinds of selected CatAttrs, and each $Cat_\diamond (1 \leq \diamond \leq n)$ is formed as $\{Cq_1^\diamond, Cq_2^\diamond, ...\}$, where $Cq_k^\diamond$ is the compressed attribute-cliques for the $Cat_\diamond$, satisfying $\forall i \neq j, Cq_i^\diamond \cap Cq_j^\diamond = \emptyset \wedge \bigcup_k Cq_k^\diamond = Set_{CatValue}^\diamond$. Then $\forall elem \in Cat_1 \times Cat_2 \times ... \times Cat_n$, we call it a Virtual-Entity.*

3.1   First Stage Clustering

In this part, we mainly aim at the *Virtual-Subjects* (Virtual-Entity for Subject) and after this stage, we can get some clusters called *First-Stage-Cluster (FSC)*. From Definition 1, we can find that if we do not compress the attributes, $|Cat_\diamond|$ and $|Cat_1 \times Cat_2 \times ... \times Cat_n|$ may be very large, leading to big cost for computing and storage in first stage! That's why we do the attribute-processing work in Sect. 2.2, based on which, we've got the similarity relation at clique-level for each $Cat_\diamond$. It's easy to extend the relation to Virtual-Subject-level by formula (8).

$$S_{vir}(VS_1, VS_2) = S_c(Cq_{x_1}^1, Cq_{y_1}^1)w_1 + ... + S_c(Cq_{x_n}^n, Cq_{y_n}^n)w_n \qquad (8)$$

VS_1 and VS_2 are both Virtual-Subjects, $Cq_{x_k}^k$ is an attribute-clique for $CatAttr_k$. The array of $\{w_i\}$ is the weight for each $CatAttr$, satisfying $\sum_{i=1}^n w_i = 1$.

To do the first stage clustering, many approaches can be adopted [20–25]. Since the similarity for the Virtual-Subjects is *pairwise-correlated*, we believe the *k-medoids* algorithm [20–22,24] is more fit rather than the *k-means* used by [6]. Based on the Virtual-Subject-Level similarity, let $d(VS_i, VS_j) = \frac{1}{S_{vir}(VS_i, VS_j)}$ be the *distance* used by the *k-medoids*. Here, an off-the-shelf *simple and fast k-medoids* method [20] is employed, which tends to select k most middle objects as the initial medoids and during iterations, and for each cluster $FSC_\triangleleft$, selects a VS_i with $min\{\sum_{VS_j \in FSC_\triangleleft - \{VS_i\}} d(VS_i, VS_j)\}$ to update the medoid. Particularly worth mentioning, it requires the distance between each pair only once and has a better performance $(O(nk))$ than the *k-means* clustering as well as a significantly reduced computation time than PAM [21,24]. Because this paper doesn't highlight the clustering algorithm, for more details, please refer to [20].

Since the first stage clustering, in fact, is based on the predesigned *CatAttrs*, it's independent of the entities in the real world. Thus, in this stage, the work for obtained clusters is qualitative.

3.2   Transition Work

In this part, we mainly complete two tasks—assign the *approximate-applicable* policies and corresponding *real* subjects to each *FSC*. Since the clustered objects in the two stages are different and the attributes involved are also different for different stages, the original policies need be adjusted to adapt to different entities. To complete the assignment task, some transition work is needed.

Firstly, we do the formal deduction and make an evolution for the policies. As described in Sect. 2.1, any rule can be expressed as a boolean function $f_r(X) = Target_r(X) \wedge Cond_r(X)$, but in X there may be some attributes such as the *non-selected CatAttrs* and *other-types* attributes (rfc822Name, x500Name,

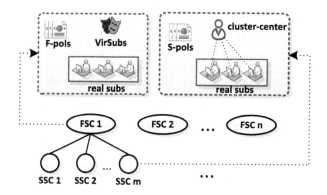

Fig. 3. Schematic diagram for FSC and SSC

etc.), which are meaningless for clustering, so we need to eliminate them and relax the corresponding matching requirements to do *approximate-matching*. In *first-order-logic* [11,38], formulas (9-10) can be reasoned by *conjunction-elimination*($\land^-$) and *disjunction-introduction*($\lor^+$) respectively, where T means *true* and $\vdash$ means *implying*. We treat the attributes that need be eliminated as B in formula (9-10): if the attribute is in *conjunction-sequence* (AND), delete its match-logic directly as (9); if it is in *disjunction-sequence* (OR), delete the whole *disjunction-sequence* as (10), that's why we count the f_{or} for a certain *CatAttr* in Sect. 2.2.1 (If f_{or} is not taken into account, there may be many *all-applicable* policies after relaxing.). Upon the two operations, we can relax a rule's boolean expression step by step—in each step, an atomic predicate is relaxed by (9) or (10), and the outcome of each step is *implied* ($\vdash$) by the former step one. So according to the *transitivity* of the formal deducibility we can finally obtain a new $f_{r'}(X')$, satisfying $f_r(X) \vdash f_{r'}(X')$, $X' \subseteq X$. Likewise, let $f_p(X) = Target_p(X)$ be the policy's Target boolean function and do the same work on $f_p(X)$ to get $f_{p'}(X')$, then $f_p(X) \vdash f_{p'}(X')$ and finally extend to (11).

$$A \land B \vdash A \tag{9}$$

$$A \lor B \vdash T \tag{10}$$

$$f_p(X) \land [f_{r_1}(X) \lor ... \lor f_{r_m}(X)] \vdash f_{p'}(X') \land [f_{r'_1}(X') \lor ... \lor f_{r'_m}(X')] \tag{11}$$

Definition 2 (Approximate-Matching). *For a policy $p(r_1, r_2, ..., r_m)$ and a request rq, without caring about its combining-algorithm, if $f_{p'}(X') \land [f_{r'_1}(X') \lor f_{r'_2}(X') \lor ... \lor f_{r'_m}(X')] = True$ after p is relaxed, then rq is said to approximate-match p and p is approximate-applicable to rq, denoted by $rq \succeq p$.*

Ensured by (11), for a request rq, its completely applicable policies Set_{pol}^{rq} and its *approximate-applicable* policies $Set_{pol'}^{rq} = \{p \in Set_{pol} | rq \succeq p\}$ satisfy $Set_{pol}^{rq} \subseteq Set_{pol'}^{rq} \subseteq Set_{pol}$. Thus, by *approximate-matching* we can get a super-set for the original complete applicabe policies. Thus, for the two-stage pre-process, there is no accuracy loss for the final evaluation. Besides, the rewritten policies just serve the preprocess, rather than the real evaluation. Let $Func(pol, \{\bullet\}_{preserved}, \{\bullet\}_{eliminated}) : pol(X) \rightarrow pol'(X')$ be the realization for relaxation. Upon $Func$, we define a family of policies in formulas (12-14), where Set_{pol} is all the policies extracted from all policysets, Cat_{all}, Num_{all} are the universal set of $CatAttrs$ and $NumAttrs$, Cat_{sect} is the set of selected $CatAttrs$, and $Others$ are the Other-types (eg. rfc822Name) or unrelated (eg. $Resource$, $Action$ and $Environment$) attributes. Focusing on $Subject$, in Appendix A, let $Region$ and $Major$ be the eliminated attributes, then (a) (b) (c) are the elements of $Set_{pol}^0, Set_{pol}^1, Set_{pol}^2$ respectively. The $Func$ can be recursively realized by DOM4J [37], according to the two operations (9-10).

$$Set_{pol}^0 = \{p \in Set_{pol} | Func(p, Cat_{all} \cup Num_{all} \cup Others, \emptyset)\} = Set_{pol} \quad (12)$$
$$Set_{pol}^1 = \{p \in Set_{pol} | Func(p, Cat_{sect}, (Cat_{all} - Cat_{sect}) \cup Num_{all} \cup Others)\} \quad (13)$$
$$Set_{pol}^2 = \{p \in Set_{pol} | Func(p, Cat_{sect} \cup Num_{all}, (Cat_{all} - Cat_{sect}) \cup Others)\} \quad (14)$$

Secondly, regardless of the structure and combining algorithm of Policy-Sets, we assign the *approximate-applicable* policies to each $FSC_\triangleright$ and get $F\text{-}pols = \{ID | \exists vir \in FSC_\triangleright, s.t.\ req(vir, \bullet, \bullet) \succeq pol_{ID}, pol_{ID} \in Set_{pol}^1\}$, where $req(vir, \bullet, \bullet)$ is a *multi-valued* request for Virtual-Subject due to many values contained in *attribute-cliques*, res and ac are arbitrary since in Set_{pol}^1 both of them are *any-labeled* except subjects. The process can be done by Sun XACML [3].

Finally, we assign the *real* subjects to each FSC. For a $sub(v_1, v_2, ... v_n, ...)$ where $v_1 \in Cat_1, v_2 \in Cat_2, ..., v_n \in Cat_n$ and a $vir(Cq_{x_1}^1, Cq_{x_2}^2, ..., Cq_{x_n}^n)$, if it's satisfied that $(v_1 \in Cq_{x_1}^1) \wedge (v_2 \in Cq_{x_2}^2) \wedge ... \wedge (v_n \in Cq_{x_n}^n)$, then sub is said to conform to vir, denoted by $sub \trianglerighteq vir$. Thus, for a sub and a certain $FSC_\triangleright$, if $\exists vir \in FSC_\triangleright$ s.t. $sub \trianglerighteq vir$, then we add the $sub.ID$ to $FSC_\triangleright$. After these transition work, we can get a series of $FSCs$, each one consists of a set of Virtual-Subjects by first-stage-clustering, a set of *real* subjects, and a set of *approximate-applicable* policies ($F\text{-}pols$), as the Fig. 3 shows. After the first stage clustering, the real entities are assigned to each FSC and the policies or entities of a FSC are independent of other $FSCs$. So we can easily treat each FSC as an independent unit and do a completely parallel work.

3.3   Second Stage Clustering

In this part, we mainly aim at the *real* subjects assigned to each FSC. For each FSC, we cluster its subjects based on the $NumAttrs$. As is well-known, the traditional *k-means* has a poor performance while faced with big data. Given that, a *web-scale k-means clustering* [8] is employed. It is a variant for traditional *k-means*, but it could yield excellent clustering results with low computation cost for big data. For each subject, we transform the $NumAttr$ values into a vector. For these vectors, we can also weight and standardize each dimension. The employed approach is based on stochastic gradient descent (SGD), additionally introducing the mini-batches that have lower stochastic noise than SGD without suffering increased computational cost ($O(kbt)$), and uses per-center learning rates for fast convergence. For more details, please refer to [8].

With the method, we can parallelly do the second stage clustering for each FSC and obtain the *Second-Stage-Clusters* (SSC). Subsequently, do transition work— for each FSC, we will assign the policies in F-$pols$ to its affiliated $SSCs$ as Sect. 3.2. Specifically, for a certain $FSC_\triangleright$, let $A = \{p|p \in Set^2_{pol} \wedge p\cdot ID \in FSC_\triangleright\cdot F\text{-}pols\}$, traverse A, if $\exists sub^* \in SSC^*, SSC^* \in FSC_\triangleright$ s.t. $req(sub^*, \bullet, \bullet) \succeq p^*(p^* \in A)$, then add $p^*\cdot ID$ to $SSC^*\cdot S\text{-}pols$. S-$pols$ is like F-$pols$, but more accurate and smaller.

After these work, we can get a series of $SSCs$ for each FSC as the Fig. 3 shows. Each SSC consists of a set of subjects that are assigned from its *parent-FSC*, a set of policies (S-$pols$) *approximate-applicable* to some subjects, and a vector which is the *cluster-center* for SSC. The second stage clustering is based on the *NumAttrs* of the *real* subjects, so it's quantitative and more fine-grained.

3.4   Other Work

The main tasks have been done by far, but there's still some remaining work. Firstly, we use IDs of policies and subjects to build *mapping-tables*: $VirSub \rightarrow FSC$ and $RealSub \rightarrow SSC$. The former is a many-to-one mapping from Virtual-Subject to corresponding FSC, which does great help to the new unhandled subjects for maintenance, and the latter is a many-to-one mapping from real subjects to corresponding SSC, which contributes to the fast realtime evaluation. All above are focusing on subjects, likewise, we can do the two-stage clustering for the *resources* and build *mapping-tables*, too. For the *Actions* in policies, due to the predesigned actions always in small scale, we can directly traverse Set^0_{pol} and build a set of applicable policies for each action-type.

Upon those mappings, we introduce a set-optimization to further cut down the scale of policies involved in the evaluation. Easy to observe that for a request $rq(sub^*, res^*, ac^*)$, if a policy p is completely applicable to rq, then it implies that sub^*, res^* and ac^* must respectively satisfy the constrains of *subjects, resources* and *actions* in p at the same time. Motivated by that, before the evaluation, we can do an intersection that $pols_{sub^*} \cap pols_{res^*} \cap pols_{ac^*}$ to earn us a more accurate set of *approximate-applicable* policies, where $pols_{sub^*}, pols_{res^*}, pols_{ac^*}$ are the ID-sets of *approximate-applicable* policies for certain sub^*, res^* and ac^*. This *Set-Optimization* is shown in Fig. 1. The intersection is a real-time operation, and we need to do it for every request. There are many highly efficient methods for the intersection between two sets, such as binary-search, hash table and so on.

Next, we talk about some details for implementation throughout. Firstly, in attribute-selecting, formula (3) isn't the only choice. Secondly, as mentioned before, in second stage clustering, the task for each FSC can be treated as an independent unit and do a completely parallel work. In experiments, we finished the preprocess work without parallel computing within 400s, which is efficient for an offline job. When the Entity-scale goes lager, using computing-clusters is inevitable. Thirdly, it's the IDs of polices that are stored in the mapping-tables to reduce the memory cost. In our implementation, the framework is built on *policy-level* while in reality it's easy to extend the framework to *rule-level*. If there're too many rules in a policy, techniques in [27, 28] can be employed to decompose a large policy to small equivalent ones. Fourthly, for multi-valued requests, we can decompose them into equivalent single-valued ones like [5] or use the original engine to evaluate them directly. Finally, we build a *Valid-Table* for the policies, each policy corresponds to a certain *valid-bit*, after set-optimization, the bits of *approximate-applicable* policies will be set valid and others invalid. When evaluating, for the invalid ones, the engine directly treats them as the *NotApplicalbe* ones without evaluating them.

4   Experimental Results

In this section, we make a *proof-of-concept* implementation using Java 1.6 based on Sun PDP [3] and Enterprise PDP [4] respectively. All these are performed on Windows 7 with 4G RAM and Intel i7 core 3.4 GHz. Experiments are conducted on different scales of *real-life-extended* policies. The reasons why *real-life* policies are not used need be addressed: Firstly, as [2,6] mentioned, it's difficult to access large real world policies that are publicly available, due to the confidential information these policies typically carry. Secondly, our framework are constructed based on the attributes, which calls for both policies and the corresponding database. Thirdly, our original intention is to achieve a improvement for large policies. Thus, it makes no sense for the small policies like *CodeA-CodeD* and *Continue-a* used in [6,10] without a corresponding database of attributes. For these concerns, a university resources sharing platform is adopted, against which we extend its policies to two large suites — one is a dense suite where entities have a large number of *applicable* rules, the other is a general suite with relatively small *applicable* rules—for both suites, the policies are organized in different scales vary from 50 to 10^4 with randomly 3 to 10 rules in each policy. After attribute-processing, each subject has 11 kinds of attributes—5 *CatAttrs* and 6 *NumAttrs*, each resource has 6 kinds of attributes—3 *CatAttrs* and 3 *NumAttrs*. We use about 10^4 users and 3×10^3 resources to participate in the experiments, and after the two-stage Clustering, we obtain 7 *FSCs* and 33 *SSCs* for subjects, 5 *FSCs* and 18 *SSCs* for resources.

Firstly, Fig. 4 describes the probability density distribution of the *policy-scale* for different stage-clusters against the two data suites using 10^4 policies.

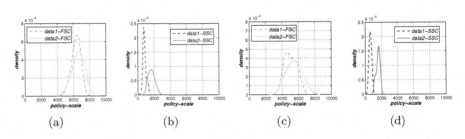

Fig. 4. The probability density distribution diagram of policy-scale—(a)(b) are for sub, (c)(d) are for res. (data1 is general suite, data2 is dense suite)

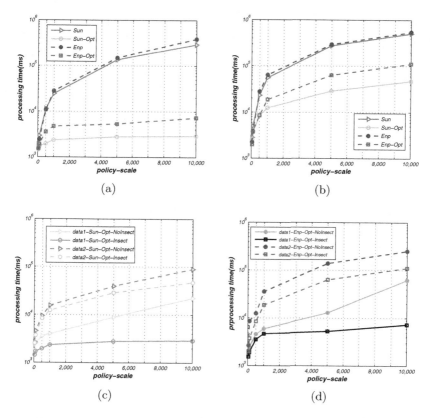

Fig. 5. The experimental results—(a)(b) test the whole framework for general suite and dense suite; (c)(d) test the effect of set-optimization. (-Opt means optimized by the framework, -NoInsect and -Insect means whether apply set-optimization or not.)

Figure 4(a) and (c) show that in both suites, the policy-scale of *FSCs* mostly range from 5×10^3 to 7×10^3. Figure 4(b) and (d) show that in general suite, the policy-scale of each *SSC* is mostly under or around 1×10^3, while in dense suite that is around 2×10^3. All these indicate that in general, our first stage clustering can effectively cut off nearly half of the policies that need to be evaluated, and the second stage clustering can further cut down the policy-scale and achieve about 10 times' downsizing. For dense suite, although the relationships are complex and subjects or resources are related to many policies, our framework can also cut down the policy-scale effectively.

Next, we test our framework with different scale of *real-life-extended* policies to prove the high performance and scalability it provides. For each test, 10^4 *single-valued* requests are randomly sent. Figure 5 shows the experimental results. Figure 5(a) and (b) are the results for the whole framework based on general and dense suite respectively. In Fig. 5(a) and (b), we can observe that for the small policy-scale, whether the data set is dense or not, the both optimized engines don't perform more superiorly than the original ones. While with the policy-scale growing larger, the engines optimized go more and more faster. Especially, in Fig. 5(a) when the policy-scale reaches 10^4, Sun-Opt can process all the requests in 2,889 ms whereas the Sun PDP need 292,988 ms; likewise, Enterprise-Opt can process all the requests in 7,321 ms but the original one need 388,560 ms. For general suite, our approach can achieve a promotion for efficiency by nearly two orders of magnitude. Although it's not better as [6], please notice that in [6] the number of clusters is almost equal to its subjects, while in our case this size is far less than the subjects. Meanwhile, [6] is identity-based without caring about the complex attributes, so it achieve a higher efficiency. In Fig. 5(b), for dense suite, when the policy-scale reaches 10^4, Sun-Opt can finish the job in 48,312 ms while the original PDP need 511,237 ms; Enterprise-Opt can finish it in 111,766 ms whereas the original one need 546,633 ms. Although the applicable policies are more in dense suites, the optimized engines can also achieve about 5 to 10 times' promotion. It is worth noting that in our experiments Enterprise PDP performs equally or a little worse than Sun PDP, for which we summarize some reasons: On the one hand, for Enterprise PDP [4], there're two ways to retrieve policies—*default-attribute-index* and *value-index*, the former is fit for non-equality match-functions while the latter is fit for equality match-functions, and both of them are mapping to a policy-list and constructed by preprocessing. When a request is coming, the Enterprise PDP firstly retrieve policies—for each attribute of the request, it gets the policy-lists by default-attribute-index and value-index one after another, then merge these lists. During this process, the value-index must be prior treated as default-attribute-index and processed before its own turn, that is to say, these two index have done many repetitive work. Due to many attributes maybe involved, the total times for merging are many, in addition, [4] uses a $O(n^2)$ method to do the merging process. Since these retrieve methods are coarse-grained, always there are still many policies in the final list. In other words, the policy-scale is not cut down efficiently but the indexing cost is too much. On the other hand, the requests are randomly sent, the hit-ratio of result-cache in Enterprise PDP is not so high. Besides, it has to manage the result-caches, even if there are only a few requests hitting the target. Similarly, this case is also encountered by others' work, such as [9]. Nevertheless, our approach can also achieve a great improvement for Enterprise PDP.

In Fig. 5(c) and (d), we test the effect of the set-optimization. Different from Fig. 5(a) and (b), the comparison is drawn between the *complete-optimized* engine and the optimized engine without set-optimization. As for the small scale of policies, the intersection doesn't work very well, but with the policy-scale growing up, it can bring about a pretty good effect on the performance. Especially, when the policy-scale goes to 10^4, in general suite, the completely optimized PDP can finish all the requests in 2,889 ms while the optimized PDP without intersection need 22,386 ms; for dense suite, the completely optimized PDP can finish all the requests in 48,312 ms while the optimized PDP without intersection need 89,771 ms. The trend is similar for Enterprise PDP. By the set-optimization, we can achieve about 2 to 10 times' promotion averagely in efficiency.

By experiments, it turns out that our framework can generally promote 1 to 2 orders of magnitude for 10^4 policies. Meanwhile, the framework can apply to most XACML based engines and preserve the good characteristics of the original ones, that is to say, our approach is relatively independent of engine-type, since we cut down the policy-scale before the original engine's evaluation and the original ones just work like a black box.

5   Related Work

With the XACML [1] popular around the world, there have been many researches about it in verification and semantics [10,11,13,19,39], detecting the conflict and redundancy [26], policy integration [27,28,31,32], policy management [12] and some other aspects [7,9,33]. However, the work on the performance of PDP is very limited. The most notable ones are [4–6].

Enterprise XACML [4] features the policy-index and result cache. As we mentioned in the experiment, once the policy-scale grows up and the control logic in the policies turns complicated, the Enterprise PDP may be more time-consuming for the policy-index process, which will make barriers for Enterprise XACML in the era of big data. XEngine [5] features the numerical ranges and its extremely high efficiency in evaluation. It converts the attributes to numbers and the matching predicates to numerical ranges. However, in big data, there are many limitations for XEngine: On the one hand, once the attribute-scale is increasing, it will directly lead to the rapid rise in ranges and their combinations, which will cause the memory's shortage and lower efficiency. On the other hand, it can't support many characteristics in XACML standard, such as the *only-one-applicable* combining algorithm, *obligations*, many complicated matching functions and so on. As is well-known, the XACML is intended to overcome the disunities in the open environment, so many limitations in [5] are not helpful. For other works, for example, in [9], it proposes an framework employing many components such as Redis server [14], Node.js framework [15] and JSON [16] Policy Language. [9] emphasizes the improvement in the implementations. [29] also enhances the performance by cache like [4]. All these means for optimization in work [4,5,9,29,30] are traditional, while the work based on Artificial Intelligence is very few, the most notable one is [6], which also categorizes the policies by clustering. However, [6] is an identity-based work, there are some drawbacks: firstly, in real policies for open environment, it's not very often to directly make constraints by certain identities, instead, constraints on attributes are more commonly used. Thus, in practice, [6] is difficult to be implemented without caring about the details in attributes. Secondly, [6] achieves a high performance using $N \sim N/10$ policy-clusters, which work like the ACLs [34]. When the user-scale growing large, it'll cost too much memory, so there're many limitations for [6] to apply to big data, but it's undeniable a great work since it opened up the new thinking to apply Artificial Intelligence to access control. To enhance the usability and robustness, we propose a framework based on a two-level structure. To build the framework, we mainly divide the attributes into two kinds —*CatAttrs* and *NumAttrs*

[7], based on which we do a two-stage clustering. In first stage, a simple and fast k-medoids [20] is employed, which is more fit for the *pairwise-corelated* similarity while [6] adopt traditional k-means that is not fit for *CatAttrs* and sensitive to the outliers [20–22]. In second stage, a *web-scale k-means* [8] is adopted to deal with the *NumAttrs*, which has a great strength on performance when the entity-scale grows larger and the dimensions go higher. Between the two stage, based on the first-order-logic [38], we do the relaxation for the policies step by step, and obtain a family of policies to serve the transition work. After the two-stage clustering, we build some mappings and proposed a set-optimization to cut down the policy-scale further. When it comes to implementation, the two-level structure has many advantages: firstly, in transition work, when assigning the *approximate-applicable* policies to *SSCs*, we can simply test the filtered policies in *F-pols* without testing other policies. Secondly, by that structure, we can easily divide the work of second stage into many independent tasks, which helps for using parallel computing or multi-thread.

Altough our work and [6] are both based on clustering, but in essence, our work is focusing on the similarity between the entities, by which we cluster the *approximate-applicable* policies; while [6] emphasizes the percentage of the common *applicable* policies for subject-pair. Additionally, [6] is identity-based that is coarse-grained, while we do a more fine-grained work based on attributes. Meanwhile, we have to mention that in our work, we consider the policy semantics for XACML and propose and prove the policy evolution methods while [6] do very little about that and only stay at the simple matching level. By the way, it's worth noting that our work is applicable to most XACML-based engines, so some techniques like cache [4,29], reordering [6] and implementation skills [9,30] can also be applied into our framework. The experimental results are not necessarily the best, since some other excellent techniques [21–25] maybe play better results and the techniques in [4,9,29,30] may earn us further improvements. In our work, the whole design is based on static work, actually, maintenance work should be considered. For some cases such as the attributes are modified or new entities join, that framework can also has great strengths. For example, when a new user is coming, there is no handled result about it in mapping-tables. As Fig. 1 shows, with that framework, the matched Virtual-Subject can be extracted for the new comer, and by the $Virsub \rightarrow FSC$ mapping, the matched FSC can be founded. Since the FSC is obtained by the Virtual-Subjects that is only related to the attribute design schema, so the first stage process is not influenced by the new comer. Then we can use the *F-pols* to cut down the policy scale for the new comer and at the same time some maintenance work for it can also be processed in background. That is just an example and not a practical method. In the future, we'll devote ourselves to designing a resilient maintenance mechanism and try to deploy our framework in the real world as well as test the practical effects it provides.

6   Conclusion

In this work, we propose a framework based on a two-level structure. To build this framework, we do a two-stage clustering work based on categorical-attributes and numerical-attributes respectively, using different clustering techniques. Besides, we propose the set-optimization to further cut down the policy-scale. We do the experiments using Sun XACML and Enterprise XACML with large dataset, and the results show that our framework can greatly enhance the performance in general and can be applied to most XACML based engines just like a black box flexibly. In the future, we'll try to search much better algorithms for clustering and matching work, and try to construct the whole framework including a resilient maintenance mechanism and test the practical effects it provides.

Acknowledgment. This work is Supported by the National Natural Science Foundation of China under Grant No. 61202476, the Strategic Priority Research Program of the Chinese Academy of Sciences, Grant No. XDA06010701, XDA06040502. Last but not least, we thank all the reviewers for the valuable advices sincerely!

A   Appendix

In essence, the Set_{pol}^0 is the original policies directly extracted from the PolicySets without caring the structure and combing-algorithm; Set_{pol}^1 and Set_{pol}^2 are the modified versions for the transition work of two stage clustering respectively. The policies in Set_{pol}^1 are the ones that *delete* match conditions of the non-selected *CatAttrs*, all the *NumAttrs* and other-type attributes abiding by the relaxation rules. The policies in Set_{pol}^2 are the ones that regain all the match conditions of *NumAttrs* based on the policies in Set_{pol}^1. The policies above are the ones in Set_{pol}^0, Set_{pol}^1 and Set_{pol}^2 respectively, focusing on subjects.

```
<Policy PolicyId=p1, RuleCombiningAlgID="Deny-overrides">
   <Target>
      <Subjects>
         <Subject> Role : TechStaff </Subject>
         <Subject> School : University of Florida </Subject>
      </Subjects>
      <Resources>
         <Resouce> Owner : Academic Library </Resouce>
      </Resources>
      <Actions> <AnyAction/> </Actions>
   </Target>
   <Rule RuleId=r1, Effect=Deny >
      <Target>
         <Subjects>
            <Subject>
               <SubjectMatch>  Role : FinancialStaff   <SubjectMatch>
               <SubjectMatch>  Age ≤ 25           <SubjectMatch>
               <SubjectMatch> Region : Gainesville <SubjectMatch>
            </Subject>
            <Subject>
               <SubjectMatch>  Role : Student     <SubjectMatch>
               <SubjectMatch>  Credits ≤ 50       <SubjectMatch>
               <SubjectMatch> Major : Computer Science <SubjectMatch>
            </Subject>
         </Subjects>
         <Resources>
            <Resouce> FileType : Documentation </Resouce>
         </Resources>
         <Actions>
            <Action> Read </Action>
         </Actions>
      </Target>
      <Condition>
            9 : 00 ≤ Time ≤ 18 : 00
      </Condition>
   </Rule>
   <Rule RuleId=r2, Effect=Permit >
      <Target>
         <Subjects>
            <Subject> Role : TechStaff </Subject>
         </Subjects>
         <Resources>
            <Resouce> FileType : Codes </Resouce>
         </Resources>
         <Actions> <AnyAction/> </Actions>
      </Target>
      <Condition>
            FileSize ≤ 5 MB
      </Condition>
   </Rule>
</Policy>
```

(a)

```
<Policy PolicyId=p1, RuleCombiningAlgID="Deny-overrides">
   <Target>
      <Subjects>
         <Subject> Role : TechStaff </Subject>
         <Subject> School : University of Florida </Subject>
      </Subjects>
      <Resources> <AnyResource/> </Resources>
      <Actions> <AnyAction/> </Actions>
   </Target>
   <Rule RuleId=r1, Effect=Deny >
      <Target>
         <Subjects>
            <Subject> Role : FinancialStaff </Subject>
            <Subject> Role : Student       </Subject>
         </Subjects>
         <Resources> <AnyResource/> </Resources>
         <Actions> <AnyAction/> </Actions>
      </Target>
   </Rule>
   <Rule RuleId=r2, Effect=Permit >
      <Target>
         <Subjects>
            <Subject> Role : TechStaff </Subject>
         </Subjects>
         <Resources> <AnyResource/> </Resources>
         <Actions> <AnyAction/> </Actions>
      </Target>
   </Rule>
</Policy>
```

(b)

```
<Policy PolicyId=p1, RuleCombiningAlgID="Deny-overrides">
   <Target>
      <Subjects>
         <Subject> Role : TechStaff </Subject>
         <Subject> School : University of Florida </Subject>
      </Subjects>
      <Resources> <AnyResource/> </Resources>
      <Actions> <AnyAction/> </Actions>
   </Target>
   <Rule RuleId=r1, Effect=Deny >
      <Target>
         <Subjects>
            <Subject>
               <SubjectMatch>  Role : FinancialStaff  <SubjectMatch>
               <SubjectMatch>  Age ≤ 25           <SubjectMatch>
            </Subject>
            <Subject>
               <SubjectMatch>  Role : Student     <SubjectMatch>
               <SubjectMatch>  Credits ≤ 50       <SubjectMatch>
            </Subject>
         </Subjects>
         <Resources> <AnyResource/> </Resources>
         <Actions> <AnyAction/> </Actions>
      </Target>
   </Rule>
   <Rule RuleId=r2, Effect=Permit >
      <Target>
         <Subjects>
            <Subject> Role : TechStaff </Subject>
         </Subjects>
         <Resources> <AnyResource/> </Resources>
         <Actions> <AnyAction/> </Actions>
      </Target>
      <Condition> FileSize ≤ 5 MB </Condition>
   </Rule>
</Policy>
```

(c)

Fig. 6. An Example of Policy.

References

1. OASIS, eXtensible Access Control Markup Language (XACML). http://www.oasis-open.org/committees/tc_home.php?wg_abbrev=xacml
2. Kolovsk, V., Hendler, J., et al.: Formalizing xacml using defeasible description logics. Technical Report TR-233-11, University of Maryland, USA (2006)
3. Sun's XACML implementation (2005). http://sunxacml.sourceforge.net
4. Enterprise XACML (2012). http://code.google.com/p/enterprise-java-xacml/
5. Liu, A.X., et al.: Designing fast and scalable XACML policy evaluation engines. IEEE Trans. Comput. **60**(12), 1802–1817 (2011)
6. Marouf, S., et al.: Adaptive reordering and clustering-based framework for efficient XACML policy evaluation. IEEE Trans. Serv. Comput. **4**(4), 300–313 (2011)
7. Lin, D., et al.: A similarity measure for comparing XACML policies. IEEE Trans. Knowl. Data Eng. **25**(9), 1946–1959 (2013)
8. Sculley, D.: Web-scale K-Means clustering. In: Proceedings of the 19th International Conference on World Wide Web (WWW 2010). ACM (2010)

9. Griffin, L., et al.: On the performance of access control policy evaluation. In: IEEE International Symposium on Policies for Distributed Systems and Networks (2012)
10. Fisler, K., et al.: Verification and change-impact analysis of access-control policies. In: Proceedings of the 27th International Conference on Software Engineering. ACM (2005)
11. Halpern, J.Y., et al.: Using first-order logic to reason about policies. In: Proceedings of the 16th IEEE Computer Security Foundations Workshop (CSFW 2003) (2003)
12. Han, W., et al.: Collaborative policy administration. IEEE Trans. Parallel Distrib. Syst. **25**(2), 498–507 (2014)
13. Philip, W.L., et al.: A white-box policy analysis and its efficient implementation. In: Proceedings of the 18th ACM Symposium on Access Control Models and Technologies (2013)
14. Lerner, R.M.: At the forge: Redis. Linux J. **197** (2010)
15. Node.js: Evented IO for V8 javascript. https://github.com/joyent/node
16. Crockford, D.: JSON: the fat free alternative to XML. In: 15th International World wide Web conference (WWW 2006). ACM (2006)
17. Yuan, E., et al.: Attributed based access control (ABAC) for web services. In: Proceedings of the IEEE International Conference on Web Services (ICWS 2005) (2005)
18. Hu, V.C., et al.: Guide to attribute based access control (ABAC) definition and considerations (draft). NIST Special Publication 800-162 (2013)
19. Ahn, G.-J., et al. Representing and reasoning about web access control policies. In: IEEE 34th Annual Computer Software and Applications Conference (2010)
20. Park, H.S., et al.: A simple and fast algorithm for K-medoids clustering. Expert Syst. Appl. **36**(2), 3336–3341 (2009)
21. Kaufman, L., Rousseeuw, P.J.: Finding Groups in Data: An Introduction to Cluster Analysis. Wiley, New York (1990)
22. Zadegan, R., et al.: Ranked k-medoids: A fast and accurate rank-based partitioning algorithm for clustering large datasets. Knowl.-Based Syst. **39**, 133–143 (2013)
23. Grabmeier, J., Rudolph, A.: Techniques of cluster algorithms in data mining. Data Mining Knowl. Disc. **6**(4), 303–360 (2002)
24. Han, J., et al.: Spatial clustering methods in data mining: a survey. In: Miller, H.J., Han, J. (eds.) Geographic Data Mining and Knowledge Discovery. Taylor & Francis, London (2001)
25. Kamvar, K., et al.: Spectral learning. In: International Joint Conference of Artificial Intelligence. Stanford InfoLab (2003)
26. Hu, H., Ahn, G.J., et al.: Discovery and resolution of anomalies in web access control policies. IEEE Trans. Dependable Secure Comput. (TDSC) **10**(6), 341–354 (2013)
27. Lin, D., et al.: Policy decomposition for collaborative access control. In: Proceedings of the 13th ACM Symposium on Access Control Models and Technologies (2008)
28. Rao, P., et al.: An algebra for fine-grained integration of XACML policies. In: Proceedings of the 14th ACM Symposium on Access Control Models and Technologies (2009)
29. Borders, K., et al.: CPOL: high-performance policy evaluation. In: Proceedings of the 12th ACM Conference on Computer and Communications Security (CCS 2005) (2005)
30. Durham, D., et al.: The COPS (common open policy service) protocol (2000)
31. Mazzoleni, P., et al.: XACML policy integration algorithms. ACM Trans. Inf. Syst. Secur. (TISSEC) **11**(1), 4 (2008)

32. Li, N., et al.: Access control policy combining: theory meets practice. In: Proceedings of the 14th ACM Symposium on Access Control Models and Technologies (2009)
33. Karjoth, G., et al.: Implementing ACL-based policies in XACML. In: Proceedings of Annual Computer Security Applications Conference (ACSAC 2008) (2008)
34. Sandhu, R.S., Samarati, P.: Access control: principle and practice. IEEE Commun. Mag. **32**(9), 40–48 (1994)
35. Moore, B., Ellesson, E., Strassner, J., et al.: Policy core information model-version 1 specification. RFC 3060, February 2001
36. Ngo, C., et al.: Multi-data-types interval decision diagrams for XACML evaluation engine. In: Proceedings of 11th Annual International Conference on Privacy, Security and Trust PST 2013) (2013)
37. Dom4J Group: Dom4J API Project. http://www.dom4j.org/
38. Smullyan, R.M.: First-Order Logic, vol. 6. Springer, Heidelberg (1968)
39. Hughes, G., et al.: Automated verification of access control policies using a SAT solver. Int. J. Softw. Tools Technol. Transfer **10**(6), 503–520 (2008)

Artifact-Metric-Based Authentication for Bottles of Wine (Short Paper)

Reina Yagasaki[✉] and Kazuo Sakiyama

Department of Informatics, The University of Electro-Communications,
Tokyo, Japan
{r.yagasaki,sakiyama}@uec.ac.jp

Abstract. The authentication system is an effective measure in avoiding counterfeit products. In order to enhance the security and to reduce the cost, artifact-metrics authentication is expected to replace the existing cryptography-based authentication system since it utilizes the intrinsic individual differences of products. In this paper, we propose a new artifact-metric-based individual authentication system that authenticates bottles of wine. The proposed system takes the light pattern of a light emitting diode transmitted through a bottle of wine as an image, and uses it as a fingerprint. Based on experimental results, we show that the system distinguishes different bottles of wine correctly, and that the authentication is sufficiently tolerant to dirt and flaws on the surface of the bottle.

Keywords: Authentication system · Artifact metrics · Phase-only correlation · LED

1 Introduction

Nowadays, counterfeit bottle of wines are circulating all over the world [1]. For instance, the counterfeit bottle of wines are produced by replacing the label with one copied or peeled from a genuine bottle of wine. And furthermore, the genuine bottle of wines are assumed not to be uncorked and replaced their contents. Moreover, manufactures of genuine bottle of wines are expected that they do not replace the labels maliciously. In this case, buyers need some way to detect that the label was replaced, which is often bothersome and infeasible in a practical sense In order to verify the matching between the label and the bottle of wine, we focus on using a transmitted light pattern through bottles and wines. This is because labels may be exchanged, the bottle itself cannot be as easily replaced. In addition, it is expected that the transmitted light will contain features of the bottle of wine, e.g., color and ingredients, as well as the physical characteristics of the bottle itself. The idea is based on the artifact-metrics technique that was introduced in 2000 [2].

The artifact-metric system uses individual physical differences of products as a fingerprint in authenticating multiple products efficiently. It is known that

© Springer International Publishing Switzerland 2015
K. Tanaka and Y. Suga (Eds.): IWSEC 2015, LNCS 9241, pp. 335–344, 2015.
DOI: 10.1007/978-3-319-22425-1_20

Fig. 1. Proposed authentication system

such physical differences cannot be reproduced even by legitimate producers since they are inherently determined by the manufacturing process. An example of an existing study is paper authentication. This system makes use of the arrangement of small fiber articles as a fingerprint of paper [3]. The idea is shown to be effective for official certificates. There are several related studies discussing the feasibility and security of the paper authentication; however, they do not refer to three-dimensional features in detail. On the other hand, optical Physical Unclonable Functions (PUFs) extract three-dimensional physical characteristics of permeable objects using transmitted light [4–6]. However, to our best knowledge, conventional optical PUFs have not been applied to bottles of wine. In this paper, in order to maximize the three-dimensional physical characteristics, the LED (Light Emitting Diode) light transmitting through the bottles of wine is used. Under the assumption that LED light is differently and uniquely transmitted through each bottle of wine, we propose a new authentication system based on artifact-metrics, and examine its feasibility by some experiments. The authentication system used for the experiments has an LED light bulb and digital still camera as shown in Fig. 1. The system is based on [7]. The transmitted light from the bottom of the bottle is recorded by a still camera, and treated as a fingerprint image. When authenticating a bottle of wine, the authentication algorithm performs two processes: feature extraction and image matching. In the feature extraction process, binarization of the image is applied to extract the shape of the light pattern from the image. In the matching process, Phase-Only Correlation [8] is employed considering the advantage of high precision by using the whole image. The experimental results show that the proposed authentication system successfully verifies the matching between the label and the bottle of wine. Furthermore, we confirm that the proposed system is tolerant to dirt and flaws on the surface of the bottle.

1.1   Our Contributions

The noteworthy contribution of the proposed system lies in the fact that it can verifies the matching between the label and the bottle of wine without causing

Fig. 2. Example of hollow of the bottle

Fig. 3. Mini-krypton bulb type LED light

damage to the bottle and using complicated devices or operations. The system comprises only two components, an LED light and a still camera, and users only need to set a bottle of wine on the system for authentication. It is not necessary to attach any tag for authentication. We use LED light as the illumination source, which can avoid deterioration of bottle of wine caused by heat since LEDs do not generate heat waves as much as filament lamps.

2 Design of Authentication System

2.1 Construction of Authentication System

We construct an authentication system as shown in Fig. 1. The system comprises mainly two components: LED light (mini-krypton bulb type, 600 [lm], TOSHIBA), and still camera (DSC-RX100, SONY(configured image size, 1920 × 1920 [pixel].)) The most notable feature of the system is that the LED bulb is inserted in the hollow at the bottom of the bottle as shown in Fig. 2. The mini-krypton bulb type LED light (Fig. 3) is an appropriate size for this hollow. The taken images of the transmitted light are analyzed using Matlab. It is worth mentioning that the image contains information regarding features of the bottle and the wine. More precisely, the image may include features of the wine (not the bottle), e.g., color and ingredients stemming from the difference in the kinds of grapes, way of winemaking, year of production, etc.

2.2 Authentication Algorithm and Preparation

The authentication algorithm is organized as follows.

1. Before authentication, images of the transmitted light are taken with the camera and stored in a database for each bottle of wine as template images.
2. When authenticating a bottle of wine, an image is taken and a search of the database is performed to make a comparison to the template image.
3. The feature extraction process is performed on both the image and stored templates.

4. The correlation between the extracted features is calculated in the image matching process.
5. The system outputs the authentication results after checking whether or not the correlation is sufficiently high based on a comparison to the predetermined threshold value.

In the step 1 of the authentication algorithm, multiple images are taken as template images using the proposed system. The reason why we store multiple images is that we try to reduce the number of false rejections. Namely, when a bottle of wine is placed on the system, it is preferable to match the direction of the bottle based on the label position. Although we try to match the direction of the bottle, there is a slight rotation displacement whenever the bottle is set. In this case, there is the possibility that an image of a genuine bottle of wine is rejected. Multiple template images are taken where each time the bottle is rotated slightly. In our experiments we assume that nine template images could reduce the number of false rejections. In the next section, the details of feature extraction and image matching process are discussed.

3   Feature Extraction and Image Matching

3.1   Feature Extraction Process

The feature extraction process consists of segmentation and binarization image processing. At first, the image is segmented into 769×769 pixels to focus on the transmitted light pattern. Then the segmented image is binarized in accordance with the following procedures.

1. Convert image to monochrome.
2. Set a certain value for binarization, k ($0 \leq k \leq 255$)
3. Binarize the monochrome image according to threshold value k.

The reason for binarization is to reduce the image noise such as unrelated light and images reflected onto the surface of the bottle. By doing so, we can address problems in which many images taken of genuine bottle of wine are rejected in error due to noise. Figure 4 represents the examples of segmented and binarized image. The pairs of image at either end of them are derived from the pairs of the bottles of wine, which have the same label. Here, the image is binarized by setting $k = 200$.

3.2   Image Matching Process

The existing technique called the POC is well studied for biometric fingerprints [8]. This method can match the entire image and is highly accurate. Moreover, it can match images even if there are a few changes in contrast. To evaluate whether the image of a transmitted light could be used as a fingerprint of a bottle of wine, we decide to apply this well-studied method to the proposed

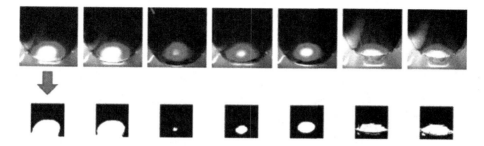

Fig. 4. Example of characteristic extraction process of seven bottles of wine

system. This method is not robust to image rotation; however, the target images are only slightly rotated in the proposed system because of the fixed camera. Namely, the POC function is derived from a normalized cross spectrum between two feature-extracted images. In this section, we introduce the theory of the POC and describe the procedures in the system.

Phase-Only Correlation. The POC function, which is derived from a normalized cross spectrum, plays a main role in the POC procedure. Thanks to the effect of the normalization, POC works well even when the brightness levels of two images are different [8]. Here, how to derive the POC function is explained. Suppose that we have two segmented and binarized images whose sizes are both $M \times N$ pixels, where M and N are odd numbers. The images are stored in two-dimensional arrays, $f(m,n)$ and $g(m,n)$, where $m \in \{-\frac{M-1}{2}, \frac{M-1}{2}\}$ and $n \in \{-\frac{N-1}{2}, \frac{N-1}{2}\}$. Hence, the center of the images can be expressed as $f(0,0)$ and $g(0,0)$, respectively. First, the two-dimensional Discrete Fourier Transform (DFT) is calculated as

$$F(u,v) = \sum_{m=-p}^{p} \sum_{n=-q}^{q} f(m,n)e^{-2\pi jmu/M}e^{-2\pi jnv/N}, \tag{1}$$

$$G(u,v) = \sum_{m=-p}^{p} \sum_{n=-q}^{q} g(m,n)e^{-2\pi jmu/M}e^{-2\pi jnv/N}, \tag{2}$$

where $p = \frac{M-1}{2}$ and $q = \frac{N-1}{2}$. Terms $F(u,v)$ and $G(u,v)$ represent a frequency-domain representation of $f(m,n)$ and $g(m,n)$, respectively. Next, the normalized cross spectrum between $f(m,n)$ and $g(m,n)$ is derived as

$$R(u,v) = \frac{F(u,v)}{|F(u,v)|} \frac{\overline{G(u,v)}}{|\overline{G(u,v)}|}, \tag{3}$$

where $\overline{G(u,v)}$ is a conjugation of $G(u,v)$. Finally, the Inverse Discrete Fourier Transform (IDFT) is performed on $R(u,v)$ in order to derive the POC function, $r(m,n)$, as

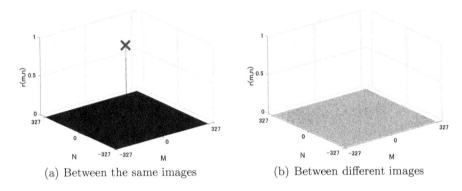

(a) Between the same images   (b) Between different images

Fig. 5. Graphs of POC function

$$r(m,n) = \frac{1}{MN} \sum_{u=-p}^{p} \sum_{v=-q}^{q} R(u,v)e^{2\pi jmu/M}e^{2\pi jnv/N}. \tag{4}$$

The POC function has a very sharp peak in a two-dimensional graph only when two waves are similar. If the two waves are not similar, it has no outstanding peaks. In our experiments, we regard the maximum value of the absolute POC function as similarity between the two images. Figure 5(a) and (b) is example of POC function between two images (655×655 pixels) plotted in three-dimensional space. These two images are pictures of a window and a chair for a test purpose. Figure 5(a) shows a graph of the POC function when the two images are the same (picture of a window) and Fig. 5(b) shows that when the two images are totally different (picture of a window and that of a chair).

3.3   Band-Limited POC Function

The high frequency band of a frequency domain representation of an image often includes noise that is not related to the characteristic of the image. This noise degrades the accuracy of the image matching. To reduce this noise and improve the accuracy, a band-limited POC function was proposed [8]. The band-limited POC function is derived by filtering the high frequency band of a normalized cross spectrum before employing the IDFT. In order to derive the the the low frequency band of a normalized cross spectrum, the variables M and N in Eq. (eq4) are replaced with S and T, respectively, where $S < M, T < N$. In our experiments, $S = 401$ and $T = 401$.

4   Evaluation of Feasibility

This section describes an experiment to examine the occurrence of false rejection and false acceptance. We prepare seven bottles of red wine and name them B1, B2, B3, B4, B5, B6, B7. Among these bottles, B1 and B2, and B6 and B7 are the same product. For the remaining bottle of wines, the product is different from each other, and hence the depth and shape of the hollow of the bottle differs. First of all, nine template images are taken from each of the bottlesand stored in a database as mentioned in Sect. 2.2. Secondly, forty images are taken from each of the bottles of wine. For the examination of false rejection of a bottle, six images are selected from the forty images and compared with its template. For example, when testing the authenticity of B1, six images are taken from forty images of B1, and they are compared with the template of B1. Finally, the following operations are performed for each of the six images.

1. After segmenting the image, brightness value k is set and the image is binarized (feature extraction process).
2. The maximum absolute value of the band-limited POC function between taken image and each of the nine template images is calculated as $R_1, R_2, ..., R_9$ (image matching process). Then the maximum value of $R_1, R_2, ..., R_9$ as R_{max} is selected.
3. If the R_{max} exceeds a certain threshold, ρ, the authentication is deemed successful. Otherwise, it is regarded as a false rejection then count it.

After performing the examination for each of the seven bottle of wines, the total number of false rejections is derived.

On the other hand, for the examination of false acceptance, six images taken from the different bottles are compared. For example, one image selected from forty images of B2, B3, B4, B5, B6, B7 is compared with the templates of B1. For the selected six images, the same operation is performed as the case for false rejection except the last step. In the last step, if the R_{max} exceeds a certain threshold, ρ, the authentication is regarded as a false acceptance. After performing the examination for each of the seven bottle of wines, the number of false acceptances is added. Finally, the total number of false rejections and false acceptances is obtained as the total number of errors. The number of examination of the false rejections and false acceptance are 42, respectively.

4.1   Experimental Result

The experimental results are shown in Fig. 6. This figure denotes the number of errors at each threshold and brightness value. The errors appearing in the left side represent the false acceptances. Those appearing on the right side represent the false rejections. The range of brightness values is selected as $196 \leq k \leq 205$ empirically. In this figure we find an area where there is no error in this experiment at the threshold is within the parameters of 0.060 to 0.070 under the brightness value condition is higher than 202. The expected maximum number of errors are 84. This result infer that the authentication performs successfully in regards to seven bottles of wine. However, there are some errors that reduce the size of this area. We consider the reason for this to be that the appropriate brightness value is different for each bottle of wine due to the level of bottle shading.

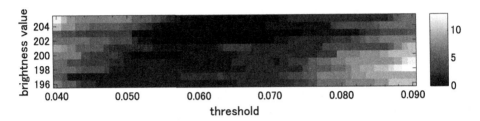

Fig. 6. Summation of errors for seven bottle of wines

5   Evaluation of Tolerance Against Dirt and Flaws

We conduct an additional experiment to evaluate the tolerance against changes in the surface conditions of the bottle. For practical use, situations in which there are some dirt or flaws on the surface should be taken into account. This experiment is separate from the previous experiment. Some parameters such as the exposure angle and the photographic sensitivity settings are different. The following are procedures for this experiment. First, the bottle of wine named as B3 is prepared Next, nine template images are taken by B3 before authentication with no dirt or flaws. Then images are taken in each of five situations: no dirt or flaw, some dirt (levels 1 and 2), and some flaws (levels 1 and 2). Two levels of dirt and flaws are established. Level 2 represents the state of the surface with more dirt or flaws than level 1. Figure 7(a) and (b) shows the surface of a bottle with dirt at levels 1 and 2, respectively. Figure 8(a) and (b) shows that of a bottle with flaws at levels 1 and 2, respectively. In addition, an image taken from one different brand bottle of wine named as B4 is prepared when there is no dirt or flaw on the surface. Finally, R_{max} is calculated corresponding to each of the six images (five situational and one different bottle of wine), where the brightness value is set to $k = 200$.

5.1   Experimental Result

Table 1 gives the calculated R_{max} for each of the six images. The situation represents the state of the surface of the bottles. The proposed system calculates

Table 1. Experimental result for dirt and flaws

Verified Bottle	Template bottle	Situation	R_{max}
B3	B3	No dirt or flaw (base line)	0.1265
	B3	Dirt (level 1)	0.1228
	B3	Dirt (level 2)	0.1699
	B3	Flaws (level 1)	0.0599
	B3	Flaws (level 2)	0.0870
B4	B3	No dirt or flaws	0.0375

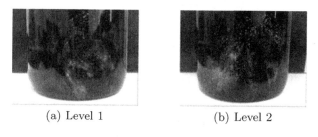

(a) Level 1   (b) Level 2

Fig. 7. Surface of bottle with dirt

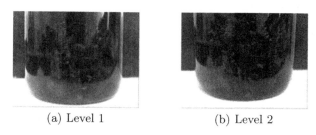

(a) Level 1   (b) Level 2

Fig. 8. Surface of bottle with flaws

correlations as high as that of the image in the situation with no dirt even if there is some dirt on the surface. On the other hand, the flaws degrade the value of correlations regardless of its level. Moreover, these correlations are obviously higher compared to that derived from the different bottle of wine. This means that the dirt and flaws do not affect the correlation drastically.

6   Conclusions and Future Work

This paper proposed a new artifact-metric-based authentication system. Based on experimental results, the proposed system successfully verify the matching between the labels and bottles of wine without false acceptance or false rejection under a certain brightness value and threshold. In addition, the system distinguish the bottles which have the same labels. Moreover, we confirmed that the proposed system provided a high level of tolerance to dirt and flaws on the bottle. We conclude that the image of transmitted light could be used for authentication.

The future work includes a refinement of the parameters, e.g., brightness value and setting of camera, the number of template images. Further investigation is required into the aging effect of the content of the wine and the bottle.

References

1. The drinks business: FAKE WINE NOW ACCOUNTS FOR 20% OF MARKET. http://www.thedrinksbusiness.com/. Accessed 2 Mar 2015

2. Matsumoto, H., Matsumoto, T.: Artifact-metric Systems. In: Workshop Record of ISEC, vol. 100, no. 323, pp. 7–14 (2000)
3. Fukuda, Y., Matsumoto, T.: An artifact-metric system which extracts values unique to individual paper. In: Workshop Record of ISEC, vol. 110, no. 113, pp. 41–46 (2010). (In Japanese)
4. Pappu, R., Recht, B., Taylor, J., Gershenfeld, N.: Physical one-way functions. Sci. AAAS **297**(5589), 2026–2030 (2002)
5. Tuyls, P., Škorić, B.: Strong authentication with physical unclonable functions. In: Petković, M., Jonker, W. (eds.) Security, Privacy, and Trust in Modern Data Management. Data-Centric Systems and Applications, pp. 133–148. Springer, Heidelberg (2007)
6. Škorić, B., Tuyls, P., Ophey, W.: Robust key extraction from physical uncloneable functions. In: Ioannidis, J., Keromytis, A.D., Yung, M. (eds.) ACNS 2005. LNCS, vol. 3531, pp. 407–422. Springer, Heidelberg (2005)
7. Yagasaki, R., Sakiyama, K.: A study of artifact-metrics based on the transmitted light through wine bottles. In: Proceedings of the IEICE General Conference, A-7-9, p. 143 (2015). (In Japanese)
8. Ito, K., Nakajima, H., Kobayashi, K., Aoki, T., Higuchi, T.: A fingerprint matching algorithm using phase-only correlation. IEICE Trans. Fundam. **E87A**, 682–691 (2004)
9. Jain, A.K.: Fundamentals of Digital Image Processing. Prentice Hall Information and System Science Series. Prentice Hall, Englewood Cliffs (1989)
10. Brigham, E.O.: The Fast Fourier Transform. Prentice Hall, Englewood Cliffs (1974)

Security in Hardware

Bit Error Probability Evaluation
of Ring Oscillator PUF (Short Paper)

Qinglong Zhang[1,2,3], Zongbin Liu[1,2], Cunqing Ma[1,2(✉)], and Jiwu Jing[1,2]

[1] Data Assurance and Communication Security Research Center, Beijing, China
{qlzhang,zbliu,cqma,jing}@is.ac.cn
[2] State Key Laboratory of Information Security,
Institute of Information Engineering, CAS, Beijing, China
[3] University of Chinese Academy of Sciences, Beijing, China

Abstract. Physically unclonable functions (PUFs) utilize the intrinsic process variation inside an integrated circuit to generate unique secret keys for cryptographic modules. PUFs can eliminate the risk that the keys stored in non-volatile memory is easy to be attacked by physical invasive attacks. Ring oscillator PUF (RO PUF) is popular for its nice property and easy implementation, and the frequencies of ROs are compared pairwise to generate one bit response. However, the frequency measurement is affected by environmental noise and the comparison of two frequencies may lead to bit error. To date, there is only a qualitative conclusion that high counting value is chosen to reduce noise's influence. In this paper, we quantitatively analyze the relationship between the frequency measurement counting value and the bit error probability. On the observation of our experiments' data, we describe a comprehensive model to estimate the bit error probability of RO PUFs and present other factors to influence the bit error probability, such as the stages of ROs, the manufacturing techniques and so on. The results calculated from our model and those from measured data achieve high consistency. Our work contributes to the evaluation scheme for RO PUFs and it is available as a guide for people to design RO PUFs with an acceptable bit error rate.

Keywords: PUFs · FPGA · Ring oscillator · Bit error probability

1 Introduction

Physically unclonable function is becoming more and more popular as PUF aims to solve a challenging problem, how to protect the private keys efficiently [1–3]. The concept of PUF is to take advantage of the uncontrolled random process variations during the manufacturing process and the process variation can not even be reproduced by the manufacturers. For a particular PUF instance

The work is supported by a grant from the National Natural Science Foundation of China (No. 61402470).

K. Tanaka and Y. Suga (Eds.): IWSEC 2015, LNCS 9241, pp. 347–356, 2015.
DOI: 10.1007/978-3-319-22425-1_21

for key generation, there is a requirement that PUF should have the ability to reproduce the key multiple times because if the key has some bit errors, this leads to the invalidity of cryptographic modules. But this requirement is not easy to be solved by PUF itself because the Gaussian noise exists in the circuit and slight measurement error may also result in the bit errors of PUFs' response. RO PUF is one of the most popular and widely used PUFs. The RO PUF's implementation is easy on both FPGA and ASIC platforms and the responses of RO PUF come from the comparison of different pairwise ROs. The bit errors happen as the environmental variables change and slight measurement random error exists. Therefore, there are researches to estimate the effects of temperature and supply voltage on the bit error probability. In CHES 2012, Katzenbeisser et al. [4] point out that RO PUF achieves almost all desired properties of a PUF: the bit error rate does not change significantly under different operating conditions and the entropy of the PUF responses is high. In the practical experiments, there comes a problem that when we compare two ROs' frequencies, are there some requirements for the reference counting values? The reference counting value is the number of the reference crystal oscillator's oscillations in a fixed interval. There are some publications [5–7] involved with the discussion on the bit error probability of RO PUFs. These two publications, [5,7], discuss the effect of the reference counter on the bit error probability. Delvaux et al. [7] give the result between repeatability and oscillation time, but they do not describe how to deduce the result theoretically. Komurcu et al. [5] list several factors to affect the bit error probability, but present the measure time's effect on the bit error probability roughly.

Hiller et al. [6] describe the relationship between the number of sample elements and the bit error probability, and present that based on multiple frequency measurements, the parameters for the distribution of ROs can be achieved and the bit error probability can be estimated from each measured frequency and these parameters. The analysis in [6] is mainly focused on the influence of the number of frequency measurements on the bit error probability, and after a practicable number of frequency measurements, their analysis can give an extremely precise estimation of the bit error rate. However, the reference counting value in every frequency measurement is not discussed and the reference counting value used in frequency measurement can affect the precision of the measured frequency. The work of our paper aims to elaborate the relationship between reference counting values and the bit error probability. In this paper, combining the process variation and noise effect, we show the different bit error probabilities with different reference counting values. In order to demonstrate the validity of our analysis, both model simulations and FPGA board experiments are conducted and the consistency of the results from simulations and experiments can support our analysis. In a word, this paper detailedly presents the relationship between reference counting values and the bit error probability from theory to practice.

Contributions. In this paper, we conduct experiments to give out a rough quantitative conclusion of the reference counting values and the bit error probability of RO PUFs. Besides the existing Gaussian noise, there are some disturbance signals when we experiment on commercial FPGA boards, and based on this comprehensive model, we estimate the bit error probability from both model calculation and data statistics.

In summary, we make the following contributions.

– Based on a comprehensive model, we quantitatively analyze the relationship between the frequency measurement counting value and the bit error probability.
– The results from data statistics and model calculation achieves high consistency. Meanwhile, the model calculation is relatively precise to predict bit error probability of the untested case.
– We conduct experiments to evaluate the effect of the number of RO's stages and the manufacturing technique.
– Our work contributes to the implementation of the evaluation scheme on RO PUFs.

Structure of the Rest Paper. In Sect. 2, we present basic concept on RO PUFs. Section 3 describes our evaluation system for our experiments. In Sect. 4, we make a description of the modified model, and give the experiment results and analysis on the estimation of the bit error probability. We conclude our paper in Sect. 5.

2   Preliminaries

A ring oscillator consists of an odd number of inverters. The frequency of the ring depends on the propagation delay of all the inverters. During the period of manufacture, slight difference of an inverter's propagation delay appears among different rings. The slight difference is called process variation and it cannot be avoidable.

2.1   Ring Oscillator PUF

Ring oscillator architecture is a typical method to construct PUFs. The ring oscillator PUF is first proposed by Suh and Devadas [8], and an RO PUF is composed of n identical ROs, RO_1 to RO_n, with frequencies, f_1 to f_n, respectively. Generally, RO PUF also contains two counters and two n-to-1 multiplexers that control which ROs are currently applied to both counters. Due to process variation, the frequencies of these two selected ROs, f_i and f_j, tend to be different from each other. One bit response r_{ij} can be extracted from two ring oscillators by using a simple comparison of their frequencies as follows.

$$r_{ij} = \begin{cases} 1 & if \ \ f_i > f_j, \\ 0 & otherwise. \end{cases} \tag{1}$$

Since all the process variation and other noises have influence on the frequencies, the resulting comparison bit will be random and device-specific. The above comparison operation is a basic form of compensated measurement, which is proposed by Gassend et al. [9]. The compensated measurements based on the ratio of two frequencies is particularly effective because the environmental changes and systematic noises can affect both frequencies of two ring oscillators simultaneously.

3   Experimental Setup

This section describes our experimental setup for the data collection. The hardware platform is Virtex-5 XC5VLX110T-1ff1136 and the FPGA boards are programmed by Xilinx ISE 14.6. An array of 100 ROs has been implemented on every FPGA boards. Every RO is composed of 15 inverters and 1 AND gate with one of the inputs used as an enable signal for the RO. One RO is implemented in two CLBs, and 15 of the 16 LUTs serve as inverters and 1 of the LUTs serves as AND gate. In order to make all the ROs identical, Hard Macro technique is used for the deployment of RO PUFs.

To measure ROs' frequencies, a 50 MHz crystal oscillator on board and a 32-bit counter are used as a counting component for reference. The ROs' outputs are connected to a multiplexer which selects one RO's output to the clock input of a 32-bit counter. In the architecture of RO PUF there are two multiplexers. When the reference counter counts to N, we can record the other two 32-bit counters' value y_1 and y_2 and calculate two frequencies respectively, $50 * y_1/N$ MHz and $50 * y_2/N$ MHz. In the application of RO PUFs, there is no need to calculate the frequencies of ROs. One response of two ROs can be extracted by just comparing y_1 and y_2. Figure 1 shows the basic architecture for our experiments. In order to analyze the measured the bit error probability elaborately, we collect all the 32-bit counters' values via UART to a PC.

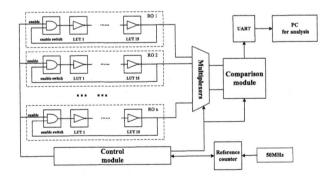

Fig. 1. The basic architecture of our evaluation systems

4   Model and Evaluation

The total delay of an RO is affected by the process variation, the environmental noise and the systematic variation. Systematic variation is mainly correlated with location. It is proposed by Maiti et al. [10] in J. Cryptol. 2011 and to compare adjacent ROs can efficiently decrease the effect of systematic variation. Therefore except systematic variation, the delay model can be simply presented as follows.

$$d_{LOOP} = d_{AVG} + d_{PV} + d_{NOISE} \tag{2}$$

Where d_{AVG} is the nominal delay value and it is the same for all the identical ROs. d_{PV} is the delay variation resulted from random process variation. d_{NOISE} is the delay variation due to the environmental noise.

For one RO, the process variation is fixed, so when we measure the frequency of this RO multiple times, the measured frequency is mainly affected by the environmental noise. So there is an assumption that the for an RO, RO_a, the distribution of the measured counting value V_a is a Gaussian distribution as follows.

$$V_a \sim N(\mu_{RO_a}, \sigma^2_{RO_a}) \tag{3}$$

4.1   Model Calibration

We conduct RO PUF experiments with 15 stages ROs and record the distributions of the measured counting values with different reference counting values. The distribution of measured counting values are as shown in Fig. 2. The Fig. 2 (a) is records the measured counting values with reference counting value is 2^{10}. For Fig. 2 (b) and (c), the reference counting values are 2^{13} and 2^{20} respectively. Figure 2 (a) indicates that when the reference counting values is 2^{10}, the distribution of the measured counting values is not a Gaussian distribution.

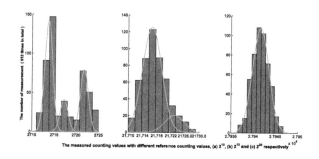

Fig. 2. The distribution of the measured counting values

From Fig. 2 (a), the distribution of the measured counting values is not just affected by the noise and the diagram indicates that there are some other disturbance signals which will have influence on the measurement of the counting

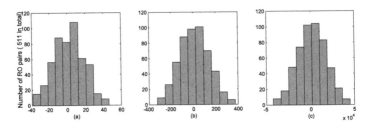

Fig. 3. The distribution of process variation with different reference counting values. (a), (b) and (c) are for 2^{10}, 2^{13} and 2^{20} respectively

values. Because the schematic diagram of FPGA boards is complex and especially the power circuit may have some disturbance on the surrounding components, the measurement of RO PUFs may have some disturbance from the power circuits. Meanwhile, from Fig. 2 (b) and (c), with the increasing of the reference counting values, the influence of these disturbance signals decreases. Rather, the magnitudes of noise's accumulation becomes larger than that of disturbance signal's accumulation with the increasing of the reference counting values. When the reference counting values is less than 2^{13}, the magnitudes of both these accumulations is almost the same. We assume that the disturbance signal is a signal with some period and limited amplitude, and the amplitude of the disturbance signal can be positive and negative, so the accumulation of the disturbance signal is limited with the reference counting value increasing.

To describe the relationship between the reference counting values and the bit error probability, when we deploy RO PUFs in the commercial FPGA boards, the effect of other components for the RO PUFs should not be ignored. Therefore, the impact factors for the measured counting values is presented as follows.

$$d_{LOOP} = d_{AVG} + d_{PV} + d_{NOISE} + d_{Disturbs} \qquad (4)$$

Assume that the effect of disturbance signals and that of noise is independent.

$$V_a \sim N(\mu_{noiseRO_a}, \sigma^2_{noiseRO_a}) + F1(a, disturb1, t) + F2(a, disturb2, t) + \cdots \quad (5)$$

$$V_a - V_b \sim N(\mu_{noiseRO_a} - \mu_{noiseRO_b}, \sigma^2_{noiseRO_a} + \sigma^2_{noiseRO_b}) \\ + \Delta F1(ab, disturb1, t) + \Delta F2(ab, disturb2, t) + \cdots \qquad (6)$$

In order to estimate the distribution of process variation between two ROs, we deploy 512 ROs and record these ROs' measured counting values 512 times. According to that the PV_{ab} is almost close to $\mu_{RO_a} - \mu_{RO_b}$, we achieve the following histograms shown in Fig. 3 to approximately describe the distribution for process variation under different reference counting values. From Fig. 3, the distribution of process variation is almost a Gaussian distribution as our assumption.

With the data statistics, we also get the histograms for the distribution of any RO pair's standard deviation as formula (6). Figure 4 shows the distribution of

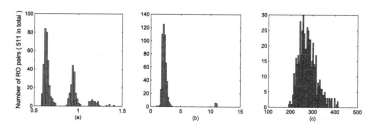

Fig. 4. The distribution of standard deviation with different reference counting values. (a), (b) and (c) are for 2^{10}, 2^{13} and 2^{20} respectively

RO pair's standard deviation under different reference counting values. If there are no disturbance signals, the distribution of the standard deviation should be almost a Gaussian distribution. But in practice because of the existing disturbance signals, there are multiple peaks in the histogram and with the reference counting values increasing, the number of peaks decreases. It can be explained that the disturbance signals have limited bounds and period, and effect of them can be positive or negative, so disturbance signals' accumulation also have limited bounds. However, noise's accumulation lead to the endless increase of the standard deviation. Therefore, with the reference counting values increasing, the effect of disturbance signals decreases.

In order to calculate the bit error probability, we divide the standard deviation values into some parts. For example, in Fig. 4 (a), there are three main peaks and divide the distribution into three parts. Every peak seems to be a normal distribution and calculate the mean of each part, σ_1, σ_2 and σ_3. The distribution of process variation is as follows.

$$PV_{ij} \sim N(\mu_{PV}, \sigma_{PV}^2) \tag{7}$$

For each σ_i, we can calculate the bit error probability $P_{error,i}$ as follows.

$$P_{error1,i} = \int_0^{+\infty} \{\{1 - \int_{-\infty}^{x} [\frac{1}{\sqrt{2\pi}\sigma_i} * exp(-\frac{y^2}{2\sigma_i^2})] dy\} * \frac{1}{\sqrt{2\pi}\sigma_{PV}} * exp(-\frac{x^2}{2\sigma_{PV}^2})\} dx \tag{8}$$

$$P_{error2,i} = \int_{-\infty}^{0} \{\{\int_{-\infty}^{x} [\frac{1}{\sqrt{2\pi}\sigma_i} * exp(-\frac{y^2}{2\sigma_i^2})] dy\} * \frac{1}{\sqrt{2\pi}\sigma_{PV}} * exp(-\frac{x^2}{2\sigma_{PV}^2})\} dx \tag{9}$$

$$P_{error,i} = P_{error1,i} + P_{error2,i} \tag{10}$$

For each σ_i, we can calculate the $P_{error,i}$ and according to the proportion of each σ_i, each $P_{error,i}$ contributes corresponding proportion to the final P_{error}, as formula (11) shows.

$$P_{error} = \sum_{i=1}^{k} (\frac{pf(\sigma_i)}{\sum_{i=1}^{k}(pf(\sigma_i))} * P_{error,i}) \tag{11}$$

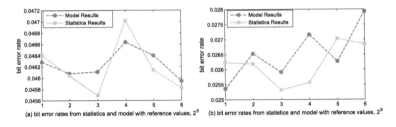

Fig. 5. The bit error rates from data statistics and model calculation

Where pf is a function to calculate the corresponding proportion of σ_i for the final P_{error}. The form of pf can be given roughly by the evaluation of the measured data with reference counting values, 2^8 and 2^9. In Fig. 5, we calculate the corresponding proportion by utilizing the function, $pf(\sigma_i) = \sigma_i^2$. In Fig. 5 (a), the rounded dot is for the bit error rate from model calculation and the square dot is for that from data statistics. The average bit error rate of data statistics is 0.0462 and that of model calculation is 0.0462. In each test, both of the bit error rates have some variation, but the variation is extremely slight. In Fig. 5 (b), the average bit error rate of data statistics is 0.0262 and that of model calculation is 0.0265. The results from Fig. 5 shows that the bit error rate computed from our modified model is nearly close to the bit error rate from data statistics.

4.2 Relationship Between Bit Error Rate and Reference Values

In order to estimate the relationship between bit error probability and the reference counting value, we conduct experiments with different reference counting values. During the experiments, we collect the test data on the same FPGA board and keep the environmental conditions to reduce the effect of other factors except the reference counting values.

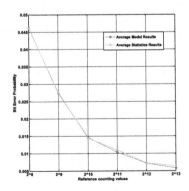

Fig. 6. The bit error rates with different reference counting values from 2^8 to 2^{13}

Figure 6 shows the average bit error rates from model calculation and data statistics with different reference counting values. The bit error rate of model calculation is fairly close to that of data statistics with the reference counting values varying from 2^8 to 2^{13}. When the reference counting value is 2^8, the bit error probability is about 4.62 %, and when the reference counting value increases to 2^{13}, the bit error probability decrease to about 0.6 %. However, Fig. 6 indicates that when the reference counting value is relatively small, the consistency of our model calculation and the practical value is extremely high. When the reference value increases with the bit error probability decreasing below 1 %, the consistency of our model calculation and the practical value has some fluctuation. The reason may be that when the bit error probability is below 1 %, just only a slight variation of measurement may result in a relatively large variation on the bit error probability. Over all, the bit error probability resulted from our model can approach to the practical bit error rate well.

Figure 6 roughly describes the relationship between the reference counting values and bit error probability. In order to verify this rough relationship, first we make a curve-fitting based on the six points (the reference counting value, the bit error rate), just like the first point, $(2^8, 0.0461)$. We utilize the least square method to get a fifth order polynomial $P_n(x)$ as shown in formula (12) and $P_n(x)$ makes the minimum of the variable I in formula (13).

$$P_n(x) = \sum_{k=0}^{n} a_k x^k \tag{12}$$

$$I = \sum_{k=0}^{n} [P_n(x_i) - y_i]^2 = \sum_{k=0}^{n} [\sum_{k=0}^{n} a_k x^k - y_i]^2 \tag{13}$$

Then, we select the reference counting value as $2^{10.5}$ and calculate the one result from the fitting curve. Meanwhile, we also get the bit error rate from data statistics and model calculation with reference counting value, $2^{10.5}$. The bit error rate from the fitting curve is 0.013688. We conduct the experiments 10 times with the reference counting value $2^{10.5}$. The average bit error rate from data statistics is 0.013781 and the average bit error rate from model calculation is 0.013669. Therefore the polynomial $P_n(x)$ can reflect the variation of bit error rate with different reference counting values accurately.

5   Conclusion

In this paper, we quantitatively describe the relationship between the bit error probability and the reference counting values. Based on the practical experiments on commercial Virtex-5 FPGA boards, the results from data statistics and model calculation achieve high consistency. Then we roughly analyze the effects of RO's stage and the manufacturing technology on RO PUF's bit error probability. Up to now, there is only qualitative conclusion that with the reference counting values increasing, the bit error rate decreases, and the reference counting value has not been considered as one factor which may influence the evaluation of RO PUFs. Our paper detailedly shows the effect of reference counting value on bit error probability. Our work is a guide for designers to construct RO PUFs with acceptable bit error rate and contributes to the implementation of the evaluation scheme on RO PUFs.

References

1. Bhargava, M., Mai, K.: An efficient reliable PUF-based cryptographic key generator in 65nm CMOS. In: Proceedings of the Conference on Design, Automation & Test in Europe, DATE (2014)
2. Delvaux, J., Verbauwhede, I.: Attacking PUF-based pattern matching key generators via helper data manipulation. In: Benaloh, J. (ed.) CT-RSA 2014. LNCS, vol. 8366, pp. 106–131. Springer, Heidelberg (2014)
3. Areno, M., Plusquellic, J.: Securing trusted execution environments with puf generated secret keys. In: Trust, Security and Privacy in Computing and Communications (TrustCom), pp. 1188–1193 (2012)
4. Katzenbeisser, S., Kocabaş, U., Rožić, V., Sadeghi, A.-R., Verbauwhede, I., Wachsmann, C.: PUFs: myth, fact or busted? a security evaluation of physically unclonable functions (PUFs) cast in Silicon. In: Prouff, E., Schaumont, P. (eds.) CHES 2012. LNCS, vol. 7428, pp. 283–301. Springer, Heidelberg (2012)
5. Komurcu, G., Pusane, A.E., Dundar, G.: Analysis of ring oscillator structures to develop a design methodology for RO-PUF circuits. In: Very Large Scale Integration (VLSI-SoC), pp. 332–335 (2013)
6. Hiller, M., Sigl, G., Pehl, M.: A new model for estimating bit error probabilities of ring-oscillator PUFs. In: Reconfigurable and Communication-Centric Systems-on-Chip (ReCoSoC), pp. 1–8 (2013)
7. Delvaux, J., Verbauwhede, I.: Fault injection modeling attacks on 65nm arbiter and RO sum PUFs via environmental changes. IEEE Trans. Circ. Syst. **61**, 1701–1713 (2014)
8. Suh, G.E., Devadas, S.: Physical unclonable functions for device authentication and secret key generation. In: Proceedings of the 44th Annual Design Automation Conference, pp. 9–14 (2007)
9. Gassend, B., Clarke, D., Van Dijk, M., Devadas, S.: Silicon physical random functions,. In: Proceedings of the 9th ACM Conference on Computer and Communications Security, pp. 148–160, 2002
10. Maiti, A., Schaumont, P.: Improved ring oscillator PUF: an FPGA-friendly secure primitive. J. Cryptol. **2**, 375–397 (2011)

Author Index

Printed in the United States
By Bookmasters